Springer Series on Environmental Management

More information about this series at http://www.springer.com/series/412

Udo Nehren • Sabine Schlüter • Claudia Raedig
Dietmar Sattler • Helga Hissa
Editors

Strategies and Tools for a Sustainable Rural Rio de Janeiro

Springer

Editors
Udo Nehren
Institute for Technology and Resources
Management in the Tropics and
Subtropics – ITT
TH Köln – University of Applied Sciences
Köln, Germany

Sabine Schlüter
Institute for Technology and Resources
Management in the Tropics and
Subtropics – ITT
TH Köln – University of Applied Sciences
Köln, Germany

Claudia Raedig
Institute for Technology and Resources
Management in the Tropics and
Subtropics – ITT
TH Köln – University of Applied Sciences
Köln, Germany

Dietmar Sattler
Institute of Geography
Leipzig University
Leipzig, Germany

Helga Hissa
Programa Rio Rural
Secretaria de Agricultura e Pecuária
Fonseca, Niterói, Rio de Janeiro, Brazil

ISSN 0172-6161
Springer Series on Environmental Management
ISBN 978-3-030-07822-5 ISBN 978-3-319-89644-1 (eBook)
https://doi.org/10.1007/978-3-319-89644-1

Printed on acid-free paper

This Springer imprint is published by the registered company Springer International Publishing AG part of Springer Nature.
The registered company address is: Gewerbestrasse 11, 6330 Cham, Switzerland

Foreword

In recent years, the state of Rio de Janeiro has experienced an alternation of intense floods and droughts, whose unparalleled magnitude leaves little room for doubt that such phenomena derive from climate change. As a matter of fact, in a geographically diverse territory such as Rio's, climate change impacts can be multifaceted, ranging from sea-level rise to landslides in mountainous areas. To face this reality, the state policy on climate change has established an ambitious set of mitigation and adaptation targets to be achieved by 2030. The latter is inextricably intertwined with the expansion of Rio Rural, a sustainable development program for rural areas under the auspices of the State Secretary for Agriculture and Livestock (SEAPEC). The centrality of this program for the state's adaptation agenda is an example of its crucial importance for a wide array of topics that are decisive for Rio's sustainable development, such as land restoration, integrated water management, and ecosystem services, among others.

The program Rio Rural perfectly illustrates the key role regional governments usually play in sustainable development. On top of their multiple legal responsibilities, policy tools, and financial mechanisms, regional governments are ideally suited to deal with issues that trespass local boundaries but are not necessarily nationwide, which is frequently the case of many topics related to sustainable development. Among these are ecosystem services and rural-urban linkages, for instance. While Rio de Janeiro is allegedly the most urbanized federated state in Brazil, traditional statistics often fail to translate the strategic value of its rural areas. The metropolitan region, where three out of every four persons live in the state, is highly reliant on surrounding rural areas for its water supply and food provision, for example. In this sense, the potential benefits of Rio Rural extend far beyond the targeted rural areas it covers. In fact, its territorial planning approach, which is focused on microwatersheds and several interventions it encompasses, such as water quality recovery, soil conservation, reforestation, or infrastructure investments, can virtually benefit the entire state.

Bearing in mind the importance that Rio Rural has for the state of Rio de Janeiro, one can imagine how meaningful a project focused on offering integrated ecotechnology and services solutions to this program could be. This is the exact reason why INTECRAL should be praised. As a jointly developed scientific cooperation supported by the German Federal Ministry of Education and Research (BMBF) and SEAPEC, this project aims to improve the competitiveness of the rural economic sector in due consideration of a sustainable watershed management, including environmental risk mitigation and protection of natural resources.

Considering that both Rio Rural and INTECRAL present innovative outcomes that are probably replicable in rural areas of other developing regions, their "strategies and tools for a sustainable rural Rio de Janeiro" will certainly be inspirational for all those interested in exploring the multiple interlinkages of the 2030 Sustainable Development Agenda that flourish in rural areas.

Joan França
International Relations Officer
for the State of Rio de Janeiro and former
Project Manager of the Regions Adapt initiative

Acknowledgments

We cordially thank the German Federal Ministry of Education and Research (BMBF) for funding the German-Brazilian research project INTECRAL – Integrated eco technologies and services for a sustainable Rural Rio de Janeiro – and the staff of the Rio Rural program for their continuous support in the preparation and in conducting the fieldwork. Moreover, we are very grateful to the following experts who volunteered their time and knowledge to review the chapters in this book. Alphabetically, our sincerest thanks go to the following:

Eduardo Roberto Alexandrino (Wildlife Ecology, Management and Conservation Lab – Department of Forest Sciences – ESALQ, University of São Paulo – USP, São Paulo, Brazil);

José Carlos de Araújo (Department of Agricultural Engineering, Federal University of Ceará – UFC, Fortaleza, Brazil);

Mariana Beauclair (Coordination of Territory Management and Geospatial Information, Instituto Estadual do Ambiente do Rio de Janeiro [Rio de Janeiro Environmental State Agency], Rio de Janeiro, Brazil);

Rachel Bardy Prado (Empresa Brasileira de Pesquisa Agropecuária [Brazilian Agricultural Research Corporation] – Embrapa Solos, Rio de Janeiro, Brazil);

Henrik Beermann (Fraunhofer Center for International Management and Knowledge Economy, Leipzig, Germany);

Jan Börner (Center for Development Research – ZEF, University of Bonn, Bonn, Germany);

Davi José Bungenstab (Embrapa Gado de Corte [Brazilian Agricultural Research Corporation], Campo Grande, Brazil);

Diogo de Carvalho Cabral (Instituto Brasileiro de Geografia e Estatística [Brazilian Institute of Geography and Statistics] – IBGE, Rio de Janeiro, Brazil);

Hugo Rosa da Conceição (Center for Development Research – ZEF, University of Bonn, Bonn, Germany);

Marcelo Eduardo Dantas (Department of Territorial Management – DEGET, Geological Service of Brazil – CPRM, Rio de Janeiro, Brazil);

Nadir Ahmed Elagib (Institute for Technology and Resources Management in the Tropics and Subtropics – ITT, TH Köln – University of Applied Sciences, Cologne, Germany);

Leonardo Esteves de Freitas (Institute of Geosciences, Department of Geography, Federal University of Rio de Janeiro – UFRJ and OTSS – Oswaldo Cruz Foundation, Rio de Janeiro, Brazil);

Gerhard Gerold (Institute of Geography, Department of Landscape Ecology, Georg-August-University of Göttingen, Göttingen, Germany);

Friederike Grüninger (Department of Physical Geography, University of Passau, Passau, Germany);

Eliane Guedes (Department of Geology and Palaeontology, Federal University of Rio de Janeiro – UFRJ, Rio de Janeiro, Brazil);

Claudia Jimenez Escobar (Faculty of Veterinary Medicine and Animal Husbandry, National University of Colombia, Bogotá, Colombia);

Antonio Jose Teixeira Guerra (Institute of Geosciences, Department of Geography, Federal University of Rio de Janeiro – UFRJ, Rio de Janeiro, Brazil);

Johannes Hamhaber (Institute for Technology and Resources Management in the Tropics and Subtropics – ITT, TH Köln – University of Applied Sciences, Cologne, Germany);

Darya Hirsch (International Centre for Sustainable Development – IZNE, Department of Management Sciences, Bonn Rhein-Sieg University of Applied Sciences, Sankt Augustin, Germany);

Antonio Carlos Oscar da Silva Júnior (Institute of Geography, Department of Physical Geography, State University of Rio de Janeiro – UERJ, Rio de Janeiro, Brazil);

Peter Krause (Thüringer Landesanstalt für Umwelt und Geologie, Gewässerkundlicher Landesdienst, Hochwassernachrichtenzentrale [Thuringian State Institute for Environment and Geology, hydrological services, flood protection center], Jena, Germany);

Georg Lamberty (Institute for Technology and Resources Management in the Tropics and Subtropics – ITT, TH Köln – University of Applied Sciences, Cologne, Germany);

Sven Lautenbach (Institute of Geodesy and Geoinformation - IGG, University of Bonn, Bonn, Germany);

Toa Loaiza (Centre for International Migration and Development [CIM], Eschborn, Germany);

Eduardo Luiz Longui (Department of Wood and Forest Products, Instituto Florestal [Forest Institute], São Paulo, Brazil);

Luiz Firmino Martins Pereira (Câmara Metropolitana de Integração Governmental [Metropolitan Chamber of Governmental Integration], Rio de Janeiro, Brazil);

Carlos Rogerio de Mello (Department of Engineering, Federal University of Lavras, Lavras, Brazil);

Marcela de Abreu Moniz (Institute of Humanities and Health, Federal Fluminense University, Rio das Ostras, Brazil);

Geoffrey Ndegwa (Independent Researcher and Consultant, Nairobi, Kenya);

Rodrigo Penna-Firme (Department of Geography and the Environment, Pontifícia Universidade Católica do Rio de Janeiro – PUC, Rio de Janeiro, Brazil);
Leandro Angelo Pereira (Federal Institute of Paraná – IFPR, Câmpus Paranaguá, Brazil);
Stefan Porembski (Department of General Botany and Systematic Botany, University of Rostock, Rostock, Germany);
Hugo Portocarrero (Institute of Geography, Department of Physical Geography, State University of Rio de Janeiro – UERJ, Rio de Janeiro, Brazil);
Fabrice G. Renaud (School of Interdisciplinary Studies, University of Glasgow, United Kingdom);
Thiago Libório Romanelli (Department of Engineering and Biosystems, University of São Paulo – USP, São Paulo, Brazil);
Eduardo Salgado (School of Agronomy, Pontifical Catholic University of Valparaíso, Valparaíso, Chile);
Simone Sandholz (Institute for Environment and Human Security, United Nations University – UNU, Bonn, Germany);
Fabio Rubio Scarano (Fundação Brasileira para o Desenvolvimento Sustentável -FBDS [The Brazilian Foundation for Sustainable Development], Rio de Janeiro, Brazil);
Gerhard Schiefer (International Center for Food Chain and Network Research, University of Bonn, Bonn, Germany);
Britta Schmalz (Department of Engineering Hydrology and Water Management, Technical University of Darmstadt, Darmstadt, Germany);
Karl Schneider (Institute of Geography, University of Cologne Cologne, Germany);
Zita Sebesvari (Institute for Environment and Human Security, United Nations University – UNU, Bonn, Germany);
Maxime Souvignet (Institute for Environment and Human Security, United Nations University – UNU, Bonn, Germany);
Katharina Stein (Department of Botany and Botanical Garden, University of Rostock, Rostock, Germany);
Karen Sudmeier-Rieux (Faculty of Geosciences and Environment, Institute of Earth Sciences, University of Lausanne, Lausanne, Switzerland)
Yuri Tavares Rocha (Department of Geography, University of São Paulo – USP, São Paulo, Brazil);
Gabriela Teixeira Duarte (Department of Biology, Federal University of Minas Gerais – UFMG, Belo Horizonte, Brazil);
Wiltrud Terlau (International Centre for Sustainable Development – IZNE, Dept. of Management Sciences, Bonn Rhein-Sieg University of Applied Sciences, Sankt Augustin, Germany);
Ana Paula Turetta (Empresa Brasileira de Pesquisa Agropecuária [Brazilian Agricultural Research Corporation] – Embrapa Solos, Rio de Janeiro, Brazil);
Raul Sanchez Vicens (Department of Geography, Federal Fluminense University, Niterói, Brazil);
Carmem Lucas Vieira (Department of Geography, Federal Fluminense University, Niterói, Brazil);

Martin Volk (Department of Computational Landscape Ecology, Helmholtz-Zentrum für Umweltforschung GmbH – UFZ, Leipzig, Germany);
Thomas Zumbroich (Department of Geography, University of Bonn and Planungsbüro Zumbroich, Bonn, Germany).

We are very grateful to Laura Noriega from the Institute for Technology and Resources Management in the Tropics and Subtropics at TH Köln for her valuable editorial support during the preparation of this book as well as to Birgit Pawelzik, Karen Schneider, Hugh Eva, Harrhy James, and Nicolas Locke for English proof-reading. Furthermore, we thank Laura Noriega and Sandra Alfonso for proofreading of the Spanish abstracts as well as Giuliano M. Locoselli, Joyce Monteiro, Antonio Soares da Silva, and Ana Paula Turetta for proofreading several Portuguese abstracts. (Spanish and Portuguese abstracts are only available in the online-version, but not in the hardcover). Finally, we thank all project partners and stakeholders for their great work and trusting cooperation.

Contents

Contributors

Nelson T. Alves Filho is a lawyer and senior officer from Rio Government, acting as the Superintendent of Sustainable Development at the Secretary of Agriculture, Livestock, Food Supply, and Fisheries of the State of Rio de Janeiro, Brazil. He holds the executive director of the Sustainable Rural Development Program – Rio Rural Program and is the head of rural development policies and projects for small farmers' sustainability with the support of the Global Environment Facility, World Bank, Inter-American Development Bank, and the United Agency for Food and Agriculture. He used to lead several projects in the area of land tenure and rural settlements in the State.

Dieter Anhuf studied Geography between 1976 and 1982 at the universities of Giessen, Munich, and Bonn. He successfully finished his university education with a diploma in Geography in 1982 and a PhD degree in 1987, both from the University of Bonn, Germany. Since then he continuously worked as a scientific assistant at the Universities of Bonn and Mannheim, Germany, where he terminated his professional dissertation in 1994. Between 1994 and 2001, he substituted the Chair of Physical Geography at Mannheim University. For more than 30 years, he has been focusing on climate and vegetation interaction in tropical Africa and later South America with a close link to the University of São Paulo, Brazil, where he was appointed the renowned von Martius Chair of Ecology in 2001. Since 2004, he has been holding the Chair in Physical Geography at the University of Passau, Germany.

Oscar Manuel Baez Villanueva is a civil engineer with a master's degree in Water Management. He is currently a PhD candidate at the Institute for Technology and Resources Management in the Tropics and Subtropics (ITT) at TH Köln and TU Dortmund, Germany. He worked as a professor in the Civil Engineering Program of the University of San Luis Potosí (UASLP), Mexico. He also worked on projects related to water management and structural design. His current research interests are hydrology, spatiotemporal analysis of precipitation, satellite products, and application of remote sensing techniques for hydrological modeling.

Rachel Bardy Prado graduated in Biological Sciences from the Federal University of São Carlos, Brazil. She holds a master's and a doctoral degree in Science in Environmental Engineering from the University of São Paulo, Brazil, specializing in planning and water management from the Federal University of the Amazon, Brazil. She has been working as a researcher at Embrapa Soils since 2003, especially in monitoring and planning of ecosystem services with a focus on water resources. She leads the Ecosystem Service Embrapa Research Network. Currently, she attends the postgraduate program in Engineering of Biosystems of the Fluminense Federal University in Rio de Janeiro, Brazil.

Lauro Bassi is an agronomist and holds a master's degree in Soil Sciences. For several years, he acted at EPAGRI/SA – Agricultural Research and Rural Extension of Santa Catarina State: At the international level as a consultant of the Food and Agriculture Organization of United Nations (FAO) projects in sustainable agriculture, soil and water conservation, and participatory planning in El Salvador, Mexico, Costa Rica, Guatemala, Honduras, Nicaragua, and Equatorial Guinea; and at the national level in projects of natural resource management in the States of São Paulo, Rio de Janeiro, Santa Catarina, and Rio Grande do Sul. Moreover, he advised and supported the preparation and development of six Global Environment Facilities projects focusing on monitoring and evaluation in different States of Brazil.

Martin Becher is a technical advisor for Gesellschaft für Internationale Zusammenarbeit (GIZ) GmbH in Brazil. He holds a diploma in Geography, Political Sciences, Agronomy, and Urban Planning from the University of Tübingen, Germany. His work focuses on ecosystem-based adaptation to climate change, disaster risk management, and the development of participative risk assessment tools. His current tasks have an emphasis on mainstreaming climate change adaptation into public policies and land-use planning instruments, framed into bilateral cooperation projects "Biodiversity and Climate Protection in the Mata Atlântica and Supporting Brazil to implement its National Agenda to Adapt to Climate Change (BMUB-IKI)."

Diana Birnbaum holds a bachelor's degree in Biology from Radboud University Nijmegen, the Netherlands, and a master's degree in Natural Resources Management and Development from the Institute for Technology and Resources Management in the Tropics and Subtropics, TH Köln – University of Applied Sciences, Germany. Her research focuses on river ecology.

Alexander Brenning has been a full professor of Geographic Information Science at Friedrich Schiller University Jena, Germany, since 2014. After studying Mathematics and Geography at the University of Erlangen-Nuremberg, Germany, Catholic University of Chile, and Technical University of Freiberg, Germany, he earned a PhD degree in Geography from Humboldt-Universität zu Berlin, Germany, and held a faculty position at the University of Waterloo, Canada, from 2007 to 2015. His research focuses on spatial statistical techniques and open-source geocomputing tools and their application to a variety of environmental issues related to mountain geomorphology, natural hazards, and environmental remote sensing.

Alicia Bustillos Ardaya is a PhD candidate in the geography department of the University of Bonn, Germany, and a researcher at the Center for Development Research (ZEF), Bonn, Germany. She holds an engineering degree from the Zamorano University for Agricultural studies in Honduras and a Master of Sciences degree from TH Köln - University of Applied Sciences, where she worked on a trilateral cooperation between Brazil, Mozambique, and Germany doing the Buzi River basin modeling for flood forecasting. Her research focuses on interdisciplinary aspects of disaster risk reduction. Based on flood models, she focuses on communication and perception of the population living in risk areas, and participatory mechanisms for risk governance, particularly in Latin American countries.

Gabriel Campos Pereira da Costa is a master's student in Geography at the State University of Rio de Janeiro (UERJ) and holds a bachelor's degree in Geography of the same institution. Moreover, Gabriel studies Environmental Management at the Estácio de Sá University in Rio de Janeiro. His academic interests are mainly linked to environmental management and rehabilitation of degraded lands. His current research emphasizes on pasture degradation and hydric erosion processes.

Jaime Cardona is an environmental manager and PhD candidate in Engineering with more than twelve years of working experience driven by the development of environmental economics, climate change abatement, and water resources management projects throughout the world. His professional experiences have been developed among others in the fields of decentralized wastewater treatment and reuse solutions, water, water treatment process engineering, technical and management assistance to water utilities, project cycle management, and capacity development. He brings extensive experience in managing large- and medium-scale development programs, mostly related to the rural urban water resources management in Middle East, i.e., Palestine, Jordan, Israel; Central Asia, i.e., Mongolia; and Latin America, i.e., Costa Rica, Colombia, Panamá, and Brazil. In-depth experience includes development of decision support methodologies, environmental economics, and Geographic Information Systems for orienting decision-making processes for environmental management.

Pedro Castro is a geographer and PhD candidate at the Institute for Geodesy and Geoinformation, Faculty of Agriculture, University of Bonn Germany. He holds a master's degree in Geography from the Fluminense Federal University, Brazil, with an emphasis on environmental planning. He worked as a consultant in several projects in cooperation with Fluminense Federal University, Rio de Janeiro Federal University, Rio de Janeiro State University, and Environmental Agency of the State of Rio de Janeiro – Brazil (INEA-RJ). His research topics include Land Use Land Cover Classification (LULCC) using optical satellite images and spatial dynamics modeling.

Elaine Cristina Cardoso Fidalgo graduated in Agronomy from São Paulo University with a master's degree in Remote Sensing at the Brazilian National Institute for Space Research (INPE), and a PhD degree in Agricultural Engineering at Campinas University, Brazil. She is a researcher of the Brazilian Agricultural Research Corporation, and based at the National Centre for Soil Research (Embrapa Soils). She has experience in GIS and remote sensing applied to studies of land-use change, environmental planning, and ecosystem services.

Marcelo Costa is an agronomist engineer with a specialization in Irrigation Engineering (UFRRJ) and GIS (UFRJ). He is currently an MSc student in Biosystems Engineering at Federal Fluminense University, Brazil (UFF), also acting as information management coordinator of the Sustainable Rural Development Program – Rio Rural. He used to be a former rural extensionist from EMATER RIO, acting as a regional coordinator at Serrana region, Brazil.

Ana Paula Dias Turetta holds a bachelor's degree from the Federal University of Rio de Janeiro (1997), a master's degree in Agronomy (Soil Sciences) from the Federal Rural University of Rio de Janeiro (2000), a PhD in Agronomy (Soil Sciences) from the Federal Rural University of Rio de Janeiro (2004) and a postdoctoral degree from the University of Wageningen, The Netherlands. She has experience in environmental management with emphasis in the subject of sustainability assessment and environmental services of soils in agricultural landscapes. Ana Turetta is a permanent professor of the Graduate Program in Territorial Development of the Federal Rural University of Rio de Janeiro (UFRRJ).

Utz Dornberger is the director of the International Small Enterprise Promotion and Training (SEPT) Program of the Leipzig University, Germany. He is a biologist with an MBA degree in Small Business Studies and a PhD degree in Natural Sciences. Since 2006, he has been the director of the start-up initiative SMILE (Self-Management Initiative Leipzig) at the Leipzig University, the director of the MBA program at the Hanoi University of Science and Technology (HUST), and of the Vietnamese-German University (VGU). Since 2015, Utz Dornberger has been the head of Unit for Entrepreneurship and Innovation for Development Cooperation at the Fraunhofer Center for International Management and Knowledge Economy (IMW) in Leipzig.

Maria Luisa Espinel is a veterinarian from the National University of Colombia and holds a master's degree in Natural Resources Management from the Institute for Technology and Resources Management in the Tropics and Subtropics (ITT), TH Köln – University of Applied Sciences, Germany. Her field research areas include best farming practices in dairy systems and farm system analysis.

Mariele Evers is a geographer and works as a professor at the University of Bonn, as head of the Eco-Hydrology and Water Resources Management Group, and as adjunct professor at the Centre for Climate and Safety at Karlstad University,

Sweden. She has national and international expertise in IWRM and human-water research, the management of hydrological risks, the development and implementation of sociotechnical instruments, and collaborative modeling and transdisciplinary research.

Marco Follador is a senior scientific officer at the European Commission and consultant at ALSO Consulting. He received a European PhD (France-Italy) in Geography at Toulouse University, France and in Engineering at Bologna University, Italy. Since 2005 he has been working on climate change related issues with a focus on supporting policy and decision making through robust evidence-based planning. He led several projects in Europe and Brazil related to adaptation issues and climate risk assessment, promotion of climate-proof investments, resilience of productive and socio-environmental systems as well as avoidance of potential losses and damages.

Leonardo S. Fernandes is a PhD student in the graduate program in Public Policies and Development Strategies at the Institute of Economics of the Federal University of Rio de Janeiro and is a biologist at the Institute of the Environment of the State of Rio de Janeiro. He received a degree in Biological Sciences from the Federal University of Juiz de Fora and a master's degree in Forestry Sciences from the Federal University of Viçosa, both in Minas Gerais, Brazil.

José Márcio Ferreira is a researcher at the Agricultural Research Company of the State of Rio de Janeiro, Brazil – PESAGRO-RIO; at the State Center for Research in Agroenergy and Waste Utilization, located in the municipality of Campos dos Goytacazes, where he is the coordinator of the Development Program Sustainable Rural – Rio Rural, in the Northern region of the State of Rio de Janeiro. He also leads projects in the area of sustainable agriculture and ecosystem services with support from the Global Environment Facility, the World Bank, the Inter-American Development Bank, and the United Food and Agriculture Agency. The main focus of his work is to support smallholder farmers and family farms for environmental conservation agriculture, ecosystem services and integrated environment-agriculture policies, and the sustainability of smallholder agriculture.

Manfred Fink is a senior researcher at the Department of Geography, Friedrich Schiller University Jena, Germany. He received a diploma in Geography from Kiel University and a PhD degree in Geography from Friedrich Schiller University Jena, Germany. His research focuses on the development and application of ecohydrological models, the assimilation of primary data in the field by designing and operation monitoring programs, and the development and application of geographic information processing methods to analyze and prepare data for model input. He led and participated in several international projects. Currently, he works on the communication of the implications of climate change in water management practice.

Silvia Berenice Fischer is a junior researcher the Center for Development Research of the University of Bonn. Her current research focuses on drought resilience and health. She holds a degree in Forestry Restoration Engineering from Chapingo University, Mexico, and a master's degree in Natural Resources Management from the TH Köln – University of Applied Sciences, Germany. She worked in several multidisciplinary and international teams, focusing on participatory research and sustainable land management in Germany, Brazil, and Mexico.

Wolfgang-Albert Flügel is retired head of the Department of Geoinformatics, Hydrology, and Modeling (DGHM) at Friedrich Schiller University Jena, Germany. His expertise is about hydrological system analysis, rainfall-runoff modeling, climate change impact assessment and analysis in large river basins, and Integrated Land and Water Resources Management (ILWRM). He acquired and fundraised 93 interdisciplinary research projects in Southern Africa, Brazil, Europe, and South Asia funded by the European Commission, the German Research Association (DFG), and the Federal Ministry of Education and Research (BMBF). In addition, he has supervised 96 MSc theses and 24 PhDs and has authorship in more than 120 publications.

Rosa Maria Formiga-Johnsson's broad research interests and experiences comprise water resources management and water policy. Since 2006, she has been an associate professor at the State University of Rio de Janeiro (UERJ) in the Department of Environmental and Sanitary Engineering (Brazil). Rosa obtained her bachelor's degree in Civil Engineering (1987), MSc degree (1992), and PhD degree (1998) in Environmental Sciences and Techniques from Université de Paris II-Val de Marne, France. She was the director of Water Management of the Environmental Agency of the State of Rio de Janeiro (INEA-RJ, Jan 2009–Jan 2015), and Water Security Manager of the State Secretariat of the Environment (Feb– Nov 2015).

Joana Fraga is a PhD student in the graduate program at the Department of Geography and Environment of Pontifical Catholic University of Rio de Janeiro, Brazil, with an emphasis on landscape changes and socioecological systems. She received a degree in Geography at the same department and a master's degree in Geography at Federal University of Rio de Janeiro, Brazil, with a focus on environmental planning and management.

Beate Frank's main research areas comprise water resources management, flood control management, and disaster risk management. She was a professor at University Blumenau, Brazil, from 1979 to 2008. Her main experience was formed in the process of the creation and development of the Itajaí River Basin Council, which she headed from 1996 to 2011, as part of Santa Catarina's water resources management system. She obtained her bachelor's degree in Physics (1976) and MSc degree in Physics (1981) at the Universidade Federal do Rio Grande do Sul, Brazil, and a PhD degree in Production Engineering (1995) at the Universidade Federal de Santa Catarina, Brazil.

Carl-Friedrich Gaese is a PhD candidate at the Chair of Agricultural Systems Engineering at TUM School of Life Sciences Weihenstephan Technical University of Munich, Germany, and a researcher at the Institute of Technology and Resources Management in the Tropics and Subtropics (ITT) at TH Köln – University of Applied Sciences, Germany. He holds a degree in Agricultural Sciences from the Bonn University, Germany. His research focuses on the interface of agricultural machinery and economics.

Daniel Gieseler is an engineer for Urban and Industrial Water Management. He graduated from Technical University of Dresden and currently works for the Training and Demonstration Centre for Decentralised Sewage Treatment – BDZ e.V. in Leipzig, Germany, in capacity development as well as in research and development. His research focuses on wastewater management solutions for rural areas.

Lorena Valeria Guzmán Wolfhard is an environmental engineer from Technical University of Bolivia and holds a master's degree in Natural Resources Management and Development from the Institute for Technology and Resources Management in the Tropics and Subtropics (ITT), TH Köln – University of Applied Sciences, Germany. She works currently as a research assistant in the Centre for Development Research (ZEF), Bonn, in the Bolivian-German cooperation project "Inclusion of biodiversity in the process of environmental impact assessment."

Anja Hebner is an environmental engineer who has been working for more than 15 years as a project manager at Vita 34 AG, business unit BioPlanta, Germany. She develops feasibility studies, environmental risk assessments, and approval planning for remediation of water and soil using biotechnologies. In addition, she manages R&D projects focusing on bio- and phyto-technologies as well as bioengineering. Besides national projects, she has developed projects in Asia and Latin America with international consortia.

Jürgen Heinrich is a physical geographer and earned a PhD and a habilitation (venia legendi) in Geography from the Johann Wolfgang Goethe University Frankfurt/Main, Germany. He holds the professorship for Physical Geography and Environmental Research at the Institute of Geography at the Leipzig University, Germany. His research focuses on geomorphology, especially of Sub-Saharan Africa, Quaternary Geology, Geoecology, and genetic questions of landscape development. For more than a decade, he addressed this research to subjects in North and Southeast Brazil within the framework of international research and development projects.

Mark Hickie earned his master's degree in Geography from the Department of Geography and Environment of Pontifical Catholic University of Rio de Janeiro. He previously obtained another master's degree for Transportation and Logistics from the Massachusetts Institute of Technology, USA. Consequently, his research interests are varied and far-ranging, primarily to understand environmental history focused on efficient modeling to solve future environmental issues. Currently, he serves as a Lieutenant Colonel in the United States Air Force as an instructor pilot of fighter aircraft.

Helga Hissa is a senior soil researcher at EMBRAPA – the Brazilian Agricultural Research Corporation working at the Secretariat of Agriculture, Livestock, Food Supply and Fisheries of the State of Rio de Janeiro, where she acts as technical coordinator of the Sustainable Rural Development Program – Rio Rural. She also leads projects in the area of Sustainable Agriculture and Ecosystems Services with support of the Global Environment Facility, World Bank, Inter-American Development Bank, and the United Agency for Food and Agriculture. She is an agronomist engineer and earned an MSc degree in Soil Sciences from the Rural University of Rio de Janeiro, Brazil. Her work focuses on conservation agriculture, ecosystem services, and environment-agriculture integrated policies aiming at sustainability of conservation agriculture by smallholders.

Elisabeth Huber-Sannwald earned a PhD degree in Range Ecology from Utah State University, USA. After a postdoc at the University of Buenos Aires, Argentina, with the Global Change and Biodiversity Program she became an assistant professor at the Chair of Grassland Science at the Technical University in Munich, Germany. In 2001, she joined the national research institute Instituto Potosino de Investigación Científica y Tecnológica, Mexico, as a researcher in the Division of Environmental Sciences. Her research interests are ecosystem services and livelihood analysis, resilience of social-ecological systems, complex system analysis of land degradation, and desertification of drylands. She is a co-coordinator of the international desertification network ARIDnet, and the International Network on Sustainable Drylands.

Silvia Marie Ikemoto is a biologist and PhD student in the Environmental Postgraduate Program of the State University of Rio de Janeiro. She holds a master's degree in Environmental Sciences from the Fluminense Federal University. She is coordinator of territory management and geospatial information at the INEA-RJ (Environmental Agency of the State of Rio de Janeiro, Brazil). Her work and research emphasize territory and water resources management, ecosystem-based management, payment of environmental services, and disaster risk management. She coordinates the Payment for Environmental Services State Program and two international projects with a focus on sustainable land use management and biodiversity conservation in Rio de Janeiro, Brazil.

Viviane Japiassú Viana holds a PhD degree in Environmental Sciences and a master's degree in Environmental Engineering from the State University of Rio de Janeiro. She is the author of several books relating to environmental topics. Her research areas include disaster risk reduction, land use management, environmental management, and water resources management. She is a faculty member of Veiga de Almeida University and other university centers in Rio de Janeiro. She also works in the environmental regularization of services and industrial activities.

Guilherme Kangussú Donagemma graduated in Agronomia from Universidade Federal de Viçosa (1997) with a master's degree in Agronomy at Universidade Federal de Viçosa (2000) and a PhD degree in Agronomy from the Universidade Federal de Viçosa (2005). He has experience in agronomy, focusing on soil physics and acts on the following subjects: nutrient dynamics, fertirrigation, soil columns, and dispersion.

Sven Kralisch is a senior researcher and lecturer in the field of Geoinformatics and Environmental Modeling at the group for Geographic Information Science of Friedrich Schiller University Jena, Germany, where he gained a diploma in Computer Science and a PhD degree in Geoinformatics. His research focuses on the development of software tools and methods for a holistic assessment of hydrological processes in river basins and their future development under changing climate and management. The application of these methods is embedded in international research activities with regional foci in Brazil, Southern Africa, Central and South Asia, and Europe.

André Kirchner holds a Junior Professorship in Geography at the University of Hildesheim, Germany. He studied Geography, Geology and History at the University of Leipzig and prepared a diploma thesis on geomorphological and pedological processes in the hyperarid Nubian Desert. During his PhD, he focused on fluvial deposits and soils as palaeoenvironmental archives in Southeast Brazil. In recent years, André worked on geoarchaeological research projects and on environmental education. Furthermore, he gives lectures, seminars and field classes on a regular basis. He is an active member of national and international working groups.

Kristina Kreter currently works as a Water, Sanitation, and Hygiene (WASH) program coordinator for an NGO in the Dominican Republic. She holds a master's degree in International Cultural and Business Studies from the University of Passau and a double master's degree in Environment and Resources Management from the TH Köln - University of Applied Sciences, Germany, and from the Autonomous University of San Luis Potosí, Mexico. Her academic field research focused on water governance with an emphasis on the Brazilian water and sanitation sector.

Annika Künne is a researcher at the group of Geographic Information Science at Friedrich Schiller University Jena, Germany. She studied Geography with a focus on hydrology and geoinformation science. Her expertise is in ecohydrological modeling as well as climate and land-use change impact assessment on river basin scale. For this purpose, she develops and applies process-based simulation models, designs, and implements hydro-climatic monitoring networks and analyses environmental data sets using geostatistical methods. Her regional foci are in Brazil and Germany.

Georg Lamberty is a senior researcher and lecturer in Integrated Water Resources Management at the Institute for Technology and Resources Management in the Tropics and Subtropics, TH Köln – University of Applied Sciences, Germany. He holds a PhD degree in Geography from the University of Bonn, Germany. His research areas include river ecosystems, hydromorphology, and environmental assessment.

Wolfram Lange works as a freelance consultant for projects of international cooperation for sustainable development, NGOs and governmental institutions in Brazil and is a PhD candidate at the Institute of Geography of the University of Cologne, Germany. His expertise is urban and social geography as well as regional planning and natural resources management focusing on land-use management in rural and urban areas and socio-spatial analysis, climate change, and disaster risk reduction. He has been working at and cooperating with several universities in Germany, Brazil, Mozambique, Chile, Mexico, and Indonesia.

Sven Lautenbach is an assistant professor of Land-Use Modeling and Ecosystem Services at the University of Bonn, Germany. His work focuses on trade-offs of land-use decisions, environmental modeling, decision support, and statistical modeling of environmental and socioeconomic data.

Alena Lepilova graduated in Water Supply and Wastewater Disposal Engineering from the Ural Federal University named after the first President of Russia B.N. Yeltzin, Russia, and continued her studies in Germany to gain a master's degree in Ecotoxicology from University of Koblenz-Landau with a focus on water pollution. After completing her studies, Alena Lepilova started to work as a research assistant supporting Training and Demonstration Centre for Decentralized Wastewater Treatment – BDZ e.V, Leipzig in research and capacity development projects at international level with an emphasis on decentralized water and wastewater management.

Silke Lichtenberg works as a lecturer at the Institute for Technology and Resources Management in the Tropics and Subtropics, TH Köln – University of Applied Sciences, Germany. She obtained her bachelor's degree in Environmental and Energy Management at the University of Applied Sciences of Pinkafeld, Austria, and her double master's degree in Environmental and Resources Management from TH Köln – University of Applied Sciences and from the Autonomous University of San Luis Potosi, Mexico. Apart from her academic background, she is a professional violinmaker with a degree from the Violinmaking School in Mittenwald. Her research focuses on social-ecological systems with an emphasis on the interconnection between environment and art.

Renato Linhares de Assis is a researcher at Embrapa Agrobiology. He graduated in Agronomy in 1993. He holds a master's degree in Soil Science and a PhD degree in Applied Economics. He works at Embrapa's research and farmers' training advanced office at Rio Serrana, Mountain Region, Brazil, and is also a teacher at UFFRJ/EMBRAPA Master of Science Course on Organic Agriculture, Brazil, and at UFRRJ's PhD binational program on Agriculture Sciende and Innovation in partnership with Universidad Nacional de Riocuarto, Argentina. He acts mainly with agroecology, mountain agriculture, family farming, and organic agriculture, especially concerning participatory knowledge transfer and agriculture sustainability indicators.

Carlos Marconi de Souza Resende is a biologist with an MBA degree in Quality, Safety, Environment, and Health from the University Castelo Branco, Brazil. He is the head and extensionist of the local office at the Company for Technical Assistance and Rural Extension (EMATER-RIO) in Italva, Northwest of the Rio de Janeiro State, Brazil.

Peter H. May received his PhD degree in Resource Economics from Cornell University, USA. He is a full professor of the Department of Development, Agriculture and Society of the Federal Rural University of Rio de Janeiro, Brazil. He is an author and editor of a number of books in the areas of ecological economics, agroforestry, payment for ecosystem services and environmental policy, including *The Subsidy from Nature*, *Pricing the Planet*, and *Natural Resource Valuation and Policy in Brazil* (Columbia University Press) and *Economia do Meio Ambiente: Teoria e Prática* (Elsevier). Peter May is the past president of the International Society for Ecological Economics-ISEE and founder and current president of the Brazilian Society for Ecological Economics-ECOECO. He previously served as program officer at the Ford Foundation's Brazil office and as forestry officer at the FAO in Rome. He is currently a visiting scholar at the Earth Institute of Columbia University, Center for Sustainable Development, USA.

Torsten Meyer graduated as an engineer from the Institute of Agricultural Machinery and Renewable Energies at TH Köln – University of Applied Sciences, Germany. From 2008 to 2013, he was research assistant at TH Köln with international activities in Brazil, South Africa, France, England, and Italy. Currently, he runs the company TM Technik in Kronberg/Taunus in Germany that develops tailor-made solutions in agriculture machine technologies.

Carlos Frederico de Menezes Veiga graduated in Agronomy, holds a master's degree in Genetics and Plant Breeding, and a doctorate in Plant Production with an emphasis on plant biotechnology. He is a researcher at the Federal Rural University of Rio de Janeiro, Brazil, and coordinated the genetic improvement program for sugarcane for the States of Rio de Janeiro, Espírito Santo, Southern Bahia, and Northeast of Minas Gerais, Brazil. He led a research project in the area of socio-economics that generated the diagnosis of the sugarcane productive chain of the State of Rio de Janeiro, identifying technological deficiencies of small and medium producers which allowed the creation of the Sugarcane Support Program in the North and Northwest of the State of Rio de Janeiro serving approximately 1,500 producers.

Joyce M. G. Monteiro graduated in Agronomy from the Federal University of Lavras, Brazil, with a master's degree in Ecology at the University of Brasilia (UnB) and a PhD degree in Environmental Planning at the Federal University of Rio de Janeiro, Brazil. She is researcher of the Brazilian Agricultural Research Corporation, and based at the National Centre for Soil Research (Embrapa Soils). Her work focuses on environmental planning: ecosystem services and strategies for adaptation and mitigation of climate change in rural area.

Friederike Naegeli de Torres is a geographer with a master's degree in Technology and Resources Management in the Tropics and Subtropics and a PhD student at the Geoinformatics and Remote Sensing Department at the Institute for Geography, Leipzig University, Germany. She has worked in different international and interdisciplinary research projects in Latin America. Her research focus is on applicable and transferable remote sensing methods for land-use classification and pasture degradation mapping as well as on GIS-based landscape and forest fragmentation analysis.

Udo Nehren is a senior researcher and lecturer in Physical Geography and Ecosystem Management at the Institute for Technology and Resources Management in the Tropics and Subtropics, TH Köln – University of Applied Sciences, Germany. He received a PhD degree in Geography from the Leipzig University, Germany, and a habilitation (venia legendi) in Geography from University of Passau, Germany. His research emphasizes landscape evolution, ecosystem-based disaster risk reduction and adaptation, and ecosystem management with regional foci in Latin America and Southeast Asia. Udo Nehren is author and editor of books and articles related to these topics, such as the book *Ecosystem-Based Disaster Risk Reduction and Adaptation in Practice* (Springer). Furthermore, he is one of the developers of the Massive Open Online Course "Disasters and Ecosystems: Resilience in a Changing Climate."

Laura Maria Noriega Gamarra is a junior researcher working on climate change and low carbon development at the Institute for Technology and Resources Management in the Tropics and Subtropics, TH Köln – University of Applied Sciences, Germany. She holds a degree in Environmental and Sanitary Engineering from the Universidad Pontificia Bolivariana in Bucaramanga, Colombia; a specialization in Integrated Management of the Environment from the Universidad de los Andes, Colombia; and a Master of Sciences degree in Technology and Resources Management in Tropics and Subtropics with an emphasis on energy management from the TH Köln – University of Applied Sciences, Germany. Her research focuses on emissions reduction potentials and climate change mitigation strategies.

Rogério Oliveira is an associate professor at the Department of Geography and Environment of Pontifical Catholic University of Rio de Janeiro, Brazil. He holds a PhD degree in Geography from the Federal University of Rio de Janeiro and a postdoc from the Institute of Social Ecology, Vienna, Austria. His research areas combine environmental history, historical ecology, and plant ecology. His research focuses on history of landscape, human dimensions of biodiversity, and forest history.

Aline Pacobahyba de Oliveira graduated in Agronomy from the Federal University of Viçosa, Brazil. She holds a master's and doctoral degree in Soil Science and Plant Nutrition from the Federal University of Viçosa. Since 2011, she has been a researcher at Embrapa Soils and has been working in Genesis, Morphology and Classification of Soil in the Pedology Research Group and Soil and Ecosystem Services in the Ecosystem Services Network at Embrapa.

Luiz Antonio Oliveira Antunes graduated in Agronomy from the Federal Rural University of Rio de Janeiro, Brazil, and holds a master's degree in Crop Sciences from the Federal University of Viçosa, Brazil. He is currently the coordinator of the participatory research group of the Sustainable Rural Development Program – Rio Rural. His work focuses on crop sciences and agroecology.

Alejandra Pedraza Luengas is a biologist from the Universidad Militar Nueva Granada, Colombia, with a double master's degree in Natural Resources Management in the Tropics and Subtropics from TH Köln – University of Applied Sciences, Germany, and in Environmental Sciences from the Universidad Autónoma de San Luis Potosí, Mexico. Currently, she co-owns a company named Danaus Consultants, dedicated to provide consultancy services and promote good agricultural practices to stakeholders in different crop production systems in Belize, while preserving the natural resources and contributing to food security for the well-being of consumers. The provision of nonformal education through seminars, workshops, and courses in a wide array of thematical areas related to agriculture, environment, and biological sciences is one of her intends.

Rui Pedroso is a senior researcher and lecturer in statistics, econometrics, optimization modeling, and nonmarket valuation methods for environmental goods/services at the Institute for Technology and Resources Management in the Tropics and Subtropics, TH Köln – University of Applied Sciences, Germany. He received a PhD degree in Agricultural Economics from the University of Bonn, Center for Development Research (ZEF). His current regional foci are in Vietnam, Kenya, and Brazil. In Brazil, the focus is on land use and land cover change in the State of Rio de Janeiro, agricultural land abandonment, and the econometric analysis of driving factors.

Santiago Penedo-Julien graduated in Chemical Engineering from the Monterrey Institute of Technology and Higher Education, Mexico, and continued his studies in Germany gaining a master's degree in Resources Management at the Institute for Technology and Resources Managements in the Tropics and Subtropics (ITT), TH Köln – University of Applied Sciences, Germany. He received a PhD from Friedrich Schiller University Jena, Germany, in Modeling the Impacts of Different Land Uses on Water Quality. He works as a researcher and lecturer at TH Köln in hydrological and water quality modeling, Ecosystem-based Adaptation (EbA), and Water-Energy-Food (WEF) Nexus.

Fatima Conceição Marquez Piña-Rodrigues is a full professor of the Federal University of São Carlos, Brazil. Her research interest lies in the area of applied seed ecology with focus on restoration of degraded areas and conservation of mainly dry zones of tropical forest. She is a forest engineer with a master's degree in Ecological Genetics. She participated in two postdoctoral programs in Tropical Ecology at the University of Missouri Saint Louis and in Conservation in the Missouri Botanical Garden. She is author and editor of many books in the areas of forest seed technology and ecology, genetic conservation, tropical silviculture, and forest restoration. Furthermore, she is the coordinator of the Atlantic Forest Seed Network and founder of the Brazilian Committee of Seed Technology, as well as a member of the Society of Ecological Restoration and of the Brazilian Committee of the International Seed Testing Association (ISTA).

Augusto João Piratelli is a professor and senior researcher in Ecology and Conservation at the Federal University of Sao Carlos, Brazil, and a member of the Brazilian Society of Ornithology. He is a biologist and holds a PhD degree in Zoology from the State University of São Paulo, Brazil, and a postdoctoral fellowship in Ecology at the University of Missouri – St. Louis, United States. His research aims at identifying and mitigating the anthropogenic effects, e.g., agriculture and urbanization, on neotropical birds. He investigates their trophic ecology, closely linked with ecosystem functions and services. Augusto J. Piratelli is author and editor of more than 30 articles in peer-reviewed journals, seven book chapters, and three books.

Claudia Raedig gained her PhD degree in Biology from Leipzig University, Germany. She works as a senior researcher and lecturer in the field of Biodiversity and Connectivity Conservation at the Institute for Technology and Resources Management in the Tropics and Subtropics at TH Köln – University of Applied Sciences, Germany. Her research focuses on connectivity conservation management, particularly on the development of strategies for enhancing connectivity for biodiversity conservation in tropical and coastal ecosystems in Southeast Asia and Latin America.

Juan Luis Ramirez Duval is a civil engineer with a master's degree in Integrated Water Resources Management (IWRM) from the TH Köln – University of Applied Sciences, Institute for Technology and Resources Management in the Tropics and Subtropics (ITT). He worked as a scientific staff at the ITT and was responsible for implementing and coordinating water monitoring activities within the research project INTECRAL in Rio de Janeiro State (2013–2017). At present he works at the company ribeka GmbH, Germany as project manager and scientific staff in the field of water and environmental monitoring and data management.

Juan Antonio Reyes-Agüero is a biologist from the Universidad Michoacana de San Nicolás de Hidalgo, Mexico. He holds a Master of Sciences degree from the Colegio de Postgraduados, Mexico, and a PhD degree in Biological Sciences from the Universidad Nacional Autónoma de México. He was a professor at four Mexican and one foreign higher education institutions (TH Köln – University of Applied Sciences, Germany). Since 1997, he has been a professor-researcher at Universidad Autónoma de San Luis Potosí (UASLP), Mexico. He has the level I of Mexican National System of Researcher. His academic interest is the research on ecology and management of renewable natural resources; as professor, he teaches the subjects of Ecology, General Ecology and Biodiversity, and Evolution in postgraduate and undergraduate courses of the UASLP, Mexico. He published more than 50 research articles, four books, and eight book chapters.

Lars Ribbe is a professor for Integrated Land and Water Resources Management and director at the Institute for Technology and Resources Management in the Tropics and Subtropics (ITT), TH Köln – University of Applied Sciences, Germany. He obtained his PhD degree in the field of hydro-informatics and works on river basin assessment, modeling, and management. His research is targeted to develop knowledge systems that help decision makers to cope with prevailing water resources related challenges like water scarcity and drought, floods, and pollution. Current research projects are implemented in various river basins in Asia, Africa, and Latin America.

Ronny Richter is a field-platform manager at the German Centre for Integrative Biodiversity Research (iDiv), Halle-Jena-Leipzig, and currently conducts his PhD research at the Department of Geoinformatics and Remote Sensing, Institute for Geography, Leipzig University, Germany. His research focuses on the assessment of scale-dependent biodiversity patterns in forests and their implications for ecosystem functions and services. His methodological background includes the acquisition and analysis of airborne hyperspectral and thermal imagery data as well as tree canopy research at the Leipzig Canopy Crane.

Vanesa Rodríguez Osuna is a senior researcher and project manager at the CUNY Advanced Science Research Center in New York, where she leads projects in the area of sustainable investments. She is an environmental engineer, and gained her MSc degree in Natural Resource Management and Technologies in the Tropics from TH Köln – University of Applied Sciences, Germany, and her PhD degree in Physical Geography from the Leipzig University, Germany. Her work focuses on ecosystem services assessments, environmental policy, environmental impact assessment, and bridging the science-policy interface on environmental topics. She is the lead author of the Intergovernmental Platform on Biodiversity and Ecosystem Services (IPBES)'s Regional Assessment and Summary for Policymakers for the American continent.

Simone Sandholz is a senior scientist at the Institute for Environment and Human Security of United Nations University (UNU-EHS). She is part of the Vulnerability Assessment, Risk Management, and Adaptive Planning (VARMAP) section and holds a PhD degree in Geography from University of Innsbruck, Austria. In her research, she focuses on sustainable urban and regional development, in particular on urban regeneration and risk reduction including nature-based solutions, with a regional focus on Asia, e.g., Indonesia, Nepal, and Latin America, e.g., Brazil. Besides other international teaching activities, Simone Sandholz is co-developer of the Massive Open Online Course "Disasters and Ecosystems: Resilience in a Changing Climate."

Juliana M. Santos is a PhD candidate of the group of Geographic Information Science at the Friedrich Schiller University Jena, Germany. She graduated in Environmental Engineering from the University of Aveiro, Portugal. Her research interests include hydrological and soil erosion processes at different spatiotemporal scales and climate and land-use change impacts on watershed processes. Therefore, she applies process-based ecohydrological models. Her regional foci are in Brazil and Portugal.

Dietmar Sattler is a biologist and senior researcher at the Institute of Geography, Leipzig University Germany. He holds a PhD degree in Botany and Urban Ecology from the Leipzig University. His research emphasizes tropical forest fragmentation and carbon stock assessment, forest conservation strategies, and the assessment and rehabilitation of land degradation. He coordinated and participated in several international research and development projects mainly in Brazil with a focus on sustainable land-use management and biodiversity conservation.

Sabine Schlüter is a professor and senior researcher in Environment and Resource Economics at the Institute for Technology and Resources Management in the Tropics and Subtropics at TH Köln – University of Applied Sciences, Germany. She holds a PhD in Agricultural Sciences from the University of Bonn, Germany. Her research areas include valuation of environmental goods/services, farm system analysis, regional sector modeling, irrigation management/governance, and food production chain analysis.

Azeneth Eufrausino Schuler graduated as a forestry engineer from the University of São Paulo, Brazil, and earned an MSc degree in Science of Environmental Engineering and a DSc degree in Sciences: Nuclear Energy in the Agriculture, both from the University of São Paulo, Brazil. From 2003 to 2005, she worked as a postdoc at the Institute for Environmental Research of Amazonia (IPAM), within the LBA project – Large-Scale Biosphere-Atmosphere Experiment in the Amazonia. Since 2006, she has been a researcher at Embrapa (Empresa Brasileira de Pesquisa Agropecuária), at the unit Embrapa Solos. She has worked in the following research lines: watershed management, hydrological modeling, small catchments, forest hydrology, and water ecosystem services.

Roman Seliger is a physical geographer and PhD student at the Institute of Geography, Leipzig University, Germany. His research focus is on landscape analysis and soil sciences in tropical and subtropical regions, with a special emphasis on assessment and rehabilitation of degraded pastures. He follows holistic and pragmatic approaches, combining remote sensing and field-work tools (ground-truth) with direct benefit to environment and target groups such as farmers and regional authorities. He has worked in several international and interdisciplinary projects, both at University and European Commission level.

Antonio Soares da Silva is a geographer and associated professor at the Institute of Geography of the State University of Rio de Janeiro (UERJ), Brazil. He holds a PhD degree in Engineering and Environmental Geology from the Federal University of Rio de Janeiro. His research emphasizes the soil science, soil erosion, and soil degradation. He coordinates a research group with a focus on tropical soils and the changes caused by deforestation, agriculture, and livestock systems. Currently, he develops a research about the influence of human activities at soil erosion and sedimentation during the last two centuries.

Alba Leonor da Silva Martins graduated in Agronomy, with a specialization in management and conservation of soil. She earned a master's degree in Agroecology from the Estadual University of Maranhão (UEMA), Brazil, and a PhD in Soil Science from the Estadual University of São Paulo (UNESP/Jaboticabal), Brazil. She has been a researcher at the Brazilian Agricultural Research Corporation (Embrapa Soils) since 2011. She is specialized in management and conservation of soil, especially no-till and ecosystem services.

Guilherme Strauch is an agronomist engineer. He earned an MSc degree in Agroecology from University of Andalucía, Spain, and has a PhD degree in Agroecology under the Natural Resources Program of University of Córdoba, Spain. He joined several working groups to elaborate public policies to support agroecology and family agriculture. He has been a member of the board of ABA – Brazilian Agroecology Association since 2012. Currently, he serves as technical advisor for training and Agroecology of the Sustainable Rural Development Program – Rio Rural Program, at the Secretary of Agriculture, Livestock, Food Supply and Fisheries of Rio de Janeiro State, Brazil.

Omar Augusto Torres Zárate is a business administrator with an MBA degree from Leipzig University, Germany. He is currently a PhD candidate in Socioeconomics, researching on the relationship between innovation and entrepreneurship within the new rurality perspective. He previously worked as a consultant in innovation projects at several German universities and research institutions, supporting the development of product and service innovations oriented to existing and new markets. Omar Torres is currently involved in the further development of a successful spin-off of the University of Applied Sciences of Leipzig (HTWK Leipzig), Germany.

Raúl Vicens gained a bachelor's degree in Geography from the University of Havana, Cuba (1991); a master's degree in Geography from Rio de Janeiro Federal University, Brazil (1997); and a PhD degree in Geography from Rio de Janeiro

Federal University, Brazil (2003). He is currently an associate professor at the Fluminense Federal University, Brazil. He has experience in geosciences, with an emphasis on remote sensing. His main research topics are as follows: monitoring land-use/land-cover changes; dynamics, evolution, and landscape stability; and landscape modeling.

Jennifer Viezzer is an environmental analyst of the Department of Ecosystems Conservation, Secretary of Biodiversity, from the Brazilian Ministry of Environment. She works on the Biodiversity and Climate Change in the Atlantic Forest Project and on the Biodiversity and Ecosystems Strategy of the National Adaptation to Climate Change Plan. She is a forestry engineer. She earned an MSc degree in Forestry Engineering, both from Federal University of Paraná, Brazil. She has experience in ecosystem-based adaptation, territorial planning, and public policies for the Atlantic Forest.

Jens Wesenberg is a biologist and senior researcher at the Senckenberg Natural History Museum Görlitz, Germany, where he works mainly on digitization projects focusing on collection-based research and on the use of digital applications for museum expositions. He holds a PhD degree in Botany and Plant Ecology from the Leipzig University, Germany. For a long time, he researched on tropical forest ecology and conservation, with a special emphasis on vegetation patterns along environmental gradients and forest fragmentation. He coordinated and participated in several international research and development projects, mainly in Brazil, with a focus on sustainable land-use management and biodiversity conservation.

Roland Wollenweber holds a master's degree in Mechanical Engineering and Environmental Technologies and works as a business field developer, project controller and team manager of the Carbon and Energy Services department at the headquarters of TÜV Rheinland in Cologne, Germany. He has been working as a project manager and expert in the area of research and development in an international environment for more than ten years. In his professional career he has acquired solid management abilities and has successfully designed and implemented various large scale multi-stakeholder projects for private clients as well as public bodies, authorities and NGOs. Roland Wollenweber has implemented the ISO 14065 accreditation for TÜV Rheinland.

Chapter 1
Rio de Janeiro: A State in Socio-ecological Transformation

Udo Nehren, Dietmar Sattler, Claudia Raedig, Helga Hissa, and Sabine Schlüter

Abstract Within the period of about 500 years since the European colonization, today's state of Rio de Janeiro has been transformed from a sparsely populated forest area to an economic and cultural hub of Brazil and the whole continent of South America. This development process is on the one hand characterized by technological progress and economic and population growth but on the other hand also accompanied by overexploitation of natural resources, land and ecosystem degradation, uncontrolled urban sprawl, and increasing social disparities. In the past few years, some efforts have been made to counteract these negative trends, but many challenges remain to keep pace with the rapidly changing demands and needs in a globalized world. In this introductory chapter, we provide an overview of the recent developments in the environmental and agricultural sectors in the state of Rio de Janeiro. We focus on the rural areas of the state that provide essential goods and services for the urban agglomerations. In this context we briefly introduce the chapters of the book that are divided into five main sections: (a) Agricultural Management; (b) Ecosystem Management and Biodiversity; (c) Integrated Water Management; (d) Land Restoration, Disaster Risk Reduction, and Climate Change Adaptation; as well as (e) Environmental Governance and Economic Instruments.

Keywords Rio de Janeiro state (RJ) · Socio-ecological transformation · Rural development

---

U. Nehren (✉) · C. Raedig · S. Schlüter
TH Köln – University of Applied Sciences, Institute for Technology and Resources Management in the Tropics and Subtropics (ITT), Köln, Germany
e-mail: udo.nehren@th-koeln.de; claudia.raedig@th-koeln.de; sabine.schluter@th-koeln.de

D. Sattler
Leipzig University, Institute of Geography, Physical Geography and Environmental Research, Leipzig, Germany
e-mail: sattler@uni-leipzig.de

H. Hissa
The Sustainable Rural Development Program in Micro-Watersheds of the State of Rio de Janeiro – Rio Rural, Niterói, Rio de Janeiro, Brazil

U. Nehren et al. (eds.), *Strategies and Tools for a Sustainable Rural Rio de Janeiro*, Springer Series on Environmental Management,
https://doi.org/10.1007/978-3-319-89644-1_1

1.1 Introduction

There are probably only a few places on Earth that are as closely associated with certain clichés and stereotypes as Rio de Janeiro. There is first of all the natural beauty and cultural diversity of the *Cidade Maravilhosa* (marvelous city), with its world famous beaches such as *Copacabana* and *Ipanema*, the spectacular rock formations with the inselberg *Pão de Açúcar* (sugarloaf mountain) and the *Corcovado*, and last, but not least, the laissez-faire lifestyle with *samba, bossa nova, and forro music*, carnival, and, of course, football. Hence, Rio de Janeiro became the epitome of a tropical paradise, a city of hope and future, which attracted both the prosperous international business world and the poor from the northeastern and interior parts of the country to the same extent as adventurers from all over the world.

Rio de Janeiro was founded in 1565 and was the capital of the Portuguese colony and later the independent state of Brazil (1763–1960). During this time, the city flourished and became Brazil's administrative, cultural, and commercial center. Even though Brasília replaced Rio de Janeiro as the capital in 1960, the city has remained one of the economic and cultural hubs of the country and the whole continent. The importance of the city is also reflected by rapid growth of the urban population, surpassing 1 million inhabitants already in 1920, reaching 3.3 million in 1960, 5.2 million in 1980, and estimated 6.5 million in 2016 (IBGE 2010). This growth, however, was accompanied by segregation and suburbanization processes with a continuous growth of marginal settlements, the so-called *favelas*. Since the abolition of slavery in the nineteenth century and severe economic crises in Rio's rural coffee and sugar cane production in the twentieth century, a strong and constant migration of the rural working population has gone toward industrial and domestic jobs in and around Rio (Abreu 2013). Additionally, recent progressing degradation of pasture and agricultural land, caused by unsustainable and inappropriate land use, makes the livelihood of smallholder farmers more and more complicated and pushes the young generation to migrate to Rio's metropolitan area. Rising inequality and a lack of prospects have led to increased poverty and have encouraged crime and violence, and insufficient urban infrastructure, such as a proper sewage disposal, has harmed the environment. This was once again visible during the 2016 Summer Olympics, when the sailing and marathon swimming competitions took place in the highly polluted Guanabara Bay.

Taking a closer look at the state of Rio de Janeiro (RJ), it is by far more than the city plus its hinterland. Located in the Southeast Region of Brazil together with the states of São Paulo (SP), Minas Gerais (MG), and Espirito Santo (ES), RJ is the third smallest of the 27 Brazilian states (Fig. 1.1). However, its area of 43,780 km^2 is comparable to the size of Denmark, Switzerland, or the Netherlands and its population of 16.64 million (estimation for 2016 provided by IBGE 2015) to that of Ecuador – and not much less than Chile, which is about 17 times larger in land surface. Despite its small size, RJ is the third most populated Brazilian state with the second highest population density (378 inhabitants per km^2) and the second largest

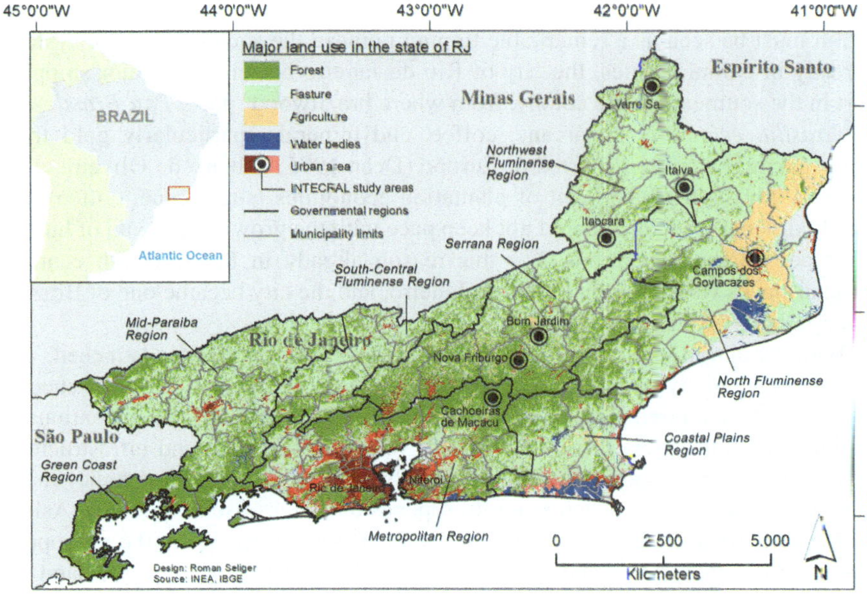

Fig. 1.1 Map of Rio de Janeiro state with main land use units and administrative boundaries (own elaboration)

economy of the country, surpassed only by SP. Furthermore, the state is characterized by enormous biological and cultural richness.

Located in the biome of the Mata Atlântica (Atlantic Forest), the state is inhabited by a high number of endemic and threatened plant and animal species (Galindo Leal and de Gusmão Câmara 2003). Within the Mata Atlântica, which is designated as one of 36 global biodiversity hotspots (CEPF; www.cepf.net), RJ is located in the "Serra do Mar Ecological Corridor," which is one of the areas with the highest biodiversity of the biome (Raedig and Lautenbach 2009). This richness can at least partly be explained with the diverse landscape patterns and topographical conditions under tropical to temperate climates that favored the formation of a variety of habitats and ecological niches. At the same time, the steep ranges of the Serra do Mar, Serra da Mantiqueira, and the coastal mountains did not allow an intensive agricultural or forestry utilization (Nehren et al. 2013). Therefore, the steep slopes of these mountain ranges are still largely covered by forests that form ecological corridors and are surrounded by highly fragmented agricultural and urban landscapes (Fig. 1.2a, b).

Human occupation in RJ can be traced back to at least 8000 years BP, when shell mound builders (*sambaquianos*) settled around the lagoon areas of today's RJ (Ribeiro de Oliveira et al. 2018). When the first Europeans arrived in the early sixteenth century, the coastal region was occupied by the agricultural-based Tupi-Guarani culture, but with the exploitation of land and human resources in the course of the European colonization, the traditional life-forms of the indigenous groups were suppressed and finally disappeared. Therefore, the European colonization must be seen as a remarkable tipping point in the socio-ecological system. Already in colonial times, the city of Rio de Janeiro became the major shipping port in the southeast of the colony, from where brazilwood (port.: *Pau-Brasil*; sci.: *Paubrasilia echinata*), sugarcane, coffee, and minerals (particularly gold from Minas Gerais) were transported to Europe (Dean 1995; Ribeiro de Oliveira et al. 2018). With the establishment of plantation economies (sugar cane, coffee), the supply of local workforce could not keep pace with the growing demand of human resources on the plantations. For this reason already in the sixteenth century, African slaves were brought to Rio de Janeiro, and the city became one of Brazil's main gateways for slave trade.

With the abolition of slavery in 1888 (*Lei Áurea*), a new era was launched. The plantation economy widely collapsed, and much of the plantations were successively transformed into grazing land and small-scale agriculture, which continue to be dominant until today (Nehren et al. 2013). The agricultural and infrastructural development of the state was strongly driven by an active immigration policy in the early twentieth century, when many Europeans and to a lesser extent also Asians immigrated to the state. At the same time, urbanization processes in the metropolis but also in smaller cities and towns increased, leading to a growing demand for natural resources and agricultural products. This in turn stimulated land use intensification and the expansion of intensive crop systems, in particular vegetable cultivation and horticulture, in the surroundings of the cities.

This briefly outlined historical development – together with the polarization of Rio de Janeiro city that attracted many immigrants from other Brazilian states in modern times – explains on the one hand the ethnic diversity and on the other hand the contrast between the vibrant metropolis and the laid-back rural hinterland of the state. However, the historical exploitation of the land, the rapid urbanization processes of the twentieth century, and recent unsustainable land use practices have also left widely visible scars in the landscape: highly fragmented secondary forests outside protected areas, deeply incised gully systems as a severe form of soil erosion (Fig. 1.2c), as well as polluted rivers and streams (Fig. 1.2d), among other forms of land and water degradation.

Today, both population and industries of the state are concentrated in the metropolitan region of Rio de Janeiro (*Região Metropolitana do Rio de Janeiro*) that surrounds the Guanabara Bay in a wide belt of urban development and includes 30 of the 92 municipalities of the state (Fig. 1.1). Among these municipalities, Rio de Janeiro (6.5 million inhabitants) and São Gonçalo (1.05 million inhabitants) are the largest, followed by Duque de Caxias, Nova Iguaçu, Niteroí, Belford Roxo, and São João de Meriti (all between 400,000 and 1 million inhabitants; IBGE 2016). Altogether, the metropolitan region covers an area of 6745 km^2 and is home to about 12.3 million people (IBGE 2016), which are about three quarters of the whole state. Together with the metropolitan regions of São Paulo and Belo Horizonte, the metropolitan region of Rio de Janeiro forms Brazil's economic backbone. However, apart from this densely urbanized area, the metropolitan region of Rio de Janeiro also includes parts of the *Serra do Mar* mountain range with the municipalities of

Fig. 1.2 Steep slopes of the Serra dos Órgãos mountain range close to Teresópolis covered with Atlantic Forest (**a**). Fragmented agricultural landscape close to Varre-Sai with plantations of vegetables and coffee and fragments of secondary forest (**b**). Massive gully erosion at a degraded sloped pasture near Cachoeiras de Macacu (**c**). Rectified tributary of the Macacu River transporting pesticides, fertilizer, and litter from agriculture to the Guanabara Bay (**d**). (Photos: D. Sattler)

Petrópolis, Teresópolis, and Magé that share one of the oldest Brazilian national parks, the NP Serra dos Orgãos, which was founded in 1939

Besides the metropolitan region, RJ is divided into five administrative mesoregions. The mesoregion *Sul Fluminense* is located in the western part of the state (Fig. 1.1) and includes 14 municipalities with a total population of 1.06 million inhabitants (IBGE 2010). The economy of this mesoregion is based on industries (metal-mechanical, automotive, energy, metal, among others), on tourism mainly along the coastline, and on livestock production and agriculture in the hinterland. Toward the east of the metropolitan region lies the smallest of the six mesoregions, *Baixadas Litorâneas* with ten municipalities and a total population of 0.68 million (IBGE 2010).

The mesoregion includes some important tourist destinations, such as Armação dos Búzios, Cabo Frio, Arraial da Cabo, and Rio das Ostras. The mesoregion of *Norte Fluminense* with nine municipalities and a total population of 0.86 million (IBGE 2010) is toward the northeast. This mesoregion is known for its oil industry but also for tourism along the coast and a high number of public universities. The mesoregion *Centro Fluminense* is located in the mountainous hinterland of the state and includes 16 municipalities, among them Nova Friburgo, an important mountainous tourism destination and location of the textile industry. This mesoregion has a population of only 0.47 million (IBGE 2010) and is in large parts dominated by livestock production and smallholder agriculture. Frequently, the term *Serrana* region is referenced, which includes Nova Friburgo and adjacent *Centro Fluminense* municipalities as well as the municipalities of Petrópolis, Teresópolis, and São José do Vale do Rio Preto in the metropolitan region of Rio de Janeiro. Finally, the mesoregion *Noroeste Fluminense* forms the northernmost part of the state and includes 13 municipalities with a total population of 0.33 million (IBGE 2010). This mesoregion is historically known for its coffee plantations, but milk production and some paper and cellulose production play a major role today. In this book, geographical directions as well as the partly contradictorily named mesoregions are used; mesoregions are written with capital letters, whereas the geographical location is written with lowercase characters.

After years of economic growths and development, the 2015–2017 economic crises severely affected the state. However, this crisis can also be seen as a chance to rethink the paradigm of neo-extractivism and exploitation of natural resources and change the direction toward a more sustainable "green" development. Not only the economic crisis but also severe disasters, such as the floods and mudslides in the Serrana region in 2011 that claimed more than 900 victims or the 2014–2017 drought event demonstrated the ecological and social vulnerability of the state's population and environment. Even though environmental degradation did not directly trigger these disasters, the impacts were intensified by land and ecosystem degradation. With the projected increase of extreme events in Southeast Brazil within the next decades (Nehren et al. 2018), the risk of floods, landslides, mudslides, and droughts will likely aggravate, so that also from the perspective of civil protection as well as food and water security, better management of land and water resources is urgently required. Therefore, we can say that the state needs a transformation toward an environment that can compensate for future disaster impact, which has probably already been started but is threatened by internal and external pressures. If this socio-ecological transformation process is successfully driven forward, the highly exploited rural landscapes will be converted into resilient landscapes that will offer various opportunities for its inhabitants.

In this book we address some of the most pressing socio-environmental challenges in the rural areas of RJ. We focus on these rural areas as they provide important ecological functions and at the same time ensure the supply of food, water, and energy for the cities. The idea of the book emerged within the German-Brazilian research project INTECRAL – Integrated eco technologies and services for a sustainable Rural Rio de Janeiro (Chap. 4) – that is conducted in cooperation with the rural development program Rio Rural (Chap. 3). It aims at compiling up-to-date knowledge and scientific advances on land, water, and ecosystem management in RJ, covering a spectrum from basic and applied research to science-policy interfaces, planning concepts, and technical tools. Thus, it has a model character for other rural areas in Latin America. All chapters are written by experts in their field including researchers and practitioners of the INTECRAL project and the Rio Rural program as well as invited scientists from Brazilian universities and research institutes. One major concern of the editors is to help bridge existing gaps between science, policies, and practice in the rural development of RJ. Therefore the book is not exclusively directed at academics but equally at rural and environmental planners, municipal representatives, policy makers, and nongovernmental actors of the civil society.

The book is organized in four introductory chapters (Chaps. 1-4), five thematic parts with four to seven chapters each, and the conclusions as Part VI. All chapters have undergone a double review process to ensure the best possible quality. Chap. 2, written by Rogério Ribeiro de Oliveira and his colleagues from the Pontifical Catholic University of Rio de Janeiro, introduces the environmental history, forests, and landscape uses in RJ state. In Chaps. 3 and 4, the editors provide an overview of the Rio Rural program (Chap. 3) and the INTECRAL project (Chap. 4), which were closely coordinated in their research and implementation activities and thus substantially facilitated this publication.

Part I is dedicated to the agricultural management in RJ. In Chap. 5, Joyce Monteiro and her colleagues from Embrapa Solos provide an overview of current soil and water conservation management in RJ and their impacts on ecosystem services (ES) based on a review of several studies. They highlight the importance of monitoring ES using multiple criteria and indicators at different scales. In Chap. 6, Pedro Castro et al. analyze the process of farm abandonment in RJ using agricultural production data, and they provide figures that show the spatial and temporal patterns for the main crops planted in the state. Interestingly, there has been a reduction of the total crop area in the last 25 years and evidence of a forest recovering process for recent past years. This development sharply differs from the agriculture expansion and forest loss trend that is observed in other parts of Brazil. Chapter 7 by Silvia Berenice Fischer and her colleagues presents a participatory appraisal for a silvopastoral system that has been implemented in the municipality of Italva. The authors analyze and discuss socioeconomic and ecological trade-offs of silvopastoral systems for small dairy farmers and formulate their lessons learned for the implementation of silvopastoral systems in the region. Also in Italva, Maria Luisa Espinel Ramos and her colleagues analyze Good Agricultural Practices (GAP) in smallholders' dairy production systems (Chap. 8). The authors focus on animal welfare concepts and sustainable production systems that at the same time contribute to increased and environmentally compatible production, access to broader markets, fair prices, and high milk quality. In Chap. 9, Carl-Friedrich Gaese and his colleagues present a new harvesting technology for small- and medium-scale sugarcane farmers, the so-called Cologne solution. They look at the implementation requirements, in particular technical, economic, and site-specific conditions, and analyze the harvest chain of sugarcane.

In *Part II*, the integration of biodiversity into ecosystem management is examined from two perspectives: how can ecosystem management aid biodiversity conservation and how can biodiversity be used for ecosystem management. Focusing on birds, in Chap. 10, Augusto João Piratelli and his colleagues use both perspectives to discuss the role of biodiversity in the agriculture-dominated matrix of rural RJ. Special attention is given to recent changes in the legal framework relevant for biodiversity conservation, mainly to the Brazilian Forest Law coupled with the introduction of an obligatory rural environmental register for landowners. In accordance with Part I, the guiding question of this chapter is how to reconcile agricultural production and biodiversity conservation. In Chap. 11, by Lorena Guzman and Claudia Raedig, the agriculture-dominated matrix provides the frame for a connectivity analysis to systematically link privately protected areas to counteract forest fragmentation and to strengthen the regional importance of protected areas. In Chap. 12, by Diana Birnbaum and Georg Lamberty, the occurrence of riparian plant species as well as abiotic factors such as flow diversity is used for the hydromorphological assessment of streams to enable the selection of adequate ecological development possibilities. In the light of an expected rise of restoration and rehabilitation measures due to recent legal changes, Chap. 13, by Dietmar Sattler and his colleagues, addresses the use of native plant species as an alternative to the cultivation of exotic species. Chapter 14, by Silke Lichtenberg et al., then focuses on the use and conservation options in RJ for the endangered national tree of Brazil, *Paubrasilia echinata* Lam., a species endemic to the Atlantic Forest.

Part III covers the topic of integrated water management from various perspectives. In Chap. 15, Lars Ribbe and Rosa Formiga provide an overview of the water resources management and water security in RJ. Chapter 16, by Juliana Santos and colleagues, analyzes hydroclimatic time series for hydrological studies, such as ecohydrological modeling. Based on the case of a mesoscale watershed, the Muriaé River, the authors provide insights into long-term hydroclimatic patterns. In Chap. 17, Annika Künne and her team present the results of the ecohydrological modeling and scenario impact assessment for the same case study area. The presented model allows better understanding of hydrological dynamics under changing environmental conditions. Chapter 18, by Jaime Cardona and his colleagues, is dedicated to reuse-oriented decentralized wastewater and sewage sludge treatment for small urbanized rural settlements. Using an integrated approach including a cost-benefit analysis, the authors show that decentralized sanitation and reuse (DESAR) solutions can contribute to the improvement of wastewater sanitation by reducing costs and improving the nutrient recovery capabilities of sanitation interventions. Finally, Santiago Penedo and his team close this section by introducing a tool to assess land use impacts on surface water quality (Chap. 19). Their findings are based on field research in the Guapi-Macacu River Basin.

Part IV includes the topics of disaster risk reduction, climate change adaptation, land degradation, and rehabilitation. In a tropical mountainous region such as RJ, the risk of disasters caused by floods and landslides is high. Chapter 20, by Udo

Nehren et al., provides an overview on the different types of natural hazards and their frequency and intensity and gives an insight on how far climate variability and human impact have affected the occurrence of hydrometeorological hazards from the early Holocene to the present. How the risk of disasters has been managed, how risk governance can be improved by participatory stakeholder involvement, and how the mega-disaster of 2011 led to a paradigm shift that includes risk assessment, prevention, mitigation, preparedness, response, and recovery are concisely described in Chap. 21 by Beate Frank et al. and in Chap. 23 by Alicia Bustillos et al. In this context, proactive approaches such as ecosystem-based adaptation (EbA) and ecosystem-based disaster risk reduction (Eco-DRR) aim at the sustainable management, conservation, and restoration of ecosystems to cope with the adverse effects of climate change and reduce the impacts of natural hazards. A comprehensive overview of climate change adaptation (CCA) and disaster risk reduction (DRR) policies at national and RJ state levels is given and discussed in Chap. 22 by Wolfram Lange and his colleagues. Besides natural hazards, degradation caused by unsustainable and inappropriate land use practices is a very critical problem compromising essential ecosystem services and rural livelihood. Chapter 24, by Antonio Soares da Silva and his colleagues, introduces the principal preconditions, drivers, and status of soil degradation in the RJ. Possibilities to counteract degradation with rehabilitation measures are presented with a case study from Northwest RJ in Chap. 25 by Roman Seliger et al., and a remote sensing-based multi-criteria prioritization of degraded land for either rehabilitation measures or afforestation is discussed in Chap. 26 by Friederike Naegeli and her colleagues.

Part V deals with environmental governance and economic instruments for land management. Chapter 27, by Peter May and his colleagues, reviews the evolution of public policies and local innovation in landscape conservation in RJ. The authors present and discuss policy interventions and instruments that are concerned with sustainable resources management and land use at multiple scales. In Chap. 28, Vanesa Rodrigues Osuna and her team present strategies and activities of the Rio Rural program that aim at promoting sustainable agriculture, boosting productivity, and enhancing climate mitigation and adaptation. Chapter 29, by Omar Zarate and Utz Dornberger, is dedicated to the promotion of alternative income diversification strategies in rural RJ. Their study is located in the economically weak northwestern region of the state, where many rural households base their income on unproductive farm activities and subsidies. Finally, Laura Noriega et al. present regional marketing strategies for sustainable production in RJ (Chap. 30). In this context, the authors refer to payment for ecosystem services (PES) for eco-friendly traditional farming, regional sustainability labels, and organic farming and carbon mitigation initiatives. The book closes with Part VI, the conclusions (Chap. 31) by Claudia Raedig et al., in which the editors summarize the main scientific outcomes and practical approaches presented in this book and sketch the way forward.

References

Abreu MA (2013) A evolução urbana do Rio de Janeiro, 4th edn. Instituto Pereira Passos, Rio de Janeiro

CEPF – Critical Ecosystem Partnership Fund (n.d.) The biodiversity hotspots. http://www.cepf.net/resources/hotspots/Pages/default.aspx. Accessed 19 Dec 2017

Dean W (1995) A Ferro e Fogo – a história e a devastação da Mata Atlântica Brasileira, Ed. Schwarcz Ltda, São Paulo

Galindo Leal C, de Gusmão Câmara (eds) (2003): The Atlantic Forest of South America: Biodiversity status, threats, and outlook (State of the Hotspots, 1), Center for Applied Biodiversity Science at Conservation International, Island Press, Washington, DC

IBGE – Instituto Brasileiro de Geografia e Estatística (2010) Sinopse do censo demográphico 2010, Tabela 1.6: População nos Censos Demográficos, segundo os municípios das capitais - 1872/2010. http://www.censo2010.ibge.gov.br/sinopse/index.php?dados=6&uf=00. Accessed 19 Dec 2017

IBGE – Instituto Brasileiro de Geografia e Estatística (2015) Estimativas da população residente no Brasil e Unidades da federação com data de referência em 1° de julho de 2015. Accessed 19 Dec 2017

IBGE – Instituto Brasileiro de Geografia e Estatística (2016) Estimativas da população residente no Brasil e Unidades da federação com data de referência em 1° de Julho de 2016. Accessed 19 Dec 2017

Nehren U, Kirchner A, Sattler D et al (2013) Impact of natural climate change and historical land use on landscape development in the Atlantic Forest of Rio de Janeiro, Brazil. Anais Acad Brasil Ciênc 85(2):311–332

Nehren U, Kirchner A, Lange W et al (2018) Natural hazards and climate change impacts in the state of Rio de Janeiro: a landscape historical analysis. In: Nehren U, Schlüter S, Raedig C, Sattler D, Hissa H (eds) Strategies and tools for a sustainable rural Rio de Janeiro. Springer International Publishing, Cham

Raedig C, Lautenbach S (2009) Broad-scale angiosperm diversity in Brazil's Mata Atlântica: using monographic data to assess prospects for conservation. In: Gaese H, Torrico JC, Wesenberg J, Schlüter S (eds) Biodiversity and land use systems in the fragmented Mata Atlântica of Rio de Janeiro, Goettingen, pp 217–243

Ribeiro de Oliveira R, Stingel Fraga J, MacLeod Hickie M (2018) Environmental history, forests and landscape uses in Rio de Janeiro state. In: Nehren U, Schlüter S, Raedig C, Sattler D, Hissa H (eds) Strategies and tools for a sustainable rural Rio de Janeiro. Springer International Publishing, Cham

Chapter 2
Environmental History, Forests, and Landscape Uses in Rio de Janeiro State

Rogério Oliveira, Joana Fraga, and Mark Hickie

Abstract Although Rio de Janeiro state (RJ) is the third smallest of the 27 Brazilian states in land area, its environmental diversity is vast; the variability of altitudes and geomorphological situations create significant ecological heterogeneity throughout the state. Human occupation in the region dates back approximately 8000 years, and when Europeans first arrived in Brazil in 1500, they discovered the agricultural practices developed by indigenous populations. The arrival of the Europeans, though, represented a sharp breaking point in land use patterns, ecosystem alteration, and erosion and soil depletion – all of which substantially altered the landscape. This chapter will focus on the historic economic eras of RJ, cycles that included sugarcane, coffee, charcoal energy production from firewood, and livestock. Each of these socioeconomic phases resulted from specific economic and historical contexts and caused distinct transformations to the landscape. As a direct consequence of its past use, only about 30% of RJ land area is presently covered by forest vegetation; the aforementioned economic eras have converted the remaining 70% of the state into pasture.

Keywords Atlantic Forest · Deforestation · Sugar production · Historical ecology · Soil use

R. Oliveira (✉) · J. Fraga
Department of Geography and Environment, Pontifical Catholic University of Rio de Janeiro, Rio de Janeiro, RJ, Brazil
e-mail: rro@pcu-rio.br

M. Hickie
Pontifical Catholic University of Rio de Janeiro, Rio de Janeiro, Rio de Janeiro, Brasil

U. Nehren et al. (eds.), *Strategies and Tools for a Sustainable Rural Rio de Janeiro*, Springer Series on Environmental Management,
https://doi.org/10.1007/978-3-319-89644-1_2

2.1 The First Uses of the Land

Although Rio de Janeiro state (RJ) is the third smallest of the 27 Brazilian states – with an area of 43,780,172 km^2 – its environmental diversity is not proportional to its size. With an extremely diversified topography, its landscape presents strong contrasts: high mountains, both on the seashore and inland; hills and valleys; bays, dunes, *restingas* and beaches, as well as rainforests in several successional stages. The variability of altitudes and geomorphological terrain are responsible for great ecological diversity. The Serra do Mar and Serra da Mantiqueira are the most prominent mountain ranges in the state, and their highest points exceed 2700 m in elevation as seen in Fig. 2.1.

Environmental diversity was also a factor influencing land use according to the specific demands of the different economic cycles and the implicit needs of territory occupation (Abreu 1992). The Serra do Mar was a mountainous and forested barrier for more than 200 years after European arrival separating the remote and "uninteresting" interior – in the eyes of the Portuguese – from the coast, where the economic activities of the colony developed (Lamego 1963). The coastal lowlands provided fertile ground for sugarcane plantations and cattle breeding; it also enabled mercantile trading efforts, mainly by maritime transport. Later, coffee cultivation occurred

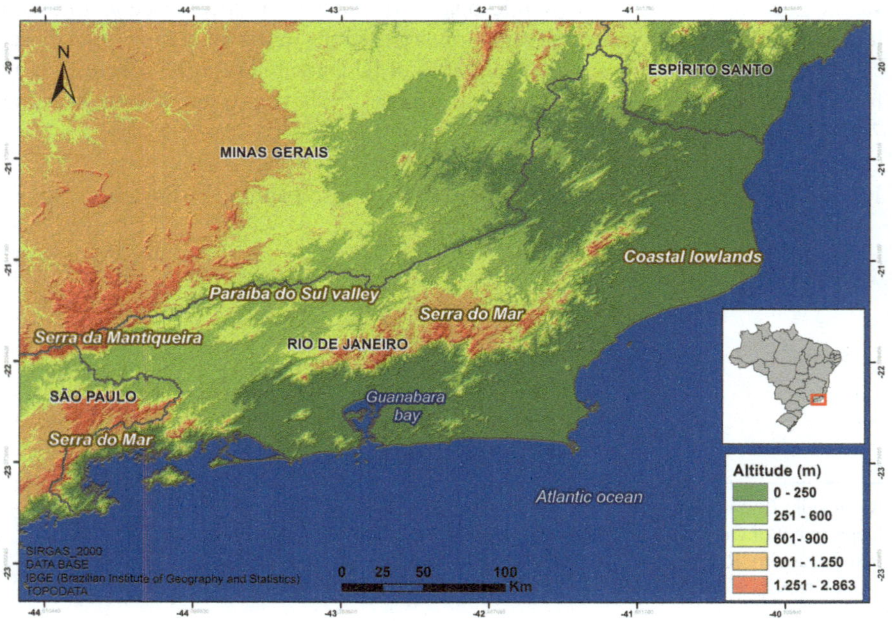

Fig. 2.1 The topography of Rio de Janeiro state depicting the coastal lowland regions, the Serra do Mar and Serra da Mantiqueira mountain ranges, and the Paraíba do Sul River Valley, which also lies in the neighboring states of São Paulo and Minas Gerais

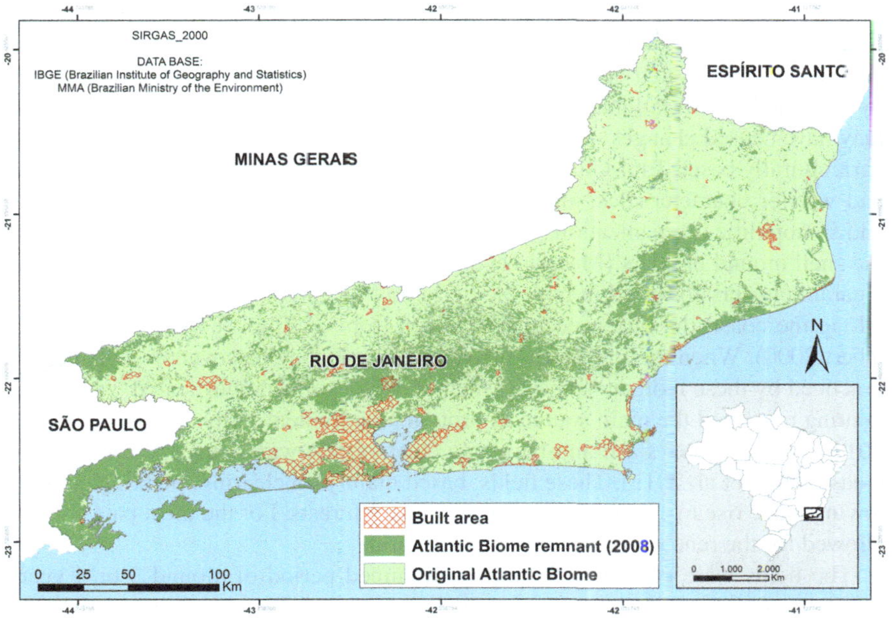

Fig. 2.2 Atlantic Forest biome remnants in Rio de Janeiro state (2008)

throughout the Paraiba do Sul valley (Fig. 2.1) due to its suitability of milder temperatures than found along the coast (Drummond 1997). In addition, the Atlantic Forest provided raw materials for a plethora of uses throughout different periods of land occupation.

Since the arrival of the Europeans, though, the Atlantic Forest biome has been reduced by approximately 93% from its original size (Strassburg 2016). As depicted in Fig. 2.2, most of the current remnants of the original Atlantic Forest biome in RJ are located in mountainous areas of the state, such as the Serra do Mar and Serra da Mantiqueira.

This demonstrates that despite the various socioeconomic influences throughout RJ's history and evolutionary ecosystem changes throughout *human* history, environmental limitations for production and human settlement were still influential. These environmental limitations are possibly more accountable for the few traces of the Atlantic Forest remaining today.

Human habitation in the region of RJ goes back about 8000 years. The first human groups were mainly hunters and gatherers (shell mound builders known as *sambaquianos*), between 6000 and 1000 cal BP (Scheel-Ybert 2000; Lima et al. 2004). With respect to their impact on land use, in general there are no trace elements (like pottery) suggesting the existence of agricultural practices. Therefore, during this era of land occupation, fire potentially had a lesser influence on landscape transformation.

According to Barbosa et al. (2004), after this period, groups migrating from the Brazil's Amazon region arrived. Unlike the *sambaquian* groups whose economy centered around fishing and gathering, the Tupi-Guarani culture was characterized by the practice of rotational agriculture in forested areas (Nehren et al. 2013). This may have caused changes in both the composition and structure of the forests. The agriculturally based Tupi-Guarani culture, along with ceramic technology advances and warrior characteristics, allowed development of new methods of subsistence and sociopolitical organization. These advances eventually led to the extinction of the shell mound builders (Macario et al. 2009) and an intense occupation by Tupi-Guarani clusters followed in both the Serra do Mar in RJ and on the littoral areas. Along the coast, the oldest occupations date from 3220 to 2840 cal BP (Scheel-Ybert 2000). When the Europeans arrived in Brazil, they discovered agriculture as practiced by these indigenous peoples, albeit lacking any domesticated livestock as hunting remained the main source of protein procurement (Oliveira and Solórzano 2014). Farming after slash-and-burn area clearing was widely used, even in sloped areas (Neves et al. 2012). These fields, based mainly on clearing, planting, and fallowing, gave rise to extensive areas of secondary forests. For the most part, this use allowed for the renewal of ecological succession.

The fields were cultivated for a predetermined period of around 3 or 4 years. After the cultivation phase, when productivity dropped due to soil impoverishment from nutrient deficiencies, the area was abandoned (Oliveira 2008a). For a period of 5–10 years, the land was left fallow, which allowed an ecological succession to occur on the site with the appearance of secondary growth. After this period, the emerging growth could be felled and burned for new planting. The fallow was therefore an integral part of their farming techniques, allowing for soil recovery exhausted by the crops (Oliveira 2008b; Ribeiro Filho et al. 2015). There are some differences of opinion regarding the ecological sustainability aspects of the slash-and-burn system. Some authors refer to the relative decline of the diversity of fallow areas (Tabarelli and Mantovani 1999; Siminski and Fantini 2007; Chazdon 2014) while others refer positively to ecological sustainability related to nutrient cycling (Oliveira 2008b; Ribeiro Filho et al. 2015; Adams et al. 2016). It is important to remember that this type of agriculture, practiced by numerous indigenous ethnic groups, was one of the legacies inherited by the European colonists. Moreover, it was one of many other important legacies from indigenous cultures, to include foods, names, and tools. Slash-and-burn agriculture is still practiced to this day in many parts of Brazil by traditional populations as a form of self-supported agriculture (Futemma et al. 2015).

In the following subsection, we will provide an overview of the main drivers of forest conversion in RJ in the last 500 years, focusing sequentially on sugar production, energy creation, and coffee growing and ranching. Our time frame is the period between the European arrival in 1500 and the beginning of the twentieth century. We will discuss how these different historical land uses altered the landscape and the ensuing consequences regarding the resiliency of forest recovery.

2.2 The Uses of Land in Colonial Brazil

The Brazilian colonial society was extremely dependent on wood as an array of technical instruments and daily activities required this raw material from the forest (Cabral 2014). Wood was also the first export product from Brazil, notably a hardwood species known as *pau brasil*[1] which was used to dye fabrics, and this exploitation of *pau brasil* lasted more than 300 years (Lichtenberg et al. 2018). In the sixteenth century, estimates indicate that approximately 2 million logs were exported; nevertheless, this activity cannot solely explain the great deforestation that the Atlantic Forest suffered after the arrival of the Europeans (Cabral 2014).

Sugar production played a predominant role in the occupation, colonization, and transformation of the Brazilian territory. Between 1530 and 1570, sugar manufacturing crystallized as a cash crop for the country becoming the first commercial enterprise of the young colony. Sugar production was also fundamental for the establishment of the donatary captaincies, land grants to Portuguese nobles, which flourished only in places where the sugar industry was successful (Schwartz 2001).

In the seventeenth century, a substantial surge in the colonization of RJ began. Between the Guanabara Bay and the mouth of the Paraíba do Sul river (Fig. 2.1), land shares were granted, and new landowners established sugarcane plantations, sugar mills, and cattle and equine breeding operations (Drummond 1997). In addition to deforestation for sugar cultivation, the operation of sugar mills prompted three more demands for forest exploitation: (1) firewood for fuel; (2) wood for routine daily use among the increasing population; and (3) the opening of pasture for cattle, mules, and horses employed in sugar production and transport (Dean 1996). According to Engemann (2005), the quantity of firewood needed to process a sugarcane crop for a medium-sized RJ sugar mill (*Engenho do Camorim*) was about 4200 m^3 of firewood per harvest. This represented the forest clearing of an area between 4 and 22 hectares per year, just to meet the firewood needs of a single mill. In addition, there were also other uses for wood besides fuel, such as the construction of houses, ox carts and plows, and boxes for transporting sugar. In order to have an idea of the impact of sugarcane activity on the Atlantic Forest, it should be noted that in the early eighteenth century, there were more than 150 mills in operation in the RJ captaincy alone (Abreu 2010). As the demand for wood increased to support operation of the sugar industry, so too did the energy needs of an ever-growing society.

2.3 Land Use for Energy: Extracting Firewood and Manufacturing Charcoal

Firewood use as a primary need energy source has been a requirement of humans throughout history. Its transformation into charcoal through sultry combustion enables an increase in calorific power with a reduction of mass, which makes it a

[1]The native tree that gave its name to the country (*Paubrasilia echinata* (Lam.) E. Gagnon, H.C. Lima & G.P. Lewis).

source of energy easily transportable over longer distances. The calorific power of charcoal per unit of weight is almost three times greater than firewood. Unlike oil, charcoal can be produced locally with a production cost of only the labor invested in it (Olson 1991).

Charcoal has a significant number of uses in industry including the driving of steam engines of a growing population. The need for charcoal for Rio de Janeiro city life starting in the eighteenth century was enormous. It is worth noting that 2.8–3.8 tons of charcoal were needed to melt and reduce iron ore in order to produce a ton of iron (Paradis-Grenouillet 2013). However, in the case of Rio de Janeiro city, possibly the most significant item of consumption in terms of volume was civil construction, specifically stonework. Working with stones, though, required a constant need to sharpen the chisels, since the stone used was very hard, e.g., granite and gneiss. The chisels could not be sharpened with emery due to losing their tensile strength. Therefore, users were required to take them to the forge for sharpening. With increasing demand, charcoal kilns were employed not only in the forests near Rio de Janeiro city but also in countless areas to support the exponential growth of construction in cities and farms in the interior of the state.

The Pedra Branca Massif in the western part of Rio de Janeiro city is a well-studied area on exploitation of charcoal between the seventeenth and the twentieth centuries. Inside its forests more than 1000 charcoal kilns (Fig. 2.3) were recently found scattered over approximately 12,000 hectares, many of them located at an

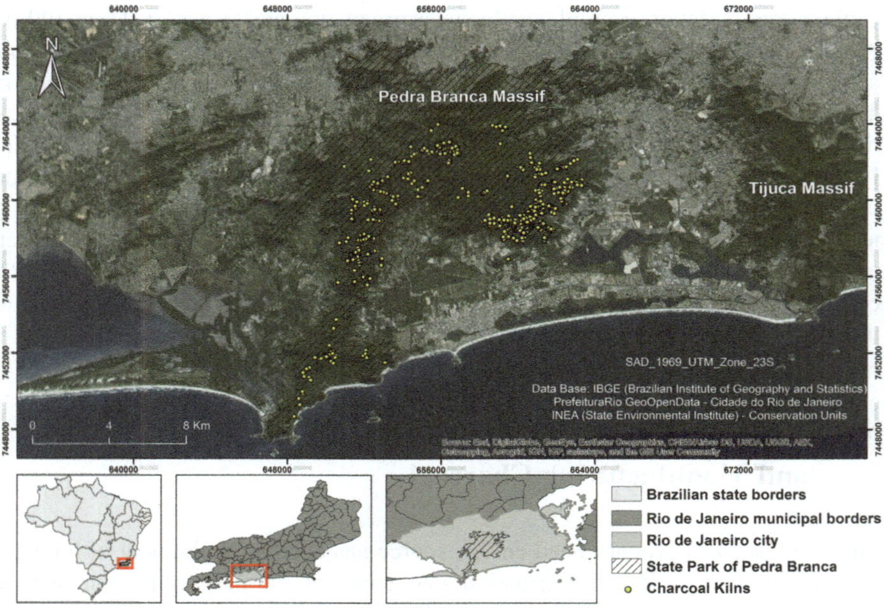

Fig. 2.3 Localization of charcoal kilns (yellow dots) in Pedra Branca Massif, Rio de Janeiro, ranging from 10 to 1000 m.a.s.l

altitude of more than 1000 meters (Oliveira and Fraga 2016). Very little is known about this process as its actors, the charcoal makers, were subject to an intense process of social invisibility (Fraga and Oliveira 2012). Despite the virtual absence of documented information, Sales et al. (2014) estimated that the average production of each charcoal kiln was about 1630 kg of charcoal. Thus, the 1000 charcoal kilns found in the Pedra Branca Massif would have provided the city of Rio de Janeiro with approximately 1600 tons of charcoal. However, the city's consumption was much higher than that. According to historical reports of that era, 28,700 tons of charcoal arrived by train between 1884 and 1886 at the Dom Pedro II train station alone (Brantes 2014). This fact points to the existence of a forest exploitation network at the end of the nineteenth century. In order to supply the city of Rio de Janeiro, this network extended beyond state borders to São Paulo and Espírito Santo and caused substantial changes in the forest landscape.

2.4 Coffee and New Uses of the Landscape

The first coffee seedlings arrived in RJ in 1760 from the northern regions where coffee was first planted in Brazil. The rapid adaptation of Arabica coffee to the climate of RJ meant that by 1787 coffee no longer needed to be imported. In fact, the widespread planting of coffee quickly allowed for its exportation to other places within the Brazilian territory and the La Plata region (Taunay 1945).

Although the mountainous areas of Rio de Janeiro city had been selectively exploited for aforementioned uses, its mountains still maintained good forest cover by the end of the eighteenth century. However, the coffee planted in the highest areas of the city developed even better than at lowlands, thanks to milder temperatures at higher altitudes during the nights (Drummond 1997). Coffee plantations in basin headwaters, fundamental for the city water supply, affected that supply. As a consequence, in the early nineteenth century, some lands were expropriated, and reforestation actions emerged by order of the Emperor.

Meanwhile, in the Paraíba do Sul Valley, coffee farms were growing at an alarming rate. According to Dean (1996), 10 million tons of coffee were produced in Brazil between 1788 and 1888. Assuming the average yield per hectare was 700 kg and the average coffee plantation was economically productive for 20 years, it would have been necessary to clear about 7200 km^2 (720,000 ha) of forest for a coffee plantation. Considering that 80% of this coffee was planted in RJ, this would yield an area of approximately 5760 km^2. This drove a critical necessity for coffee transportation from the Paraíba do Sul Valley to the littoral ports for exportation.

Serving as an ideal vessel for the transport problem presented, the mule became crucial in supporting the economic growth of coffee in the valley (Hickie 2015). Due to its durability over long distances and capability in carrying heavy loads, the mule further transformed the landscape while carrying millions of pounds of coffee on its back in the nineteenth century. Geographically, the Paraíba do Sul Valley is blocked from easy access to coastal ports in RJ and nearby São Paulo state by the Serra do

Mar and Serra da Bocaina mountain ranges. These ranges forced slaves to clear single-track trails through the dense forest at a width too narrow for use by carts drawn by horses or oxen. In addition, the *tropeiros*, or individuals who guided the mule packs, were forced to clear additional forest to provide rest areas and pasture at periodic intervals with suitable grasses for the non-indigenous mule. These rest areas would later serve as starting points for further lateral expansion into the forest. While notable, the total deforestation supporting the needs of mules was still minor in comparison to the area cleared supporting techniques toward coffee cultivation.

The deforestation of native forests was followed by the planting of coffee, carried out in vertical rows from the tops of the hills to their bases.[2] The removal of the native vegetation used in this technique led to the creation of "ramps" for sediment transport. In the more intense rain storms, these ramps would drain the surface soil to the valley bottoms. The slopes underwent a process of organic topsoil loss as well as loss of aggregation and "sealing" of the soils. The amount of sediment that converged to the bottom of the valley was much higher than the transport capacity of the channels, generating an intense continuous process of silting in the rivers. Parallel to this imbalance in hydrological dynamics, coffee production in the Paraíba do Sul Valley irreversibly changed the regional climate. In a period of 30–40 years, coffee transformed the hydrological regime of the region from a typical tropical rainforest to that of a savannah. This caused an increase in torrential rains, capable of triggering severe erosive processes, aggravating even more the environmental problems in these areas (Dantas and Coelho Netto 1996).

With the exhaustion of soil nutrients due to coffee production, the economic activity soon after dominating the region was extensive cattle breeding, which caused its own environmental implications to the region (Fig. 2.4). In addition, the use of fire was a common practice in the creation of pastures, which further hampered the natural regeneration of these areas.

2.5 Resilience of the Landscape and Results of Its Historical Uses

In general terms, RJ forest landscape is characterized by a great spatial diversity due to different altitudes, slope orientation, soil types, microclimate, and historical uses. A considerable part of the remaining forest areas, i.e., mainly in the area of the submountain dense Ombrophilous Forest, presents diverse historical vestiges such as building foundations, ruins of old sugar mills, fragments of charcoal in the soil, and exotic plant species. All of these are indicative of previous uses of secondary forests and of the Atlantic Forest's regeneration ability pervasive throughout the Serra do Mar.

[2] Marquese (2008) shows that the main objective of the vertical alignment with large row spacing of the coffee cultivation techniques was to aid the foremen in the supervision of the work of the slaves.

Fig. 2.4 Erosion and cattle breeding in the Paraíba do Sul Valley

Table 2.1 Environmental characteristics of historical cycles in Rio de Janeiro state

	Sugarcane	Coffee
Location of cultivation	Terraces, plains, valleys	Hillside slopes, "sea of hills," hill tops
Level of erosion	Low-medium	High
Fertilization	Burning of straw	None
Fallow practice	None	None
Dependency on firewood	Yes	Low
Expansion of tillage to replenish land fertility	Medium	High
Post cycle use of the land	Abandonment	Pastures

The type and intensity of historic land use is a key aspect for understanding the present landscapes. Some characteristics of the production and environmental consequences of the two great Brazilian economic cycles, i.e., sugarcane and coffee, are shown in Table 2.1 and help us understand its environmental results.

Coffee in the Paraíba Valley was planted in a monoculture system with total deforestation of native forests and techniques detrimental to soil conservation. The plantation was made at the slopes, favoring erosion processes along with deforestation. Even after the decline of coffee, extensive cattle production continued with pasture conversion (Romeiro et al. 2004). While in the Paraíba do Sul Valley pastures were converted, several areas previously used for sugarcane production, subsistence crops, firewood extraction, and even coffee plantations were abandoned and turned into secondary forest (Oliveira 1999; Oliveira 2015; Rezende et al. 2015).

Subsisting crops demonstrated the great capacity of forest recovery after abandonment for a few years (Oliveira 2008a). The forest harvesting for production of charcoal in Pedra Branca Massif resulted in a reduction of diversity but a recovery of biomass (Oliveira and Fraga 2016). Natural regeneration of the Atlantic Forest could be a significant factor to promote biome restoration, but different vectors of human pressures continue to affect its spatial dynamics (Rezende et al. 2015).

2.6 Conclusion

In this chapter we illustrated some of the main drivers of forest conversion since the European arrival in RJ, although Europeans were not the first humans to manage and change the landscape. Examining historic land use helps to understand the significant deforestation suffered by the Atlantic Forest biome over the last 500 years. In addition, it demonstrates the great recovery capacity of these forests in areas where these uses have long since stopped.

Forest conversion for coffee production in Paraiba do Sul Valley generated a varied number of environmental consequences, such as local climate change, slope erosion, and silted rivers. Significant pasture extensions took place after the exhaustion of the soils making the return of forest even more difficult.

The great number of vestiges found in some Atlantic Forest remnants showed that different kinds of land use could be replaced by secondary forest after abandonment. It is still important to recognize, however, the consequences of each use in forest quality. There are also other less-evident aspects, as the forest is both the product and the producer of RJ's landscape. As discussed, the forests have been impregnated by human presence for untold ages. Despite forest growth in many areas, some of which appear to be primordial without any trace of previous human occupation, the truth about RJ is that it represents a largely anthropogenic landscape in which the deforested areas are the most remarkable aspect.

References

Abreu M (1992) Natureza e sociedade no Rio de Janeiro. Secretaria Municipal de Cultura, Turismo e Esportes, Rio de Janeiro

Abreu M (2010) Geografia histórica do Rio de Janeiro (1502–1700). Andrea Jakobson Estúdio, Rio de Janeiro

Adams C, Rodrigues ST, Calmon M, Kumar C (2016) Impacts of large-scale forest restoration on socioeconomic status and local livelihoods: what we know and do not know. Biotropica 48(6):731–744

Barbosa M, Buarque A, Gaspar MD et al (2004) Intermittent occupation of the sambaqui builder settlements in Rio de Janeiro state, Brazil. Nucl Instrum Methods Phys Res, Sect B 223–224:695–699
Brantes C (2014) Carvoeiros e Imprensa: uma análise das publicações no Rio de Janeiro entre 1850 e 1920. Dissertation, Pontifícia Universidade Católica do Rio de Janeiro
Cabral DC (2014) Na presença da Floresta: Mata Atlântica e História Colonial. Editora Garamond, Rio de Janeiro.
Chazdon R (2014) Second growth: the promise of tropical forest regeneration in an age of deforestation. University of Chicago Press, Chicago
Dantas M, Coelho Netto AL (1996) Resultantes geo-hidroecológicas do ciclo cafeeiro (1780–1880) no médio vale do rio Paraíba do Sul: uma análise quali-quantitativa. An Inst Geociências 19:61–78
Dean W (1996) A ferro e fogo: a história e a devastação da Mata Atlântica brasileira. Companhia das Letras, Rio de Janeiro
Drummond JA (1997) Devastação e Preservação Ambiental no Rio de Janeiro. EDUFF, Niterói
Engemann C (2005) Consumo de recursos florestais e produção de açúcar no período colonial: o caso do Engenho do Camorim, RJ. In: Oliveira Rogério Ribeiro de (Org.). As marcas do homem na floresta: História Ambiental de um trecho urbano de Mata Atlântica. Editora PUC-Rio, Rio de Janeiro, pp 119–142
Fraga J, Oliveira R (2012) Social metabolism, cultural landscape, and social invisibility in the forests of Rio de Janeiro. In: Canevacci, Massimo (org) Polyphonic anthropology – theoretical and empirical cross-cultural fieldwork. InTech, Rijeka
Futemma C, Chamlian Munari LC, Adams C (2015) The afro-Brazilian collective land: analyzing institutional changes in the past 200 years. Lat Am Res Rev 50(4):26–48
Hickie M (2015) The mule as an agent of landscape transformation in Southeast Brazil. Pontifical Catholic University, Department of Geography and Environment. PUC-Rio, Rio de Janeiro
Lamego A (1963) O Homem e a Serra. No. 8. Serviço Gráfico do Instituto Brasileiro de Geografia Estatística, Rio de Janeiro
Lichtenberg S, Huber-Sannwald E, Nehren U, Reyes-Aguero A (2018) Use and Conservation of the Threatened Brazilian National Tree *Paubrasilia echinata* Lam.: A Potential for Rio de Janeiro State? In: Nehren U, Schlüter S, Raedig C, Sattler D, Hissa H (eds) Strategies and tools for a sustainable rural Rio de Janeiro, Springer International Publishing, Cham
Lima TA, Macario KD, Anjos RM et al (2004) The earliest shell mounds of the central-south Brazilian coast. Nucl Instrum Methods Phys Res 223–224:691–694
Macario KD, Buarque A, Scheel-Ybert R et al (2009) The long term Tupiguarani occupation in Southeastern Brazil. Radiocarbon 51(3):937–946
Marquese R (2008) Diáspora africana, escravidão e a paisagem da cafeicultura no Vale do Paraíba oitocentista. Almanack Braziliense 7:138–152
Nehren U, Kirchner A, Sattler D et al (2013) Impact of natural climate change and historical land use on landscape development in the Atlantic Forest of Rio de Janeiro, Brazil. An Acad Bras Ciênc 85(2):497–518. https://doi org/10.1590/S0001-37652013000200004
Neves W, Murrieta RSS, Cristina A et al (2012) Coivara: cultivo itinerante na floresta tropical. Ciência Hoje 50:26–30
Oliveira RR (1999) O rastro do homem na floresta: sustentabilidade e funcionalidade da mata atlântica sob manejo caiçara. Doctoral thesis. Universidade Federal do Rio de Janeiro.
Oliveira R (2008a) Environmental history, traditional populations, and paleo-territories in the Brazilian Atlantic coastal forest. Global Environment 1:176–191
Oliveira R (2008b) When the shifting agriculture is gone: functionality of Atlantic Coastal Forest in abandoned farming sites. Boletim do Museu Paraense Emílio Goeldi. Ciências Humanas 3(2):213–226
Oliveira RR (2015) Fruto da terra e do trabalho humano: paleoterritórios e diversidade da Mata Atlântica no Sudeste brasileiro. Revista de História Regional 20:277–299.

Oliveira R, Fraga J (2016) Fluxos de energia, matéria e trabalho na construção da paisagem do Rio de Janeiro do século XIX. In: Franco, José Luiz et al (org.). História ambiental: territórios, fronteiras e biodiversidade. Editora Garamond, Goiania, pp 35–54

Oliveira R, Solórzano A (2014) Três hipóteses ligadas à dimensão humana da biodiversidade da Mata Atlântica. Fronteiras 3:80–95

Olson SD (1991) Firewood and charcoal in classical Athens. Hesperia 60(3):411–420

Paradis-Grenouillet S (2013) Les forêts métallurgiques: analyses dendroecologiques et approches géohistoriques. Dissertation, Université de Limoges

Rezende C, Uezu A, Scarano F, Araujo D (2015) Atlantic Forest spontaneous regeneration at landscape scale. Biodivers Conserv. https://doi.org/10.1007/s10531-015-0980-y

Ribeiro Filho AA, Adams C, Manfredini S et al (2015) Dynamics of soil chemical properties in shifting cultivation systems in the tropics: a meta-analysis. Soil Use Manag 31(4):474–482

Romeiro AR, Mangabeira JA, Valladares GS (2004) Biodiversidade, reflorestamento e agropecuária no Brasil. Florestar Estatístico 16:15–22

Sales GPS, Solórzano A, Patzlaff RG, Oliviera R (2014) Resultantes ecológicas, práticas culturais e provisão de lenha para a fabricação de carvão nos séculos XIX e XX no Rio de Janeiro. Pesquisas Botânica 65:389–402

Scheel-Ybert R (2000) Vegetation stability in the southeastern Brazilian coast area from 5500 to 1400 14C yr BP deduced from charcoal analysis. Rev Palaeobot Palynol 110(1–2):111–138

Schwartz S (2001) Escravos, roceiros e rebeldes. Edusc, Florianópolis

Siminski A, Fantini AC (2007) Slash-and-burn agriculture: use of forest resources and dynamics of rural landscape in Santa Catarina state, Brazil. Ciência Rural 37(3):31–54

Strassburg B (2016) The role of natural regeneration to ecosystem services provision and habitat availability: a case study in the Brazilian Atlantic Forest. Biotropica 48:890–899

Tabarelli M, Mantovani W (1999) A regeneração de uma floresta tropical montana após corte e queima (São Paulo - Brasil). Rev Bras Biol 59(2):239–250

Taunay A (1945) Pequena história do café no Brasil. Editora UNB, Rio de Janeiro

Chapter 3
Sustainable Rural Development in Rio de Janeiro State: The *Rio Rural* Program

Helga Restum Hissa, Nelson T. Alves Filho, Marcelo Costa Guilherme Strauch, Lauro Bassi, and Renato Linhares de Assis

Abstract The eradication of hunger and poverty are part of the challenges established in this century by various countries as members of the pact to meet the global Sustainable Development Goals (SDGs). This pact is expected to strengthen agricultural production systems that can reverse environmental degradation with the adoption of agroecological technologies and greater social participation of actors historically excluded from public policies, such as family farmers, women, and rural youth. This chapter deals with the Sustainable Rural Development Program of Rio de Janeiro state – RIO RURAL – initially implemented as a pilot project and later transformed into a public policy of sustainable rural development with the support of the GEF and World Bank. The methodology adopted was built on community-based rural development in micro-watersheds, with the support of rural organizations and decision-making actors at local, municipal, and regional levels, to strengthen the beneficiaries' sense of governance and social management of the project. The financial resources are allocated to individual and collective projects with practices to encourage the agroecological transition of production systems, environmental adequacy of properties, water management, and conservation of rural roads. The autonomy of the project's actions has been built through co-investments and a varied establishment of partnerships.

H. R. Hissa (✉) · N. T. Alves Filho · M. Costa · G. Strauch
The Sustainable Rural Development Program in Micro-Watersheds of the State of Rio de Janeiro, Niterói, Rio de Janeiro, Brazil

L. Bassi
Technical Consultant, Food and Agriculture Organization of United Nations, Rome, Italy

R. Linhares de Assis
Embrapa Agrobiologia's Center for Research and Training of Farmers in Nova Friburgo – NPTA, Embrapa Agrobiologia, Empresa Brasileira de Pesquisa Agropecuária (Embrapa), Nova Friburgo, RJ, Brazil
e-mail: renato.assis@embrapa.br

U. Nehren et al. (eds.), *Strategies and Tools for a Sustainable Rural Rio de Janeiro*, Springer Series on Environmental Management,
https://doi.org/10.1007/978-3-319-89644-1_3

Keywords Micro-watershed approach · Sustainable agriculture · Local governance · Public policy · Empowerment

3.1 Introduction

The eradication of poverty and hunger through sustainable food production and consumption has been set under the framework of the global Sustainable Development Goals (SDGs). This commitment to supply the growing world population in an ecofriendly path is a challenge for several United Nations' member countries, including Brazil, within a scenario of scarce natural resources, degraded soils, loss of agro-biodiversity, and extreme climatic events (ONUBR 2016).

By 2030, signatory countries are committed to reach bold targets such as doubling agricultural productivity and income while ensuring food production systems that help maintain ecosystems and reverse land degradation with social and economic inclusion (ONUBR 2016). In order to achieve these results, the adoption of agroecological technologies will be required both in industrial and subsistence agriculture systems (IPES-Food 2016) as well as the engagement of historically excluded social stakeholders, such as smallholders, family farmers, rural women, and youth. To this end, governments need to provide technical assistance, credit, research, and infrastructure to support the transition to more sustainable systems, fostering public policies that value the territories and the role of family farmers responsible for supplying consumer markets.

In Latin America and the Caribbean, there are an estimated 17 million family production units with 60 million people, corresponding to 75–90% of all rural establishments in some countries in the region (Salcedo and Guzmán 2014; Sabourin et al. 2015). In addition to being the largest source of rural employment, they produce most of the food for domestic consumption and local markets in the region and usually develop diversified agricultural activities which give them a key role in ensuring environmental sustainability and provision of ecosystem services and supporting the conservation of biodiversity.

As in the entire Latin American and Caribbean region, family farming in Brazil is very expressive in socioeconomic terms and directly responsible for food and nutritional security of the population. It accounts for most of the basic food basket products, such as cassava, beans, milk, poultry, fruits, and vegetables. The family farmer sector is based on 4,367,902 production units which represent 84% of all rural units, with an average size of 18.4 hectares. Family farming covers 12.3 million people, making up 74.4% of the personnel employed in Brazilian agriculture. In terms of income, family-based agriculture generates a gross production value almost 90% higher than that of nonfamily agriculture (IBGE 2006).

The state of Rio de Janeiro (RJ) is the second smallest state in Brazil with only 0.5% of the country's territory. Despite its limited territorial dimension and the relatively low contribution of the agriculture sector to the state's GDP, the metropolitan region of Rio de Janeiro encompasses the country's second highest population and consumer market, i.e., about 16 million people. Overall, 75% of the agricultural production in RJ is family-based (IBGE 2006) with a key contribution to the state's population food and nutritional security. Considering the enrollment of agro-industries in the food supply chain, the role of agriculture certainly goes well beyond its low participation in the state's GDP. Besides tomatoes and sugarcane, the main agricultural products in terms of income generated are fruit and vegetables, the latter mainly produced by family farmers (Barros 2013). Most of this production supplies the families of RJ with fruit and vegetables, also with milk and its derivatives. In the last 3 years, it also supplies state and municipal schools, as well as kindergartens and asylums through its insertion in institutional markets.

Historically RJ has undergone consecutive and unsustainable cycles of land use that led to deforestation, loss of biodiversity, soil erosion, and alteration of hydrological recharge processes, culminating in the degradation of natural resources in the Atlantic Forest biome and the impoverishment of the rural population (compare Ribeiro de Oliveira et al. 2018). In the beginning of the 2000s, the state government decided to undermine and reverse this path by setting efforts to establish a rural development approach based on sustainably raising agriculture productivity and restoring ecosystem functions, thus introducing a new perspective of wealthy and environmental-friendly well-being for the state's rural population.

3.2 The RIO RURAL Program

The Sustainable Rural Development Program of RJ – RIO RURAL – is a governmental program that aims to promote transition of conventional agriculture into agroecological and climate smart systems, thus supporting the increase of smallholders' income while reversing land degradation and improving resilience and biodiversity conservation in rural landscapes with the provision of ecological services in the Atlantic Forest biome.

Developed by the Rio State Secretariat of Agriculture, Livestock, Fishery, and Supply (SEAPPA) through its Superintendence of Sustainable Development, the Program has the outcome to reverse rural poverty and environmental degradation in biodiversity hotspots. The program envisions to support family farmers in the transition to more profitable and ecofriendly systems in order to meet the new market demands for cleaner and healthier food while linking farmers to key contemporaneous global issues, such as food security, solidarity economy, good governance, environmental services, climate change mitigation, and adaptation and biodiversity conservation.

The challenge to "produce and protect" goes beyond the simple application of stereotyped technological packages pressing the environment to adapt to them. It is a matter of adapting technology to environmental conditions and users. Moreover, to achieve long-term well-being, raising productivity should impact beyond positive economic results and deliver social and ecological benefits evenly. Key social engagement to promote this path requires strategies that are simple and participatory in order to effectively address the sustainable use of natural resources. Using micro-watersheds as territorial units for planning and measure implementation allows for an attitude change in rural communities toward more profitable and environmentally sound livelihood, delivering healthier food and ecosystem services to society. In order to face this challenge, RIO RURAL adopted the "community sustainable rural development in micro-watersheds" methodology, an adaptation of the watershed sustainable management approach more suitable to rural areas. Adapted in Brazil in the early 1980s by rural technicians, the micro-watershed methodology builds upon participatory diagnosis and planning strategies and tools to address key social, economic, and environmental issues and guide territorial sustainable management while promoting community empowerment and good governance. RIO RURAL provides technical assistance, financial incentives, adaptive research, and capacity building to support adoption of agroecological technologies and natural resources conservation management practices in productive systems. It also enables farmers' social and economic inclusion through enhancing community self-management skills, productivity, and quality of agricultural products. Furthermore, RIO RURAL seeks to improve mobility, sanitation, and communication infrastructure in rural areas, aiming to facilitate farmers' access and more efficient linkages to markets.

Moreover, the program incorporates several methodological improvements, especially concerning the social dimension, such as community capacity building, environmental education, and dissemination activities, aiming to strengthening the role of social organizations in the decision-making process. The evolution of rural social groups' empowerment has been facilitated toward local sustainable development through the leverage of public and private co-investments and networking among farmers, scientists, and extensionists to improve exchanging and disseminating of agroecological knowledge. Additionally, the social engagement of farmers, women, and rural young people to local managers and technicians enables a democratic environment for a more efficient coordination of public policies and private partnerships that will provide sustainability to climate resilient, ecofriendly, gender, and economic inclusion of smallholders in the future.

3.2.1 Background, Rationale, Objectives, and Goals

RJ is entirely situated within the Atlantic Forest biome and hosts the largest portion of the Serra do Mar Corridor, considered to be one of the richest and globally most significant remnants of the original Atlantic Forest (Myers et al. 2000; Eisenlohr et al. 2015). Fortunately, some of these tracts are already being conserved in protected areas. According to Fundação SOS Mata Atlântica (2017), the RJ state territory encompasses less than 20% of native forest. Actually, within the state's territory rests a number of additional albeit smaller forest fragments, mostly on private lands. Despite the global significance of the biodiversity of the biome, these remnants are highly fragmented and threatened by (i) deforestation, primarily attributable to logging and charcoal production, i.e., in tropical moist broadleaf forests, and (ii) conventional mono-cropping agriculture, i.e., in floodplain forests and grasslands, slash and burn for agriculture and pastures, i.e., in tropical semi-deciduous forests, and irrigated horticulture in fragile *restingas*, i.e., in coastal tropical and subtropical moist broadleaf forest and mangroves.

In the beginning of the 2000s, several constraints were affecting the effectiveness of the aforementioned responses to addressing the threats to biodiversity conservation while promoting sustainable agricultural practices included (i) an absence of policies and institutional frameworks promoting the incorporation of ecological considerations in agricultural production systems, (ii) limited human and financial resources to facilitate the integration of ecological and sustainability planning principles in sector policies and institutional processes, and (iii) the lack of alternative livelihoods relevant to communities living in proximity to areas of high biodiversity.

In order to face these constrains, the Rio de Janeiro Sustainable Integrated Ecosystem Management in Production Landscapes of the North-Northwestern Project (Rio GEF Project) was approved in 2005, a US$14 million grant that became effective in 2006, funded by the Global Environment Facility (GEF) with the World Bank as implementing agency and technical support of the Food and Agriculture Organization of the United Nations – FAO.

From 2006 to 2011, the Rio GEF pilot project was carried out to introduce new approaches to overcoming critical constrains in the state's hotspots areas in terms of endangered forest remnants, land degradation, and family farming. The Rio GEF Project provided technical assistance, adaptive research, financial incentives, capacity building, and training programs engaging more than 4000 farmers, technicians, and local managers in participatory diagnosis and planning processes, raising awareness of global environmental issues, and creating democratic decision-making spaces in 48 micro-watersheds and 22 municipalities. This enabling environment also supported the design of strategies and tools to implement a broaden public program – the RIO RURAL program – to promote integrated adoption of sustainable land management and natural resources conservation practices within the state territory (World Bank 2012).

3.2.2 Upscaling from the Pilot Project to a Public Policy

Based on effective results and impacts gathered during the Rio GEF Project, a loan agreement in the amount of US$79 million was approved by the World Bank in 2009. The loan project development objective was to increase the adoption of integrated and sustainable farming systems approaches in micro-watersheds, thus contributing to the higher-order objective of increasing small-scale farming productivity and competitiveness in the state.

The RIO RURAL program focuses on enhancing the entrepreneurial capabilities and opportunities available to small farmers for sustained improvements in their socioeconomic and environmental conditions via a participatory and decentralized organizational framework. Within this approach, the program builds on the local and central institutional structures established under the Rio GEF Project to offer comprehensive and innovative support that addresses key rural sector challenges. For this purpose, the program has introduced and mainstreamed a new, integrated, and coordinated approach to public and private interventions within agricultural areas of RJ to maximize the impact and sustainability on a long-term basis.

In order to achieve these goals, RIO RURAL has established the following three pillar components (SEAPEC 2014):

- *Component 1*: *Supporting rural production and competitiveness* provides assistance to rural beneficiaries by working with community groups across local, municipal, and regional levels to increase organizational and participation skills through capacity building and planning activities. This component also aims to operate changes in rural production processes within a framework of market-driven agricultural development focused on sustainable and increased productivity of small farmers. It also helps to reestablish the productive environment of small farmers affected by the January 2011 natural disaster in the *Serrana Region*.
- *Component 2*: *Strengthening institutional frameworks* seeks to improve the state's institutional frameworks supporting market-driven agricultural development by (a) strengthening rural institutions and coordination mechanisms, (b) improving public and private financial support mechanisms, and (c) undertaking a participatory research network to facilitate agroecological knowledge transference between agriculture research and extension services and farmers.
- *Component 3*: *Project coordination and information management* supports overall project management, coordination and implementation, monitoring and evaluation, as well as dissemination of key sustainable rural development information under the project by financing (a) project coordination and (b) information management.

In the beginning of RIO RURAL implementation, in January 2011, unprecedented heavy rainfalls occurred in the *Serrana Region* of RJ with huge flooding and mudslides causing deaths and extensive damage to houses, roads, bridges, crops, and livestock affecting 11 municipalities in the region. Seeking at once for responding to the emergency needs, RIO RURAL was quickly restructured by the World Bank and Rio government teams to redirect almost US$ 19 million to support the resumption of on-farm productive endeavors and to restore farmers' mobility and access to markets. Most funding activities included rural roads restoration, recovery and rehabilitation of degraded lands, restoration of damaged houses, acquisition of productive assets and natural restoration of river banks and forests (CREA-RJ 2011).

The good performance of RIO RURAL's emergency support habilitated the state of RJ to proceed to an additional loan in an amount of US$100 million, complemented by US$40.5 million equivalent counterpart, for the additional financing of the RJ Sustainable Rural Development Program. Undergoing this additional loan credit operation, over US$213 million will be invested in RJ to benefit 85% of its

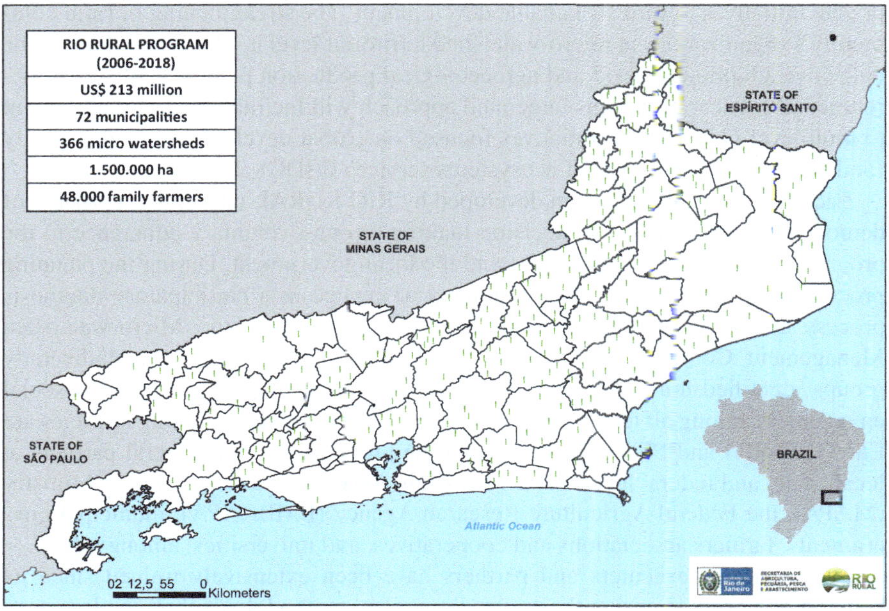

Fig. 3.1 RIO RURAL distribution throughout Rio de Janeiro State

rural population through 1.5 million hectares under improved production systems, i.e., 35% of the state area, as presented in Fig. 3.1.

By 2018, almost 48,000 family farmers living in 366 micro-catchments in 72 municipalities will be benefitted by RIO RURAL, 16,000 of which with direct financial incentives to support the transition to agroecological systems. About 34,000 km of rivers and 14,000 protected springs are expected to be more preserved by reducing organic and sediment load. Additional positive environmental effects rely on biodiversity conservation and climate change mitigation and adaptation in rural landscapes while cleaner and healthier food production will sustain food security in the state, benefiting the entire population of RJ. Most importantly, smallholders' economic inclusion and improved community empowerment will ensure good governance to sustain these results in the long term.

3.3 The Micro-Watershed Approach

RIO RURAL approaches focus on three foundation pillars – democratic, transparent, and decentralized management practices – aiming to improve governance at all territorial levels, i.e., micro-watershed, municipal, regional, and state levels. These pillars also seek to more efficiently implement and coordinate public polices and private initiatives toward sustainable development. The strengthening of rural community's organizations at micro-watershed territorial level is a building block to the collective adoption of good and agroecological production practices allied to environmental conservation. This integrated approach will facilitate farmers' alignment to multilevel policies and initiatives focused on global development goals, mainly food security and provision of ecosystems services (MDG8 2015).

Each micro-catchment basin developed by RIO RURAL undergoes a process of democratic participation and decision-making through voluntary adherence to the program that involves communities and the local government. During the planning process, rural communities are stimulated to engage in a participatory diagnosis process and to establish a steering coordination body – the Micro-watershed Management Committee (MMC) – with representatives of all social diversity groups identified during the appraisal, aiming to improve governance and social engagement among all territorial levels. The program's main executing agencies are EMATER RIO and PESAGRO RIO, but the program also has several partners at local, state, and federal level like the Brazilian Agriculture and Husbandry Ministry (MAPA), the Federal Agriculture Research Agency (EMBRAPA), municipal governments, farmers associations and cooperatives, and universities, among others.

All program's executers and partners have been extensively trained since its beginning in order to ensure the application of key methodological principles, strategies, and tools while setting the stage. Mostly, capacity building challenged the introduction of participatory and community self-management/empowerment approaches adapted by the Technological Incubator of Popular Cooperatives ITCP/COPPE of the Federal University of RJ aiming to engage rural small farmers, women, and youth in democratic decision-making and monitoring toward their own sustainable development. Training also seeks to improve the technological base for agroecological transition, by changing the behavior not only among farmers but, also importantly, by delivering rural extension and agricultural research services.

3.4 Methodological Steps

In order to raise awareness among relevant stakeholders, to engage and empower rural communities into a democratic bottom-up decision-making to identify priorities, as well as to support transition to agroecological productive systems and conservation of natural resources in micro-watersheds, the RIO RURAL approach is implemented in the following four major phases (also presented in Fig. 3.2):

- Motivation phase: (i) presentation of the project, (ii) mapping and validation of all municipal watersheds, (iii) prioritization and selection of watersheds, and (iv) community engagement
- Planning phase: (i) preparation of Participatory Rural Appraisal (PRA), (ii) development of Executive Micro-watershed Plan (EMP), (iii) selection of beneficiaries, and (iv) development of the Individual Development Plan (IDP)

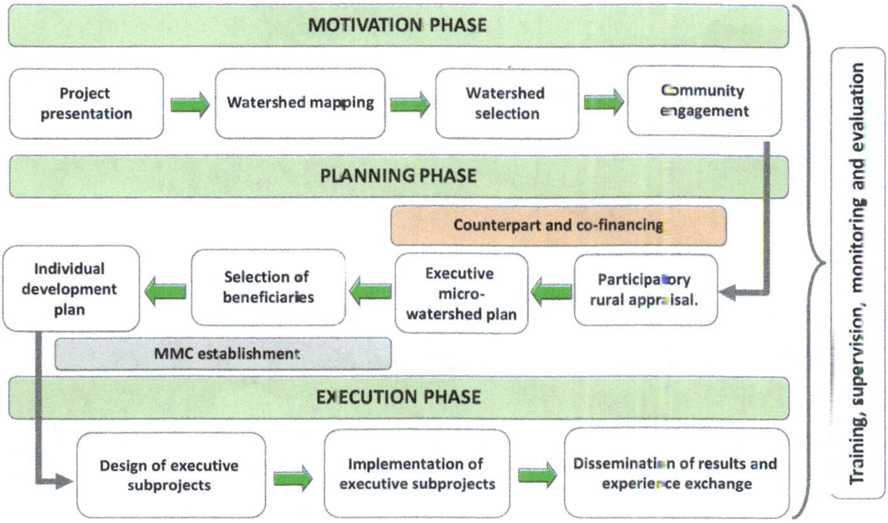

Fig. 3.2 Flow of the micro-watershed methodological approach (SEAPEC 2014)

- Implementation phase: (i) preparation of executive subprojects and (ii) implementation of executive subprojects
- Monitoring phase: (I) scientific monitoring system and (ii) participatory monitoring

During the planning process, both EMP and IDP are translated into business, individual, and group projects, the subprojects which are developed jointly between the executor technician, beneficiaries, and the steering committee (MMC). Investments in proposed subprojects are prioritized by rural communities within Rio Rural rules, boundaries, and lines of incentives agreed with the World Bank.

Each beneficiary shall voluntarily access the program's financial incentives, agreeing to adopt both sustainable productive and environmental practices. The aim of these incentives is to demonstrate that it is indeed feasible to conserve and produce simultaneously *at farm level*. RIO RURAL' s incentives also seek to stimulate the creation of more competitive, efficient, and empowered small business groups within rural farmers, women, and youth. Thus, higher amounts of financial resources are mainly canalized to collective entrepreneurship projects.

Executive subprojects are divided into the following four main categories: (a) sustainable productive subprojects, i.e., for sustainable intensification, adding value, and development of supply chains; (b) environmental subprojects, i.e., to support environmental adequacy of farms and agroecological transition; (c) rural roads; and (d) water management (Fig. 3.3a–d). Rio Rural also supports emergency interventions to combat the effects of drought and heavy rainfalls through encouraging actions that allow the prevention, mitigation, and adaptation to extreme events water conservation, i.e., protection of springs, restoration of riparian forests, and

Fig. 3.3 Examples of subprojects supported by RIO RURAL: **a**) fence insulation to protect springs and water recharge areas, **b**) agroforestry systems, **c**) diversified agroecological production, and **d**) facilitated access to institutional school feeding program and local markets. Photos: Rio Rural

conservation of water recharge areas, building catchment infrastructure, storing and waterworks, such as deep tubular wells, and the construction and maintenance of dams, soil conservation practices, i.e., oil cover, counter line, crop rotation, no till, minimum till; agroforestry and silvopastoral systems among others that enable water infiltration and rational use of water resources in agriculture and livestock.

3.5 Results and Achievements

At the time of this writing, RIO RURAL is running in 373 micro-watersheds in 72 municipalities, covering almost 80% of the whole state's territory. Currently, 45,400 families of farmers and inhabitants are committed to program activities, and over 3500 farmers, women, and youth are participating in 366 Micro-watershed Committees that have already been established. More than 350 executive micro-watershed plans are designed driving communities' efforts to achieve local sustainable development. With financial incentives of the program, about 12,000 family farmers have already adopted sustainable productive systems in almost 100,000 hectares of land. Almost 35,000 subprojects, including sanitation, sustainable production, and environmental conservation practices and rural infrastructure, have been incentivized, and 5700 km of rural roads have recovered, benefiting 28,000 rural residents. Table 3.1 shows RIO RURAL's main achievements by February 2017.

Table 3.1 RIO RURAL's main results (until June 2017)

RIO RURAL main achievements	
Result/achievement	Complementary information
32,400 smallholder farmers transitioned from traditional to sustainable agroecological farming systems	Financial incentives, capacity building, planning and research actions to promote changes in land management and adoption of good agricultural practices
More than a *100,000* ha of agricultural lands under sustainable production systems	Within *1,500,000* ha of land in micro-watersheds indirectly influenced by program actions
2400 small farmers with enhanced links to agricultural products value chains	Strategies and incentives to add value to small farmer's products and to facilitate access to new markets
373 micro-watershed committees organized and effectively implemented at community level	*360* micro-watershed executive plans and participatory rural appraisal elaborated at community level. More than **5000** rural women and youth being involved in local management decisions
More than *27,000* farmers, women, rural youth, technicians, and local stakeholders involved in capacity building events	More than *200* capacity building and training events were performed
52 participatory research fields	Participatory research trials carried out at small farmer level, adapting appropriate technologies in support of sustainable rural development
US $ 25.4 million of co-investments leveraged from various sources	This amount represents almost 15% of World Bank investments in the project. More than *150 territorial initiatives* were designed, 86 of which have successfully integrated co-investments funds
5700 km of rural roads maintenance and rehabilitation benefiting more than 30,000 people in rural areas	Rural roads rehabilitation and maintenance were carried out ensuring transport of goods, mobility of residents, and environmental protection (sediments reduction)
5000 springs protected and *4000* sanitation systems implemented in rural residences	Improvement in water quality and quantity along *34,000* km of rivers
32 soil conservation sets of equipment and machinery	Supported machinery acquisition and training to 32 farmer's associations in target micro-watersheds to improve soil conservation
Agroecological research network	The program strengthened an innovative network design involving 19 institutions, NGO's and farmers' organizations with focus on agroecology knowledge gaps
International and national *technical exchanges*	Exchange of experiences with Africa, India, Latin America, and other Brazilian federal states.

3.5.1 Capacity Building

As one of the most positive and long-term results, RIO RURAL program promoted a set of training activities to strengthen rural organizations and to improve community decision-making as well as to enhance local capacity toward transition to agroecological and environment-friendly rural landscapes.

The program's capacity building strategies focused on training of trainers, awareness raising, exchanging of experiences, field visits, and dissemination of good agricultural and agroecological practices. Main training topics have been related to the program's innovations, such as good governance, participatory approaches, community self-management, agroecological transition and environmental conservation practices, rehabilitation of rural roads, sustainability of food supply chains, and environmental education. Since 2010, more than 27,000 farmers, rural young people and women, fishermen, students, teachers, municipal managers and members of municipal councils of rural development, technicians, and local stakeholders have been democratically involved in more than 200 training events, meetings, lectures, and exchanging visits to raising awareness.

Furthermore, other relevant intangible results emerged from RIO RURAL's capacity-building actions – enhancement of community empowerment and engagement on local and regional social groups or steering committees; improved small business groups accessing markets more efficiently; gender empowerment, mainly through women participation in crafts and local markets; enhancement of organic certified groups; and leakage effect of voluntary adoption and expansion of environmental conservation practices.

3.5.2 Financial Incentives

In order to demonstrate that it is feasible to simultaneously produce safe food and protect the environment, RIO RURAL provides financial incentives and technical assistance to facilitate adoption of agroecological and conservation practices by smallholders. In return, farmers agree to implement environmental conservation and forest restoration practices in their lands with or without program's support. The overall program's results concerning financially incentivized practices are presented in Fig. 3.4.

In a range of the ten practices most demanded by farmers, i.e., 44% of the total, spring protection (fence insulation), conservation of water recharge areas (fence insulation), and recovery (afforestation) of riparian vegetation were the most accessed environmental practices making up 26% of the total. Key factors that may have contributed to raise farmers' interest in adopting sustainable and environmentally sound practices are as follows: (i) raised environmental awareness through RIO RURAL training events, (ii) increased climate concerns due to more frequent and extreme events, (iii) farmers' need to comply to environmental legislation in order to access rural credit federal polices, and/or (vi) recent market demands.

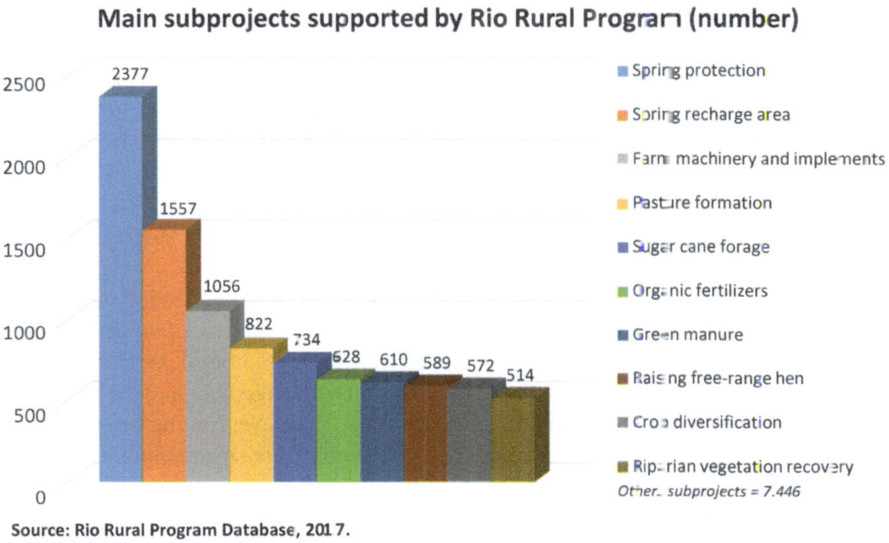

Fig. 3.4 Adoption of environmental and sustainable productive practices financed by RIO RURAL until February, 2017 (RIO RURAL Program Database 2017, unpublished)

Furthermore, a broad campaign for recovery and preservation of springs in rural areas was carried out by the RIO RURAL program on the eve of the 2016 Olympic Games.

Concerning adoption of sustainable and agroecological productive practices, pasture formation and sugarcane forage represented 9.2% of the total, which reflected the need of dairy farmers to climate adaptation after a 2-year drought period (2014–2015). Agricultural mechanization was also highly demanded, stressing a surrogate to labor scarcity and green manure and organic fertilizers, deeply related to the agroecological transition in horticulture systems – these subprojects represent 7.3% of the total. Farming diversification was also intensively accessed by beneficiaries in light of the crescent need for climate adaptation, albeit raising hens freely provided extra rural income, especially to women.

3.5.3 *Co-investments*

One of the most successful strategies implemented by the RIO RURAL program was the leverage of co-investments. These co-investments were integrated by public and private partners to enhance financial sustainability of the social, economic, and environmental services and results provided by smallholders. Furthermore, the co-investments aimed to supporting farmers with technical assistance, capacity-building, and public and private matching funds to ensure long-term results, far beyond RIO RURAL's implementation period.

These funds were identified upon community demands raised during the planning phase, recorded on the PRDs (participatory rural diagnosis) and consolidated in the MEPs (micro-watershed executive plans). From 2007 to 2016, the RIO RURAL program, through its beneficiaries, technicians, and researchers, has designed more than 150 territorial initiatives to leverage co-investment funds, 86 of which have successfully managed to raise approximately US $ 25.4 million in co-investments from various sources. This represents almost 15% of World Bank's investments, which indicates a fruitful and promising way of sustaining these actions in the future.

The cooperation of RIO RURAL with German universities and small enterprises under INTECRAL project is a successful example of leveraging of co-investments from the German Federal Republic government to undergo complementary studies and knowledge transfer to improve natural resources management and the provision of ecosystems services in the Atlantic Forest biome while overcoming research gaps to facilitate adoption of good agriculture practices in a more sustainable rural development economy.

All these partnerships reinforced the main strategy of seeking for financial sustainability to rural development activities beyond the end of the project, generating an evolutionary process through the local watershed committees, using participatory planning tools to match partnership funds and empower rural communities. However, most of these co-investments were derived from public initiatives, indicating an urgent need to a stronger commitment of the private sector.

3.5.4 Environmental Services

Undoubtedly, farmers' transition to agroecological and more environmental-friendly productive systems improves environmental services and availability of water, forests, biodiversity, safe and nutritious food, as well as climate resilience, benefiting the whole society. These services shall be compensated through public or private support to ensure its provision in a long-term basis.

Agroecological soil and crop management, agroforestry and silvopastoral systems, and forest conservation and restoration practices improve carbon immobilization in rural landscapes and reduce emissions of greenhouse effect gas by agriculture sector. In addition, besides several climate adaptation and diversification of production systems (Torres and Dornberger 2018), the program also encouraged farmers to take on preventive measures in relation to the interaction with the growing climate instability and soil degradation, reducing their vulnerability to such factors (Seliger et al. 2018).

Several rural communities engaged in the RIO RURAL program have successfully established partnerships with watershed committees, schools, hospitals, restaurants, and local markets in order to provide clean water and organic and safe food.

Moreover, RIO RURAL joins additional efforts to enhance financial sustainability of payment for ecosystems services (PES) schemes to support farmers' adoption of sustainable intensification technologies in compensation to their environment conservation actions. In the recently approved GEF Project "Recovery of Climate and Biodiversity Services in the Paraíba do Sul River Basin in the Brazilian Atlantic Forest," US$ 14.39 million grants will be invested from 2017 to 2022 in the state's northwestern and southern regions; part of the counterpart will be financed by RIO RURAL. The main objective of this new project is to pilot a PES scheme that will integrate both agriculture and environmental funds. While RIO RURAL will support farmers' environmental compliancy, PES will canalize funds to farmers' technological jump and rise in productivity and income. In counterpart, farmers will commit to preserve ecosystem services associated to biodiversity conservation and carbonization in critical rural landscapes of the Brazilian Atlantic Forest Southeastern Corridor.

3.5.5 *Agroecological Technologies Research Network*

Another challenge to long-term rural development outcome is how to maintain and expand the systematic provision of extension services and research focused on developing agroecological technologies tailored to the small farmers. These services are essential once they provide essential knowledge and key information to qualify agriculture products and raise competitive access of small farmers to markets. To this end, RIO RURAL has established an agroecological network based on partnerships among research, extension, farmers' associations, and development institutions, both nationally and internationally.

The RIO RURAL agroecological network was established to overcome the technological gap and facilitate knowledge transfer among researchers, extensionists, and farmers. Using an innovative and collaborative way of network training, the agroecological network focuses on horizontal exchanging of sustainable agriculture management practices knowledge to support and facilitate agroecological transition of production systems.

3.6 The Way Forward

The transformative seed capital nested through a long-term public program such as RIO RURAL will simply initiate a process that shall afterward rely upon solidarity, social responsibility, and collaborative networking to sustain alliances between end consumers and producers to long-term provision of goods and services. Public and private partnerships will also have to be envisioned to necessarily improve governance toward better managed and climate-resilient landscapes.

Simplified compensation mechanisms shall be designed in a near future to facilitate rural communities' access to collaborative financial schemes that will provide technical assistance and financial incentives to support their transition to a more productive, sustainable, and climate-smart agriculture.

Another challenge will be to strengthen rural infrastructure and communication networks to reduce asymmetries of information and to improve farmers' access to markets. By creating accreditation tools, solidarity markets, e-commerce, and negotiation platforms, the geographical and cultural gap shall be overcome, uniting producers and consumers, creating spaces of collaborative technical assistance, rural extension, and research approaches, attracting public policies and private companies. The main focus shall be to improve market access and knowledge sharing on practices, technologies, and sustainable agricultural systems, joining efforts with NGOs, private sector, and research institutions to ensure smallholders sustainability.

Acknowledgments The authors recognize the financial support from the government of the state of Rio de Janeiro and the World Bank and FAO's technical support to RIO RURAL program implementation. The authors also acknowledge EMATER-RIO and PESAGRO-RIO for their role and commitment to RIO RURAL program achievements. The authors also thank Thiago Rodrigues Sabatinelli and Nathália Duarte for their relevant contribution to text translation and designing of maps and figures presented in this chapter.

References

Barros GSC (2013) Dimensionamento do PIB do agronegócio do Estado do Rio de Janeiro. Relatório final. CEPEA, ESALQ/USP, Piracicaba

CREA-RJ – Conselho Regional de Engenharia e Agronomia do Estado Rio de Janeiro (2011). Tragédia na Região Serrana do Rio de Janeiro após decorridos 6 meses. Relatório No. 9/2011, Rio de Janeiro

Eisenlohr P, de Oliveira-Filho A, Prado J (2015) The Brazilian Atlantic Forest: new findings, challenges and prospects in a shrinking hotspot. Biodivers Conserv 24:2129–2133

Fundação SOS Mata Atlântica (2017). Atlas dos remanescentes florestais da Mata Atlântica. https://www.sosma.org.br/14622/divulgados-novos-dados-sobre-a-situacao-da-mata-atlantica/. Accessed 24 Apr 2017

IBGE (2006) Censo Agropecuário 2006, Agricultura familiar: Primeiros resultados. Brasil, grandes regiões e unidades da federação. IBGE, Rio de Janeiro, p 267

IPES – Food (2016) From uniformity to diversity: a paradigm shift from industrial agriculture to diversified agroecological systems. International Panel of Experts on Sustainable Food systems. Report 2. http://www.ipes-food.org/images/Reports/UniformityToDiversity_FullReport.pdf. Accessed 23 Jun 2016

MDG8 – Millenium Development Goals 8 (2015) Taking stock of the global Partnership for Development - MDG gap task force report executive summary. United Nations, New York

Myers N, Mittermeier RA, Mittermeier CG et al (2000) Biodiversity hotspots for conservation priorities. Nature 403:853–845

ONUBR – Nações Unidas no Brasil (2016) Objetivos de desenvolvimento sustentável. https://nacoesunidas.org/pos2015/agenda2030/. Accessed 23 Jun 2016

Ribeiro de Oliveira R, Stingel Fraga J, MacLeod Hickie M (2018) Environmental history, forests and landscape uses in Rio de Janeiro state. In: Nehren U, Schlüter S, Raedig C, Sattler D, Hissa H (eds) . Springer International Publishing, Strategies and tools for a sustainable rural Rio de Janeiro, Cham

Sabourin E, Samper M, Massardier G (2015) Políticas públicas para as agriculturas familiares: existe um modelo latino-americano? In: Grisa, C. e Schneider, S. (org.) Políticas públicas de desenvolvimento rural no Brasil. Editora da UFRGS, Porto Alegre, p 624

Salcedo S, Guzmán L (eds) (2014) Agricultura familiar en América Latina y el Caribe: recomendaciones de política. FAO, Santiago, p 497

SEAPEC – Secretaria de Estado de Agricultura e Pecuária (2014) Manual Operacional: Projeto Desenvolvimento Rural Sustentável em Microbacias Hidrográficas do estado do Rio de Janeiro. Superintendência de Desenvolvimento Sustentável, Rio de Janeiro

Seliger R, Sattler D, Soares da Silva A et al (2018) Rehabilitation of degraded sloped pastures – lessons learned in Itaocara. In: Nehren U, Schlüter S, Raedig C, Sattler D, Hissa H (eds) Strategies and tools for a sustainable rural Rio de Janeiro. Springer International Publishing, Cham

Torres OA, Dornberger U (2018) Promotion of alternative income diversification in rural Rio de Janeiro: making use of innovation resources. In: Nehren U, Schlüter S, Raedig C, Sattler D, Hissa H (eds) Strategies and tools for a sustainable rural Rio de Janeiro. Springer International Publishing, Cham

World Bank (2012) Implementation completion and results report (tf-54399) on a grant from the global environment facility in the amount of us$ 6.75 million equivalent to the state of Rio de Janeiro for a Rio de Janeiro sustainable integrated ecosystem management in production landscapes of the north-northwestern fluminense (GEF) project. Washington, USA

Chapter 4
The INTECRAL Project

Sabine Schlüter, Udo Nehren, Dietmar Sattler, and Claudia Raedig

Abstract INTECRAL (Integrated Eco Technologies and Services for a Sustainable Rural Rio de Janeiro) is a German-Brazilian research project (2013–2017) funded by the German Federal Ministry of Education and Research (BMBF) in cooperation with the Rio de Janeiro State Secretariat of Agriculture and Livestock Program RIO RURAL (SEAPPA-PRR). It contributes to a sustainable development of the rural areas of Rio de Janeiro state by (a) enhancing the competitiveness of the agricultural sector, (b) providing environmentally friendly and climate-adapted solutions for land and water management, and (c) adopting appropriate green technologies. In this chapter, we provide an overview of the project and outline the main outcomes as well as the way forward.

Keywords INTECRAL project · Sustainable rural development · Rio de Janeiro state (RJ)

4.1 Background and Partners

INTECRAL (Integrated Eco Technologies and Services for a Sustainable Rural Rio de Janeiro) is a scientific cooperation project funded by the German Federal Ministry of Education and Research (BMBF) in the program FONA (Research for

S. Schlüter (✉) · U. Nehren · C. Raedig
TH Köln – University of Applied Sciences, Institute for Technology and Resources Management in the Tropics and Subtropics (ITT), Köln, Germany
e-mail: sabine.schlueter@th-koeln.de; udo.nehren@th-koeln.de; claudia.raedig@th-koeln.de

D. Sattler
Leipzig University, Institute of Geography, Physical Geography and Environmental Research Johannisallee 19, Leipzig, Germany
e-mail: sattler@uni-leipzig.de

U. Nehren et al. (eds.), *Strategies and Tools for a Sustainable Rural Rio de Janeiro*, Springer Series on Environmental Management,
https://doi.org/10.1007/978-3-319-89644-1_4

Sustainability) in collaboration with the Rio de Janeiro State Secretariat of Agriculture and Livestock Program RIO RURAL (SEAPPA-PRR). The project, with a funding period from 2013 to 2017, contributes to providing the scientific basis to improve the competitiveness of the rural economy in Rio de Janeiro state (RJ) in due consideration of sustainable use of land and water resources and support of local family farmers. It builds on previous German-Brazilian cooperation projects in RJ, in particular the BMBF projects BLUMEN (biodiversity in integrated land use management for economic and natural system stability in the Mata Atlântica of RJ (2002–2005)) and DINARIO (Climate Change, Landscape Dynamics, Land Use, and Natural Resources in the Atlantic Forest of RJ (2009–2012)).

The aim and the scope of the project were jointly defined by the German INTECRAL consortium and RIO RURAL, a large-scale program to advance sustainable development in rural areas of RJ (Hissa et al. 2018). The INTECRAL consortium consists of the Institute for Technology and Resources Management in the Tropics and Subtropics (ITT) at the TH Köln (project lead), the Institute of Geography and the SEPT (small enterprise promotion and training) program at Leipzig University, the Geographic Information Science Group at the Institute of Geography at Friedrich Schiller University Jena, and the Training and Demonstration Centre for Decentralized Wastewater Treatment – BDZ. These academic institutions closely cooperated with the following German Small-Medium Enterprises (SME) as implementation partners: codematix GmbH (http://www.codematix.de), ribeka GmbH (http://www.ribeka.com), Seba Hydrometrie (www.seba-hydrometrie.com), Tilia GmbH (http://www.tilia.info), TM Technik (http://tm-technik.de), TÜV Rheinland (https://www.tuv.com/germany/de), and Vita34/bioplanta (https://www.vita34.de).

4.2 Objectives and Study Regions

INTECRAL helps contribute to a sustainable development of rural areas in RJ by integrating scientific and technological knowledge and tailor-made solutions into land use planning, watershed and ecosystem management, and agricultural production systems. The action plan was elaborated in collaboration with Brazilian partners and German institutions and enterprises in a participative planning process, from the analysis to the definition of development goals and strategies. With respect to rural development, several bottlenecks were identified based on the experience from the RIO RURAL program and the previous research projects of the German-Brazilian consortium. From these bottlenecks, the following complexes of specific demands were derived:

Demand Complex 1

The first complex includes analyzing, planning, and monitoring the RIO RURAL activities at different temporal, spatial, and administrative scales. The participative planning processes, from the analysis to the definition of development goals and strategies at the micro-watershed level, are improved by a stronger integration of relevant social and environmental system components and the inclusion of projected scenarios (environmental risks, climate change).

Demand Complex 2

The second complex refers to specific measures and investments in the RIO RURAL program to improve ecological quality and economic performance of rural RJ. In some cases, the local institutions are confronted with a limited availability of expertise and suitably adopted processes and technologies (e.g., restoration of degraded areas, community sanitation, and agricultural techniques).

Demand Complex 3

The third complex comprises the provision of alternative income sources for farmers, including strategies of income diversification and access to green markets.

From these demands, the following three overall goals were defined:

1. *Competitiveness of Rural RJ*

Increasing the capability of family farms and the rural administration to compete in globalized markets through improved farm practices and appropriate technological solutions, sustainable management of natural resources, and income diversification at the farm and landscape levels and through benefiting specifically from green market opportunities.

2. *Environmental and Climate Protection*

Enhancing the environmental quality and carbon-reduction potentials through (a) the introduction of environmental friendly technologies; (b) the support of governance processes; (c) the development of strategies and tools for landscape and ecosystem conservation, land restoration and rehabilitation, and sustainable land and water management; and (d) the identification of viable certification services in carbon markets for the family farming sector.

3. *Adapting German Green Technologies*

Applying and further developing German know-how and technology for monitoring the state of water, soil, and vegetation designing solutions for rural wastewater treatment and for sugarcane harvesting according to the demands in rural RJ.

The activities were carried out on different temporal and spatial scales depending on the specific research questions. Specific pilot measures were implemented at the farm level in the municipalities of Italva for the implementation of silvopastoral systems (Fischer et al. 2018) and in Itaocara for the rehabilitation of degraded pasture lands (Seliger et al. 2018). Ecohydrological modeling was carried out in three pilot river basins, namely, the Guapi-Macacu River Basin, the Dois Rios River Basin, and the Muriaé River Basin.

4.3 Project Design and Activities

The design of project activities follows the need to respond to the three overall goals mentioned in Sect. 4.2 through achieving the targets for each project component. To this end, five work packages (WPs) were defined (see Fig. 4.1). WP 1 includes project management and stakeholder integration, while WP 2 deals with good agricultural practices and participatory planning in the agricultural sector. WP 3 is dedicated to land restoration, rehabilitation, and conservation, while WP 4 has two main components: (a) ecohydrological modeling, scenario impact assessment, and development of web-based software tools for river basin management and monitoring and (b) sanitation and watershed management and monitoring. Finally, WP 5 is a highly integrative package that focuses on ecosystem services including the services for disaster risk reduction and climate change adaptation, as well as alternative markets and governance aspects. Table 4.1 summarizes the main activities of each work package and lists the German and Brazilian partners involved.

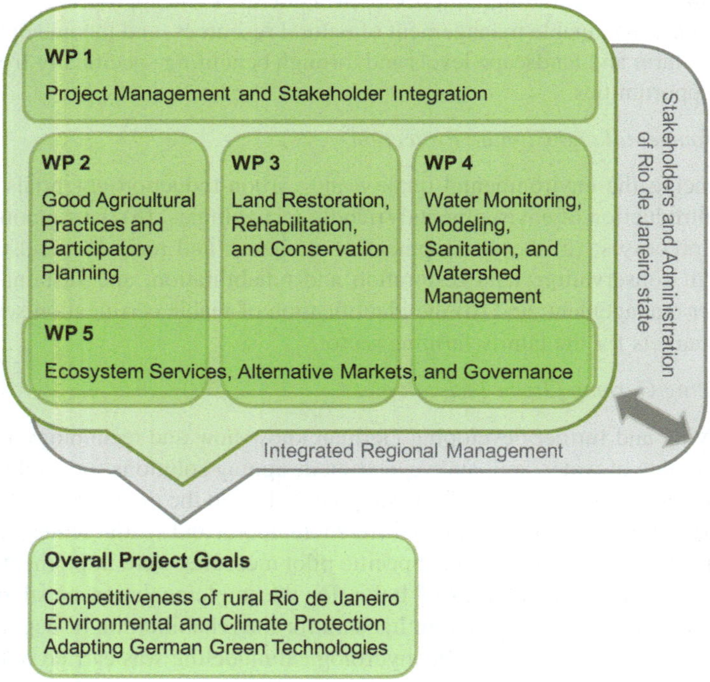

Fig. 4.1 Design of the INTECRAL project

Table 4.1 INTECRAL main activities and partners involved

Work package (WP)	Main activities	Partners involved[a]
1. Project Management and stakeholder integration	- Project coordination including logistics and budgeting - Stakeholder involvement	**TH Köln**, Leipzig University, University of Jena, BDZ
2. Good agricultural practice and participatory planning	- Development of the BIOSS tool for management of silvopastoral systems - Implementation of a silvopastoral pilot measure (fodder bank) in the municipality of Italva - Elaboration of procedures and training programs in good dairy practices for small-scale farmers and rural extensionists - Agricultural value chains analysis - Optimization of an environment-friendly sugarcane harvester	**TH Köln**, TM Technik Universidade Federal Rural do Rio de Janeiro – UFRRJ, RIO RURAL program
3. Land restoration, rehabilitation, and conservation	- Assessment of land degradation and multi-criteria analysis - Multi-criteria site prioritization for rehabilitation and restoration of degraded pasture lands - Elaboration of concepts for rehabilitation planning - Implementation of a pilot rehabilitation measure on a degraded pasture - Scientific monitoring of degradation and rehabilitation measures - Development of training programs and participative degradation assessments	**Leipzig University**, bioplanta/Vita34, Universidade do Estado do Rio de Janeiro – UERJ, TH Köln, University of Jena, RIO RURAL program
4. Water monitoring, modeling, sanitation, and management	4.1 Water monitoring and integrated water resources management (IWRM) - Construction of a pilot system for water quality monitoring at Rio Dois Rios/Muriaé River basin - Design of a strategy for community-based water quality monitoring at the micro-watershed level - Development of a functional river zoning methodology - Development and implementation of a water monitoring and information system - Integrated water resources governance and planning - Capacity building and training measures	**TH Köln**, SEBA, ribeka, RIO RURAL program

(continued)

Table 4.1 (continued)

Work package (WP)	Main activities	Partners involved[a]
	4.2 Ecohydrological modeling, scenario impact assessment, development of web-based software tools and LULC classification - Implementation and extension of the INTECRAL River basin information system (RBIS) - Development of two web-based tools for (1) delineation of ecohydrological modeling entities (webHRU) and (2) land use classification (webIMX) - Development of a client-server environment (JAMS Cloud) and an automated modeling workflow for web-based modeling - Data collection, correction, and trend analysis of time series data LULC classification - Design and application of computer models to analyze and predict integrated ecohydrological systems - Development of methods and tools for scenario impact assessment within the region - Evaluation of future development scenarios and the quantification of their impacts on hydrology - Development of a model parameter library for land use, soil, and geology for the whole RJ - Capacity building and training	**Friedrich Schiller University Jena**, codematix
	4.3 Sanitation - Status quo sanitation analysis and identification of suitable treatment technologies - Development of scenarios for decentralized wastewater treatment - Case study for suitability analysis of decentralized wastewater treatment solutions in a rural community (Campos, municipality of Nova Friburgo) - Capacity building and training measures	**BDZ**, TH Köln, RIO RURAL program

(continued)

Table 4.1 (continued)

Work package (WP)	Main activities	Partners involved[a]
5. Ecosystem services, alternative markets, and governance	- Assessment and valuation of ecosystem services as a basis for the creation of incentive systems (payment for ecosystem services, PES) as well as the implementation of ecosystem-based approaches for climate change adaptation and disaster risk reduction - Development of ecological corridors in a selected municipality (Varre-Sai) - Creation of the project design document "improvement of agricultural practices in small- and medium-size sugarcane farming, Brazil" - Design of a toolkit for carbon projects applicable methodologies - Analysis of economic efficiency of agricultural production and possibilities of formal and informal income-generating strategies for small farmers - Capacity building and training workshops	**TH Köln**, TÜV Rheinland, Leipzig University – SEPT

[a]Leading project partners of work packages in bold

4.4 Scientific Outputs

The INTECRAL project sets a clear focus on applied research and technology development, the elaboration of strategies and tools for sustainable land and water management, and the implementation of local solutions to support rural livelihoods and foster a "green" development in the rural areas of RJ. Therefore, the results of many activities were disseminated though workshops and training measures, as shown in Table 4.1. In addition, training materials were developed in English and Portuguese to be directly used by local authorities and stakeholders. These include, among others, manuals for "Mapping and Monitoring Protected Areas with Unmanned Aerial Vehicles (UAVs)" and "Private Reserves of Natural Heritage (RPPNs): Manual for Awareness Building and Motivation of Landowners," as well as a toolkit for the identification of suitable carbon projects in RJ state and guidelines for participative pasture degradation assessment. Besides this, research results of the project in the field of disaster risk reduction and climate change adaptation were used for a case study and exercise sourcebook (Nehren et al. 2014) as well as the massive open online course "Disasters and Ecosystems; Resilience in a changing climate" (TH Köln and UNEP 2015/2017) and the global online course "Environmental Education: Trans-disciplinary Approaches to Addressing Wicked Problems" (Cornell University 2016). A compilation of the project activities, additional information, and promotion material such as video-clips and internal publications is presented on the INTECRAL project homepage at http://intecral-project.web.th-koeln.de.

Many of the scientific results of the project are presented in this book, but beyond this, eight peer-reviewed articles, six book chapters, and seven conference proceedings have been published up to 01/2018 – and several more are in print. Furthermore, results of the INTECRAL project have been presented in international conferences and trade fairs, such as the Latin American Symposium on Climate Change Adaptation (10–12 November 2016, conference proceedings, Leal Filho and Esteves de Freitas 2017) and the trade fair "Green Rio" (2–4 June 2016) in RJ.

In addition to scientific achievements, INTECRAL project allowed for research-oriented teaching and studying and thus enabled students to carry out their PhD, master, and bachelor research in the course of the project. To date (01/2018), 2 PhD, 20 MSc, and 2 bachelor students have completed their theses, and 5 PhD and 2 master students will most probably be graduated by the end of 2018.

Most important of all, the INTECRAL project created a needed space for scientists, decision-makers, students, farmers, landowners, and many other stakeholders to explore options for a more sustainable RJ jointly, which is the best guarantor of the consistent and successful further implementation of the project's results.

4.5 The Way Forward

After the official end of the project in October 2017, ongoing project activities will be pursued by PhD students that are jointly supervised by German and Brazilian researchers. INTECRAL and the previous German-Brazilian projects BLUMEN and DINARIO have laid a solid and trustful foundation, so that several activities can be continued. Moreover, the close cooperation of scientists and decision-makers aided in identifying new areas for which scientific research is needed in order to develop effective and integrated solutions. For these areas, German and Brazilian researchers will jointly acquire funds to initiate new activities based on the results of the INTECRAL project. The first successful applications include the project DRYTIP (social-ecological tipping points in drought-prone transition zones of the Mata Atlântica and the Cerrado), which is funded by the BMBF until 2018 with a possible extension until 2021, and the project PEDES (soil-related ecosystem services in the Atlantic Forest of Rio de Janeiro), funded by the German Academic Exchange Service (DAAD) and the Brazilian Coordination of Improvement of Personal Higher Education (CAPES).

References

Cornell University (2016) Global Online Course: Environmental Education: Trans-disciplinary Approaches to Addressing Wicked Problems; http://www.globalee.net/. Accessed 30 Sept 2017

Fischer SB, Pedraza Luengas A, Schlüter S, Oliveira Antunes LA (2018) From design to implementation: a participatory appraisal for silvopastoral systems. In: Nehren U, Schlüter S, Raedig

C, Sattler D, Hissa H (eds) Strategies and tools for a sustainable rural Rio de Janeiro. Springer International Publishing Cham, Switzerland

Hissa H, Teixeira N, Costa M et al (2018) Sustainable rural development in Rio de Janeiro state: the Rio Rural Program. In: Nehren U, Schlüter S, Raedig C, Sattler D, Hissa H (eds) Strategies and tools for a sustainable rural Rio de Janeiro. Springer International Publishing Cham, Switzerland

Leal Filho W, Esteves de Freitas L (eds) (2017) Climate change in Latin America – managing vulnerability, fostering resilience, Climate change management series. Springer International Publishing, p 537 Cham, Switzerland

Nehren U, Sudmeier-Rieux K, Sandholz S et al (eds) (2014) Creating resilient landscapes in the state of Rio de Janeiro. In: Nehren et al: the ecosystem-based disaster risk reduction case study and exercise source book. CNRD/PEDRR http://postconflict.unep.ch/publications/DRR_CASE_STUDIES_&_EXERCISES.pdf. Accessed 30 Sep 2017

Seliger R, Sattler D, Soares da Silva A et al (2018) Rehabilitation of degraded sloped pastures – lessons learned in Itaocara, Rio de Janeiro. In: Nehren U, Schlüter S, Raedig C, Sattler D, Hissa H (eds) Strategies and tools for a sustainable rural Rio de Janeiro. Springer International Publishing Cham, Switzerland

TH Köln, UNEP UN – Environment (2015/2017) Massive Open Online Course 'Disasters and Ecosystems; Resilience in a changing climate'. https://courses.adpc.net/courses/course-v1:CNRD+EDRR001+2017_T1/about. Accessed 30 Sept 2017

Part I
Agricultural Management

Chapter 5
Soil and Water Management for Ecosystem Services Provision in Agricultural Landscapes: The Challenge of Monitoring

Joyce Maria Guimarães Monteiro, Azeneth Eufrausino Schuler, Rachel Bardy Prado, Elaine Cristina Cardoso Fidalgo, Ana Paula Dias Turetta, Alba Leonor da Silva Martins, Aline Pacobahyba de Oliveira, and Guilherme Kangussú Donagemma

Abstract Soil and water conservation management includes practices that help to preserve the quality of soil, water, and biodiversity, to promote sustainable agriculture and to provide ecosystem services (ES). In the state of Rio de Janeiro, some agriculture conservation practices such as minimal tillage, crop rotation, agroforestry systems, rotational grazing, and fallow can contribute to value and provide ES and also increase productivity and income of the family farmers. However, it is necessary to identify the impacts of these practices on the provision of ES, and monitoring is an important means of analysis. This chapter presents a review of studies on the conservation management practices in the state of Rio de Janeiro and their impacts on ES. Moreover, it highlights aspects to be considered in monitoring of ecosystem service indicators. The main challenges of the monitoring approach include participatory efforts and multidisciplinary criteria used at appropriate scales in order to provide adequate and useful answers to society.

Keywords Good agricultural practices · Agricultural family farming systems · Soil and water quality · Ecosystem services assessment

J. M. Guimarães Monteiro (✉) · A. E. Schuler · R. Bardy Prado · E. C. Cardoso Fidalgo
A. P. D. Turetta · A. L. da Silva Martins · A. Pacobahyba de Oliveira · G. Kangussú Donagemma
Brazilian Enterprise for Agricultural Research – Embrapa Soils, Rio de Janeiro, Brazil
e-mail: joyce.monteiro@embrapa.br; azeneth.schuler@embrapa.br; rachel.prado@embrapa.br; elaine.fidalgo@embrapa.br; ana.turetta@embrapa.br; alba.leonor@embrapa.br; aline.oliveira@embrapa.br; guilherme.donagemma@embrapa.br

U. Nehren et al. (eds.), *Strategies and Tools for a Sustainable Rural Rio de Janeiro*, Springer Series on Environmental Management,
https://doi.org/10.1007/978-3-319-89644-1_5

5.1 Introduction

Ecosystems provide ecosystem services (ES), welfare, and goods to people. The Millennium Ecosystem Assessment has estimated that some 60% of the ES are being used unsustainably (MEA 2005). ES are closely related to agricultural systems. On the one hand, ecosystems provide the services necessary to maintain agricultural systems such as climate regulation, water resources supply, soil formation, erosion control, carbon storage and sequestration, and nutrient cycling, among others. On the other hand, agricultural systems supply provisioning services such as food, fiber, and energy and, depending on the type of management, supporting services such as soil and water conservation and regulating services, for example, carbon storage (MEA 2005; FAO 2015).

Power (2010) discussed the relationship between ES and the type of management in agriculture. The author mentioned inadequate farming practices including deforestation, burning, and overfishing, which can result in serious damages, also called disservices, such as pollution, silting of rivers, and loss of habitats to conserve biodiversity. However, this situation can be reversed by the provision of ES through proper management of property and landscape, e.g., crop rotation and integration, no-tillage, level planting, mulching, soil and water conservation, optimization of agricultural inputs, conservation of riparian zones, and connecting forest fragments.

In Rio de Janeiro state (RJ), the Atlantic Forest ecosystems have been degraded and fragmented by deforestation due to successive economic cycles and the pressure from population growth (Ribeiro de Oliveira et al. 2018). The intensive exploitation has caused damages such as biodiversity loss, erosive processes, reduction of productivity, rural poverty, and decreased water quality. Current agriculture in RJ, though not very expressive when compared to other states, plays a fundamental role in the cultural, economic, social, and environmental aspects of local and regional communities, mostly practiced by family farmers (Rodríguez Osuna et al. 2014).

Considering the presence of agricultural lands alongside Atlantic Forest remnants, it is crucial to develop synergetic strategies to value and provide ES and increase productivity and income of the family farmers based on soil and water conservation management. These strategies are usually policies, programs, and other initiatives that encourage farmers to adopt adequate management practices. To evaluate how these practices succeeded in promoting ES provision, it is necessary to monitor the changes in the agroecosystems and the impacted ecosystems.

This chapter reviews studies regarding conservation management practices in RJ, their impacts on selected ES, and some key aspects of monitoring: joint design, multiscale approaches, selection of indicators, and participatory monitoring.

5.2 Soil and Water Conservation Practices and Their Impacts on Ecosystem Services

Changes in agricultural and forestry systems management practices tend to contribute to the provision of ES. Several studies show how conservation management practices in agriculture systems improve soil conditions, which may reduce water erosion and surface runoff, and can contribute to improve water quality, nutrient cycling, carbon storage, biodiversity, and pest control (Hengsdijk et al. 2005; Verhulst et al. 2011; Grace et al. 2012; Coutinho et al. 2014; Lal 2014). Some conservation management practices used in the family farming systems in RJ include fallow, no-tillage systems, green manure, organic fertilization, agroforestry systems, crop rotation, and rotational grazing.

Fallow consists of the cultivation of an area during consecutive years followed by a period of land use suppression. This area may or may not be cleared and cultivated thereafter. In an experiment carried out in Bom Jardim (RJ) with annual crop rotation followed by fallow for 5 years, the soil parameter stable aggregates index (SAI) has shown significant differences among the vegetation cover types with an average of 93% under fallow and 59% in soil samples under crop rotation (Coutinho et al. 2006). The soil losses varied from 30 to 40 $Mg.ha^{-1}$ in annual crop plots, from 10 to 11 $Mg.ha^{-1}$ in perennial banana crop plots, and from 6 to 9 $Mg.ha^{-1}$ in 1 and 2 years fallow crops. The average runoff in these areas varied from 250 to 300 $m^3\ ha^{-1}$ (Mendes 2006).

Other common conservation management practices in RJ are green manure and organic management. In the case of green manure, the soil is covered with plants, generally legume crops, which can capture atmospheric nitrogen by their symbiosis with Rhizobium bacteria. In organic management, farmers use either animal manure or vegetal matter such as crop residues and compost. Their impacts on soil chemical properties are usually greater than impacts on the physical ones. However, effects on the latter have been also observed in organic systems, such as the improvement of soil aggregation and bulk density reduction, leading to increase of soil porosity and better root growth (Cardozo et al. 2008).

Crop rotation consists of different crop species planted successively in the same area, with root systems that absorb nutrients from different soil depths. This practice has been often used together with no-tillage and the use of cover crops. Duarte (2006) showed that no-tillage crop rotation of corn, beans, and millet in the municipality of Campos dos Goytacazes (RJ) showed a productivity rate that was 43% higher than the conventional tillage of corn and beans without cover crops between successive harvests.

Agroforestry systems are characterized by a combination of trees or shrubs that grow together with agricultural crops; some systems also include animal husbandry. Planting trees and agricultural crops together allow better productivity by exploiting different soil layers and thus, reducing the competition among species. According to Rodríguez Osuna et al. (2014), local farmers reported an increase in agricultural productivity in banana-producing areas under agroforestry systems in Cachoeiras de Macacu (RJ).

Rotational grazing consists of subdivided pastureland in more than three parts protected by fences – each part is grazed in sequence by one or more lots of animals. Cesário et al. (2014) monitored rotational grazing in São Francisco de Itabapoana (RJ) and registered higher values for the aggregate stability index (ASI) of rotational pasture and forest (92% and 91%, respectively), while the succession of pineapple-cassava had lower average values of 58%. The carbon content was higher in the microaggregated class (ranging from 1 to 0.5 mm diameter) in forest and rotational pastures, respectively, 9.7 and 9.6 g.Kg^{-1}, in comparison to pineapple-cassava that showed 4.9 g.Kg^{-1}. These results indicate that this conservation practice has favored soil aggregation and carbon cycling and contributed to reduce erosion and improve pasture quality.

Soil and water conservation practices play an important role for the maintenance and restoration of ES. Considering some conservation practices in RJ, we list some key ES and relate them to reported practices and their potential beneficial impacts in Table 5.1.

Table 5.1 Selection of ecosystem services (ES), main conservation practices adopted in Rio de Janeiro state and their impacts on ES (based on various sources)

Ecosystem services (ES)	Conservation practices	Impacts on ES	References
Nutrient cycling	Fallow, no-tillage system, agroforestry, crop rotation, organic management, green manure, rotational grazing	Improved soil organic matter, biological nitrogen fixation	Araújo et al. (2006); Coutinho et al. (2006); Cardozo et al. (2008); Rangel et al. (2011)
Climate and GHG regulation	Fallow, no-tillage system, agroforestry, crop rotation, organic management, green manure	Carbon storage, improved soil organic matter	Coutinho et al. (2004); Duarte (2006); Cardozo et al. (2008)
Soil formation and erosion control	Fallow, no-tillage system, agroforestry, crop rotation, organic management, green manure, rotational grazing	Soil surface protection, restoration of degraded land	Araújo et al. (2006); Coutinho et al. (2006); Rangel et al. (2011)
Water supply, quality, and regulation	Fallow, no-tillage system, agroforestry, crop rotation, organic management	Improvement of soil physical properties	Cardozo et al. (2008); Rangel et al. (2011); Prado et al. (2013)
Food and fiber supply	No-tillage system, agroforestry, crop rotation;	Increments in productivity rates	Duarte (2006); Rodríguez Osuna et al. (2014)
Biodiversity habitat	Fallow, agroforestry, crop rotation, organic management, green manure	Increase of biological diversity	Coutinho et al. (2004); Rodríguez Osuna et al. (2014)
Control of pests and pathogens	Fallow, agroforestry, crop rotation, organic management Green manure	Increase of biological diversity, improved soil organic matter, improved chemical and physical soil properties	Coutinho et al. (2004); Rodríguez Osuna et al. (2014)

Even though the relationship between conservation practices and the impacts on ES has been reported, there are still gaps in the knowledge of ecological dynamics in agroecosystems. Thus, monitoring conservation practices and related ecosystem processes is essential to understand and evaluate their impacts on ES provision.

5.3 Guidelines to Monitor Conservation Practices

5.3.1 Joint Monitoring Design

According to van Oudenhoven et al. (2012), the analysis of ES to support land management decisions faces a number of challenges. These include (1) identifying comprehensive indicators to measure the capacity of ecosystems to provide services, (2) dealing with the complex dynamics of the link between land management and ES provision, (3) quantifying and modeling the provision of ES by connecting ecological processes to ES, and (4) accounting for the multiple spatial and temporal scales of ecological processes and ES provision.

We add two major challenges to be considered (i) the need for multidisciplinary studies to understand how land management affects ES with respect to the set of biotic and abiotic parameters involved and (ii) a suitable approach to analyze them together. These conditions should contribute to identify and to comprehend processes of environmental degradation or restoration, as well as the sites where they have occurred and their intensity.

Thus, monitoring the effects of land management on the provision of ES requires an integrated sample scheme. This requirement is crucial to understand the different levels of restoration of the environmental functionality. This is why the Comperj[1] monitoring of the revegetation project (Prado et al. 2014) adopted a landscape division on three different levels as a sampling scheme to consider geology, geomorphology, and pedology aspects. The first level classified the landscape into two sections: flat lands and slopes; the second level classified the slopes into two environments, depending on their origin, sedimentary rocks of Neogene and Paleogene periods and metamorphic rocks (gneisses) of the Proterozoic; the third level sectioned the upper, middle, and lower part of the slopes. The sampling scheme that was adopted aimed to obtain data on a set of themes (soil, water, climate, land use, flora, and fauna), an integrated analysis of them, and the spatialization of the results (Curcio et al. 2014; Fidalgo et al. 2014).

Other important factors to consider when defining sampling sites include the access conditions throughout the year and the land owner agreement for data collection on their farms. Sometimes sampling design needs to be adapted due to access limitations.

[1] Rio de Janeiro Petrochemical Complex (Complexo Petroquímico do Rio de Janeiro) is a Petrobras enterprise under construction that revegetated part of the area of its surroundings.

Before developing a monitoring program, it is essential to define and prioritize the need for information, to understand the current types and sources, and to understand the ones that could be used in the future besides identifying critical information gaps and opportunities to fill them in (Higgins and Zimmerling 2013). It is also vital to optimize monitoring linking sampling schemes, themes, and indicators for the monitoring objectives and intended answers.

While analyzing payment for ecosystem services (PES) schemes in watersheds, Porras et al. (2013) claimed that all monitoring efforts have to deal with the obstacles in establishing causal links with any degree of certainty due to patchy or short-term data collection; stochastic events masking local impacts, e.g., weather patterns; and uncertainty regarding groundwater movements. The key issue is to understand whether factors other than the target actions can affect water quality and quantity over time and how they do so.

Measuring ES response also requires a substantial baseline as reference to demonstrate both qualitative and quantitative environmental changes. According to Porras et al. (2013), very few PES schemes have developed measuring programs built on a sound baseline. The lack of reference regarding former land management in early PES schemes was an obstacle in monitoring the environmental changes and the benefits brought by the schemes, the so-called environmental additionality.

The study by Meals et al. (2010) on changes in water quality due to management practices highlights that monitoring efforts may not show definitive results if the monitoring period, program design, and sampling frequency are not sufficient to address the lag between treatment and response. The main components of lag time include (a) the time required for an installed practice to produce an effect, (b) the time required for the effect to be delivered to the water resource, (c) the time required for the water body to respond to the effect, and (d) the effectiveness of the monitoring program to measure the response.

The authors highlight that the frequency of monitoring also depends on the availability of teams, financial resources, and equipment. We can stress that even when monitoring is based on previous planning, disruptions and lack of continuity in monitoring programs and activities often occur, and these are a major cause of the lack of robust information on the effects of management practices on the environment and ES.

5.3.2 Multiscale Approach

To assess and monitor the impacts of the agricultural and forestry production systems on ES, a temporal and spatial multiscale approach must be used. A spatial multiscale approach is necessary to support decisions in different geographic scopes. For example, the rural producer has to provide elements at the farm level to support crop, pasture, and forest management. But it is also important to obtain information regarding the rural landscape to support planning and actions in the regional decision-making context.

It is also necessary to consider the different time scales in which ecosystem processes take place. For example, soil formation and erosion occur at the geological scale, but there are intense erosive processes with rapid effects. These processes can cause changes in water quality regarding sediment supply and increase the discharge of pollutants from point or diffuse sources, which compromise on water supply and aquatic biodiversity. Thus, a well-defined monitoring frequency must take the real-time scale into account.

Studying ES in the multiscale approach has been an international trend already reflected in Brazilian research. Power (2010) presents the importance of the multiscale approach in the assessment and the monitoring of services and disservices provided by soil and water management in agroecosystems (Fig. 5.1).

Minella et al. (2010) mention that erosion monitoring and modeling is a challenge to be met from the experimental plot to landscape scale with catchments as a study unit. At the landscape level, it is important to consider the fragmentation, the recharge of aquifers, the vulnerability of relevant plant species, and the pollinating and seed dispersing species among others to subsidize policies (De Groot et al. 2010; Power 2010).

The multiscale approach is also important in heterogeneous landscapes with their mosaics of crops, pastures, and forests, as it enables identifying the complexity of ecosystem responses. In terms of soil biodiversity, heterogeneous landscapes

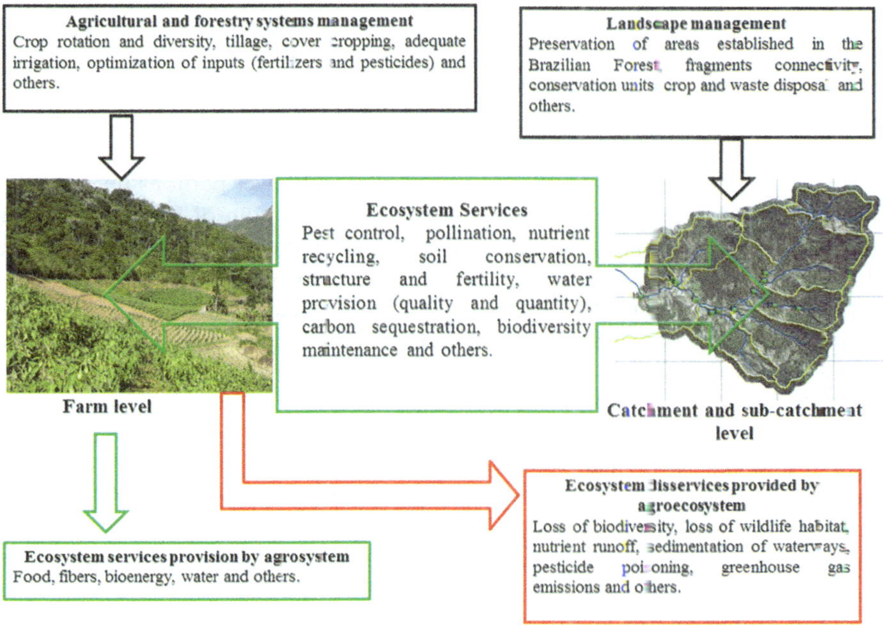

Fig. 5.1 Ecosystem services and disservices due to the management of agroecosystems in a multiscale approach. (Source: Adapted from Power (2010))

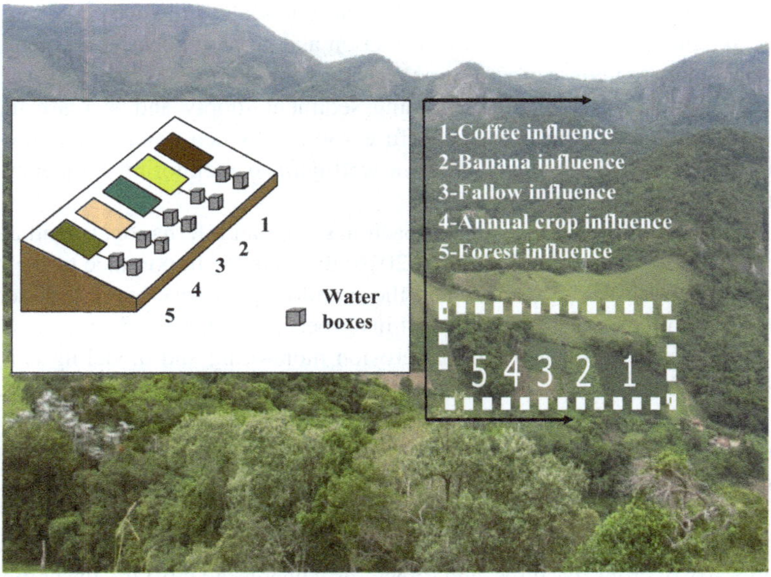

Fig. 5.2 Plots where soil and water were monitored (1, 2, 3, 4, and 5) at the local scale, Pito Aceso catchment, RJ. (Photo: Vinícius Benites adapted from Mendes (2006))

generally have greater biodiversity than homogeneous ones (Tscharntke et al. 2005; Rundlöf and Smith 2006).

In the Pito Aceso catchment, located in the Bom Jardim municipality, the soil and water quality was monitored at the local scale, to understand how different uses and conservation management influence the ecosystem. Soil samples were taken in five plots (25 m x 70 m) from the upper, middle, and lower portions once a year. Water samples were collected in water boxes at the bottom of each plot, which stored runoff water and suspended soil particles after each precipitation event (Fig. 5.2) (Coutinho et al. 2006). The landscape structure and fallow-based agricultural production system reduce the negative impact, so that the water that drains out of the system does not show significant deterioration of its quality, when compared with the water that entered the system from undisturbed areas.

At the landscape scale, 60 pedo-environmental units (UPs) were defined as areas that have the same soil, land use, and cover characteristics. The UPs were used as sampling units to assess and monitor soil quality. Pedo-environmental units may be also useful to estimate other environmental parameters (Prado et al. 2013). To assess water quality parameters at the landscape scale, samples were collected at Pito Aceso sub-catchments outlets, and the results were related to land use and management in each sub-catchment (Fig. 5.3).

When monitoring and assessing the impacts of conservation management on ES, very subtle processes of land use and management change deserve special attention. Small but significant alterations are caused by changes in agricultural management,

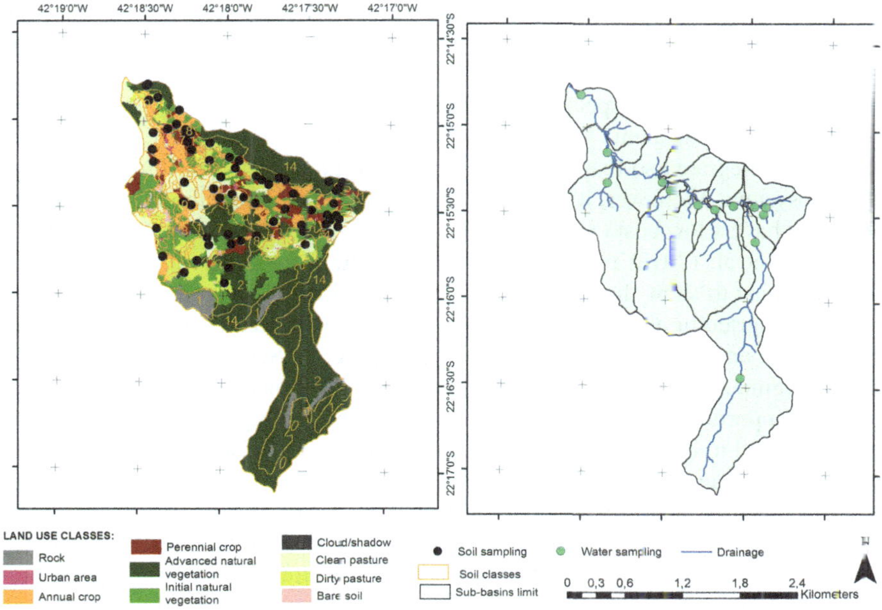

Fig. 5.3 Soil and water sampling points at landscape scale in the Pito Aceso catchment

including input levels, fallow cycles, rotation systems, and secondary vegetation. In such cases, the information obtained from remote sensing images requires field visits and data sampling in situ to characterize the farming systems and analyze the relationship between land use dynamics and ES (Lambin et al. 2000).

In addition, other aspects are important for successful monitoring, such as the selection of indicators able to measure the impact of land use and management in ES.

5.3.3 Indicators

A key point concerning agroecosystems and their potential to provide ES is to define indicators to evaluate the impact of soil management on ES. Indicators and indexes to evaluate different kinds of impacts are powerful tools, capable of expressing complex phenomena. The indicator must also take into account clear communication, which gives it the ability to express the change in a simple, concise, and reliable way. It also helps to improve the perception of different actors on the main impacts of a real event (Smeets and Weterings 1999).

Dale and Polasky (2007) state that the challenge to select ecological indicators to link agroecosystems and ES provision is to identify key features that represent the compositional, structural, and functional components of the system important for

ES provision. This idea would use a minimum set of indicators to show ES provision changes influenced by land use and soil management or an index (Karlen and Scott 1994; Chaer and Tótola 2007). This is a central recommendation as it is not possible to monitor all aspects of the interactions between ES provision and agroecosystems management. Considering soil ES, this minimum set of indicators should be integrated so that it reflects changes in the various soil functions in the agroecosystem. Moreover, it should be easy to measure, be methodologically reliable, and have a low cost (Stenberg 1999).

Turetta et al. (2016) proposed a set of indicators for different agroecosystem managements to show their influence on soil functions. Some criteria that are correlated to agricultural practices were selected and validated by representatives of agricultural entities, producers, and the research team. Selected criteria such as a no-fire regime, rational use of agrochemicals, and integrated agricultural crops represented opportunities to improve the ES provided by each agroecosystem. A matrix was produced to demonstrate the relationship among the criteria for the establishment and management of the agroecosystems and the ES types, soil functions, potential soil indicator, ES benefits, and policy relevance. The proposition of simple and easy indicators was prioritized, and all indicators are soil parameters validated by soil science and can be easily found in the literature. The authors suggest biomass stock as one of the most appropriate soil parameters used as indicators to monitor the impact of agroecosystem on ES provision.

In Bom Jardim (RJ), Benites et al. (2010) found that areas under agricultural use showed minor contents of carbon, nitrogen, and humic acids, although they present higher fertility, indicating that fallow or nonintervention in the soil (i.e., preparation, pH correction, or fertilization) is an important management component, especially for soil organic matter conservation.

In this same area, Mendes (2006) found that areas under short-cycle crops presented the highest values of soil losses and flows in the surface compared to the fallow system, indicating low soil protection against erosion. Miranda et al. (2008) emphasized that porosity, macroporosity, and saturated hydraulic conductivity are parameters sensitive to differentiate the impact of agricultural management.

Soil microbiology is also sensitive to land use systems, and its information has potential to be used as indicators of ES provision by soil and agroecosystems. Grazing, stocking rate, tillage, and other soil practices may affect soil microbial activity and its role in the transformation of nutrients. Coutinho et al. (2004) presented results that show that microbial diversity was sensitive to soil management. The authors found similar results between forest and 5-year fallow that presented higher microbial diversity when compared to crops, pasture, and 1- and 3-year-old fallow.

A good strategy to obtain robust analyses on how soil management affects ES provision is to integrate soil and water indicators. For example, Menezes et al. (2012) studied a watershed in a rural area in the northwest of Rio de Janeiro (São Domingos watershed) and found that the most frequent nonconformity parameters of water quality were associated with agricultural land.

Prado et al. (2015) state that the intense conversion of natural vegetation into cropland increased the pollution potential in a water reservoir in the Mata Atlântica biome, with a direct impact on water supply ES.

In the face of the many possibilities to use soil and water parameters as ES indicators, it is important to define clearly the study's goal. Simple questions should be answered before defining the set of indicators. These include indicators for what (objective)? Indicators for whom (users)? What are the resources (human and financial) available to monitor these indicators? Other questions that can help choose the indicators are the following: What is the level of feasibility to obtain an indicator? What is the relevance of the indicator as representative of increment or monitoring of environmental service (Fidalgo et al. 2017)?

5.3.4 Participatory Monitoring

Participatory monitoring (PM) aims to empower all actors involved, such as farmers, community leaders, scientists, rural extension technicians, and decision-makers. According to Danielsen et al. (2014), participation within monitoring should be present from the planning phase up to the application of the results into decision-making processes.

The participatory approach in the agroecosystem monitoring may have other goals:

- To promote the local stakeholders' knowledge on the ecosystems and the impacts of land management on the ES
- To contribute to research on socioeconomic and environmental issues of communities and their territories to build solutions to their needs
- To translate technical results into a more accessible language for a target group
- To subsidize policies to meet suitable solutions for ES provision
- To facilitate the implementation of ES policies and the reduction of their costs

Constantino et al. (2016) presented a PM proposal in which the guidelines targeted the local actors (stakeholders) involvement. These authors listed questions that have to be considered, such as:

(a) How does local participation affect conservation and how do communities learn and change with PM?
(b) How can data reliability be verified?
(c) How can PM be successfully maintained and integrated to scales which present more detailed information?
(d) How can PM influence political decision-making?
(e) How can institutional arrangements from PM initiatives be established?
(f) How can resources for collaborative projects be obtained?
(g) How can information about the challenges and the solutions related to the work of partners in PM be exchanged?

Choosing the correct participatory method depends on the goals. The participative approach may be applied, e.g., to assess and characterize farming systems, which is desirable before the participatory monitoring. Rodríguez Osuna et al. (2014) and Martins et al. (2014) adopted different strategies to evaluate farming systems in Cachoeiras de Macacu (RJ). Rodríguez Osuna et al. (2014) used structured forms to assess farm household data to estimate the opportunity costs of abandoning current land uses to drive conservation investments for the recovery of the natural vegetation. Martins et al. (2014) applied guided interviews to collect information on soil management and farming systems and used the results to develop a strategy for participatory training in soil and water management.

In some cases, local individuals can be trained to participate in the monitoring done on the field. Prado et al. (2005) described the use of analytical kits by students and teachers to monitor water quality in São José de Ubá municipality (RJ). This experience also led to improved environmental education. Another participatory approach can be developed by training residents to assess data based on field observations, e.g., presence of living organisms, sediment deposition, and soil surface crusting. These participatory approaches can reduce costs, such as the need to transport specialized technicians or equipment that could be damaged or stolen.

To facilitate communication with stakeholders, some PM can integrate field parameters observed, such as the riparian zones' quality index used to monitor the restoration of these zones and the river channel quality (Bjorkland et al. 2001). Barrios et al. (2011) proposed a methodology that integrates soil quality knowledge from farmers, extension technicians, and scientists and generates a basis to promote the adoption of sustainable land use and management technologies. This approach allows monitoring ES while improving environmental perception and management.

5.4 Final Comments

When considering the relevant socioeconomic, cultural, and environmental role of agriculture and the incipient conservation management in RJ, investments in rural extension, PES, agrotourism, certifications of organic products, and fair trade among other alternatives are necessary to provide support for farmers promoting an increase in income and provision of ES.

Different conservation management practices occur in RJ, and their impacts on ES have been reported. However there are still gaps in the knowledge on the relationship between these practices and ecosystem dynamics. Monitoring them is essential to evaluate their impacts on ES provision.

An important aspect of monitoring is the systematization and dissemination of the data obtained to facilitate communication between decision-makers and rural producers. The main challenges of the monitoring approach include participatory efforts and multidisciplinary criteria used at appropriate scales in order to provide adequate and useful answers to society. In addition, monitoring is also useful to support policies related to water and soil conservation.

References

Araújo ES, Andrello AC, Appoloni CR et al (2006) Avaliação do manejo agrícola adotado em propriedade familiar da região serrana Fluminense quanto ao potencial de conservação do solo. Seropédica: Embrapa Agrobiologia Circular Técnica 15:4

Barrios E, Coutinho HLC, Medeiros CAB (2011) InPaC-S: integração participativa de conhecimentos sobre indicadores de qualidade do solo: guia metodológico. Nairobi: World Agroforestry Centre (ICRAF), Embrapa, Ciat, p 178. http://www.worldagroforestry.org/downloads/Publications/PDFS/B17293.pdf. Accessed 01 Jul 2016

Benites VM, Moutta RO, Coutinho HLC, Balieiro FC (2010) Análise discriminante de solos sob diferentes usos em área de Mata Atlântica a partir de atributos da matéria orgânica. Revista Árvore, Viçosa-MG 34(4):685–690

Bjorkland R, Pringle CM, Newton B (2001) A stream visual assessment protocol (SVAP) for riparian landowners. Environ Monit Assess 68:99–125

Cardozo SV, Pereira MG, Ravelli A, Loss A (2008) Caracterização de propriedades edáficas em áreas sob manejo orgânico e natural na região serrana do Estado do Rio de Janeiro. Semina: Ciências Agrárias, Londrina 29(3) 515–528

Cesário FV, Donagemma GK, Macedo JR et al (2014) Índices de agregação e teor de carbono intra-agregados como indicadores de qualidade do solo. Boletim de Pesquisa. Embrapa Solos, Rio de Janeiro, p 24

Chaer GM, Tótola MR (2007) Impacto do manejo de resíduos orgânicos durante a reforma plantios eucalipto sobre indicadores qualidade solo. Rev Bras Ciênc Solo 31:1381–1396

Constantino PAL, Buening JK, Silvius K et al (2016) Monitoramento participativo da biodiversidade e dos recursos naturais: Seminário Internacional e Formação da Rede Internacional de Monitoramento e Manejo Participativo. Biodiversidade Brasileira 6(1):18–33

Coutinho HLC, Aboim MCR, Rosado AS, Barbosa JC (2004) Impacto à diversidade microbiana (IDM): uma estratégia para incorporação de resultados de análises moleculares de biodiversidade em estudos integrados de qualidade do solo. - Rio de Janeiro: Embrapa Solos. Embrapa Solos. Boletim de Pesquisa e Desenvolvimento 62, p 27

Coutinho HLC, Prado RB, Donagemma GK et al (2006) Qualidade de solo e água como indicadores de recuperação de áreas degradadas submetidas a manejo agroflorestal. Documentos 100. Embrapa Solos, Rio de Janeiro, p 38

Coutinho HLC, Noellemyer E, Balieiro FC et al (2014) Impacts of land-use change on carbon stocks and dynamics and Central-Southern South American Biomes: Cerrado, Atlantic Forests and Southern Grasslands. In: Banwart SA, Noellemyer E, Milne E (eds) Soil carbon, science, management, and police for multiple benefits. CABI, Wallingford, pp 243–265

Curcio GR, Bonnet A, Resende AS (2014) Compartimentação das paisagens. In: Prado RB, Fidalgo ECC, Bonnet A (eds) Monitoramento da revegetação do COMPERJ: etapa inicial. Embrapa, Brasília, pp 23–36

Dale VH, Polasky S (2007) Measures of the effects of agricultural practices on ecosystem services. Ecol Econ 64:286–296

Danielsen F, Pirhofer-Walzl K, Adrian TP et al (2014) Linking public participation in scientific research to the indicators and needs of international environmental agreements. Conservation Letters, 7(1):12–24

De Groot RS, Alkemade R, Braat L et al (2010) Challenges in integrating the concept of ecosystem services and values in landscape planning, management and decision making. Ecol Complex 7:260–272

Duarte Jr JB (2006) Avaliação agronômica da cana-de-açúcar, milho e feijão em sistema de plantio direto em comparação ao convencional em Campos dos Goytacazes – RJ. Tese. Campos dos Goytacazes, p 284. Doutorado (Produção Vegetal) – Universidade Estadual do Norte Fluminense (UENF)

FAO (2015) Healthy soils are the basis for healthy food production, FAO. Available at: http://www.fao.org/3/a-i4405e.pdf

Fidalgo ECC, Prado RB, Resende AS et al (2014) Estratégia metodológica para o monitoramento. In: Prado RB, Fidalgo ECC, Bonnet A (eds) Monitoramento da revegetação do COMPERJ: etapa inicial. Embrapa, Brasília, pp 37–47

Fidalgo ECC, Prado RB, Turetta APD, Schuler AE (2017) Manual para pagamento por serviços ambientais hídricos: seleção de áreas e monitoramento. Brasília, Embrapa, p 80

Grace PR, Antle J, Aggarwal PK et al (2012) Soil carbon sequestration and associated economic costs for farming systems of the Indo-Gangetic plain: a meta-analysis. Agric Ecosyst Environ 146(1):137–146

Hengsdijk H, Meijerinkb GW, Mosuguc ME (2005) Modeling the effect of three soil and water conservation practices in Tigray. Ethiopia. Agric Ecosyst Environ 105:29–40

Higgins JV, Zimmerling A (eds) (2013) Um compêndio para o monitoramento de fundos de água. The Nature Conservancy, Arlington

Karlen DL, Stott DE (1994) A framework for evaluating physical and chemical indicators of soil quality. In: Doran JW, Coleman DC, Bezdicek DF, Stewart BA (eds) Defining soil quality for a sustainable environment. Soil Science Society of America - Special Publication 35, Madison, pp 53–72

Lal R (2014) Soil conservation and ecosystem services. Int Soil Water Conserv Res 2(3):36–47

Lambin E, Rounsevell M, Geist HJ (2000) Are agricultural land-use models able to predict changes in land-use intensity? Agric Ecosyst Environ 82(1–3):321–331

Martins ALS, Schuler AE, Fidalgo ECC et al (2014) O enfoque sistêmico no diagnóstico participativo dos sistemas de produção da comunidade Faraó. Rio de Janeiro: Embrapa Solos, p 42 (Embrapa Solos. Documentos, 178)

Meals DW, Dressing SA, Davenport TA (2010) Lag time in water quality response to best management practices: a review. J Environ Qual 39:85–96. https://doi.org/10.2134/jeq2009.0108

Mendes CAR (2006) Erosão Superficial em encosta íngreme sob cultivo perene e com pousio no Município de Bom Jardim – Rj. Ph.D. thesis, COPPE- Universidade Federal do Rio de Janeiro, Rio de Janeiro, RJ

Menezes J, Prado RB, Silva GC Jr et al (2012) Qualidade da água superficial em microbacias do Noroeste Fluminense. Caderno de Estudos Geoambientais CADEGEO 3(1):32–43

Millennium Ecosystem Assessment MEA (2005) Ecosystems and human well-being: synthesis. Island Press, Washington, DC, p 155

Minella JPG, Merten GH, Reichert JM, Cassol EA (2010) Processos e modelagem da erosão: da parcela à bacia hidrográfica. In: Prado RB, Turetta AP, Andrade AG (eds) Manejo e Conservação do Solo e da Água no Contexto das Mudanças Ambientais. Embrapa Solos, Rio de Janeiro, pp 105–122

Miranda JPL, Silva LM, Lima RL et al (2008) Influência do pousio na recuperação de propriedades físicas de um Cambissolo em ambiente agrícola serrano – Bom Jardim (RJ). In: Reunião Brasileira de Manejo e Conservação do Solo e da Água, 17. Rio de Janeiro. Manejo e conservação do solo e da água no contexto das mudanças ambientais. Rio de Janeiro: SBCS: Embrapa Solos: Embrapa Agrobiologia, p 4 (Embrapa Solos. Documentos, 101)

van Oudenhoven APE, Petz K, Alkemade R et al (2012) Framework for systematic indicator selection to assess effects of land management on ecosystem services. Ecol Indic 21:110–122

Porras I, Alyward B, Dengel J (2013) Monitoring payments for watershed services schemes in developing countries. International Institute for Environment and Development, London

Power AG (2010) Ecosystem services and agriculture: tradeoffs and synergies. Philos Trans Royal Soc B: Biol Sci 365:2959–2971

Prado R, Macedo JR, Perez D et al (2005) Monitoramento de indicadores de qualidade da água como subsídio à gestão de recursos hídricos em microbacia do município de São José de Ubá – RJ. In: IV Workshop Do Projeto Gestão Participativa da Sub-Bacia do Rio São Domingos/ RJ – Rio de Janeiro – GEPAR- MBH. Rio de Janeiro, RJ

Prado RB, Chaer GM, Balieiro FB et al (2013) Proposta metodológica para amostragem de solo e água visando o monitoramento e avaliação de serviços ambientais em unidades de paisagens rurais. Comunicado Técnico 69. Embrapa Solos, Rio de Janeiro, p 16

Prado RB, Fidalgo ECC, Bonnet A (eds) (2014) Monitoramento da revegetação do COMPERJ: etapa inicial. Embrapa, Brasília, p 350

Prado RB, Fidalgo ECC, Balieiro FC et al (2015) Serviços ambientais no bioma Mata Atlântica no Estado do Rio de Janeiro: lições e desafios. In: Parron LM, Garcia JR, Oliveira EB, Brown GG, Prado RB (eds) Serviços ambientais em sistemas agrícolas e florestais do Bioma Mata Atlântica. Embrapa Florestas, Brasília: Embrapa, Colombo-PR, pp 202–218

Rangel LA, Donagemma GK, Macedo JR et al (2011) Estabilidade de agregados em solo sob pastejo rotacionado em São Jose de Ubá-RJ. In: Solos nos biomas brasileiros: sustentabilidade e mudanças climáticas. Anais do Congresso Brasileiro de Ciência do Solo, 33 Uberlândia, MG: SBCS: UFU, ICIAG

Ribeiro de Oliveira R, Stingel Fraga J, MacLeod Hickie M (2018) Environmental history, forests and landscape uses in Rio de Janeiro state. In: Nehren U, Schlüter S, Raedig C, Sattler D, Hissa H (eds) Strategies and tools for a sustainable rural Rio de Janeiro. Springer International Publishing, Cham

Rodríguez Osuna V, Prado RB, Monteiro JM (2014) Perfil dos sistemas de produção agropecuária: uma Visão do Produtor Rural na Sub-bacia de Batatal em Cachoeiras de Macacu, Rio de Janeiro, RJ Embrapa Solos: Rio de Janeiro Boletim de Pesquisa e Desenvolvimento 247, p 49

Rundlöf M, Smith HG (2006) The effect of organic farming on butterfly diversity depends on landscape context. J Appl Ecol 43:1121–1127

Smeets E, Weterings R (1999) Environmental indicators: Typology and overview. European Environmental Agency Technical report, Copenhagen, no. 25

Stenberg B (1999) Monitoring soil quality of arable land: microbiological indicators. Soil Plant Sci 49:1–24

Tscharntke T, Klein AM, Kruess A et al (2005) Landscape perspectives on agricultural intensification and biodiversity ecosystem service management. Ecol Lett 8:857–874

Turetta APD, Tonucci R, Mattos LM et al (2016) An approach to assess the potential of agroecosystems in providing environmental services. Pesq Agrop Brasileira 51(9):1051–1060

Verhulst N, Nelissen V, Jespers N et al (2011) Soil water content, maize yield and its stability as affected by tillage and crop residue management in rainfed semi-arid highlands. Plant Soil 344:73–85

Chapter 6
Spatial Patterns of Farmland Abandonment in Rio de Janeiro State

Pedro Castro, Rui Pedroso, Sven Lautenbach, Oscar Manuel Baez Villanueva, and Raúl Vicens

Abstract While land use change in Brazil is characterized by strong cropland expansion, the federal state of Rio de Janeiro, located in Southeast Brazil, shows an opposite trend: a significant amount of cropland has been abandoned. The temporal and spatial distribution of farmland abandonment in Rio de Janeiro state was analyzed at the municipal level from 1991 to 2013. Developments differ strongly between the different regions. The Norte Fluminense Region, which holds the highest share of cropping area in the state, showed the highest cropland abandonment. This decrease was mainly due to a reduction of the areas used for the cultivation of banana, maize, and sugarcane. In the Baixadas and Metropolitan regions, the reduction of areas for orange plantations played an important role, while the reduction of areas for coffee production was important for mountainous areas. Another transition

P. Castro (✉)
TH Köln – University of Applied Sciences, Institute for Technology and Resources Management in the Tropics and Subtropics (ITT), Köln, Germany

Institute for Geodesy and Geoinformation, Faculty of Agriculture, Bonn University, Bonn, Germany

Departament of Geography, Fluminense Federal University, Niterói, Rio de Janeiro, Brazil
e-mail: pbastosd@th-koeln.de

R. Pedroso · O. M. Baez Villanueva
TH Köln – University of Applied Sciences, Institute for Technology and Resources Management in the Tropics and Subtropics (ITT), Köln, Germany
e-mail: rui.pedroso@th-koeln.de; obaezvil@th-koeln.de

S. Lautenbach
Institute for Geodesy and Geoinformation, Faculty of Agriculture, Bonn University, Bonn, Germany
e-mail: sven.lautenbach@uni-bonn.de

R. Vicens
Departament of Geography, Fluminense Federal University, Niterói, Rio de Janeiro, Brazil
e-mail: rsvicens@id.uff.br

U. Nehren et al. (eds.), *Strategies and Tools for a Sustainable Rural Rio de Janeiro*, Springer Series on Environmental Management,
https://doi.org/10.1007/978-3-319-89644-1_6

seems to be the substitution of sugarcane cultivation with pineapple cultivation. Areas for coconut crop production increased mainly in the Quissamã municipality. Areas for manioc crop production remained stable throughout the analysis. The patterns identified in this paper will provide an important background for the policy-makers in implementing spatially explicit plans for the agricultural sector.

Keywords Farmland abandonment · Agricultural dynamics · Spatial analysis · Time series analysis

6.1 Introduction

Farmland abandonment is a process in which croplands are no longer maintained. The abandoned croplands are either used to keep livestock or are left unmanaged which allows a natural succession; sometimes they make also place for plantations (FAO 2006; Keenleyside and Tucker 2010; Lambin and Meyfroidt 2010; Rudel et al. 2010). Typically, this process is a result of agricultural intensification associated with the abandonment of marginal areas (Baumann et al. 2011; Hostert et al. 2011). One immediate cause for farmland abandonment is the low viability for smallholders, underlined mostly by agricultural policies (Aide and Grau 2004; Keenleyside and Tucker 2010; Renwick et al. 2013; Terres et al. 2015).

Farmland abandonment takes place in different geographical regions and is triggered by different geopolitical developments (Munroe et al. 2013). In Europe, it is particularly evident in Eastern Europe, Southern Scandinavia, and in mountain regions (Baumann et al. 2011; Hostert et al. 2011; Griffiths et al. 2013; Munroe et al. 2013; Prishchepov et al. 2013; Estel et al. 2015). Examples of cropland abandonment in Latin America can be found in Western Honduras (Southworth and Tucker 2001), Northern Argentina (Izquierdo and Grau 2009), and Southern Chile (Díaz et al. 2011).

This study focuses on farmland abandonment in the Rio the Janeiro state (RJ) in Southeast Brazil. The agricultural activity in RJ is characterized by the production of sugarcane, banana, coffee, manioc, maize, and orange, which account for more than 90% of the cultivated areas (IBGE 2017). The definition of farmland used there includes orchards and plantation used for fruit production and excludes animal husbandry.

From 1988 to 2016, the state lost 70% of its areas used for crop cultivation. The developments in RJ are in sharp contrast with the nationwide agriculture expansion. This national trend contributed to maintaining Brazil as one of the leading producers and exporters of agricultural commodities worldwide (Gasques et al. 2010; Maranhão 2015) – however, at the cost of biodiversity, especially in the Cerrado and Amazon biomes (Brannstrom et al. 2008; INPE 2017).

6.2 Relevant Policies

Environmental policies play an important role in the farmland abandonment and forest recovering (García-Barrios et al. 2009; Renwick et al. 2013). In Brazil, the Forest Act (Federal Act. 12,651/2012), the Private Nature Reserves Act (RFPN) (Federal Act. 5746/2006), and the Payment for Ecosystem Services (PES) Act (Federal Act. 12,512/2011) addressed essential issues related to environmental protection at a national level.

The Forest Act (Federal Act. 12,651/2012) promoted the Environmental Rural Cadaster (CAR) that aims to create spatial database regarding the land uses in the farms for monitoring and promoting an environmentally friendly management. The Federal Law 5746/2006 regulated and promoted the private natural reserves (RPPN), and the Federal Law 12,512/2011 implemented the Payment for Ecosystem Services (PES) at national level.

Those policies encourage the recovering of natural habitats and create income to the farmers either through nature-oriented tourism or PES. However, these policies are still at the implementation level, i.e., their effects can currently not be sufficiently evaluated. Some study cases related to the implementation of those policies are discussed in more detail by Guzmán Wolfhard and Raedig (2018), May et al. (2018), and Noriega Gamarra et al. (2018) in this book.

The substantial reductions in agricultural areas in RJ might be at least partially due to government policies. Therefore, since 2000, the state's government implemented several state-level acts aiming to foster different spheres of agriculture production for reestablishing the balance between the various economic activities in RJ.

The State Law 34,335/2003 aimed to promote the cultivation of ornamental and medicinal plants by providing differentiated credit. The State Law 34,015/2003 intended to strengthen the organic agriculture, stimulating the already existent productions as well as promoting the establishment of new ones. The State Law 26,278/2000, also known as the Frutificar program, increased the offer of credits and projects for fruits production in the Norte and Noroeste Fluminense Administrative Regions.

Understanding the patterns and driving factors of farmland abandonment is essential due to its different socioeconomic and environmental impacts (Foley 2005; Griffiths 2013). Farmland abandonment is clearly associated with socioeconomic consequences. In RJ, the agricultural activity is the main source of income of approximately 170 thousand inhabitants (IBGE 2009, 2017); cropland abandonment is likely to lead to increasing unemployment rates in rural regions or migration to cities (Aide and Grau 2004).

When based on crop genetic engineering, those previous intensive crop areas become dependent on genetically modified species, generating substantial environmental and socioeconomic impacts (Mcafee 2005). Areas with past agriculture intensification show reduced capacity for a recovery of the original vegetation (Cramer et al. 2008). Abandoned croplands converted into grassland have led to increasing runoff rates in other regions (Lasanta et al. 1995; Algeet-Abarquero et al.

2015) and posed threats to ecological high value species inhabiting croplands (Katayama et al. 2015; Zakkak et al. 2015).

In order to understand consequences of current and future developments, it is essential to identify drivers of change. Within its scope, this chapter uses administrative data to understand the spatial distribution of farmland abandonment in RJ. The objective is to identify the regions where farmland abandonment is most critical and to identify the crop types that are most affected by farmland abandonment. Those results will be used to set up a research design for an econometric analysis to identify the driving factors for farmland abandonment in RJ. Furthermore, the spatial pattern identified in this study establishes the basis for the assessment of the effects of public policies in RJ.

6.3 Material and Methods

6.3.1 Study Site

RJ (Fig. 6.1) is the second biggest economy in the country, with a share of 11% in the national gross domestic product (GDP). The state is the fourth most populated in the country with nearly 16 million inhabitants. Its population density is only surpassed by the Federal District (IBGE 2011).

The economy of the state is based on the services sector, which accounts for 69% of the state GDP (CEPERJ 2017). The importance of agricultural activities has been continuously decreasing: the share of the agricultural production of the federal state GDP has decreased from 0.4% in 2002 to 0.26% in 2015 (CEPERJ 2017). The share of the livestock activity has also decreased from 0.2% to 0.17% (CEPERJ 2017).

In RJ there is evidence for a rural out-movement: while the urban population is increasing, rural areas are characterized by an opposite trend. From 1991 to 2010, the number of inhabitants in RJ grew by 36%, from 9.5 million to 12.9 million (IBGE 2011). In the meantime, the rural population decreased by 5%, from 470,222 to 447,694 (IBGE 2011).

6.3.2 Environmental Factors of Land Use Change

It is well documented that steep topographic conditions play a substantial role regarding farmland abandonment and subsequent forest transitions in Southeast Brazil (Silva et al. 2016, 2017). In RJ, the most likely direct consequence is the conversion from crop farms to low-maintenance activities such as cattle raising. In Argentina, for example, this trend occurs differently. In its northern parts, livestock activity is not common in highlands because the cattle-raising activity relies on extensively flat areas (Izquierdo and Grau 2009).

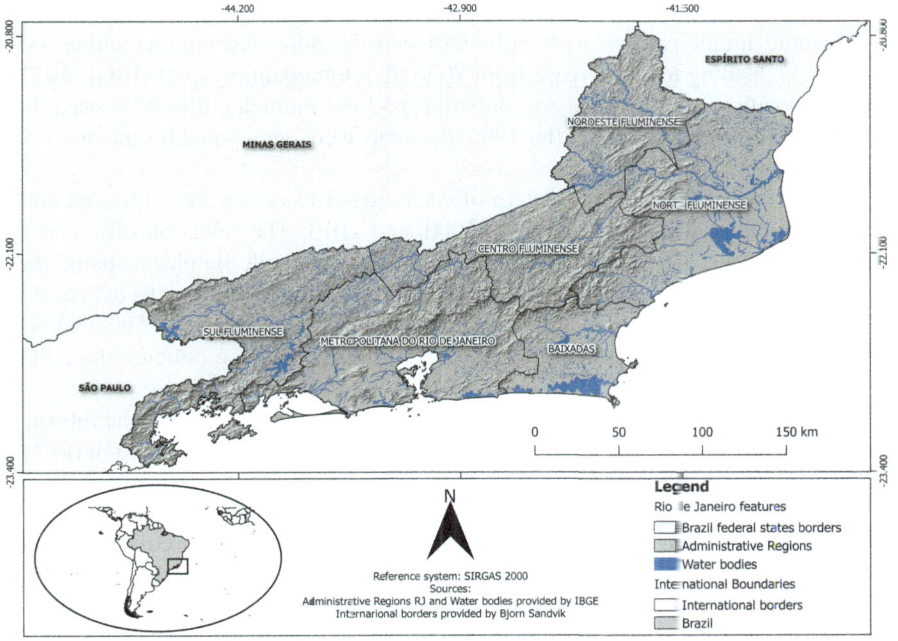

Fig. 6.1 Study site location

Topographic conditions are also seen as an important driving factor in farmland abandonment in other parts of the world. In the Swiss mountains, soil and topographic conditions associated with an underdeveloped road infrastructure triggered substantial farmland abandonment, mostly impacting landscapes of traditional agriculture as well as the habitats of high ecological value species (Gellrich and Zimmermann 2007).

6.3.3 Methods and Data

In order to assess changes in cropland areas in RJ, the authors used the Municipal Agricultural Production (PAM) survey, from the Brazilian Institute of Geography and Statistics (IBGE). Since 1988 the PAM has been providing annual data per municipality about the cultivated area, harvested area, crop yields, and production values for 29 different temporary crop types and 33 permanent crop types (IBGE 2002).

The time series of each of the main crops in RJ was analyzed to uncover specific temporal and spatial dynamics. The eight crops with the largest cropland share in the federal state were selected: sugarcane, bananas, coffee, manioc, maize, oranges, coconuts, and pineapples. For the analysis of spatial pattern, administrative statistics were linked to the RJ federal state municipality administrative divisions. After 1991 some municipalities were split into two or more new municipalities (see Table 6.1), leading to an increase from 70 to 92 municipalities in RJ (IBGE 2017). In order to allow a proper analysis, only the pre-1991 municipalities were used. For municipalities that were split after 1991, the areas were aggregated for the pre-1991 municipalities.

The analysis of the spatial pattern of each crop considering the cultivated areas per municipality was done for 1991, 2000, and 2013. The selection of a starting period from 1991 intended to match the tabular results with the oldest spatial data containing the municipality boundary available. The year 2000 was selected because it was the time where several agricultural policies were implemented. The final year of 2013 was selected to observe the possible impacts of those policies since their implementation.

The information on the amount of livestock was analyzed as well. The information on livestock was obtained through the Municipal Livestock Production (PPM). Pasture was not considered in the definition of farmland because the data for pasture was not available for the period of analysis.

Information about pasture uses was obtained by the IBGE agricultural census, which covers a different period, from 1975 to 2006. In this survey, the information was provided only at a federal state level. The agricultural census considered crop farms and husbandry farms, denominated agriculture establishments. This concept of farm differs from the one it is included on the farmland abandonment in this chapter, which considers only crop farms. Bearing in mind that limitation, the agricultural census data was used only to evaluate if the abandoned farms are being converted into areas for cattle raising.

6.4 Results

Cropland areas (excluding livestock activities) in RJ decreased by 301,906 hectares (70%) from 1988 to 2016 (IBGE 2017). This development was in sharp contrast to the national trends (Fig. 6.2), where crop areas have increased by around 20 million hectares from 1988 to 2016 (IBGE 2017). The amount of livestock also decreased. The figures show a 15% reduction from 1974 to 2015 (IBGE 2017).

The total area of agricultural establishments (crop farms and husbandry farms) in RJ decreased from 3,181,385 ha to 2,059,462 ha, from 1980 to 2006 (IBGE 2009). In 1980, pasture accounted for the biggest share within the establishments with 1,744,614 ha, i.e., 54% of the total area. Despite having the biggest absolute loss of 454,252 ha from 1980 to 2006, pasture still accounted for the highest share in the establishments. In 2006, the pasture uses accounted for 63% of the uses within the establishments. Meanwhile, the share of crops decreased from 19% to 14%.

In 2013, pineapples, bananas, coffee, sugarcane, coconut, oranges, manioc, and maize accounted for 92.91% of the cultivated areas in RJ (Fig. 6.3). From those, sugarcane accounted for 57%, followed by bananas (11%), coffee (7%), manioc

Table 6.1 Municipalities subdivisions in Rio de Janeiro state from 1980 to 2013

1980	1991	2013
Resende	Resende	Resende
		Porto Real
	Itatiaia	Itatiaia
Barra Mansa	Barra Mansa	Quatis
		Barra Mansa
Piraí	Piraí	Pinheiral
		Piraí
Itaguaí	Itaguaí	Itaguaí
		Seropédica
Nova Iguaçu	Nova Iguaçu	Nova Iguaçu
		Queimados
		Japeri
		Mesquita
		Belford Roxo
Vassouras	Vassouras	Paty do Alferes
		Vassouras
Três Rios	Três Rios	Areal
		Três Rios
		Comendador Levy Gasparian
Magé	Magé	Guapimirim
		Magé
Itaboraí	Itaboraí	Tanguá
		Itaboraí
Petrópolis	Petrópolis	Petrópolis
	São José do Vale do Rio Preto	São José do Vale do Rio Preto
Cabo Frio	Cabo Frio	Cabo Frio
		Arraial do Cabo
		Armação dos Búzios
São Pedro da Aldeia	São Pedro da Aldeia	São Pedro da Aldeia
		Iguaba Grande
Casimiro de Abreu	Casimiro de Abreu	Casimiro de Abreu
		Rio das Ostras
Cordeiro	Cordeiro	Cordeiro
		Macuco
Macaé	Macaé	Macaé
		Carapebus
	Quissama	Quissama
Santo Antônio de Pádua	Santo Antônio de Pádua	Santo Antônio de Pádua
		Aperibe
Cambuci	Cambuci	São José de Ubá
		Cambuci

(continued)

Table 6.1 (continued)

1980	1991	2013
Natividade	Natividade	Natividade
		Varre-Sai
Campos dos Goytacazes	Campos dos Goytacazes	Campos dos Goytacazes
	Italva	Italva
		Cardoso Moreira
São João da Barra	São João da Barra	São João da Barra
		São Francisco do Itabapoana

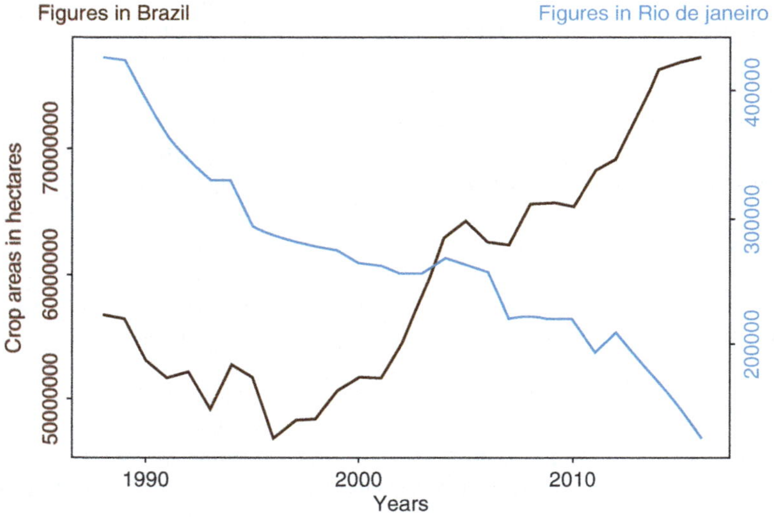

Fig. 6.2 Cropland areas in Brazil (black) and the federal state of Rio de Janeiro (blue) from 1988 to 2016

(7%), maize (3%), coconuts (2%), oranges (2%), and pineapples (2%). Areas used to cultivate traditional crops decreased during the study period, while fruit production increased, especially from the 2000s.

The areas used to cultivate sugarcane, bananas, maize, and oranges showed a clear decrease over time. The areas used to farm manioc were relatively stable with periods of oscillation. The areas used to cultivate coffee showed a sharp decrease at the beginning of the 1990s but recovered in the following decade. Coconuts and pineapples showed a strong increase in cultivated areas, especially during the 2000s.

The spatial patterns for the different crops in RJ are shown in Figs. 6.4 and 6.5. Cultivated areas for sugarcane were concentrated in the Norte Fluminense Region (Fig. 6.4). The municipalities of Quissamã (13,000 ha), São João da Barra (23,200 ha), and Campos de dos Goytacazes (56,650 ha) hosted most of the cultivated sugarcane area of the state. Although most sugarcane cultivation areas kept

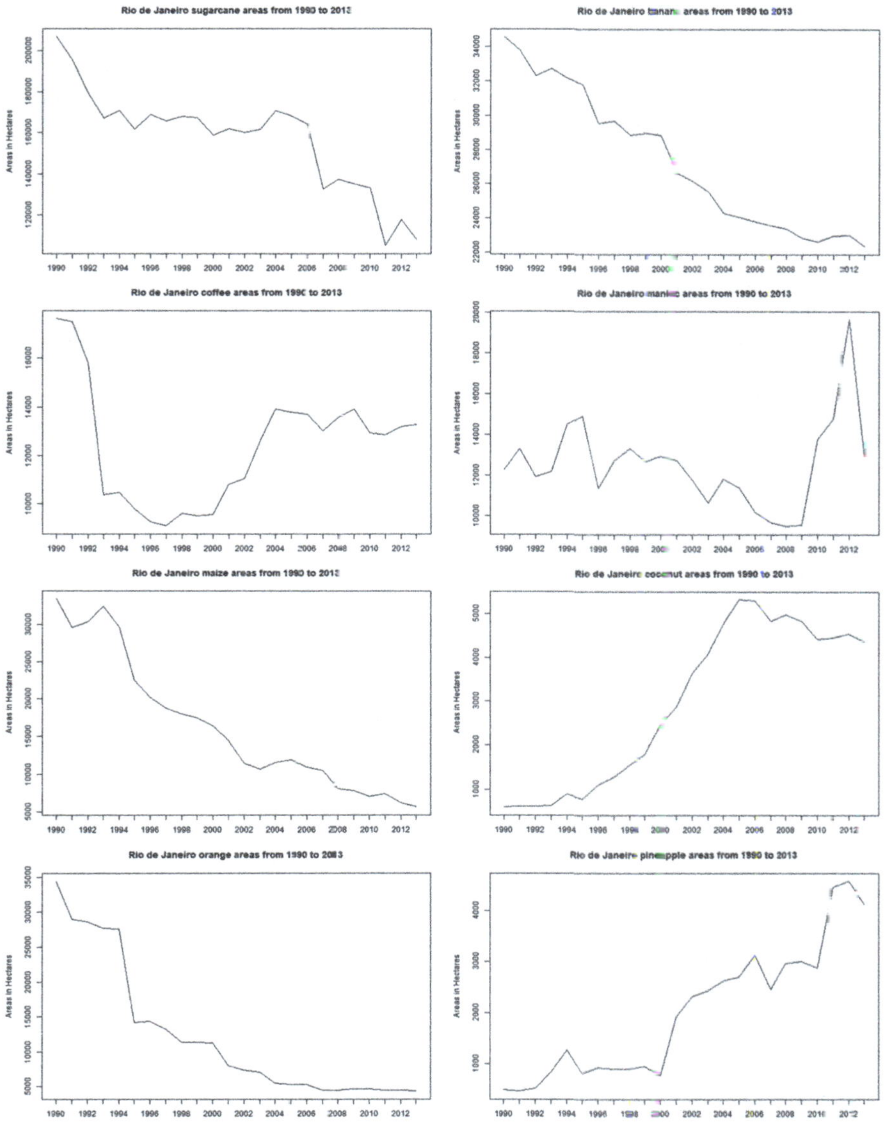

Fig. 6.3 Temporal dynamics of main crops' acreages in Rio de Janeiro Federal state

concentrated in the Norte Fluminense, the cultivated areas were reduced by 45% in the federal state and by 46% in the Norte Fluminense Region between 1991 and 2013. The municipalities with the highest reduction were Itaocara, São Fidélis, Macaé, and Conceição de Macabu; they are also characterized as areas with extensive pasture occupation.

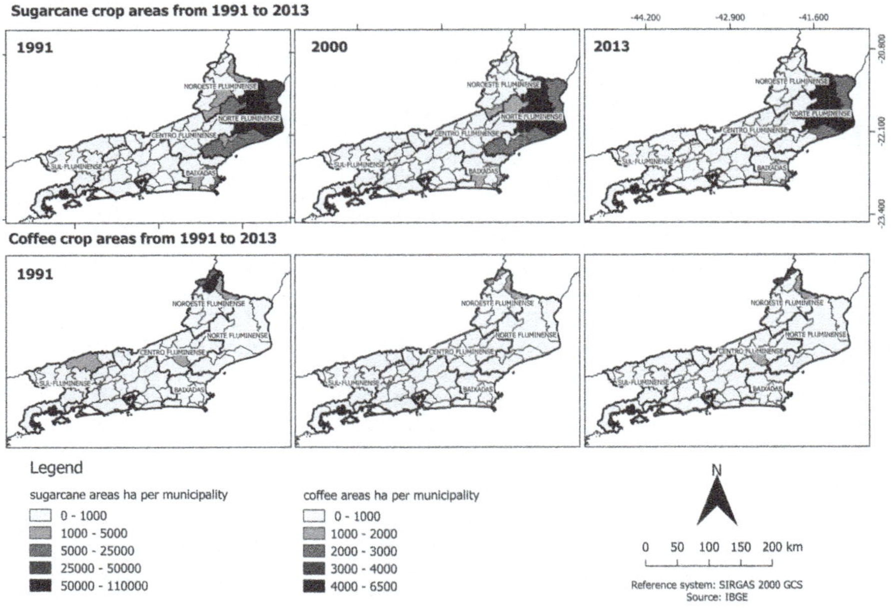

Fig. 6.4 Spatial dynamics of sugarcane and coffee in Rio de Janeiro state in 1991, 2000, and 2013

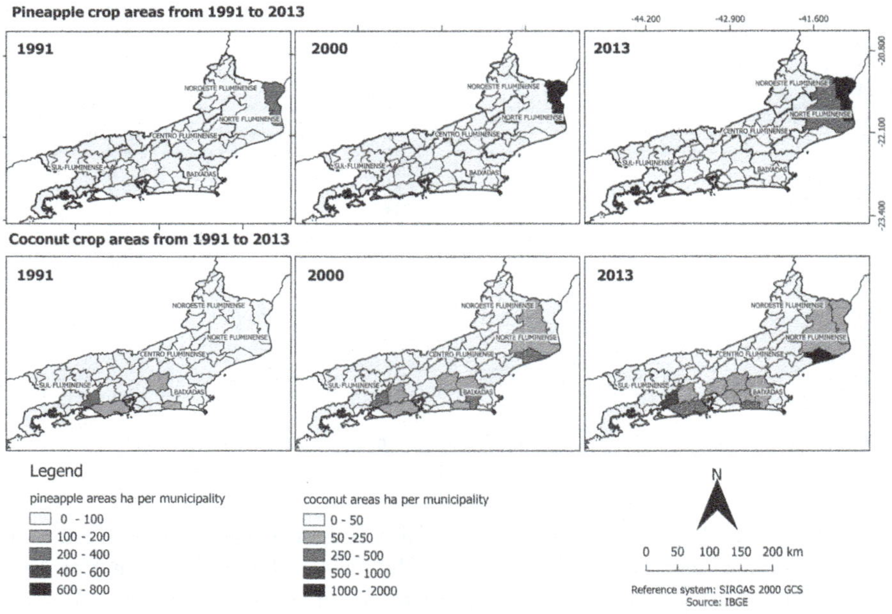

Fig. 6.5 Spatial dynamics of pineapple and coconut in Rio de Janeiro state in 1991, 2000, and 2013

Besides sugarcane, maize, bananas, and oranges showed a remarkable reduction in cultivated areas. Maize reduction occurred along its main occupation area in the Paraíba do Sul River Basin, which also includes the Centro Fluminense and the Noroeste Fluminense. The abandonment occurred in the mountainous areas of the state.

The cultivation of oranges was concentrated in the Macacu River Basin, which includes the Metropolitan and the Baixadas region. This is where the strongest decrease in cultivated areas happened. The bananas production was spread along the federal state mostly southward of the mountain ridge due to a more humid climate. The reduction of bananas acreages was strongest in the Norte Fluminense Region.

Manioc cultivation areas were spread across the federal state. The spatial pattern was fundamentally different from sugarcane. Approximately 81% of the manioc crop production in the state was focused in the Norte Fluminense (42%) and the Metropolitan Region (39%). Between 1991 and 2000, the cultivated areas for manioc in RJ showed a slight reduction from 13,307 to 12,948 hectares, which represents a 3% decrease in cultivated area. In the period from 2000 to 2013, the cultivated area of manioc increased slightly by 1%.

From 1991 to 2013, areas used to cultivate coffee decreased by 24% (Fig. 6.4). The period from 1991 to 2000 faced an important reduction of 7921 ha (45%). From 2000 to 2013, about 3727 ha of coffee were partially recovered. Coffee production occupied the headwaters of the Paraíba do Sul River Basin, in the mountainous landscapes located in the Sul, Centro, and the Noroeste Fluminense. From 1991 to 2000, there was a withdraw of the traditional coffee plantations in the municipality of Valença, in the Sul Fluminense Region. The municipality lost 1446 of its initial 1450 ha. During the following period from 2000 to 2013, coffee crop production migrated toward the Muriaé River Basin, in the Noroeste Fluminense Region.

The areas used to cultivate pineapples were concentrated in the Norte Fluminense Region (Fig. 6.5). The town of São João da Barra, in the Norte Fluminense, hosted 91% of the cultivated areas. Land use to cultivate pineapples increased between 1991 and 2000 at a rate of 66% and between 2000 and 2013 at a rate of 434%. In total, the pineapple area boomed by 784%, starting in 1991 with 466 ha and reaching 4121 ha in 2013.

An even stronger increase by 600% was observed for the areas used to cultivate coconuts with a 297% increase between 1991 and 2000. The municipality of Quissamã, in the Norte Fluminense, pushed this growth with an increase of 1391 ha of cultivated areas from 1991 to 2013 (Fig. 6.5).

From the 240,836 hectares areas used as cultivated land that was lost in RJ between 1991 and 2013, sugarcane accounted for 87,348 hectares (34%). Coconuts and pineapples were the crops that had the highest increase in cultivated areas – 3751 and 3655 ha, respectively.

The figures show that the reduction of sugarcane crop production of 87,348 ha was considerably higher than the increase of area used to cultivate pineapples (3655 ha). Considering also the 3751 ha increase in land used to cultivate coconuts, the sum of both crops does not reach 10% of the decrease of cultivated areas for sugarcane. Despite the crop substitution from sugarcane to pineapples or coconuts in some farms in São João da Barra, the figures reinforce that most of the sugarcane fields are being abandoned.

Sugarcane holds 57% of the cultivated areas in RJ. From those sugarcane areas, 85% are concentrated in the Norte Fluminense. Therefore, the farmland abandonment observed in RJ was pushed mostly by sugarcane and took place mainly in the Norte Fluminense Region.

6.5 Discussion

The open access policy from IBGE and its wide range of data originated from its various survey programs provide essential information for research. The survey program agriculture production per municipality considers temporary and permanent crops. The PAM did not provide information regarding livestock activities in its survey. For livestock activity, IBGE has another survey: Municipal Livestock Production (PPM), providing only data for the amount of livestock, but not for the size of pasture areas.

For pasture uses, data from the agricultural census was considered; it provided data at the federal state's level only. The agricultural census considered both animal crop farms and husbandry farms. The census showed the total areas of farms decreased. However, the share of pastures had increased within the remaining farms. This means while some farms were abandoned, others switched from crop to livestock activities.

Despite its advantages, the main disadvantage of the PAM data is its limited spatial resolution that does not allow an analysis at the sub-municipal scale. Moreover, PAM itself does not provide elements to explain the driven factors for changes in agricultural areas. Other socioeconomic factors such as demographic and production factors, e.g., the value of production, labor, fertilizers, and transportation costs, and environmental factors should be incorporated in the analysis to understand in which level the policies reduced or aggravated the current farmland abandonment process. Nonetheless, the agriculture production per municipality data allowed understanding the patterns of agriculture production in RJ.

The production of maize seems to be affected by the steep topography of the region. The abandonment is associated with the high-cost maintenance of the farm activity in those regions. Farmlands might be converted into extensive grassland for cattle raising or being simply abandoned in those areas (Alentejano 2006).

Coffee, which occupies mainly the headwaters of the Paraíba do Sul River Basin, became more concentrated in the Noroeste Fluminense. The trend in RJ followed a nationwide trend, which appears to be related to price fluctuations in the international market, especially after the fall of the International Coffee Agreement (ICA) (Jarvis 2005; Mehta and Chavas 2008; Russell et al. 2012).

The ICA, established in 1979, created quotes for export and importing country members as well as mechanisms to control prices fluctuations. The objective was to offer guarantees and protect the local producers (Jarvis 2005). In 1989, the ICA fell apart, leading to a more competitive market. Outside of this shield, some countries became more susceptible to price fluctuations (Russell et al. 2012).

From 1987 to 1995, the coffee farm gate prices in Brazil were particularly low. Fluctuations mainly impacted those coffee producers who failed to increase either productivity or quality (Mehta and Chavas 2008). The impact of the price fluctuations on those farmers could be a reason for the decrease of coffee productions in the case study region.

Possible impacts of Frutificar (State Act. 26,278/2000) can be seen in the Campos dos Goytacazes (Bahiense 2014; Bahiense et al. 2015). The municipality located in the Norte Fluminense Region shows an intriguing development. Areas used for the cultivation of pineapples decreased by 13% between 1991 and 2000. Between 2001 and 2013, the production area trend reverted with an increase of 508%, from 57 to 290 hectares. This boom matched the start of the Frutificar program.

The increase in areas for the cultivation of pineapples was followed by a decrease in sugarcane-cultivated areas, which suggests a crop substitution in the São João da Barra. The availability of new pineapple varieties with a shorter phenological cycle and, thereby, increased productivity enabled this substitution. Nonetheless, the increase in areas for the cultivation of pineapples and the decrease for sugarcane were not balanced.

The high labor cost and a small farm size structure, non-suitable for mechanization, may have affected the sugarcane farms. Nevertheless, sugarcane still has the biggest share in the crop areas of the state. Besides, sugarcane still accounts for 85% of the areas cultivated in the Norte Fluminense, which indicates the absence of any fundamental transformation in the spatial distribution of the sugarcane in RJ (Bahiense et al. 2015).

In the Norte Fluminense Region, the topography is known to be suitable for crop production due to the Paraíba do Sul River Basin extensive floodplain. However, the past overuse of soil resources might have created a dependency on fertilizers, seeds, and machinery to make the farm activity profitable.

Despite the increasing importance of pineapples, the figures indicate that the state-level acts aiming to promote fruits production (State Act. 26,278/2000), flowers production (State Act. 34,335/2003), and organic agriculture (State Act. 34,015/2003) could not avoid an evident loss of cropland in RJ. The Frutificar program (State Act. 26,278/2000) might have reduced the impact of farmland abandonment by promoting the pineapples and coconuts (Bahiense 2014; Bahiense et al. 2015). Nonetheless, the effects are locally restricted to the Norte Fluminense.

Low crop productivities led to land fragmentation and conversion to low-maintenance extensive grassland occupation in the federal state of RJ. In some cases, those abandoned croplands were converted to pasture for raising the cattle in extensive areas. In other cases, those areas were simply converted to fallow lands, for speculation purposes (Alentejano 2006). The most affected area was the Norte Fluminense Region.

6.6 Conclusions

Agriculture in RJ showed a diverging trend from the rest of Brazil. The importance of agricultural production in RJ decreased during the study period. The Norte Fluminense Region, which holds the highest share of cropping areas in the state, also showed the highest crop area reduction. The decrease of bananas, maize, and sugarcane had a dominant impact on farmland abandonment in the Norte Fluminense Region.

Cultivated areas of oranges and coffee also decreased. Oranges were mainly produced in the Baixadas and Metropolitan regions and coffee mainly in mountainous areas. In the latter, low-cost maintenance could have been a trigger factor for switching from crop to grassland. Croplands were either converted to pasture for raising the cattle in extensive areas or simply converted to fallow land for speculation purposes.

The increase in the area used to cultivate pineapples accompanied by a shrinkage in sugarcane-cultivated fields in the Norte Fluminense Region indicates a crop substitution in the region. Areas for coconut crop production also increased in the federal state, pushed mainly by the Quissamã municipality. From 2000 to 2013, incentives for fruits such as pineapples and coconuts might have triggered a substitution of crops, avoiding an even more extensive farmland abandonment. The trends in fruits crops matched with the period of implementation of the Frutificar program (State Act. 26,278/2000), suggesting the program had a positive impact on the production of fruits in RJ, particularly in the Norte Fluminense.

In this exploratory research, a general description of the agricultural sector in RJ was performed. The patterns of spatial distribution were described, and the trend of the most important crops was identified. Within the analysis, it was possible to observe interrupted trends in the time series probably associated with farmland abandonment. Throughout the spatial distribution, it was possible to observe where the farmland abandonment was most prominent. This analysis could be used as a basis to develop an econometric model to identify the factors that drive and trigger farmland abandonment.

References

Aide M, Grau R (2004) Globalization, migration, and latin american ecosystems. Science 80(300):1725–1727. https://doi.org/10.1126/science.1103788

Alentejano P (2006) Um breve balanço da política agrária no estado do Rio de Janeiro nas últimas décadas, pp 1–14

Algeet-Abarquero N, Marchamalo M, Bonatti J, Fernández-Moya J, Moussa R (2015) Implications of land use change on runoff generation at the plot scale in the humid tropics of Costa Rica. Catena 135:263–270. https://doi.org/10.1016/j.catena.2015.08.004

Bahiense D (2014) Análise do desenvolvimento da fruticultura e seus efeitos sobre a produção agrícola no Estado do Rio de Janeiro. Universidade Estadual do Norte Fluminese

Bahiense D, Souza P, Ponciano N (2015) Incentivos à produção de frutas e as mudanças na agricultura da Região Norte do estado do Rio de Janeiro. Revista Brasileira de Fruticultura, Jaboticabal – SP 37:387–395

Baumann M, Kuemmerle T, Elbakidze M, Ozdogan M, Radeloff V, Keuler N, Prishchepov A, Kruhlov I, Hostert P (2011) Patterns and drivers of post-socialist farmland abandonment in Western Ukraine. Land Use Policy 28:552–562. https://doi.org/10.1016/j.landusepol.2010.11.003

Brannstrom C, Jepson W, Filippi A, Redo D, Xu Z, Ganesh S (2008) Land change in the Brazilian Savanna (Cerrado), 1986–2002: Comparative analysis and implications for land-use policy. Land Use Policy 25:579–595. https://doi.org/10.1016/j.landusepol.2007.11.008

CEPERJ (2017) Produto interno bruto do Estado do Rio de Janeiro. http://www.ceperj.rj.gov.br/ceep/pib/pib.html. Accessed 12 Dec 2017

Cramer V, Hobbs R, Standish R (2008) What's new about old fields? Land abandonment and ecosystem assembly. Trends Ecol Evolut 23:104–112. https://doi.org/10.1016/j.tree.2007.10.005

Díaz G, Nahuelhual L, Echeverría C, Marín S (2011) Drivers of land abandonment in Southern Chile and implications for landscape planning. Landsc Urban Plan 99:207–217. https://doi.org/10.1016/j.landurbplan.2010.11.005

Estel S, Kuemmerle T, Alcántara C, Levers C, Prishchepov A, Hostert P (2015) Mapping farmland abandonment and recultivation across Europe using MODIS NDVI time series. Remote Sens Environ 163:312–325. https://doi.org/10.1016/j.rse.2015.03.028

FAO (2006) The role of agriculture and rural development in revitalizing abandoned/depopulated areas. Food and Agriculture Organization of the United Nations. Rome

Foley J (2005) Global Consequences of Land Use. Science 80(309):570–574. https://doi.org/10.1126/science.1111772

García-Barrios L, Galván-Miyoshi Y, Valsieso-Pérez I, Masera O, Bocco G, Vandermeer J (2009) Neotropical forest conservation, agricultural intensification, and rural out-migration The Mexican experience. Bioscience 59:863–873. https://doi.org/10.1525/bio.2009.59.10.8

Gasques J, Filho J, Navarro Z (2010) A Agricultura Brasileira: desempenho, desafios e perspectivas. Brasília

Gellrich M, Zimmermann N (2007) Investigating the regional-scale pattern of agricultural land abandonment in the Swiss mountains: a spatial statistical modelling approach. Landsc Urban Plan 79:65–76. https://doi.org/10.1016/j.landurbplan.2006.03.004

Griffiths P (2013) Utilizing the depth of the Landsat archive to reconstruct recent land change in the Carpathian ecoregion. Humboldt-Universität zu Berlin – Geographisches Institut

Griffiths P, Müller D, Kuemmerle T, Hostert P (2013) Agricultural land change in the Carpathian ecoregion after the breakdown of socialism and expansion of the European Union. Environ Res Lett 8:45024. https://doi.org/10.1088/1748-9326/8/4/045024

Guzmán Wolfhard L, Raedig C (2018) Chapter 12: Connectivity conservation management: connecting private protected areas. In: Nehren U, Schlüter S, Raedig C, Sattler D, Hissa H (eds) Strategies and tools for a sustainable rural Rio de Janeiro. Springer Series on Environmental Management

Hostert P, Kuemmerle T, Prishchepov A, Sieber A, Lambin E, Radeloff V (2011) Rapid land use change after socio-economic disturbances: the collapse of the Soviet Union versus Chernobyl. Environ Res Lett 6:45201. https://doi.org/10.1088/1748-9326/6/4/045201

IBGE (2002) Pesquisas Agropecuárias. Rio de Janeiro

IBGE (2009) Censo agropecuário 2006. Brasília

IBGE (2011) Censo Demográfico 2010

IBGE (2017) Sistema IBGE de Recuperação Automática – SIDRA. https://sidra.ibge.gov.br/pesquisa/censo-agropecuario/censo-agropecuario-2006/segunda-apuracao#series-temporais. Accessed 4 Aug 2017

INPE (2017) PRODES – Monitoramento da Floresta Amazônica Brasileira por Satélite. http://www.obt.inpe.br/OBT/assuntos/programas/amazonia/prodes. Accessed 12 Dec 2017

Izquierdo A, Grau H (2009) Agriculture adjustment, land-use transition and protected areas in Northwestern Argentina. J Environ Manag 90:858–865. https://doi.org/10.1016/j.jenvman.2008.02.013

Jarvis L (2005) The rise and decline of rent-seeking activity in the Brazilian coffee sector: Lessons from the imposition and removal of coffee export quotas. World Dev 33:1881–1903. https://doi.org/10.1016/j.worlddev.2005.06.003

Katayama N, Osawa T, Amano T, Kusumoto Y (2015) Are both agricultural intensification and farmland abandonment threats to biodiversity? A test with bird communities in paddy-dominated landscapes. Agric Ecosyst Environ 214:21–30. https://doi.org/10.1016/j.agee.2015.08.014

Keenleyside C, Tucker G (2010) Farmland abandonment in the EU: an assessment of trends and prospects. Report prepared for WWF. Institute for European Environmental Policy (IEEP), London

Lambin E, Meyfroidt P (2010) Land use transitions: socio-ecological feedback versus socio-economic change. Land Use Policy 27:108–118. https://doi.org/10.1016/j.landusepol.2009.09.003

Lasanta T, Pérez-Rontomé C, García-Ruiz J, Machín J, Navas A (1995) Hydrological problems resulting from farmland abandonment in semi-arid environments: the central Ebro depression. Phys Chem Earth 20:309–314. https://doi.org/10.1016/0079-1946(95)00042-9

Maranhão R (2015) Análise da dinâmica da produção de soja, cana de açúcar, milho e bovina no território brasileiro. Universidade de Brasília

May P, Fernandes LS, Rodríguez Osuna V (2018) Evolution of public policies and local innovation in landscape conservation in Rio de Janeiro. In: Nehren U, Schlüter S, Raedig C, Sattler D, Hissa H (eds) Strategies and tools for a sustainable rural Rio de Janeiro. Springer International Publishing, Cham

Mcafee K (2005) Gegraphies of risk and difference in crop genetic engineering. Geogr Rev 94:80–106

Mehta A, Chavas J (2008) Responding to the coffee crisis: what can we learn from price dynamics? J Dev Econ 85:282–311. https://doi.org/10.1016/j.jdeveco.2006.07.006

Munroe D, van Berkel D, Verburg P, Olson J (2013) Alternative trajectories of land abandonment: causes, consequences and research challenges. Curr OpinEnviron Sustainability 5:471–476. https://doi.org/10.1016/j.cosust.2013.06.010

Noriega Gamarra LM, Raedig C, Schlüter S (2018) Regional marketing strategies for sustainable production in Rio de Janeiro state. In: Nehren U, Schlüter S, Raedig C, Sattler D, Hissa H (eds) Strategies and tools for a sustainable rural Rio de Janeiro. Springer International Publishing, Cham

Prishchepov A, Müller D, Dubinin M, Baumann M, Radeloff V (2013) Determinants of agricultural land abandonment in post-Soviet European Russia. Land Use Policy 30:873–884. https://doi.org/10.1016/j.landusepol.2012.06.011

Renwick A, Jansson T, Verburg P, Revoredo-giha C, Britz W, Mccracken D (2013) Policy reform and agricultural land abandonment in the EU. Land Use Policy 30:446–457. https://doi.org/10.1016/j.landusepol.2012.04.005

Rudel TK, Schneider L, Uriarte M (2010) Forest transitions: an introduction. Land Use Policy 27:95–97. https://doi.org/10.1016/j.landusepol.2009.09.021

Russell B, Mohan S, Banerjee A (2012) Coffee market liberalisation and the implications for producers in Brazil, Guatemala and India. World Bank Econ Rev 26:514–538. https://doi.org/10.1093/wber/lhr055

Silva R, Batistella M, Moran E (2016) Drivers of land change: human-environment interactions and the Atlantic forest transition in the Paraíba Valley, Brazil. Land Use Policy 58:133–144. https://doi.org/10.1016/j.landusepol.2016.07.021

Silva R, Batistella M, Moran E (2017) Socioeconomic changes and environmental policies as dimensions of regional land transitions in the Atlantic Forest, Brazil. Environ Sci Policy 74:14–22. https://doi.org/10.1016/j.envsci.2017.04.019

Southworth J, Tucker C (2001) The influence of accessibility, local institutions, and socioeconomic factors on forest cover change in the mountains of Western Honduras. Mt Res Dev 21:276–283. https://doi.org/10.1659/02764741(2001)021[0276:TIOALI]2.0.CO;2

Terres J, Scacchiafichi L, Wania A, Ambar M, Anguiano E, Buckwell A, Coppola A, Gocht A, Källström H, Pointereau P, Strijker D, Visek L, Vranken L, Zobena A (2015) Farmland abandonment in Europe: identification of drivers and indicators, and development of a composite indicator of risk. Land Use Policy 49:20–34. https://doi.org/10.1016/j.landusepol.2015.06.009

Zakkak S, Radovic A, Nikolov S, Shumka S, Kakalis L, Kati V (2015) Assessing the effect of agricultural land abandonment on bird communities in southern-eastern Europe. J Environ Manag 164:171–179. https://doi.org/10.1016/j.jenvman.2015.09.005

Chapter 7
From Design to Implementation: A Participatory Appraisal for Silvopastoral Systems

Silvia Berenice Fischer, Alejandra Pedraza Luengas, Sabine Schlüter, and Luiz Antonio Oliveira Antunes

Abstract Small-scale dairy farming in rural areas of Rio de Janeiro state is characterized by low levels of productivity, lack of technological innovation, and land degradation particularly due to overgrazing. A field study was conducted to identify the potentialities and constraints to adopt integrated systems pasture-tree-livestock (silvopastoral systems, SPS) to improve small-scale dairy farming in Italva municipality in the Northwest of the state. Qualitative and quantitative information on dairy production regimes as well as the willingness to adopt SPS was collected, based on semi-structured interviews. The present dairy farming system was analyzed and characterized. Additionally, an on-farm fodder bank trial plot, adapted to the local socio-environmental conditions, was implemented. The barriers for adoption and socio-environmental benefits for small dairy farmers were identified and analyzed. Lessons learned from this research such as the reluctance to adopt these systems due to the long-term investment return and the high costs for initial implementation show the need of expanding the dissemination of knowledge and implementation of SPS in the region. Research and trainings were integrated in participatory capacity building of farmers to provide adequate tools and concepts and support decision-making.

S. B. Fischer (✉) · S. Schlüter
TH Köln – University of Applied Sciences, Institute for Technology and Resources Management in the Tropics and Subtropics (ITT), Köln, Germany
e-mail: sabine.schlueter@th-koeln.de

A. Pedraza Luengas
Danaus Consultants, Belmopan, Belize
e-mail: apedraza@danausconsultants.com

L. A. Oliveira Antunes
The Sustainable Rural Development Program in Micro-Watersheds of the State of Rio de Janeiro – Rio Rural, Niterói, Rio de Janeiro, Brazil

U. Nehren et al. (eds.), *Strategies and Tools for a Sustainable Rural Rio de Janeiro*, Springer Series on Environmental Management,
https://doi.org/10.1007/978-3-319-89644-1_7

Keywords Agroforestry practices · Dairy farming · Small-scale farming · Sustainable land management · Farmer's participation

7.1 Introduction

There is no simple explanation about the causes and effects of tropical deforestation due to the complexity and dynamicity of their nature; the overexploitation of forest resources, the land use change to carry on anthropogenic activities, and the governmental promotion for agricultural expansion are some of the identified causes of deforestation in the Mata Atlântica (Tabarelli et al. 2005; Barretto et al. 2013).

According to De Sy et al. (2015), 80% of the total deforestation in Brazil was associated with conversion to pastureland, while in the Mata Atlântica, the changes in land use are attributed to a mix of pasture, commercial cropland, and tree crops expansion.

The prevalent land uses in the biome are agriculture, pasture, and urban areas (Scarano and Ceotto 2015). Throughout the Brazilian history, projects were implemented to relocate farmers to tree-covered hinterlands favoring the conversion of forests into cropland and pasture (Southgate 1994). In Rio de Janeiro state (RJ), the dominant land use is pastureland for livestock production, i.e. meat and dairy, occupying 47.2% of the territory of the state (IBGE 2016), concentrating in the North and Northwest regions of RJ with approximately 45% of the total production of the state (IBGE 2012).

Forest conversion to pastureland has the following substantial environmental and socioeconomic impacts: depletion of natural resources; modification of hydrologic, carbon, nitrogen, and phosphorus cycles; increase of greenhouse emissions; increase of soil erosion; and loss of biodiversity rates (Conant et al. 2001; Vendramini et al. 2007; Herrero et al. 2009). Hence, there is an urgent need on finding sustainable land use system alternatives able to offset these impacts and to improve the current situation.

Agroforestry systems are likely to constitute a suitable approach to tackle this challenge. These systems provide an ecologically and economically sound alternative to the current production systems, increasing and improving the productivity of the farming systems, which can specifically improve low-input systems (Nair and Fernandes 1984; Kang and Akinnifesi 2000). One specific type of agroforestry system that covers the ecological, socioeconomic, and structural composition of livestock production is classified as silvopastoral systems (SPS). SPS emerge as an alternative to the challenges mentioned above, being characterized by the incorporation of trees and/or bushes in the animal production systems. These systems combine woody perennial production with animals and pastureland in the same period of time or in a temporal sequence. SPS can be classified according to the components and interactions as protein banks, living fences, trees and shrubs on pastures, and integrated production of animals and wood products (Nair 1985, 1987; Franke and Furtado 2001).

Trees in these systems play a significant role at the farm as well as landscape level. They not only provide wood and non-wood products for farms' consumption and external sale but also function as shelter for several wild animal species, being in situ biodiversity conservation spots, regulating the soil and water conservation processes as well as the microclimate, among other ecosystem services (Anderson and Sinclair 1993; Schroth et al. 2004; Shibu 2009).

In the last 30 years, one focus of agroforestry research was on providing conclusive results of environmental and socioeconomic benefits of silvopastoral land use systems. The scope of one of the applied research pillars is the adoption of agroforestry practices by main users on grassroots level for the replication and successful dissemination of these practices. This research is guided by the principle that the design and implementation of agroforestry systems need to be planned based on the ecological and socioeconomical characteristics of the study area in order to guarantee a successful adoption and dissemination. The active participation of local farmers in identification of priorities and constraints for possible implementation of agroforestry systems is a success-promising strategy in comparison to implementing preconceived systems since farmers' in-depth knowledge of their production system is considered right in the beginning (Clarke 1991;Weber et al. 2001). One of the main axes of agroforestry practices scale-up is the engagement of farmers through the whole process of agroforestry innovation research (Franzel et al. 2001).

This paper describes and analyzes the determinants of the early stages of participatory adoption of silvopastoral systems in Italva, RJ, using a bottom-up approach. The objectives of this chapter are as follows: (i) to identify characteristics of local dairy farming systems and major types of production systems, (ii) to assess the potential and constraints for adoption of SPS as an alternative to current dairy farming systems in the Northwest of RJ based on a participatory process from learning to adoption, as well as (iii) to analyze the potential socioeconomic costs and benefits for small dairy farmers implementing SPS in a pilot project in the municipality of Italva.

7.2 Methods

7.2.1 Study Area

The municipality of Italva is located in the Northwest mesoregion (*Noroeste Fluminense*) of RJ. The municipalities bordering Italva are Campos dos Goytacazes, Cambucí, São Fidélis, Itaperuna, and Bom Jesús de Itabapoana. The municipality has a total area of 296 km^2, 64.2 km^2 of which are arable lands and 224.7 km^2 are pastures (EMATER-RIO 2013).

The topography is characterized mainly by hilly terrains. The soils are classified as red-yellow argisols, presenting a textural B horizon with dominating low activity clays or with high activity clays and low base saturation. Pasturelands are of low productivity dominated by *Brachiaria* spp. (Gomes Tôsto and Santos Brandão 2006; Muniz Benedetti et al. 2011).

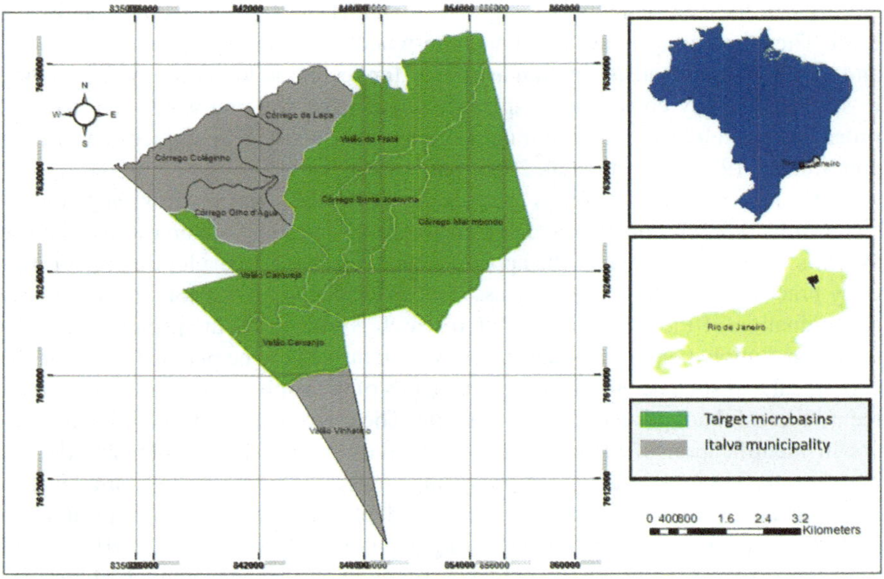

Fig. 7.1 Location of Italva municipality. (Source: Own elaboration)

The mean annual precipitation is 1200 mm and the mean annual temperature 23.6 °C. According to the Köppen climate classification, the region corresponds to an Aw climate with a distinct wet season and a dry winter (Guerreiro Martorano et al. 2003). The Northwest region displays the highest indices of poverty in the state featuring a lack of development in agriculture, rural exodus to nearby industrial cities, and environmental degradation.

The municipality of Italva is divided into nine administrative subunits referred to as micro-basins (Fig. 7.1). For micro-basins self-organizing management committees, the Comitês Gestores da Microbacias Hidrográficas (COGEMs) were established, representing producers and inhabitants of each micro-basin. The COGEMs are bottom-up organizations, self-managed and recognized by the town hall, and public institutions of the area (Prefeitura Municipal de Italva 2002; EMATER-RIO 2013; Pedraza 2014).

Overexploitation, low productivity, and lack of technological innovation are typical for dairy farming, which has led to an expanded degradation of the pastures.

7.2.2 Methodological Approach

Participatory methods have profound implications on the success of adoption and dissemination of SPS. In a dynamic approach, these methods intent to cover the whole range of stakeholders' involvement in the analysis, design, as well as implementation and monitoring of these systems (Pretty 1995; Cardoso et al. 2001). The

Table 7.1 Participatory appraisal on silvopastoral systems in Italva, Rio de Janeiro

Steps	Pre-diagnostic	Diagnostic	Farmers field school	Implementation of trial plot
Objectives	Analysis of agroecosystems Identification of sustainable agricultural alternatives Status quo of agricultural production	Identification of opportunities and constraints Information on the dairy farming system	Trainings on SPS Exchange of knowledge among farmers Consolidation of key concepts	Design and planning of the candidate SPS Practical examples of implementation
Methods	Literature review Experts consultation Secondary data analysis	Semi-structured interviews On-farm visits Characterization of farming system (FEAST analysis)	Participatory workshops, field days and trainings, hands-on mapping	On-farm implementation Economic analysis

presented research follows a four step participatory approach (Table 7.1) in order to build and strengthen the capacities for adoption and implementation of SPS.

7.2.2.1 Pre-diagnostic

The pre-diagnostic phase of this research identified the land use systems of the study area and the type of land use the research should be focused on. The information was collected by consultation of local experts, extension agents, and a set of individual farming characteristics information, using the individual development plans (IDPs), a participatory diagnostic tool developed by the Rio Rural program (see Hissa et al. 2018). A total of 35 IDPs were analyzed; the selection was done randomly, after consultation with the local rural extension office.

7.2.2.2 Diagnostic

After the pre-diagnostic stage, an in-depth household survey was carried out based on semi-structured interviews of targeting 20 key local farmers that were beneficiaries of the Rio Rural program. The survey questionnaire focused on general aspects of dairy production, presence and use of trees in pasturelands, as well as the perception of farmers about SPS. Consultations with key local actors were carried out based on open interviews. Additionally, secondary information was obtained from government representatives and extension agents from the Technical Assistance and Rural Extension Company (Empresa de Assistência Técnica e Extensão Rural-EMATER). Characterization of dairy farms was carried out using the FEAST tool (https://www.ilri.org/feast) that assesses farming systems with particular emphasis on livestock feed aspects.

7.2.2.3 Farmers Field School

A participatory training program was designed and implemented in Italva, using the approach of farmer field schools adapted for SPS (Feder et al. 2004; Braun et al. 2006; FAO 2016). This approach promotes farm-based experimentation, group organization, and decision-making. A combination of different extension methods such as field days, participatory training workshops, on-farm visits, and on-farm demonstration trials was implemented as shown in Table 7.2.

Table 7.2 Farmers field school contents

Method	Theme	Content
Introductory meeting	Introducing SPS	Presentation of stakeholders and trainers Exploration of farmers' knowledge about SPS Farmers interest and commitment to participate in the farmers' field schools
Workshop I	Basic concepts on SPS	Brainstorming on silvopastoral systems concept Identification of farmers' background on silvopastoral systems Summary presentation by the trainers Participatory mapping of the farms
On-farm survey	Dairy farming characterization/trees on farm	Semi-structured interviews with dairy farmers Evaluation of learned concepts in previous workshop On-farm transects using participatory mapping
Field day I	Established SPS	Field visit to a local established SPS Farmers exchange of knowledge Identification of barriers and constraints for implementation and management
Workshop II	Types of SPS	Presentation on different types of SPS Case studies presentation and best practices analysis Evaluation of knowledge transfer Identification of most suitable SPS for the study area
Workshop III	Nursery techniques	Propagation of tree species Nursery techniques Establishment of participatory tree nurseries
Field day II	Implementation of a fodder bank	Implementation of a fodder bank Environmental benefits of fodder banks Economic benefits of fodder banks
Field day III	Post-planting visit	Identification of problems during and after implementation and possible solutions Planning post-planting management activities
Workshop IV	Training the trainers	Concepts and implementation of SPS for rural extension workers Hands-on extension tools for the planning and implementation of SPS

The series of participatory activities were attended by a total of 20 key local farmers and members of the local rural extension office. Strategies of facilitation included a preset of topics for discussion allowing participants to take the lead on the conversations and to be supported, only when necessary, by the rural extension workers.

An analysis of potentialities and constraints of implementation as well as the adoption of SPS was carried out as an output of all participatory activities.

7.2.2.4 Implementation of a Pilot Study

On-farm trials provide data on the biophysical performance, economic feasibility and profitability, and adoption and adaptation of SPS. The trials can also be viewed as means of providing feedback to improve research and to document the process under local conditions from the farm to the research station (Pinney 1991; Franzel et al. 2001; Franzel and Scherr 2011).

A silvopastoral pilot plot was established as part of the farmers' field schools. This plot was done based on identified needs and priorities for farmers in order to improve productivity in their farms. The post-establishment assessment is still in process, and results will support the decision-making on expanding the implementation of these systems on a landscape level.

Data was analyzed using the FEAST tool, evaluating primarily cash flow dynamics and financial performance. An analysis of the costs and revenue structures of a current dairy farming system as well as a proposed silvopastoral system was carried out, a net cash flow was constructed for a period of 5 years, and investment decision indicators were calculated in order to explore the financial feasibility of the implementation of a silvopastoral trial plot (net present value and internal rate of return).

7.3 Results and Discussion

7.3.1 Characterization of the Dairy Farming System in Italva, RJ

After a secondary information analysis on farming systems and problems in the study area, research gaps were identified, specifically in terms of need in alternative farming systems that could enhance the socioeconomic and environmental state of the region. Since dairy farming is the most practiced activity in the study area, a characterization of the system was carried out in order to understand its trade-offs and opportunities.

The dairy farming system in Italva is carried out on a small scale. It is a year round extensive system with milking once a day, mostly by hand (80%). Milk sale represents the major source of income (46%), followed by the sale of cash crops,

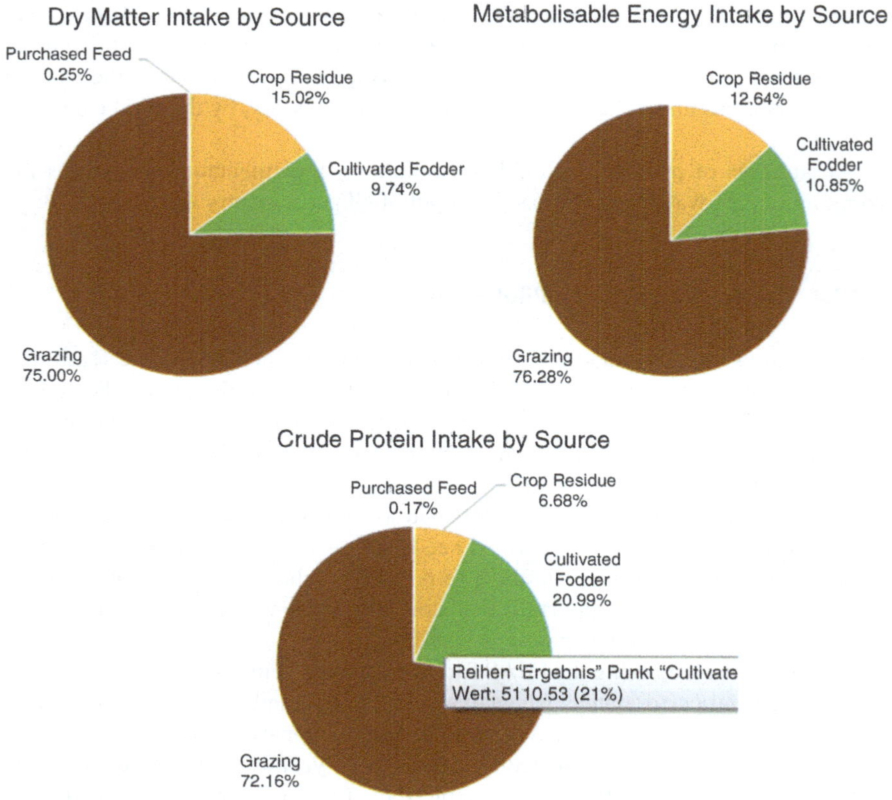

Fig. 7.2 Feed resources of dairy livestock in Italva, RJ. (Source: Own elaboration)

green peppers (*Capsicum annuum*), common beans (*Phaseolus vulgaris*), and rice (*Oryza sativa*) representing 32%, and non-farm sources, i.e., pension schemes and salaries, represent 22% of the total household income. The average daily milk production per cow is 8.5 l per day^{-1}; the average milk price paid to farmers during this research ranged from R$0.90 to R$1.20.[1]

The herds are composed entirely by a Brazilian hybrid breed, Girolando (*Bos indicus* × *Bos taurus*), with an average size of 30 animals per farm, which is composed of dairy cows, dry cows, and replacement heifers. Herd feeding is based in its majority on *Brachiaria* spp. pastures (71%), complemented by collected forage of sugar cane (*Saccharum officinarum*) as an energetic supplement (Fig. 7.2) and crop residuals of corn (*Zea mais*). Farmers expressed to have more difficulties to feed the herd during the dry season, i.e., May to September, due to the lower availability of pasture. During the dry season, crop residues are bought in a larger quantity (Fig. 7.3).

The protein intake relies mostly on the grazing of *Brachiaria* spp. (74%), while the energetic intake is shared between the collected crops and the grazing. According

[1] 1 R = 0.453 dollars (June 2014).

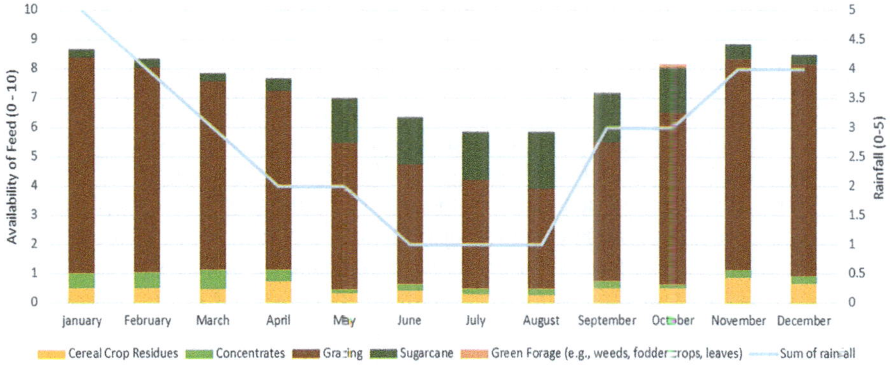

Fig. 7.3 Available dairy livestock feed resources and average rainfall pattern in Italva, RJ. (Source: Own elaboration)

to Campos Paciullo et al. (2003), the feed quality of *Brachiaria* spp. in degraded pastures is very low in terms of its nutritional value, when compared, for example, with some species of legumes.

The average stocking rate in the pastureland is of 2.4 cow ha^{-1}, and pastureland shows several stages of degradation. Pastureland management activities consist primarily of control of primary forest regrowth, done using chemical herbicides or manual mowing as well as fence and corral maintenance. Sanitation controls include vaccination against food and mouth disease (two applications per year), rabies (one application per year), and brucellosis (only for females in reproductive age). The secretary of agriculture and the sanity control office of the municipality assists with the application of the vaccines along the year; a deworming against ecto-endo parasites is made once per year.

The dairy farming system in the study area is characterized by its low technological and productivity levels. The farm household income relies mainly on the cash flow from agricultural activities, and further recommendation should be done oriented to increase the productivity of the farming system.

7.3.2 *Farmers' Field School on Silvopastoral Systems*

The design and objectives of the famers' field school were done based on the information gathered through the survey and the pre-diagnostic on the study area. The objective was to enhance the knowledge on silvopastoral systems as an alternative land use system, promoting the local appropriation of sustainable alternatives tested and designed specifically for the current conditions in the area.

One of the main success factors of this research was the already existing social organization structure of farmers in the area, e.g., COGEM (micro-basin management committees) which enabled extension services and researchers to easily identify target farmers who participated in the farmers' field school. This participatory approach provides additional recognition and potential to become multipliers in their communities (Holguín and Mora-Delgado. 2007; Murgueitio et al. 2010). Therefore, it is necessary to understand the farmers' perspectives with participatory tools that facilitate the open dialogue and co-learning by researchers and producers (Cardoso et al. 2001; Holguín and Mora-Delgado 2007; Murgueitio et al. 2010).

The participatory activities had the objective of facilitating the concept of SPS, their characteristics, advantages and disadvantages, implementation problems, and proposed solutions. The perceptions were identified, and participants were asked to perform a field map of their properties (Geilfus 1997) to identify their actual productive areas and spatial distribution of their farms. Finally, they were asked to point an area where future SPS could or should to be established. With the results collected in the first workshop, modifications and new ideas were performed and improved in the following meetings with the target group of farmers followed by visits to the farms, where semi-structured interviews were held. Questions about SPS were asked about their awareness on these systems and their knowledge about them, and the arboreal component of their properties was assessed; through on-farm transects, farmers were asked to provide information about common names and uses of the trees to promote self-reflection on the different services that trees provide. The interviews were designed to evaluate the knowledge transfer after the first workshop. All interviews were carried out at the farmers' properties, with the participation of the farmer, family members present at the time of the visit, the research group, and a rural extension agent.

The main objective of the second workshop was to identify the concepts and ideas the producers had learnt and to clarify doubts. A presentation with SPS images was the main activity; therefore, they were able to analyze and discuss about what they had observed. From this process, main conclusions, clarification of doubts, and main concerns of farmers as well as an assessment of their acquisition of knowledge and the farmers "perception" could be derived.

The contents of the farmers' field school curricula were adjusted based on the identified needs during the first workshops. The following topics were included: (i) the different types of SPS, (ii) a visit to a local implemented SPS, and (iii) the implementation of a SPS demonstration plot. The implementation of a fodder bank was found to be the most suitable SPS according to the needs and preferences of the target group as well as the environmental conditions of the study area, since it boosts increased in the milk productivity. In this way it contributes to the household income as well as enhances rehabilitation of degraded pastures.

The participatory implementation of the pilot trial of a fodder bank was carried out in 3 training days: (i) demonstration techniques of fodder tree species, (ii) propagation and implementation of a tree nursery, and (iii) field day to provide insights on the implementation of a fodder bank, soil preparation, spatial and temporal arrangements, as well as environmental and socioeconomic benefits of fodder banks, such as soil enrichment erosion reduction, nitrogen fixation, improvement in milk productivity and weight gain of cattle, increase in the household income, and reduction on the use of fertilizers in the pastureland. A post-planting workshop was held 6 months later in order to identify preliminary status and changes in the trial plot. Farmers' reactions were positive. Therefore, a greater interest on implementation of SPS in their own farms was achieved. The concluding workshop targeted the rural extension officers of the study area in order to build capacities for further dissemination and implementation of SPS.

At a first stage, farmers showed reluctance to the concept of having more trees on their farms, especially on pastures. However, during the program lifetime reluctance levels changed. This can be attributed mostly to the implementation of the trial plot and the exchange of information with farmers who practice this system. The analysis of potentialities and constraints shows that main constraints for adoption are related to capital limiting investment in such systems and, in general, the lack of knowledge about practices. The identified opportunities are related to a general improvement of environmental conditions as well as an increase in the productivity of the farming system (Table 7.3).

Table 7.3 Opportunities and constraints for adoption and dissemination of silvopastoral systems in Italva, RJ

Dimensions	Opportunities	Constraints
Environmental	Ecosystem services: Effect on local and catchments hydrology, nutrient recycling, nitrogen fixation, pest-predator relationships, biodiversity increase, habitat for diverse biodiversity of fauna, flora and microbes, maintenance of soil fertility, windbreaks, among others Use of local materials and energy Carbon sequestration Improved natural resources management More stable, robust, and resilient systems	Extreme weather events Introduction of exotic species
Higher annual income (cash flow) Diversity of income from various products Net present value positive Cost-benefit advantages	Lack of loan funding sources Long-term investment, some of the income effects are only tangible after various years	Higher annual income (cash flow) Diversity of income from various products Net present value positive Cost-benefit advantages
Social	Household food security and nutrition Community participation Local people involvement in all the stages Farmer's willingness to adopt the systems Use of existing or easily adaptable skills More training options for farmers	Shift of paradigm – Cultural barriers Diverse perception of SPS Lack of knowledge Doubts on the effect of the trees on pasture and animals development Shortage of qualified personnel and local training centers Lack of young people in the farms

(continued)

Table 7.3 (continued)

Dimensions	Opportunities	Constraints
Political	Promote regional development Development of diverse marketing structures Development of a regulatory framework (voluntary or mandatory) Possible compensation for leaving trees in their pastures (payments for ecosystem services, e.g., tree tenure ship)	Communication disconnection between national, states, district, and community levels Regulatory framework needs to be developed: Supporting policies that include aid schemes and instruments to pay producers Logging regulations More resources allocation for research Integration of animal, crop, and environmental sciences in teaching, research, and development planning
Agroecological	Diversification of agroecosystem (fruit trees, fuelwood, shade, etc.) Identification of suitable species and arrangements for implementation according to the region Potential to use social species and thus contribute to biodiversity conservation. Continuous feed availability for the animals. Balanced and better quality diets to improve production	Availability of tree seedlings Farm size: Space or land availability Seasonal occurrence of plant and animal diseases Seasonal distribution of rains – Irrigation systems must be implemented to ensure survival of seedlings Meet the demands of local geography, particularly climate and topography Lack of knowledge on management of local species within SPS

7.3.3 Investment Analysis: Implementation of a Fodder Bank

A fodder bank trial plot (Fig. 7.4) was established in one of the farms of the participants in the farmers' field school (micro-basin Còrrego Marimbondo, Italva, RJ – 21° 24´ S, 41° 35 W). The size of the plot was 0.8 ha with three different arrangements (legume tress *Gliricidia sepium* and *Leucaena leucocephala* and in the lower strata *Arachis pintoi*). Legume trees were pruned two times during a period of 8 months in order to allow the less developed trees to prune and to form a homogeneous stem and to control the level of shadowing. A survival rate of 80% was observed in *Gliricidia sepium*, while the survival rate in *Leucaena leucocephala* was 73%.

The higher survival rate of *Gliricidia sepium* can be attributed to the fact that *Gliricidia sepium* was originated from stakes, while *Leucaena leucocephala* was from seeds. Persistence of *Arachis pintoi* in association with *Brachiaria* spp. showed less growth and survival. Further studies in the area are recommended to analyze the behavior of the association of these two species under different management scenarios.

Fig. 7.4 Plot prior implementation of the fodder bank (a) and implemented fodder bank (b)

Table 7.4 Cost of implementation of a 1 ha fodder bank in Italva, RJ

Description	Category	Unit	Quantity	Unit value R$	Total value R$
Soil preparation (ox plowing)	Services	Hour	24	40	960
Planting	Services	Day	5	50	250
Reposition	Services	Day	2	50	100
Fertilization	Services	Day	4	50	200
Seedlings transport	Services	Hour	4	50	200
Fencing	Services	Day	18	50	900
Seedlings (*Gliricidia sepium*, *Leucaena leucocephala*)	Input	Unit	2000	2	4000
Seedlings (*Arachis pintoi*)	Input	kg	6	300	1800
Dolomitic limestone	Input	50 kg	20	16	320
Wire	Input	Roll (250 m)	4	135	540
Fence clamps	Input	kg	2	10	20
Piles 8–10 cm/2.20 m	Input	Dozen	4.5	110	495
Total R$					9785.00

All input and output costs of implementation (Table 7.4) and the first year of management were recorded to perform an evaluation of economic indicators.

7.3.4 Investment Analysis Indicators

The net present value (NPV) and the internal rate of return (IRR) were calculated for a discount rate of 16% for a 5-year period. The NPV of a dairy farming with a 1 ha fodder bank was R$ 27903.39 and shows that the investment in this system is economically feasible and recommended.

The internal rate of return indicates that revenues coming from an increment on milk yield are 53% showing the expected profitability of the invested money over the calculated period.

These results are based on the assumptions of an increase of 20% of milk production (Camero Rey and Ibrahim 1995; Barragán Hernández et al. 2016; Rivera-Herrera et al. 2017) in a 1 ha fodder bank, under a 2-h/day foraging system, maintaining a stocking rate of 2 animal units/ha. The usage of the fodder bank, besides incrementing milk production, will replace expenses related to purchase of feeding concentrates.

The analysis of these indicators shows the financial feasibility of the implementation of a fodder bank in Italva, RJ.

7.4 Lessons Learned

Regardless the enthusiasm and understanding of the systems, the willingness to establish SPS and to increase the trees in the pastures was evidenced; but there are still a lot of barriers and reluctances.

Farmers identified the low productivity, degraded pastures, and lack of resources to buy dietary supplements as the main problems for dairy production in the study area. Most of the farmers had already some knowledge regarding the benefits of trees on pastures; however, the concept of SAS was proved to be innovative for them.

One of the major concerns of the farmers for adoption is the long-term return of the investment (Arnold 1987). Farmers will not risk and change their traditional production system to a new technology if they are not very sure about the benefits that can be achieved by farmer – farmer exchange (Barrow 1991; Beer 1991). As a consequence, most of the farmers involved in this research manifested against disposition to invest money or to borrow some for the SPS implementation in their farms, because they did not have the economic solvency for this investment. The importance of governmental incentives or external investments is essential for appropriate adoption of these technologies (Murgueitio et al. 2010). Additionally, policy-makers should consider designing hybridized schemes in order to integrate conservation and development (Petheram and Campbell 2010).

It was observed that the adoption success of SPS will be determined by the level of response to the farmer's needs and problems of the SPS. It is necessary to raise awareness on SPS among all farmers in the community, tested with locally available species with consideration of minimal costs of implementation and labor.

According to a study carried out in Costa Rica (Beer 1991), farmers were completely committed to a 2-year period of implementation with no financial return. Further research has to be oriented toward the biophysical performance of different SSPs technologies under current socio-environmental conditions. An essential opportunity for SPS is the occurrence of suitable local species. Without developing routines on how to integrate them into pasture management, this opportunity cannot be tapped.

The authors consider that SPS have a large potential to increase dairy production in the study area; the implementation of a two-stage SPS might facilitate the adoption of these systems. The first stage should focus on implementing fast-growth tree species in order to increase productivity in a shorter term; the second stage should focus on planting a higher diversity of local species with different spatial and temporal arrangements to diversify. The participatory approach adopted here required introducing familiar practices including nonlocal species recognizing the risks to the local ecosystems. Substantial research is still needed to identify suitable local species. This research will support the understanding of the environmental and socioeconomic benefits of SPS.

Acknowledgments We gratefully acknowledge the contribution of Dr. Juan Carlos Torrico for his contribution to the conceptualization and operationalization of this research. We thank our colleagues from Rio Rural program and EMATER Italva who provided insight and supported this research during all stages.

References

Anderson LS, Sinclair FL (1993) Ecological interactions in agroforestry systems. Agroforestry Abstracts. http://r4d.dfid.gov.uk/Output/11654/Default.aspx. Accessed 23.11.2016

Arnold MJE (1987) Economic considerations in agroforestry. In: Steppler HA, Nair PKR (eds) Agroforestry: a decade of development. International Council for Research in Agroforestry, Nairobi, Kenia, pp 173–190

Barragán Hernández W, Mahecha-Ledesma L, Cajas-Girón Y (2016) Efecto de sistemas silvopastoriles en la producción y composición de la leche bajo condiciones del valle medio del río Sinú, Colombia. Revista Colombiana de Ciencia Animal 8(2):187–196

Barretto AGOP, Berndes G, Sparovek G, Wirsenius S (2013) Agricultural intensification in Brazil and its effects on land-use patterns: an analysis of the 1975–2006 period. Glob Chang Biol 19(6):1804–1815. https://doi.org/10.1111/gcb.12174

Barrow EGC (1991) Evaluating the effectiveness of participatory agroforestry extension programmes in a pastoral system, based on existing traditional values – a case study of the Turkana in Kenya. Agrofor Syst 14(1):1–21. https://doi.org/10.1007/BF00141594

Beer J (1991) Implementing on-farm agroforestry research: lessons learned in Talamanca, Costa Rica. Agrofor Syst 15(2–3):229–243. doi: https://doi.org/10.1007/BF00120190

Braun A, Jiggins J, Röling N et al (2006) A global survey and review of Farmer Field School experiences. Wageningen, The Netherlands. https://www.researchgate.net/profile/J_Jiggins/publication/228343459_A_Global_Survey_and_Review_of_Farmer_Field_School_Experiences/links/0046353bd1e61ab7f7000000/A-Global-Survey-and-Review-of-Farmer-Field-School-Experiences.pdf. Accessed 12 Apr 2017

Camero Rey A, Ibrahim M (1995) Bancos de proteína de poro (Erythrina berteroana) y madero negro (Gliricidia sepium). Agroforestería en las Américas 2(8):31–33

Campos Paciullo DS, Magalhaes Aroeira LJ, Maurilío JA, Mesquita Carvalho M (2003) Características produtivas e qualitativas de pastagem de braquiária. Pesquisa Agropecuária Brasileira 38(3):421–426

Cardoso I, Guijt I, Franco F et al (2001) Continual learning for agroforestry system design: university, NGO and farmer partnership in Minas Gerais, Brazil. Agric Syst 69:235–257

Clarke J (1991) Participatory technology development in agroforestry: methods from a pilot project in Zimbabwe. Agrofor Syst 15(2–3):217–228. https://doi.org/10.1007/BF00120189

Conant R, Paustian K, Elliot E (2001) Grassland management and convertion into grassland effects on soil carbon. Ecol Appl 11(2):343–355

De Sy V, Herold M, Achard F et al (2015) Land use patterns and related carbon losses following deforestation in South America Land use patterns and related carbon losses following deforestation in South America. Environ Res Lett 10:1–15. https://doi.org/10.1088/1748-9326/10/12/124004

EMATER-RIO – Empresa de Assistência Ténica e Extensão Rural do Rio de Janeiro (2013) Relatório de atividades da EMATER-RIO. Escritório local de Italva, Rio de Janeiro, p 25

FAO (2016) Farmer Field School guidance document planning for quality programmes. Rome, Italy. http://www.fao.org/nr/land/sustainable-land-management/farmer-field-school/en/. Accessed 12 Apr 2017

Feder G, Murgai R, Quizon JB (2004) Sending farmers back to school: the impact of farmer field schools in Indonesia. Rev Agric Econ 26(1):45–62. https://doi.org/10.1111/j.1467-9353.2003.00161.x

Franke IL, Furtado SC (2001) Sistemas Silvipastoris: Fundamentos e Aplicabilidade. Embrapa Acre – Documentos 74:51

Franzel S, Scherr S (eds) (2011) Trees on the farm: assessing the adoption potential of agroforestry practices in Africa. CAB International – World Agroforestry Centre (ICRAF), Nairobi

Franzel S, Cooper P, Denning GL (2001) Scaling up the benefits of agroforestry research: lessons learned and research challenges. Dev Pract 11(4):524–534. https://doi.org/10.1080/09614520120066792

Geilfus F (1997) 80 Herramientas para el Desarrollo Participativo: diagnóstico, planificación, monitoreo, evaluación. Prochalate–IICA, San Salvador, el Salvador, p 208

Gomes Tôsto S, Santos Brandão E (2006) Avaliação de impactos ambientais com a utilização do método "AMBITEC – produção Animal" – Projeto RADEMA: Recuperação de Áreas Degradadas com Pastagens. Boletim de pesquisa e desenvolvimento 98:26

Guerreiro Martorano L, Pereyra Rossiello R, do Amaral Meneguelli N et al (2003) Aspectos climáticos do Noroeste Fluminense, RJ. Documentos 43:28

Herrero M, Thornton P, Gerber P, Reid R (2009) Livestock, livelihoods and the environment: understanding the trade-offs. Curr Opin Environ Sustain 1:111–120

Hissa H, Teixeira N, Costa M et al (2018) Sustainable rural development in Rio de Janeiro state: the Rio Rural Program. In: Nehren U, Schlüter S, Raedig C, Sattler D, Hissa H (eds) Strategies and tools for a sustainable rural Rio de Janeiro. Springer International Publishing, Cham

Holguín VI, Mora-Delgado M (2007) El aprendizaje participativo como base de un cambio positivo del uso del suelo en fincas ganaderas de Costa Rica. (19 ed). http://www.lrrd.org/lrrd19/4/holg19053.htm. Accessed 14 Feb 2017

IBGE (2012) Produção da Pecuária Municipal - PPM. https://www.ibge.gov.br/estatisticas-novoportal/economicas/agricultura-e-pecuaria/9107-producao-da-pecuaria municipal.html?&t=publicacoes. Accessed 15 Nov 2017

IBGE (2016) Estado do Rio de Janeiro cobertura e uso da terra 2016 No Title.. ftp://geoftp.ibge.gov.br/informacoes_ambientais/cobertura_e_uso_da_terra/uso_atual/mapas/unidades_da_federacao/rj_uso.pdf. Accessed 15 Nov 2017

Kang BT, Akinnifesi FK (2000) Agroforestry as alternative land-use production systems for the tropics. Nat Res Forum 24(2):137–151. https://doi.org/10.1100/2012/350258

Muniz Benedetti M, Curi N, Sparovek G, de Carvalho Filho A, Silva SH, G (2011) Updated Brazilian's Georeferenced soils database – an improvement for international scientific information exchanging. In: Özkaraova Güngör E (ed) Principles, application and assessment in soil science. InTech. Available from: https://mts.intechopen.com/book/principles-application-and-assessment-in-soil-science/updated-brazilian-s-georeferenced-soildatabase-an-improvement-for-international-scientific-informat

Murgueitio E, Uribe F, Zuluaga AF et al (2010) Sistemas Silvopastoriles establecidos en las fincas de difusión tecnológica (FDT). In: Murgueitio E, Uribe F, Zuluga AF, Galindo WF, Valencia LM, Giraldo C, Soto R (eds) Reconversión ganadera con sistemas silvopastoriles en la provincia de Chiriquí, Panama. Panamá. Feriva S.A. Chap 2, pp 35–79

Nair PKR (1985) Classification of agroforestry systems. Agrofor Syst 3(2):97–128. https://doi.org/10.1007/BF00122638

Nair PKR (1987) Agroforestry systems inventory. Agrofor Syst 5(3):301–317. https://doi.org/10.1007/BF00119128

Nair PKR, Fernandes E (1984) Improved Production systems as an alternative to shifting cultivation. FAO Soils Bulletin 53

Pedraza A (2014) Assessment of silvopastoral systems establishment in Italva, Rio de Janeiro, Brazil. Thesis submitted to the Cologne University of Applied Sciences, Institute for Technology and Resources Management in the Tropics and Subtropics, Cologne, Germany

Petheram L, Campbell BM (2010) Listening to locals on payments for environmental services. J Environ Manag 91:1139–1149

Pinney A (1991) Farmer-augmented designs for participatory agroforestry research. Agrofor Syst 15:259–274

Prefeitura Municipal de Italva (2002) Plano Municipal de Desenvolvimento Rural do Municipio de Italva 2002–2005. Rio de Janeiro. p 230

Pretty JN (1995) Participatory learning for sustainable agriculture. World Dev 23(8):1247–1263. https://doi.org/10.1016/0305-750X(95)00046-F

Rivera-Herrera JE, Molina-botero I, Chará-orozco J (2017) Sistemas silvopastoriles intensivos con Leucaena leucocephala (Lam.) de Wit: alternativa productiva en el trópico ante el cambio climático. Pastos y Forrajes 40(3):171–183

Scarano FR, Ceotto P (2015) Brazilian Atlantic forest: impact, vulnerability, and adaptation to climate change. Biodivers Conserv 24(9):2319–2331. https://doi.org/10.1007/s10531-015-0972-y

Schroth G, da Fonseca GAB, Harvey CA et al (eds) (2004) The Role of Agroforestry in Biodiversity Conservation in Tropical Landcapes. In Schroth G, da Fonseca GAB, Harvey CA et al (eds) Agrofrestry and biodiversity conservation in tropical landscapes. Island Press, Washington, DC. pp 1–12

Shibu J (2009) Agroforestry for ecosystem services and environmental benefits: an overview. Agrofor Syst 76:1–10

Southgate D (1994) Tropical deforestation and agricultural development in Latinamerica. In: Brown K, Pearce D (eds) The causes of tropical deforestation: the economic and statistical analysis of factor giving rise to the loss of tropical forests. UCL Press Limited, British Columbia, Canada

Tabarelli M, Pinto LP, Silva JMC et al (2005) Challenges and opportunities for biodiversity conservation in the Brazilian Atlantic forest. Conserv Biol 19(3):695–700. https://doi.org/10.1111/j.1523-1739.2005.00694.x

Vendramini J, Silveira M, Dubeux J Jr, Sollenberger L (2007) Environmental impacts and nutrient recycling on pastures grazed by cattle. Rev Bras Zootec 36(suplemento especial):139–149

Weber JC, Montes CS, Vidaurre H et al (2001) Participatory domestication of agroforestry trees: an example from the Peruvian Amazon. Dev Pract 11(4):425–433. https://doi.org/10.1080/09614520120066710

Chapter 8
Towards Good Agricultural Practices in Smallholder Dairy Production Systems from an Animal Welfare Perspective

Maria Luisa Espinel, Sabine Schlüter, and Carlos Marconi de Souza Resende

Abstract Good agricultural practices (GAP) are norms and procedures developed to obtain agricultural products in a sustainable way. The specific GAP for dairy farms are good dairy practices (GDP) which objective is sustainable milk production. Animal products obtained with environmentally responsible practices, joined with increased awareness for animal well-being, concern modern consumers. This fact challenges the livestock sector and opens opportunities for products from animals maintained under high welfare standards. The measures to achieve high states of animal welfare are complementary with GDP. Animals that live in enhanced environmental conditions, with optimal health status, reach higher wellness and have a better performance. Therefore, animal well-being is strongly related with efficiency and profitability in dairy production. In the case study area of Italva, Rio de Janeiro state, Brazil, the traditional milking practices of small dairy farmers, the low milk yield, the environmental conditions, the degradation of natural resources, and poor infrastructure hamper good quality and a high and constant milk production. The potentials and constraints of small milk producers of Italva to adopt GDP through animal welfare concepts are analyzed to support farmers to achieve sustainability, to increase production, and to have access to broader markets and fairer prices for high-quality milk.

Keywords Good agricultural practices (GAP) · Good dairy practices (GDP) · Animal welfare · Dairy production · Rio de Janeiro

M. L. Espinel (✉) · S. Schlüter
TH Köln – University of Applied Sciences, Institute for Technology and Resources Management in the Tropics and Subtropics (ITT), Köln, Germany
e-mail: sabine.schlueter@th-koeln.de

C. Marconi de Souza Resende
Empresa de Assistência Técnica e Extensão Rural do Estado do Rio de Janeiro-EMATER-Rio, Italva, RJ, Brazil

U. Nehren et al. (eds.), *Strategies and Tools for a Sustainable Rural Rio de Janeiro*, Springer Series on Environmental Management,
https://doi.org/10.1007/978-3-319-89644-1_8

8.1 Context of Dairy Production in Brazil

Dairy farming has a strong influence on the Brazilian agricultural sector. Brazil is the world's fifth biggest milk producer after India, the USA, China, and Pakistan (FAO 2010), and milk is the sixth most important agricultural product in the country (Cruz 2013). During the period between 2000 and 2013, total milk production had a progressive increase from 20.379 million tons/year to 34.25 million tons/year resulting in a raise of production of 40%. The number of milking cows increased as well, from 18 million heads to 24 million heads, and the productivity average per animal raised from 1139.5 lit/cow/year in 2000 to 1492.3 lit/cow/year in 2013 (FAOSTAT 2016), showing a tendency of milk production intensification almost doubling the production in a period of 13 years (Figs. 8.1 and 8.2).

Around 35% of the Brazilian milk production is concentrated in the Southeast Region where Minas Gerais state (MG) is the largest producer with a share of 80% of the total production in the region. MG is also the biggest producer of the country reporting around 9 million tons/year of milk and has the highest annual productivity with about 2400 lit/cow/year. In this state, dairy farming is characterized by large-scale farms and intensive production (Cruz 2013). In contrast, Rio de Janeiro state (RJ), part of the Southeast Region too, has on average a total milk production of 0.5 million tons/year and an average productivity of 1200 lit/cow/year. The difference lies in a dairy production based on traditional grazing systems (Gomes Tôsto et al. 2003).

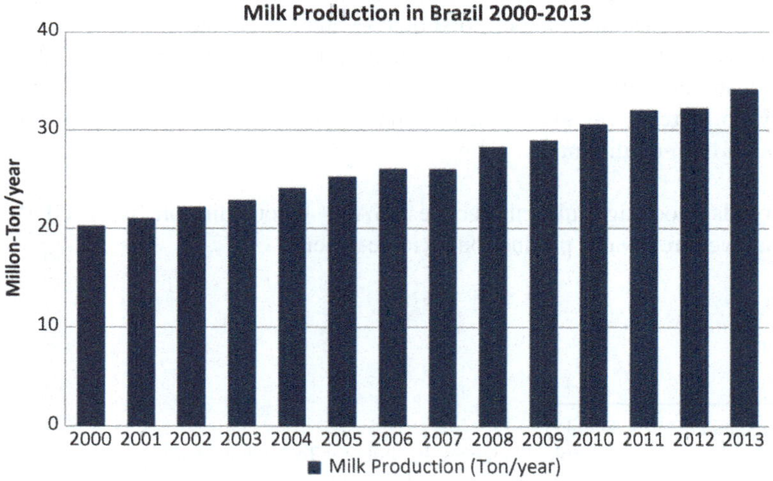

Fig. 8.1 Total cow milk production in Brazil in the period 2000–2013. (Source: Elaborated with data from FAOSTAT (2016))

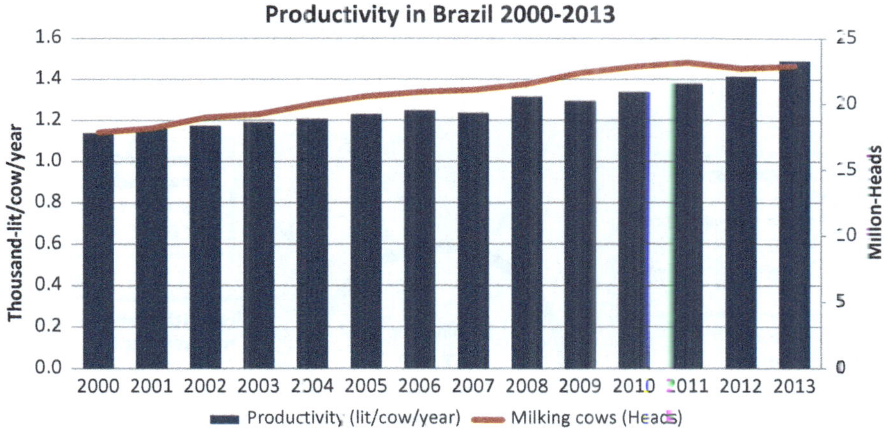

Fig. 8.2 Productivity and milking cows in Brazil in the period 2000–2013. (Source: Elaborated with data from FAOSTAT (2016))

8.2 Consumers, Production Trends, and Legislations

Good agricultural practices (GAP) are a set of preventive and proactive methods oriented to avoid potentially upcoming critical circumstances rather than to solve problems. They are applicable to every component and through the entire productive process of the farm. The final aim of GAP is to obtain agricultural products in a sustainable way. Their application in dairy farms enhances farm management, making it more efficient in terms of infrastructure, inputs, labor, and animal handling (Uribe et al. 2011). The specific GAP for dairy farms are good dairy practices (GDP) which are grounded in aspects as the safety of animals and stockpersons, animal welfare, sanitary measures, preservation of the environment and profitability of production (FAO – International Dairy Federation 2011).

The importance of GDP for small dairy farmers consists of the nine main activities described in Fig. 8.3. These actions have major implications on animal welfare and vice versa. Furthermore, these actions are essential for efficient, sustainable, productive, and profitable dairy farm (Doyle and Moran 2015). Animals in good health and mental status have a better performance, and their health and well-being have a direct impact on growth, reproduction, and production rates. This means higher milk output with better sanitary characteristics and nutritional content, hence its importance for producers and consumers (Humane Society International 2008).

Additionally, the rising awareness for animal welfare and ethical animal handling concerns consumers and renders new pressure for the livestock sector (Humane Society International 2008). Animal well-being became a requirement to satisfy consumer's expectations. It pushes producers to embrace sustainable and animal welfare measures to keep access to the market (Humane Society International 2008).

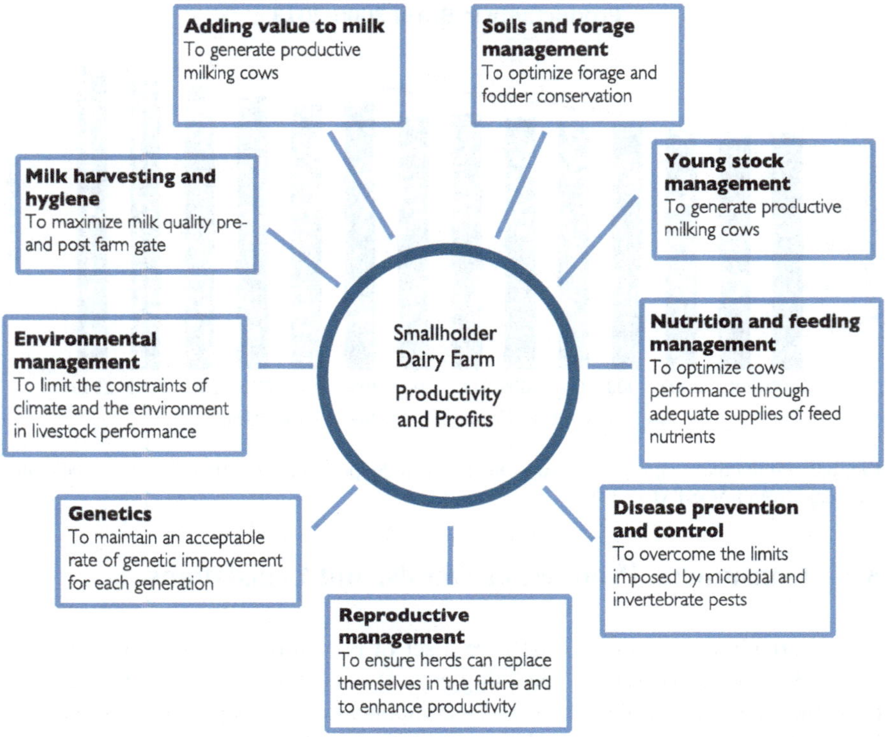

Fig. 8.3 Farm main activities to improve performance and sustainable production (Doyle and Moran, 2015)

Another aspect that influences milk production and market are the sanitary aspects. Milk is a highly perishable product that should be collected according to several hygienic norms and refrigerated quickly after the milking process to prevent its bacterial content increase and spoilage (Knips 2005). The evaluation of microbiological quality is an important parameter to determine the hygiene conditions during milk harvesting and storage. Some milk quality standards were developed to guarantee the product safety for human consume and health (Knips 2005). There are legal limits for the content of microorganisms and somatic cells (SC) in milk, i.e., cells that are from the cow's immune system and increase in milk in case of udder infection (see Table 8.1 comparing the basic hygiene requirements in the EU, USA, and Brazil). In Brazil the Ministry of Agriculture, Livestock and Supply developed the Normative Instructions (NI) 51/2002 and 62/2011 establishing an initial maximum of 1,000,000 CFU/ml and 1,000,000 CSC/ml for raw milk for the year of 2002. These amounts should reduce gradually to reach a limit of total bacterial account of 100,000 CFU and 300,000 CSC/ml by 2017 (see Table 8.2). In order to trade milk and its by-products, it is mandatory to fulfill these standards. Therefore, small milk producers have to overcome many obstacles to satisfy the regulations and requirements established by the sanitary institutions, the market, and consumer's preferences.

Table 8.1 General milk hygiene requirements in the EU, USA, and Brazil – 2004

Parameter	EU	USA	Brazil
CSC / ml	400,000	750,000	1,000,000
CFU / ml	100,000	100,000	1,000,000

Source: Hillerton and Berry 2004; Olival et al. 2004

Table 8.2 General milk hygiene requirements in NI 51/2002 and 62/2011

Regions	Parameter	Normative instruction 51/2002			Reform with normative instruction 62/2011			
South, southeast, and Midwest	**Date**	**01/07 2005**	**01/07 2008**	**01/07 2011**	**01/07 2008**	**01/07 2012**	**01/07 2014**	**01/07 2016**
	CSC / ml	1,000,000	750,000	400,000	750,000	600,000	500,000	400,000
	CFU / ml	1,000,000	750,000	100,000 individual 300,000 collective	750,000	600,000	300,000	100,000 individual 300,000 collective
North and northeast	**Date**	**01/07 2007**	**01/07 2010**	**01/07 2012**	**01/07 2010**	**01/07 2013**	**01/07 2015**	**01/07 2017**
	CSC / ml	1,000,000	750,000	400,000	750,000	600,000	500,000	400,000
	CFU / ml	1,000,000	750,000	100,000 individual 300,000 collective	750,000	600,000	300,000	100,000 individual 300,000 collective

Source: Alves Dias and Goldschmidt Antes 2014; Olival et al. 2004

8.3 Smallholder Dairy Production in Italva

Italva municipality, located in RJ, was selected as a study region for the INTECRAL research project considering that its dairy sector has a high influence for the municipal economy although it has a lower productivity than the average for the state. Additionally, it was selected because of the high environmental damage and the extreme state of land degradation posting a tremendous threat to the region and dairy production. These aspects give the region a high potential to explore the effects of GDP implementation (IBGE 2006).

Milk production in Italva is carried out mostly by smallholders. Agriculture has a land distribution based on many small-scale farms and a few large-scale farms. Land misuse and extensive cattle grazing are the predominant characteristics (Abrahão et al. 2014). The husbandry systems are established on slopes and hills, and cattle production occupies 74% of agricultural areas (Pedraza Luengas 2014). The municipality has many producers associations, cooperatives, and a medium-sized milk processing plant. Around 70% of the dairy farmers belong to a producer's association (Abrahão et al. 2014). Milk production is a strong part of the municipality's identity, and it is the main economic activity with rising importance. From 2013 to 2014, the number of producers increased from 218 to 359, while the total amount of milk production increased from 4936.6 tons to 7735.5 tons (Resende

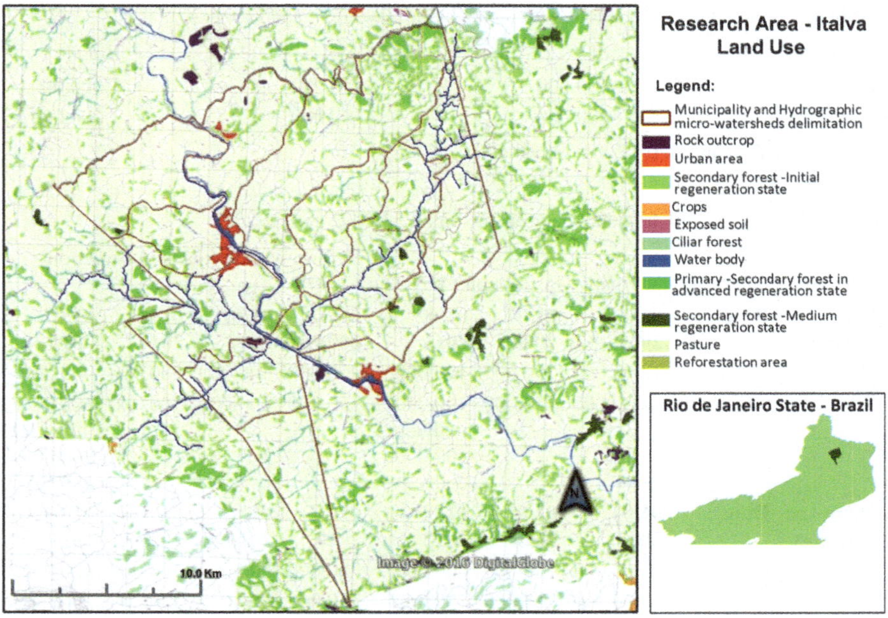

Fig. 8.4 Italva, Rio de Janeiro state, Brazil – land use. Elaborated from data of Instituto Brasileiro de Geografía e Estatística (IBGE 2002)

et al. 2013, 2014). Farmers tend to implement cattle farming as their agricultural activity due to the low initial investment required and the continuous returns obtained also in crisis moments (Abrahão et al. 2014). Figure 8.4 shows the map of Italva and its land use.

Ten representative smallholder dairy farms were selected for the study, and the characterization was made through the information collected by means of semi-structured interviews and field observation. The extension of farms varied between 4.8 ha and 42 ha, and the calculated ratios of areas for pasture over the total farm areas ranged between 29% and 98%. The number of lactating cows varied from 3 to 17; in total there were 10 to 60 heads. Performance varied significantly from 2.5 liters/cow to 13.7 liters/cow or 547.5 liters/cow/year to 3185.5 liters/cow/year. The milking process was mostly manual, once a day with the calf on foot, with the exception of two producers using mechanical milking systems. For production, Girolando is the typical race, a robust cross-bread of Holstein x Gyr.

The feeding systems were based principally in extensive grazing and supplementation with sugarcane. Reproductive management was basic, and natural mating was the main breeding strategy. The lack of veterinarian assistance was a major concern for all farmers. Corrals and milking areas for animal handling were in general small to medium sized, provided good ventilation, and were made of wood, tin roof, and earth floor or bad conditions cement. These characteristics made it difficult to keep the facilities free from dirt, and in some cases, accumulation of feces or waste of different sources was found in the exterior or inside the corral. See the Italva farms basic characterization in Table 8.3.

Table 8.3 General information of smallholder dairy farms in Italva, Rio de Janeiro state

Farm	Main economic activity	Farm total area (ha)	Pasture percentage	Grazing system	Total herd	Lactating cows	Daily production (L)	Yield (L/cow/year)	Total milk production (L/year)
1	Agriculture	17.5	29%	Continuous	10	3	25	1825.0	9125
2	Dairy farming	4.8	42%	Continuous	16	7	18	547.5	6570
3	Dairy farming	12.0	80%	Continuous	60	14	192	3185.5	70,080
4	Dairy farming	6.0	58%	Continuous	40	17	90	1825.0	32,850
5	Dairy farming	42.0	88%	Rotational	35	13	110	1911.9	40,150
6	Dairy farming	30.5	98%	Rotational	36	10	40	811.1	14,600
7	Dairy farming	40.6	79%	Continuous	37	10	37	675.3	13,505
8	Dairy farming	19.0	95%	Continuous	40	17	60	1095.0	21,900
9	Dairy farming	31.1	49%	Continuous	38	17	100	1587.0	36,500
10	Dairy farming	16.0	94%	Continuous	36	12	120	1990.9	43,800

Source: Own data - Interviews: 10 small-scale farmers, Italva, Rio de Janeiro state – Brazil, 2015

8.4 Constraints and Potentials of Small Dairy Farmers of Italva for Sustainable Production

In order to assess the information collected, a system of parameters and indicators, explained in Fig. 8.5, was developed based on the main dairy farming activities described by Doyle and Moran 2015, the GDP described by FAO – International Dairy Federation 2011 – and "The guide to the good animal welfare in dairy production" by International Dairy Federation 2008. Those factors are strongly related with the animal well-being and influence drastically the productivity and efficiency of animals in dairy farms. The purpose was to examine the farms' performances in seven different elements (see Figs. 8.5 and 8.6). This system was designed due to the need to make an analysis through subjective and qualitative data as animal

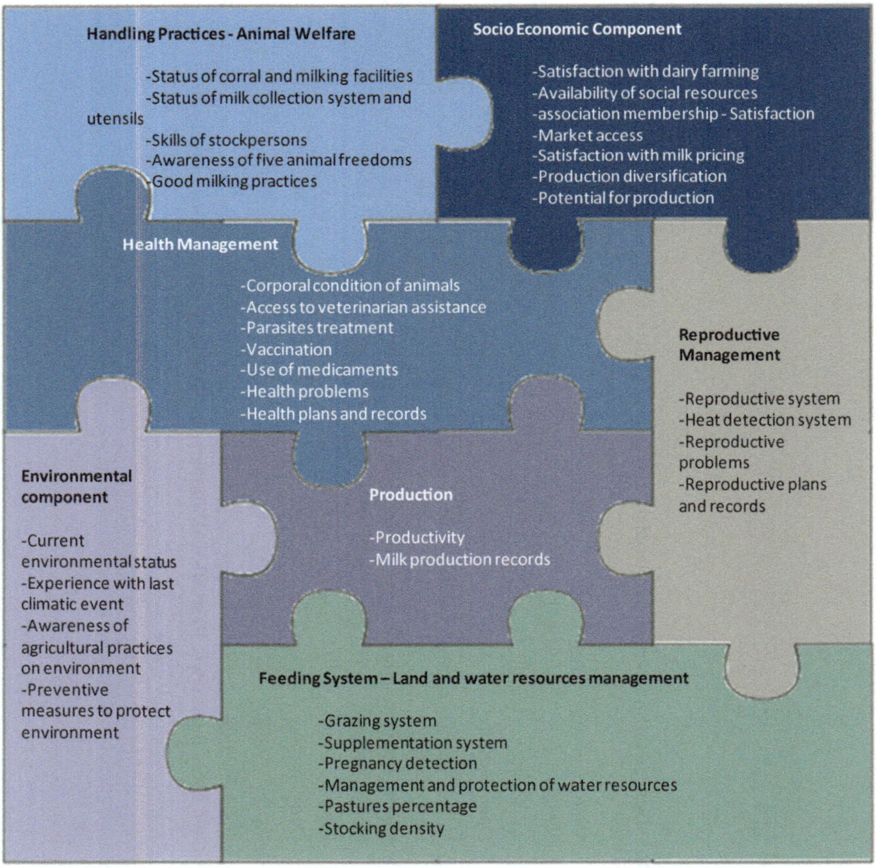

Fig. 8.5 Indicators and parameters per indicator assessed. (Source: Own data – Interviews – 10 small-scale farmers, Italva, Rio de Janeiro state, Brazil, 2015)

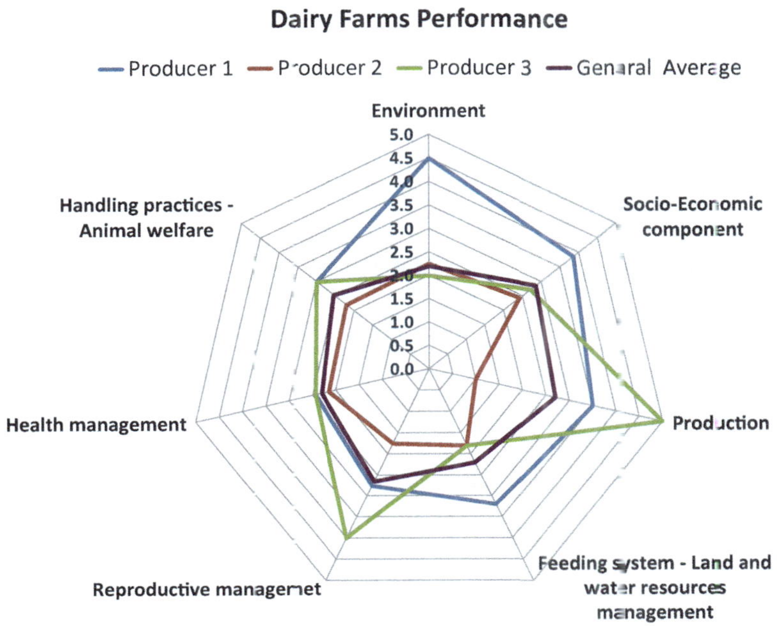

Fig. 8.6 Current performance of selected smallholder dairy farms in Italva. (Source: Own data – Interviews – ten small-scale farmers, Italva, Rio de Janeiro state, Brazil, 2015)

welfare and the variable handling practices performed by each farmer, combined with the fact that the measurable information available in health, reproductive, and productive aspects were not enough to make a deeper rigorous analysis of the farms' performances. Ranking was made on a scale of one to five, with one being the worst and five the best condition. Each indicator was composed of a list of parameters. For instance, the reproductive management component included parameters such as reproductive systems used and techniques for heat and pregnancy detection. The score was given depending on how suitable or applicable the used techniques were. The scores enabled a graphic overview of the farms' performances in the different areas allowing a comparison between the farms. Based on these indicators' analysis, critical points, constrains, and potentials for sustainable milk production were identified. For example, the current performance of three dairy farms compared with the average for all ten farms assessed is represented in Fig. 8.6. The main constraints and potentials are described in Table 8.4.

One of the most important findings was the influence of the milking facility conditions and the traditional milking practices on hygiene. Inadequate construction materials, bad floors in corrals, and deficient access to an adequate source of water observed in small dairy farms of Italva make it difficult for the producers to keep up with some important milk production requirements, such as the hygienic measures

Table 8.4 Main constraints and potentials of small-scale dairy farmers of Italva for sustainable production

Main constraints	Main potentials
Rudimentary conditions and production practices Low financial resources Low awareness of the impact of farming practices on the environment Poor knowledge of good farming and good milking practices Low awareness of animal welfare High dependence of environmental conditions Lack of veterinarian assistance Lack of farm records General deficiencies in farm management	Producers associations and cooperatives Diversification of products and organic production Environmentally important areas inside the farms Extension groups with the willingness to transfer technology Good community-based management through different associations inside the micro-watersheds Animal breed adapted to the area and climatic conditions

Source: Own data – Interviews: ten small-scale farmers, Italva, Rio de Janeiro state, Brazil, 2015

and appropriate animal handling. Additionally, these handling areas are in some cases uncomfortable and unsafe for both animals and stockpersons.

In the animal handling component, the farmers interviewed were experienced persons who had acquired their skills through experience and tradition. During the milking process, they carried out some basic norms of hygiene but had scarce notion of concepts/topics with respect to good dairy and animal welfare practices, like mastitis prevention, milking hygiene, animal healthcare, milk quality, animal reproduction management, environmental management, etc. These are key issues for dairy production, and the lack of those concepts represents a crucial obstacle for adequate performance of the dairy farms. It shows poor background about basic measures to keep milk production under safe and reliable conditions for animals, workers, and consumers.

Another important constraint was the vulnerability to erratic environmental conditions, leading to substantial losses in cases of extreme climatic events. The widely applied practice of grazing on periodically plowed pasture areas (head down) in the hilly area is the reason for extensive land degradation, decrease of soil nutrients, and poor-quality forage resulting in deficient animal nutrition, hence low efficiency and productivity. It can also raise costs for fertilizers and silages or supplementation for animals. However, there is a considerable potential for improvement, since the region is located in a rich biodiverse and environmentally important area.

Moreover, dairy farmers lacked adequate farm registers, records of costs, production, animal handling, animal health, reproductive management, etc. This means that they have no certainty and control of their farming performance, leading to misuse of resources and management problems.

In contrast, the most important strength found of dairy farmers of Italva was the social capital. They were open and willing to work in associations, allowing them to join for many common purposes, like collecting higher amounts of milk, reducing costs, negotiating better prices for their product, and having access to credits or subsidies. The strong motivation and satisfaction due to the common benefits of being associated assure the continuity of production and the enhancement of their livelihoods.

8.5 Heading to Good Dairy and Animal Welfare Practices in Italva

A decisive aspect to improve milk quality is the implementation of hygienic procedures, correct animal handling measures, and mastitis control. For small producers in RJ and especially in areas like Italva municipality that has a strong dependence on dairy production for their economic development, it is of high importance to make an appropriate guidance and implementation for such measures. There are many studies concerning the use of GDP for the improvement of small farmers' milk production conditions in rural areas of Brazil. For instance, in the research with small dairy farmers in the Pirassununga region, by Olival et al. (2004), 21.0% of the producers evaluated had a bacteria total account with values above the limit (1,000,000 CFU/ml) and 11.1% above the limit for CSC (1,000 000 CSC/ml) established by the NI 51. According to this research, the constraints to reach these quality standards were the lack of knowledge about the causes of bacterial contamination and somatic cell count in milk and their prevention, joined with the lack of mastitis control measures (Olival et al. 2004).

Some other studies with small dairy farmers in Brazil such as the ones by Vallin et al. in 2009, Domingues et al. in 2009, and Mattioda and Bittencourt in 2011 coincide that the general hygiene conditions, the herd management, and the milk quality in smallholder dairy farms required adjustments to accomplish the NI 51/2002 and 62/2011. All of them confirm that the main limitation for the milk producers is the lack of information about milking routines, the poor use of sanitary measures, and a deficient mastitis control. These observations are consistent with the findings in the study carried out in Italva. The need of training in hygienic milking routines, milk quality, mastitis control, and animal health management is a crucial aspect for dairy production improvement.

The application of simple techniques can lead to important enhancement of hygienic and microbiological milk characteristics and make it possible for small producers to overcome their difficulties to reach the milk quality standards established in the Brazilian law. In the study by Vallin et al. (2009), the application of two simple measures in the milking routine resulted in a substantial reduction of CSC and CFU. The routines implemented were as follows: to examine and discard the first streams of milk in a black cup to check for mastitis and pre-dipping or disinfection of the cow's teats with chlorine solution 750 ppm before milking. The second measure was to clean the buckets and milk cans with alkaline detergent after the milk harvesting and to leave them top down in a dry airy area. The result for manual milking was a decrease of 63% in CSC and 86% in CFU/ml (Vallin et al. 2009).

In a different work, Mattioda and Bittencourt (2011) stated that the use of measures such as disinfecting the cow's teats soon after milking, keeping animals standing a sufficient time for the sphincter of the teat to close again, and adequate cleaning of utensils contributed to the improvement and maintenance of milk quality. Furthermore, Brito et al. (2003) reported that the use of buckets with lateral superior coverage compared with buckets without superior coverage for manual milking carried out an improvement in the microbiological quality and a reduction between 26% and 73% of CFU in milk. In addition, Brito et al. evaluated the application of measures such as feces cleaning in the corral once or twice a week, the sterilization of utensils, and the pre-dipping of cow's teats. These measures led to a reduction of 99% in CFU. These practices are easy to implement; however, the effectiveness of good milking techniques depends on the strict and constant use of such procedures and the careful management of the whole process.

Additionally, it is important for small producers, as members of cooperatives with communal tanks, to pay special attention to the GDP adopted. If one producer does not preserve milk in good conditions, this can cause damage to the collected production of many farmers, thus leading to economic losses for all. A contrario, if all have good quality, the differential prices can be distributed for all. In the study by Brito et al. (2003), just 20% out of a total of 345 samples of communal milk tanks had counts below 1,000,000 CFU/ml. Only in one of the tanks the values had a mean of 320,000 CFU/ml in 10 of 12 months analyzed. In that case, the association had the commitment of all producers to keep good milk quality, to pay special attention to hygienic measures, and to deliver their yield to the tank immediately after the milk harvesting (Brito et al. 2003).

Moreover, mastitis prevention and control should be one of the main activities to enhance production. Mastitis is a disease that frequently affects dairy cattle and produces one of the largest losses in terms of decrease of production, decrease on milk quality, milk waste, costs of treatments, medicaments, and veterinarian assistance (Halasa et al. 2007). It is the infection and inflammation of the udder and a main aspect for dairy cow welfare, producing pain and chronic stress in cows. Subclinical mastitis can cause a reduction of production up to 30% with a content of somatic cells of 1,500,000 CSC/ml (Alves Dias and Goldschmidt Antes 2014).

In a different animal welfare aspect and with regard to land use, ongoing studies are being carried out in the area for the transition to rotational systems and improved pastures mixed with silvopastoral systems (SPS) by the governmental investigation and extension institutions PESAGRO and EMATER-RIO as well as the INTECRAL project. The land use change to rotational systems and SPS may have higher costs of investment but at long term could represent higher production and efficiency rates. These systems are an important contribution to the protection of the environment, the provision of adequate areas, welfare, and nutrition for the optimal development and performance of animals. According to Mahecha and Angulo 2011, the carrying capacity can increase between 87.5% and 166.6% depending on the SPS implemented and the tree species used. Milk production could increase between

20% and 35% in addition. Those SPS systems support more animals per hectare because of the high nutritional offer from pastures, trees, and shrubs included in the diet, leading to an increment in the cow's productivity (Rosales 2006). The use of SPS produces as well a reduction of environmental temperature under the tree canopy between 2 and 9 °C, compared with areas of open pasture giving the animals an adequate environment to increase production (Panadero 2010). This aspect is important within the climatic conditions of Italva where animals suffer from heat stress due to the extremely high temperatures. For the producers of Italva, the implementation of such techniques could represent an increment expectation of productivity of at least 160 to 280 lit/cow/year.

8.6 Conclusions and Outlook

The good dairy and animal welfare practices and adequate farm management measures are important tools to overcome constraints of small dairy farmers for sustainable dairy production. It puts them ahead in a leading trend of the market, satisfying the requirements from the sanitary sector and the growing interest for consumers: milk production, performed by healthy animals kept under high welfare conditions. Good dairy practices based on animal welfare comprise animal health, hygienic procedures, animal nutrition, and handling techniques, as well as land and resources management, among others. The aim is to provide animals with the best conditions to produce efficiently. These measures also put producers at the cutting edge of the market and give them a strong ability to compete and to achieve better conditions to trade with a high-quality product.

However, good dairy practices and animal welfare measures should be supported by experts. Milk quality and milking routines should be assessed. The next step to follow is the implementation of tests for mastitis. It is fundamental to identify the real status of milk production in Italva for its improvement. Therefore, the presence of veterinarians and technical assistance is essential for the development of dairy production in the region. Small producers need as well support regarding the use of resources, basics of management, animal nutrition, reproduction, and health all important fields for dairy farming. Dairy production is extremely dependent on the animal's health, reproductive performance, and nutritional status. In other words, it depends on the welfare of cows.

Another important step is the change of inappropriate habits and procedures of small dairy farmers combined with technical support and guidance along the production process. This is the basis of the implementation of good dairy practices and animal welfare measures, thus the importance of informative workshops and the distribution of educative materials. The main aim is to guarantee the application of correct practices and the continuity of the measures adopted.

Moreover, it is vital to promote the strongest potential of the small dairy farmers in Italva, i.e., the associations of producers. These associations give them the conditions to bargain and commercialize bigger volumes of milk and sub-products. It is essential to support and verify the well-functioning of these associations concerning management, technical, social, and economic aspects to assure the sustainability of their activities. Ultimately, the expectation is the enhancement of the livelihoods and lifestyles of small producers of Italva in a permanent way.

References

Abrahão J, de Lopes R.S, Melo G (2014) Plano municipal de conservação e recuperação da Mata Atlântica de Italva. Secretaria de Estado do Ambiente – Prefeitura do Município de Italva, Estado do Rio de Janeiro

Alves Dias J, Goldschmidt Antes F (2014) Qualidade físico-química, higiênico-sanitária e composicional do leite cru. EMBRAPA – Empresa Brasileira de Pesquisa Agropecuária, Brasil

Brito M, Portugal JAB, Diniz FH et al (2003) Qualidade do leite armazenado em tanques de refrigeração comunitários. Alternativas tecnológicas e políticas públicas para produção leite em bases sustentáveis. Embrapa – Gado Leite, pp 21–43

Cruz AA (2013) Indicadores de sustentabilidade: estudo de caso em propriedades produtoras de leite nas regiões sul e sudeste do Brasil utilizando a metodologia RISE. Universidade de São Paulo Escola Superior de Agricultura "Luiz de Queiroz"

Domingues F, De Freitas Trancoso JR, Urizzi E., Mattioda F, Barbosa de Morais SA (2009). A relevância de ações assistenciais e técnicas, junto a propriedades de leite dos municípios de teixeira soares e fernandes pinheiro. Apoio técnico aos pequenos produtores de leite, dos municípios de Teixeira Soares e Fernandes Pinheiro. Universidade Estadual de Ponta Grossa, Brasil

Doyle R, Moran J (2015) Cow talk: understanding dairy cow behaviour to improve their welfare on Asian farms. CSIRO PUBLISHING, Australia. ISBN: 1486301614, 9781486301614

FAO (2010) Food and Agriculture Organization of the United Nations – Dairy production and products: Farm practices. http://www.fao.org/agriculture/dairy-gateway/milk-production/farm-practices/en/#.VrNYq_EnX5c. Accessed 02 Apr 2016

FAO, International Dairy Federation (2011) Guide to good dairy farming practice, Rev. ed. FAO animal production and health guidelines. Food and Agriculture Organization of the United Nations: International Dairy Federation, Rome

FAOSTAT (2016) Food and Agriculture Organization of the United Nations – Statistics Division. http://faostat3.fao.org/download/Q/QL/E. Accessed 03 May 2016

Gomes Tôsto S, Santos Brandão E, Tinoco dos Anjos G, Pereira Said U (2003) Diagnóstico socioeconômico dos municípios da região Noroeste do estado do Rio de Janeiro (Research Report No. 41), Boletim de Pesquisa e Desenvolvimento. Empresa Brasileira de Pesquisa Agropecuária (EMBRAPA) – Ministério da Agricultura, Pecuária e Abastecimento, Rio de Janeiro, Brasil

Halasa T, Huijps K, Østerås O, Hogeveen H (2007) Economic effects of bovine mastitis and mastitis management: a review. Vet Q 29:18–31. https://doi.org/10.1080/01652176.2007.9695224

Hillerton JE, Berry EA (2004) Quality of the milk supply: European regulations versus practice. In: NMC Annual Meeting Proceedings, pp 207–214

Humane Society International (2008) Animal welfare, international trade and sustainable development improving the lives of animals, farmers and communities. Humane Society International, World Trade Organization

IBGE – Instituto Brasileiro de Geografia e Estatística (2002) Base Cartográfica Digital do Levantamento Topográfico Sistemático IBGE, Municipio de Italva, Microbacias Hidrográficas sobre Uso do solo, Rio de Janeiro

IBGE – Instituto Brasileiro de Geografia e Estatística (2006) Censo agropecuário 2006. http://www.ibge.gov.br/home. Accessed 10 Aug 2017

International Dairy Federation (2008) Guide to good animal welfare in dairy production. Brussels, Belgium

Knips V (2005) Developing countries and the global dairy sector part I – Global Overview, Working Paper No. 30. Food and Agriculture Organization (FAO) – Pro-Poor Livestock Policy Initiative
Mahecha L, Angulo J (2011) Nutrient management in silvopastoral systems for economically and environmentally sustainable cattle production: a case study from Colombia. Soil Fertility Improvement and Integrated Nutrient Management. Global Perspective 201–216
Mattioda F, Bittencourt JVM, Kovaleski JL (2011) Qualidade do leite de pequenas propriedades rurais de Fernandes Pinheiro e Teixeira Soares-PR. Rev ADMpg Gestão Estratégica 4:1–9
Olival AA, Spexoto AA, Mano G, Santos MV (2004) Avaliação das limitações para melhoria da qualidade do leite na região de Pirassununga-SP. Revista Ciência em Extensão 1:171–183
Panadero AN (2010) Importancia de los sistemas silvopastoriles en la reducción del estrés calórico en sistemas de producción ganadera tropical. Rev Med Vet:113–122
Pedraza A (2014) Assessment of silvopastoral systems establishment in Italva, Rio de Janeiro, Brazil (Master Thesis). Universidad Autónoma de San Luis Potosí – Cologne University of Applied Sciences – Institute For Technology And Resources Management In The Tropics And Subtropics
Resende CM, Neto PS, Silveira F et al (2013) Relatório de atividades da EMATER – RIO Escritório Local de Italva 2013. Empresa de Assistência Técnica e Extensão Rural do Estado do Rio de Janeiro (EMATER – RIO) – Escritório Local de Italva
Resende CM, Neto PS, Silveira F et al (2014) Relatório de Atividades da EMATER – RIO Escritório Local de Italva 2014. Empresa de Assistência Técnica e Extensão Rural do Estado do Rio de Janeiro (EMATER – RIO) – Escritório Local de Italva
Rosales M (2006). Cattle ranching and deforestation, livestock information, sector analysis and policy branch animal production and health division. Food and Agriculture Organization (FAO)
Uribe F, Zuluaga AF, Valencia LM et al (2011) Buenas prácticas ganaderas. Manual 3, Proyecto Ganadería Colombiana Sostenible. G. GEF, BANCO MUNDIAL, FEDEGÁN, CIPAV, FONDO ACCION, TNC, Bogotá, Colombia
Vallin VM, Beloti V, Battaglini APP et al (2009) Melhoria da qualidade do leite a partir da implantação de boas práticas de higiene na ordenha em 19 municípios da região central do Paraná. Semina: CiênciasAgrárias 30:181–188

Chapter 9
Adequate Harvesting Technology for Small- and Medium-Scale Sugarcane Farmers in Rio de Janeiro: The "Cologne Solution"

Carl-Friedrich Gaese, Carlos Frederico de Menezes Veiga, José Márcio Ferreira, and Torsten Meyer

Abstract Ten million hectares of sugarcane make Brazil the largest global producing country. By the year 2024, legislation will ban the burning of dry leaves before manual harvest. Solutions like the self-propelled harvester technology are available. Nevertheless, since these solutions require high investments and large cropping fields for economically feasible operational costs, they are more favorable and widely adapted for large-scale producers. However, small- and medium-scale producers are under increasing pressure to adapt and remain competitive as the Brazilian market for agricultural engines does not offer adequate harvesting technologies. The average sugarcane farm size in Rio de Janeiro state is approximately 11 ha, with a low average yield when compared to the average yield of the country. The conditions to be considered for the implementation of innovative harvesting technology, so far developed up to a classified test-bench status, involve all technical, economic, and site-specific requirements. This research defined the adequate conditions for a harvesting technology based on the sugarcane cycle analysis.

C.-F. Gaese (✉)
TH Köln – University of Applied Sciences, Institute for Technology and Resources Management in the Tropics and Subtropics (ITT), Köln, Germany
e-mail: carl-friedrich.gaese@th-koeln.de

C. F. de Menezes Veiga
Universidade Federal Rural do Rio de Janeiro, Campos dos Goitacazes, RJ, Brazil

J. M. Ferreira
Empresa de Pesquisa Agropecuária do Estado do Rio de Janeiro, Campos, RJ, Brazil

T. Meyer
TM Technik, Kronberg im Taunus, Germany
e-mail: tmeyer@tm-technik.de

U. Nehren et al. (eds.), *Strategies and Tools for a Sustainable Rural Rio de Janeiro*, Springer Series on Environmental Management,
https://doi.org/10.1007/978-3-319-89644-1_9

Keywords Sugarcane production · Adapted technology solutions for sugarcane harvesting · Socioeconomic impacts

9.1 Introduction

Sugarcane includes several species of tall perennial true grasses of the genus Saccharum, tribe Andropogoneae. It is cultivated between the 35° north and 30° south latitude. According to FAOSTAT (2017), Brazil is the largest sugarcane producer in the world with more than 10 million hectares of harvested area. On the second place is India with approximately 5 million hectares. Sugarcane came from South and Southeast Asia to the Northeast of Brazil about 500 years ago (Nastari 1983). Brazil has implemented a governmental subsidy program named "National Alcohol Program" (Pró-Alcool) since 1975. It was launched in response to the oil crisis in 1973 when sugar prices were low and oil prices were high. The overall goal of Pró-Alcool was to get more independence from the production of crude oil. This was a key decision for optimizing the value chain of sugarcane in Brazil. Meanwhile Brazil is the world's largest producer of sugarcane and ethanol as well as the largest exporter of sugar (Solano 2011).

The most important Brazilian state for sugarcane production is São Paulo (SP) where the majority of scientific institutions, industries, and associations are located. The state of Rio de Janeiro (RJ) does not play a major role in the overall Brazilian sugarcane production (IBGE 2017); however, sugarcane is an important source of income in agriculture especially for small and medium-sized enterprises (SMEs). About 90% of the companies dealing with sugarcane in RJ belong to these categories (Bastos and Moraes 2014).

Preharvest burning of sugarcane leaves and manual cutting and collecting of the stalks have been common practice for sugarcane harvesting. By the year 2002, around 75% of sugarcane in Brazil was harvested in this way, today it is only about 55% (Solano 2011). Related to the total of 10 million hectares, the average area currently being burnt down is still around 5.5 million hectares.

Since the Kyoto Protocol entered into force in 2005 preharvest burning has abolished step by step. A major obstacle for faster advancement is the fact that even 13 years later there is still no technology on the market that is suitable for small-scale farmers.

To date no adapted harvesting technologies have evolved while parts of the harvesting process chain, such as the receipt of harvest goods in sugar refineries, have been adapted to process more biomass. There are various approaches for harvesting technologies for SMEs which could not yet penetrate the market. SME are defined with a production area up to 150 ha (Ferreira 2014). In this chapter the challenges of a development of an environmental friendly harvesting technology are discussed, using the example of RJ.

State Law 2049/92, Art. 2 and the latest update of Law 5990/11 defined a reduction of burning in the crops to be harvested in mechanized harvesting areas to be as follows: 20% by 2012, 50% by 2014, 80% by 2018, and to cease by 2020. For plantations where mechanized harvesting cannot be implemented (of >12% slope), deadlines for reductions are longer: 20% by 2016, 50% by 2018, 80% by 2022, and final end of practice by 2024 (Governo do Estado RJ 1992, Lei n° 2.049/92; Governo do Estado RJ 2011, Lei no. 5990/2011).

According to Moreira (2006), this legislative factor with its emphasis on the government to mobilize the reduction of burning is a great incentive to mechanization. However, due to the many peculiar characteristics of sugarcane harvesting, the technical progress to date has not been achieved.

This chapter explains the reasons making a development of an adequate harvesting technology for sugarcane so challenging and difficult. It refers to the structure of sugarcane farms in the region, preconditions due to the pre-burning process, the physiology, the plantation, the harvest of sugarcane, and the adoption of a harvesting technology to site conditions in RJ. As a result it shows a developed technology approach, named "Cologne solution," based on these defined challenges to close the gap on an adequate environmentally friendly sugarcane harvesting technology for SMEs in RJ.

9.2 Material and Methods

The following investigations were carried out to develop a sugarcane harvesting technology suitable for small and medium-sized enterprises:

(i) Systematic literature research along the value chain of sugarcane production in Brazil and especially in RJ.
(ii) Patent research.
(iii) Semi-structured expert interviews in the years 2014 (March and April), together with J.M. Ferreira, researcher of Empresa de Pesquisa Agropecuária do Estado do Rio de Janeiro (Pesagro-Rio), shown in Table 9.1.

The results of the semi-structured expert interviews were confirmed by literature research (see References). Unpublished data, such as sugarcane farm structures in RJ (Table 9.2) were given by Asflucan during semi-structured interviews.

Thanks to the collected data an integrated mobile test bench was constructed and adapted to the conditions of RJ enabling field tests to be carried out. These field tests were carried out together with UFRRJ in Campos dos Goytacazes.

Table 9.1 Overview about semi-structured expert interviews

Date	Name	Organization	Topic
March 25 and 26; April 2, 2014	Frederico Rangel Paes	Cooperativa Agroindustrial do Estado do Rio da Janeiro Ltda. COAGRO	State of the art of sugarcane harvesting technology, analysis of process chain and weaknesses
March 25 and 26; April 2, 2014	Marcos Luiz de Vasconcelos Pereira	Cooperativa Agroindustrial do Estado do Rio da Janeiro Ltda. COAGRO	State of the art of sugarcane harvesting technology, analysis of process chain and weaknesses
March 25 and 31; April 3 and 10, 2014	Carlos Frederico Veiga	Universidade Federal Rural do Rio de Janeiro UFRRJ	Sugarcane breeding, especially with respect to targets for mechanization, reproduction of plants for planting regarding quality of stalks, physiology of sugarcane and components (condition, weight, proportions, masses), composition of biomass, requirements of the plant for growing and producing sugar juice, economic and socioeconomic situation of farms, agricultural structure in RJ
March 24, 2014	Saul de Barros Ribas Filho,	Pesagro-Rio	State of the art, row distances
March 26 and 27; April 1 and 4 and 9	SMEs in the municipality of Campos	(wanted to stay anonymous)	Cultivation and maintaining of fields, harvest chains referred to level of mechanization, transport to refineries
April 7 and 11, 2014;	Luis Marcos Barreti da Silva, Tito Lídio Campello Inojosa de Andrade,	Associação Fluminense dos Plantadors de Cana Asflucan	Adoption process of harvested sugarcane and processes within refineries, requirements to quality of harvested sugarcane

Table 9.2 Diagnosis of the sugarcane farm structures of RJ (Asflucan 2014, modified)

Extract	Calculated yield [R$/ha]	Farms [units]	Production [tons]	Total area [ha]	Average farm size [ha]
Up to 300 t	5240[a]	8738	502,249	43,690	5
>300–1000 t	17,816[a]	896	480,039	15,232	17
>1000 t – 5000 t	57,640[a]	390	821,273	21,420	55
>5000 t	125,760[a]	72	758,284	8640	120
Total		10,096	2,461,845	88,982	

[a]Ø costs of 69R$/ha

9.3 Economic Characterization of Sugarcane Production in the State of Rio de Janeiro

For the last 40 years, sugarcane was the economic backbone in the municipality for Campos dos Goytacazes (Campos) (Garcia and Silva 2009). The main sugarcane cultivation area of RJ is located in this area (Asflucan 2014).

In RJ the cultivation area has declined sharply since the 1970s. At the beginning of the 1970s, the cultivated area was still around 160,000 ha. According to the Asflucan (2014), it stagnated around 70,000 ha in 2007 and has since then risen to 88,000 ha.

From originally 24 sugar mills located in the northern region of the state, only three are currently still in operation. The radical reduction of sugar mills is an effect of structural change under competition (Asflucan 2013) due to the price decline of sugar and bioethanol.

The main evolving economic pillar of the region of Campos produces crude oil. As a result of the immigration, a considerable share of cultivation area was lost with the expansion of the city. Furthermore, the sugarcane production areas disappeared from the slopes and are mainly present in the plains. This fact would allow an easier mechanization of the harvest. According to Asflucan (2014), in farms with less than 100 ha, only 14.2% were mechanized and in farms with more than 100 ha up to 43.9%.

In RJ, 87% of farms are smaller than 10 ha, i.e., between 0.01 and 9.9 ha, with a harvest yield below 300 t/ha[1] (Asflucan 2014). Table 9.2 shows an overview of the farm size structures of SMEs in RJ.

This means that micro and small farmers operate at the subsistence level. Usually the farms have several branches of production, sugarcane being one of them (Asflucan 2013). In RJ, yields of 46–65 t/ha are generated depending of quantity of harvest (Viana et al. 2017); by comparison, SP produces on average 85 to 120 t/ha (UNICA 2016). For the year 2013, Coagro calculated average annual per hectare costs of 3,752R$ for the plantation, 1,030R$ for maintenance, and 1,570R$ for harvest (Coagro 2013). The recent purchase price of sugarcane paid by sugar mills is 91.78R$/t in RJ (Agrolink 2017). In RJ, there are five to seven harvests per plant stock. Then the field has to be replanted for maintaining or increasing the yields. Micro and small farms typically lack access to capital needed for reinvestments. Medium-scale farms normally own tractors with minimum 48 kW and a truck to transport their harvest to the sugar mill.

RJ's sugarcane sector is under pressure because of the lack of adequate harvesting technology. When the pre-burning law comes into force, small and some of the medium producers will not be able to harvest and as a consequence will be forced to cease production; depending on the scale, the operating grace of the sugar mills might not be achieved causing a closure of the business. This would result in the disappearance of the entire sugarcane sector in RJ, an important source of income with socioeconomic consequences.

Table 9.3 Most important advantages and disadvantages of the pre-burning process (ELLA 2012, modified)

Advantages	Disadvantages
Cost-effective removal of the dry leaves and reduction of transport costs (because of burned biomass)	CO_2 emission and lost opportunity to capture stocked carbon for beneficial use, e.g., energetic use
Facilitation and increased productivity of manual harvest	Harvest losses due to heat
Higher work safety in manual harvesting by removal of sharp dry leaves and animals, e.g., snakes and spiders	Negative health impacts on field hands local population caused by ashes and emissions
Destruction of pests, spores, fungi	Continuous destruction of biodiversity
Fertilization of the soil by the ash	No humus formation from burnt leaf mass

9.4 Advantages and Disadvantages of the Pre-burning System

The pre-burning prohibition will protect the environment. Emissions, especially CO_2 emissions, will be avoided; soil carbon stocks will increase and the potential to use the residual biomass will grow (ELLA 2012). But also abolishing the pre-burning means to abolish manual harvest and, connected to this, employment for unskilled labor like field workers will disappear, while at the same time higher educated labor force is demanded. This can lead to considerable social challenges considering that a traditional harvesting machine replaces around 100 workers (ELLA 2012). Although the costs of the mechanized harvest are up to 20% less than the manual harvest, according to calculations by Solano (2011), small- and medium-sized enterprises cannot provide the necessary capital to purchase a commercial harvester. Table 9.3 gives a short overview about the most important pros and cons of burning the leaves.

9.5 Adopting Technology to Sugarcane Morphology, Physiology, and Harvesting Cycle

The sugarcane belongs to the sweet grass family (Poaceae). It propagates vegetatively and is perennial.

Roots The planting of the sugarcane is usually carried out vegetatively with pieces of stalks placed in the soil. These sugarcane sticks form adventive roots at the nodes shortly afterwards and new shoots resprout quickly from the same nodes. These sprouts in turn form new roots and secondary and tertiary shoots, after harvesting (Bakker 1999).

Seedlings are planted in rows with a width and spacing (row geometry) adapted to the harvesting machine. The number of stalks and stock width depends on the number of nodes of the seedling as well as on the harvesting periods the so-called cuts (Bakker 1999). After 5 years, the rootstock is up to 50 cm wide. Vehicles driving over would heavily damage the rootstock. The cutting edges of the logs have to be clean and without fraying for the attack area for germs and pests to be as low as possible in order to maintain productivity.

Anatomy of Stalks The cane is cylindrical, standing, fibrous, and rich in various sugars. It consists of 10–14 nodes. The internode length varies between 2 and 20 cm. A bud can be clearly seen on the node from which a lateral shoot can grow, for example, in vegetative propagation, stocking in the case of growing plants, or storage on the ground (Bakker 1999).

As the sugar content is highest in the lowest part of the stalk, it has to be cut close to the ground, a challenge for man, as manual harvesting is done in a continual bent position, and also for the machine, as rotary cutters cut into the soil and the blades have to be renewed after every working day of 8 h. The upper third of the trunk shows the highest water content, thus delivery to the refinery is avoided.

Leaf The leaf margin clings tightly to the stem. The leaf begins to split from the stalks into a tubular shape and then grows wider, i.e., between 2 and 10 cm, and flat in a length of 50 cm to 1.5 m (James 2004). The green leaves at the top are very strong. The field trials showed that uncovered rotating machine parts should be avoided because the green leaves are wrapped around it and can stop the gear.

Important for the harvesting and processing of sugarcane is the avoidance of unwanted admixture of foreign materials to the crop. Also the leaf mass holds a potential value for energetic use if heavy contamination with soil during the harvesting process can be avoided.

Entire Plant The individual plant parts and requirements on their quality after the harvest show the challenges for the development of a new harvesting technique. The percentage of the plant components and their weights are shown in Tables 9.4 and 9.5. Stalks are the main crop, the other above ground components "tops and leaves" are relevant by-products as long as they can be separated and transported.

Mechanized harvest is a challenge because of the variance of length, i.e., 1620–4460 mm without tops and with tops up to 5 m long and width between 193 mm and 300 mm as well as the density to be dealt with; the straw weight per row meter is very irregular and varies between 1.53 and 19.20 kg per meter. The number of stalks per meter varies between 4 and 16 pieces. The area yields of the different cane varieties have values from 35.6 to 183.19 t/ha, only straws without head and leaves (Bakker 1999).

Previous breeding and investigations aimed at high sugar yields and the reduction of the so-called nonrecyclable residuals. Leaves, heads, fiber of the stalk, and roots were unwanted ballast which had to be reduced or removed. In a comprehensive view, especially according to modern, interdisciplinary aspects, all components

Table 9.4 Morphological composition of a 12-month-old cane plant, in each case dry material (van Dillewijn 1952)

Part of plant	Percentage (%)	Location
Tops and leaves	9.0	Above ground
Stalks	49.2	
Leaf drop	24.6	
Stumps of stalks	4.5	Below ground
Roots	12.7	

Table 9.5 Weight distribution of the cane plant [t/ha], selected cane, first cut after 12 months (Bakker 1999)

Stalks [t/ha]	Roots [t/ha]	Tops [t/ha]	Leaves [t/ha]
60	21	11	30
70	24	13	35
80	28	15	40
90	31	17	45
100	35	18	50
110	39	20	55
120	42	22	60

of the plant are re-examined. They are discussed in detail according to their values and possibilities not only for the production of sugar but also for the energy use, the protection against soil erosion, and the return of important nutrients. What used to be "waste" is now being redefined.

Planting and Harvesting Cycle Sugarcane refineries themselves plan a large part of the cultivation and harvesting, for themselves and for the suppliers to guarantee sufficient supply for maintaining a continuous process during the harvest season. In order to minimize the risk of yield loss due to abiotic and biotic factors, different cane cultivars are used to ensure that no more than 15% of the total cultivated area is planted with the same variety.

Normally, harvest lasts from April to December. The variety mix is selected for about 30% early, 50% middle, and 20% of late harvesting throughout the harvesting period (Barbosa and da Silveira 2015). Approximately 15% of the land is replanted per year. Each stock is harvested between four and six times before being replanted. The row distances vary widely, from 0.8 to 1.5 m, including combinations, e.g., 2 × 0.8 m with 1 × 1.5 m. In middle- and large-sized farms, the rows' distances have increasingly been adapted to the applied harvesting technology. Mechanical harvest is hampered by:

- Difficulties in identifying the rows in matured plantations. Absence of pre-turning areas leading to difficulties in cutting the fields.
- The absence of headland limits the opportunities for harvesting the first sugar-cane rows of a field and the space for maneuver.

9.6 Harvesting Methods and Harvest Chains

States of the art of harvesting sugarcane are two systems: the manual practice with pre-burning dry leaves and mechanized harvest.

Manual Practice The field to be harvested is burnt and then harvested manually with a special cutting device, comparable to a machete. The stalk is separated as close to the soil as possible and then the top, i.e., green leaves, is separated from the canal. The crop is put on a heap and later loaded onto vehicles transporting it to the sugar refineries.

Mechanized Practice A convergence analysis of the approved patents reveals that all the commercially available self-propelled harvesters (Fig. 9.1) have a similar technology: a harvester drives into the row, cuts the tops which fall into the harvest row, pulls the stalks into the middle of the machine, and cuts them into pieces about 30 cm long. The leaves are returned to the field with the help of a blower. Because

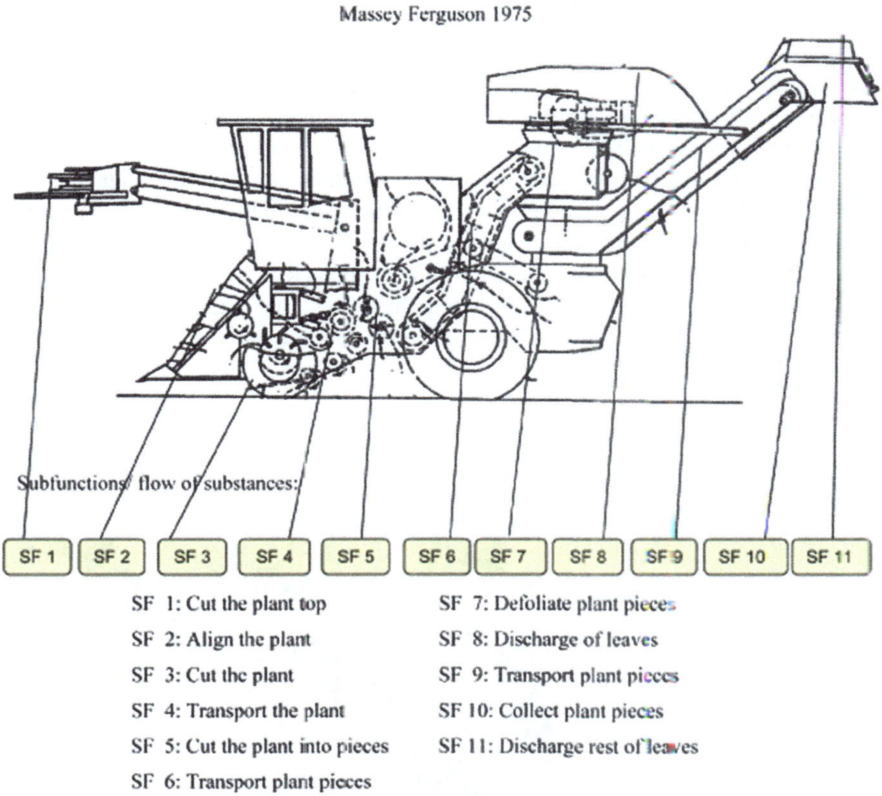

Fig. 9.1 Cross section through a sugarcane harvester from Patent Nr. DT 2630621 A1 Massey Ferguson. (Modified according to Meyer 2007)

of the deep cut, i.e., partly into the soil, soil particles are absorbed and get into the process as a whole – even for processing in the refineries. Leaves and tops are contaminated with soil. Parallel to the harvester an overloading vehicle is loaded directly. When it is full, it goes to the field edge and fills the trucks that supply the refineries (Meyer 2007).

Harvest losses of 10–15% are common (Almeida et al. 2009). Self-propelled harvesters can work with a speed of up to 7 km/h and a maximum slope of 12%.

This harvest technology has a maximum gross capacity of 110.18 t/h. The machine price is approximately R$ 600,000 by agricultural machinery manufacturers Brastoft and Santal (Garcia and Silva 2009).

The technology assumes the existence of overloaded and transporting vehicles to bring the harvested sugarcane to the mills and a low-loader to bring the harvester to the field. This kind of harvest chain is typical for large sugarcane producers. Any innovation in harvesting technology has to fit into the harvesting chain as described above.

Several field factors are relevant for the success of sugarcane production, highlighting among them: the leveling of the soil, length of the plots, productivity and homogeneity of the sugarcane, characteristics of cultivars, quality of operation, operator training, and work speed (Meneguetti et al. 2010). For the municipality of Campos dos Goytacazes (RJ), Solano (2011) defined the factors limiting the mechanization of sugarcane harvest and elaborated on suitable condition. For Campos de Goytacazes, heterogeneity is a major impeding factor; most of the plots do not comply with the standardization recommended for mechanization of sugarcane harvesting. It was observed that with only 7.18% of the recommended length of planting rows fields are by far too small to achieve the optimal operational performance of a harvester of raw sugarcane as a function of the speed (Ripoli et al. 1999).

The self-propelled harvester as shown in the Fig. 9.2 has established itself on the market. Weaknesses of the technology were defined and solutions optimized (Table 9.5).

Fig. 9.2 First concept of pulled sugarcane harvester

The "Cologne solution" approach combines the framework conditions with the weakness of the state of the art to develop an adequate harvesting technology primarily for SME but also large producers. This approach can be embedded in the state of the art of the existing manual harvest chain. The machine is pulled and propelled by an external source of energy like a tractor (Fig. 9.2). This development study has already been tested in sugarcane under real conditions (Table 9.6).

Table 9.6 Shortcomings of conventional sugarcane harvesting technology and development of a new concept

Identified shortcomings of the state of the art	"Cologne solution"
High power between 129 and 298 kW (Coelho 2007; Dalmolin 2007; Gouveia 2007)	Tractor possible from 112 kW
High weight – Up to 20 tons own weight and chassis mostly consisting of chains so that the soil fertility is adversely affected by enormous soil pressure (Magro 2005). Numbers of harvest seasons are reduced by soil pressure and injuries of rootstocks (Barreti and Inojosa 2014). This affects higher replanting rate and higher costs	Reduction of weight to maximal 5 tons and driving on wheels to reduce soil pressure and the effects such as higher replanting rate
Transport mainly on low-loaders (Barreti and Inojosa 2014)	Simplified transport from farm-to-field and field-to-field (normally without additional truck/low-loader)
Rotating knifes - Vulnerability to repairs because of the rotating blades, as they cut deep into the ground blades have to be replaced every working day - Since a lot of soil is absorbed in the process, the wear is very high so that the machines have to be completely reinforced at the end of the harvest season (Magro 2007)	Oscillating knifes Gentle harvest for material by choice of knifes and material flow without soil particles and abrasion
Leaf mass is contaminated with soil particles, so that it can be used only with difficulty or without further use (Paes 2014)	Laying down of leaves and stalks on the ground almost without soil contamination
Work safety by rotating knife systems	Higher work safety by oscillating knife systems
Social factor: High training required	Normal knowledge about tractor handling is necessary
Driving/harvesting only possible up to a slope of 12% (Meyer 2007)	Harvesting possible over a slope of 12%; limiting factor is the tractor
The rots are cut approximately 30 cm long, thereby reducing the juice loss and low storage capacity of the crop (Meyer 2007)	Harvesting of entire cane stalks; therefore minimum of juice loss
Harvested sugarcane is not appropriate for replanting (Veiga 2014)	By gentle defoliation of sugarcane stalks by propylene brushes, i.e., not vulnerable to insolation and high temperatures, acidic juice etc., and the harvest of entire stalks, the rhizomes of the stalks are not damaged and can be used for replanting

Fig. 9.3 Sugarcane harvesting test bench, pulled by a Ford tractor of the series 6610 (own picture). The test bench was tested by COAGRO and UFFRJ in Campos by ITT/TH Köln and TM Technik from Germany

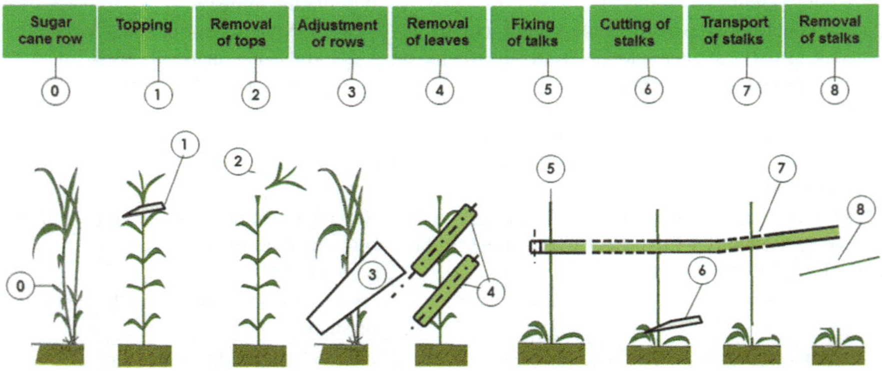

Fig. 9.4 Function and schematic diagram of "Cologne solution" concept (own schematic diagram)

The "Cologne solution" approach for harvesting sugarcane was tested in sugarcane fields in RJ for further development. Long-term field tests are planned in the state of RJ close to Campos dos Goytacazes.

The latest state of development is shown in Fig. 9.3:

Figure 9.4 shows the various sub-functions for the material flow:

The schematic shows six assemblies of the machine and eight steps of the material flow. During the harvesting process, the tops fall into the next row of plant stock and the dry leaves on the ground into the harvested row, and the stalks are transported to a bunker to build bundles.

9.7 Challenges for Developing and Implementing Adequate Harvest Technology

Mattos (1992) emphasizes the importance of planning from the choice of the plantation area to the slope of less than 12%, soil preparation, removal of physical barriers (stumps, stones, etc.), planting, and to row spacing.

Ripoli and Ripoli (2005) point out that a better planning of the area is needed, starting with the definition of the fields, their format, and the carriers, according to the relief and soil of the new or retirement area in order to achieve a better result in the mechanical harvesting of raw cane. According to the same authors, the implementation of mechanized harvest requires a new management model, initiated by soil preparation and systematization of the area. The plots should be rectangular and have a minimum of 600 m length. Care should also be taken of the parallelism between rows and spacing in the planting which cannot be less than 1.50 m to avoid trampling the rake.

Only an adjustment to the possibilities of technological limits and almost doing groundwork from the agriculture, e.g., arranging and creating fields, to the agricultural engineering is vital to make technological progress possible.

References

Agrolink (2017) Cotações. https://www.agrolink.com.br/cotacoes/diversos/cana-de-acucar. Accessed 13 Jul 2017

Almeida BR, Ripoli TCC, Neves EM (2009) Avaliação das perdas de cana-de-açúcar na colheita mecanizada em diferentes condições de sistematização de terreno. http://www.usp.br/siicusp/Resumos/17Siicusp/resumos/2773.pdf. Accessed 02 Apr 2017

Asflucan – Associação dos Plantados de Cana (2013) Informações sobre as propriedades Cadastradas. Unpublished excel data file

Asflucan Associação dos Plantados de Cana (2014) Audiência Pública: A crise do setor canavieiro Fluminense. http://www2.camara.leg.br/atividade-legislativa/comissoes/comissoes-permanentes/capadr/audiencias-publicas/audiencias-publicas-2014/audiencia-publica-10-de-dezembro-de-2014-asflucan. Accessed 23 Mar 2017

Bakker H (1999) Sugarcane cultivation and management. Springer US, New York, pp 3–8

Barbosa MH, da Silveira LC (2015) Breeding program and cultivar recommendations. In: Santos F, Borém A, Caldas C (eds) Sugarcane: agricultural production, bioenergy, and ethanol. Elsevier. pp 241–255. https://doi.org/10.1016/B978-0-12-802239-9.00011-6

Barreti LM, Inojosa TLC (2014) Employees of Asflucan. Semi-structured expert interview (14 March 2014)

Bastos AdC, Moraes MAF (2014) Perfil dos fornecedores de cana-de-açucar na região centro-sul do Brasil. http://www.iea.sp.gov.br/ftpiea/publicacoes/ie/2014/tec1-0414.pdf. Accessed 04 Mar 2017

Coagro (2013) Custos de 1 há de cana de acucar 04 de dezembro de 2013.Unpublished table
Coelho JL (2007) Inovações Tecnológicas na Colheita Mecanizada da cana-de-açúcar. Paper presented at the 24 symposium of Agroindústria da cana-de-açúcar de Alagoas
Dalmolin P (2007) Santal – Linha de produtos. Paper presented at the 24 symposium of Agroindústria da cana-de-açúcar de Alagoas
van Dillewijn C (1952) Botany of sugarcane. The Cronica Botanica Co., Book, Department, Waltham
ELLA – Evidence and Lessons from Latin America (2012) Policy Brief, from manual to mechanical harvesting: reducing environmental impacts and increasing cogeneration potential. https://assets.publishing.service.gov.uk/media/57a08a6340f0b64974000590/120907_EV_BraEthPro_BRIEF1.pdf. Accessed 26 Jul 2017
FAOSTAT (2017) Crops. http://www.fao.org/faostat/en/#data/QC. Accessed 02 May 2017
Ferreira JM (2014) Pesagro-Rio, Campos dos Goytacazes, expert interview (31.03.2014)
Garcia RF, Silva LS (2009) Avialiação do corte manual e mecanizado de cana-de-açúcar em Campos dos Goytacazes, RJ. http://www.seer.ufv.br/seer/index.php/reveng/article/viewFile/67/114. Accessed Mar 08 2015
Gouveia RP (2007) Marketing CASE IH. Paper presented at the 24 symposium of agroindústria da cana-de-açúcar de Alagoas
Governo do Estado RJ (1992) Lei n. 2049, de 22 de dezembro de 1992. http://alerjln1.alerj.rj.gov.br/CONTLEI.NSF/b24a2da5a077847c032564f4005d4bf2/485fe39fc6e8451e0325651b0068e926?OpenDocument. Accessed 25 Oct 2017
Governo do Estado RJ (2011), Lei n. 5990 de 20 junho de 2011. http://alerjln1.alerj.rj.gov.br/CONTLEI.NSF/bff0b82192929c2303256bc30052cb1c/e3a21f49c6520869832578bc005e6b97?OpenDocument. Accessed 25 Oct 2017
IBGE – Instituto Brasileiro de Geografia e Estatística (2017) LSPA – Levantamento sistemático da produção agrícola, janeiro 2017. ftp://ftp.ibge.gov.br/Producao_Agricola/Levantamento_Sistematico_da_Producao_Agricola_[mensal]/Fasciculo/lspa_201701.pdf. Accessed 20 Nov 2017
James G (2004) Sugarcane, 2nd edn. Blackwell Publishing Company, Oxford
Magro JA (2005) Perdas de cana na colheita mecanizada. Paper presented at the 23 symposium of Agroindústria da cana-de-açúcar de Alagoas
Magro JA (2007) Perda de produtividade. Necessidades para colher cana mecanicamente. Paper presented at the 24 symposium of Agroindústria da cana-de-açúcar de Alagoas
Mattos JR (1992) Planejamento da lavoura para implementação e operacionalização mecânica da cana-de-açúcar. Encontro Técnico 8. Ribeirão Preto. São Paulo
Meneguetti CC, Mezaroba S, Groff AM (2010) Fatores relacionados ao cultivo da cana-de-açúcar. http://atividaderural.com.br/artigos/56019d9740759.pdf. Accessed 03 Oct 2017
Meyer T (2007) Zuckerrohrvollernter – Stand der Technik, Diplomarbeit 959/07, Fachhochschule Köln, Köln 2007
Moreira FL (2006) Analise da viabilidade econômica da mecanização da colheita de cana-deaçucar na região centro sul do Brasil. p 77. Monografia (Engenharia de produção) – Escola Politecnica da Universidade de São Paulo. http://pro.poli.usp.br/wp-content/uploads/2012/pubs/analise-da-viabilidade-economica-da-mecanizacao-da-colheita-de-cana-de-acucar-na-regiao-centro-sul-do-brasil.pdf. Accessed 05 May 2017
Nastari PM (1983) The role of sugar cane in Brazil's history and economy. Iowa State University, Retrospective Theses and Dissertations 8948. http://lib.dr.iastate.edu/rtd/8948. Accessed 20 Nov 2017
Paes FR (2014) Director – President of COAGRO. Semi-structured expert interview (10 March 2014)
Ripoli TCC, Ripoli MLC (2005) Biomassa de cana-de-açúcar: Colheita, energia e ambiente, 2nd edn. ESALQ/USP, Piracicaba, p 302p
Ripoli TCC, Nery MS, De León MJ et al (1999) Desempenho operacional de uma colhedora em cana crua em função da velocidade de avanço. Engenharia Agrícola. Jaboticabal 19(2):199–207

Solano CS (2011) Avialicao técnica e economica dos sistemas de colheita da cana-de acucar no município de Campos dos Goytacazes. http://www.uenf.br/Uenf/Downloads/PRODVEGETAL_3434_1324572497.pdf. Accessed 03 Dec 2014

UNICA (2016) Setor sucroenergético do estado de São Paulo frente à crise econômica. http://www.unica.com.br/convidados/29063662920344564516/setor-sucroenergetico-do-estado-de-sao-paulo-frente-a-crise-economica/. Accessed 02 Jun 2017

Veiga CFM (2014) Professor de UFRRJ. Semi-structured expert interview (11 March 2014)

Viana AV, Ferreira JM et al (2017) Pesagro-Rio CANA-DE-AÇUCAR, http://www.pesagro.rj.gov.br/cana.html. Accessed 01 May 2017

Part II
Ecosystem Management and Biodiversity Conservation

Chapter 10
Integrating Biodiversity Conservation into Agroecosystem Management: Using Birds to Bring Conservation and Agricultural Production Together

Augusto João Piratelli, Fatima Conceição Marquez Piña-Rodrigues, and Claudia Raedig

Abstract The conversion of natural areas into agricultural systems for food and energy production has globally put biodiversity and ecosystem functions at risk. Recent research has focused on characterizing biodiversity and ecological functions, e.g., pollination, seed dispersal, and control of herbivorous arthropods, as well as on the best agricultural practices to consort the production and conservation of natural environments. In this context, birds can serve as models – both for the relatively good taxonomic resolution and ecological knowledge and for the range of ecosystem services that they consistently perform. We discuss the most sustainable practices for both the maintenance of biodiversity focusing on bird fauna and of ecological processes, and thus for human well-being. In order to close ranks between biodiversity conservation and agricultural production, biodiversity can be incorporated directly or indirectly into agroecosystem management, either to measure or monitor ecological parameters, to identify priority areas for conservation, or to identify suitable native plant species for ecological restoration or agricultural productivity. For the state of Rio de Janeiro (RJ), different ecosystem management solutions are developed to link biodiversity conservation with the safe operation of agroecosystem and sustainable production, taking into account Brazil's complex legal-environmental framework.

A. J. Piratelli (✉) · F. C. M. Piña-Rodrigues
Departamento de Ciências Ambientais CCTS, Universidade Federal de São Carlos, Sorocaba, SP, Brazil
e-mail: piratelli@ufscar.br; fpina@ufscar.br

C. Raedig
TH Köln – University of Applied Sciences, Institute for Technology and Resources Management in the Tropics and Subtropics (ITT), Köln, Germany
e-mail: claudia.raedig@th-koeln.de

U. Nehren et al. (eds.), *Strategies and Tools for a Sustainable Rural Rio de Janeiro*, Springer Series on Environmental Management,
https://doi.org/10.1007/978-3-319-89644-1_10

Keywords Biodiversity-friendly agricultural production · Avifauna · Fragmentation · Ecosystem services · Rio de Janeiro

10.1 Agriculture in the Anthropocene

Humans have been responsible for dramatic environmental changes threatening the ecological and climate balances and the survival of countless biological species (Foley et al. 2005). This new epoch of the Earth, under such huge human influence, has been called Anthropocene – the human age (Caro et al. 2011; Corlett 2015; Sarrazin and Lecomte 2016). There are now billions of people and a huge demand of food and energy in a scenario of significant climate change and water and food scarcity (FAO et al. 2015).

Agriculture has experienced robust expansion in the last centuries, and cultivated areas occupy nearly 38% of the planet's surface (Renwick et al. 2014). Several factors have contributed to accelerate agricultural development in Latin America, such as public policies to encourage and increase new technologies and the international demand (Barbier 2004). In this region, the environmental consequences of land use are dramatic because nearly 70% of the world's biodiversity lives in tropical forests (Johnson et al. 2011; Beston et al. 2016).

Large-scale agriculture is responsible for severe environmental changes and a direct agent in degradation, fragmentation, and destruction of remnants of native vegetation associated biological communities (Pimm and Raven 2000; Hooper et al. 2005; Klink and Machado 2005). Agroecosystems, seminatural or modified natural systems for food and agricultural production purposes (FAO 2009), often develop into monocultural landscapes. Along with global climate change, land use has led to biological simplification and homogenization (Domínguez-López and Ortega-Álvarez 2014) putting at risk the stability of ecosystems (McKinney and Lockwood 1999; Frishkoff et al. 2014) and globally affecting the respective local biodiversity (Newbold et al. 2015). Indeed, loss of biodiversity is one of the most dramatic results of human environmental interventions, largely associated with habitat loss (Gonthier et al. 2014).

The expansion of the agricultural frontier worldwide is responsible for the decline of about 60% of birds threatened by extinction (Johnson et al. 2011). Bird species are affected in different ways by agriculture, depending on their sensitivity and tolerance of habitat degradation, population dynamics, and behavioral characteristics (Barrett et al. 1994; Alexandrino et al. 2016). The responses range from those that benefit and increase their populations to those that disappear locally due to failure of finding food, refuge, or nesting sites (Sekercioglu 2012). In the first case, species are more tolerant to dry climates, have low dependence on forest habitats, and/or have a broad diet (Fig. 10.1), whereas at the other extreme, species have high dependence on forests and specific food requirements and little tolerance to open areas (Fig. 10.2) (Petit and Petit 2003; Lindell et al. 2004; Frishkoff et al. 2014).

More than half of the tropical avifauna is considered forest specialists, while only 3% are "specialists" in agricultural areas – i.e., typical species of grasslands

Fig. 10.1 Southern lapwing (*Vanellus chilensis*), a typical species of open areas (Sick 1997) whose populations tend to be large in agricultural landscapes (Photo: AJ Piratelli)

Fig. 10.2 A female spot-billed toucanet (*Selenidera maculirostris*) (**a**) and white-bearded antshrike (*Biatas nigropectus*) (**b**). Both are endemic to Atlantic forest and may disappear with the conversion of pristine forest into agriculture lands. The spot-billed toucanet nests in tree cavities; it is mainly frugivore and considered an important seed disperser (e.g., Galetti et al. 2013). The white-bearded antshrike is a rare, vulnerable species, often associated with bamboos, where it feeds on arthropods (del Hoyo et al. 2016) (Photos: Marcos A Melo)

and open areas (Sekercioglu 2012) – and a turnover from the more to the less specialized bird species is expected, with possible changes in ecological functions and, therefore, in ecosystem services (Sekercioglu 2012; Newbold et al. 2013). Yet, nearly one-third of the world's birds may be found in agricultural areas regularly or sporadically, often providing important ecosystem services such as pest control, pollination, and seed dispersal (Sekercioglu 2012; Cassano et al. 2016). Indeed the interaction of wildlife with agricultural systems can positively or negatively influence food production. On the one hand, birds and rodents are recognized for negatively impacting crop production in California, USA (Gebhardt et al. 2011). On the other hand, invertebrate pests cause an annual loss of nearly $360 million in Australia's six main grain crops (Murray et al. 2013). Birds provide many ecosystem services such as pest control and pollination (Wenny et al. 2011), and it has been shown that they contribute positively to the coffee production in the region of

Kilimanjaro, Tanzania (Classen et al. 2014). There is still a large gap in understanding how these processes occur together. Some studies have addressed cost-benefit trade-offs in the same context, but most of them neglected the costs imposed by loss of birds or insects and the associated ecosystem services (Peisley et al. 2015 and references herein).

In practice, only a small part of the world can – or could – be legally protected in large pristine areas. Hence, conservation efforts need to go further, focusing on the ecological value of farming systems, thoughtful of agricultural areas as habitats for wildlife (Petit et al. 1999). Thus, agricultural alternatives have been proposed to reduce environmental impacts and convert farmlands into biodiversity-friendly areas, where the idea is to merge food production and protection of biodiversity (Johnson et al. 2011). The search for efficient conservation strategies involves (1) decreasing management intensity while reducing pesticides and fertilizers at local level and (2) incorporating natural or seminatural areas into the farming systems at regional level (Johnson et al. 2011; Wenny et al. 2011).

Such multiple-scale conservation strategies should be jointly managed because they affect biodiversity differently, depending on the taxonomic group (Gonthier et al. 2014). Nearly 63% of all animal species living in agricultural areas depend on seminatural habitats, and the presence of native forests near plantations, e.g., riparian forests, plays an important role for regional diversity and abundance of birds (Duelli and Obrist 2003; Domínguez-López and Ortega-Álvarez 2014) and provision of additional ecosystem services (Douglas et al. 2014). Small shade coffee plantations may harbor bird communities very similarly to communities in patches of secondary native vegetation, acting as corridors among them, contributing to landscape conservation at regional level (Hernandez et al. 2013). In Ethiopia, where *Coffea arabica* is native (Buechley et al. 2015), many similarities among bird communities of shade coffee plantations and native forests were found. In Central America such corridors of shade coffee plantations and riparian forest corridors were assessed as valuable for bird conservation, while sugarcane and Caribbean pine plantations were of no value for biodiversity conservation (Petit et al. 1999).

Agroforestry systems (AFSs), i.e., mixed arrangements of forest species and agricultural crops that provide biological interactions among their components (Somarriba 1992), house controllers of natural pests such as insects, birds, and bats (Nyffeler and Benz 1987; Morrison and Lindell 2012; de la Peña-Domene et al. 2014). These are mobile organisms. Therefore, the management of agricultural systems toward biodiversity-friendly landscapes should produce a more robust effect for their conservation and the maintenance of the services they deliver (Beecher et al. 2002; Cassano et al. 2016). Indeed, wild birds can control pests on farms as "functional insectivores," and this effect can be potentiated in mixed plantations rather than monocultures (Jones et al. 2005). Predators and their prey build complex networks of interactions, and predators consume not only other predators but also herbivores. Understanding this network is crucial to assess changes in ecological functions such as pest control. For instance, studies in coffee plantations, using experiments excluding birds, lead to an increase in abundance of herbivorous arthropods, which may lead to a progress in foliar herbivory. Therefore, the presence of arthropod-eating birds may benefit long-term coffee production – and other farming systems – by preventing pests (Greenberg et al. 2000; Karp and Daily 2014).

Other initiatives that will benefit biodiversity include educational and engaging policies, strengthening communication between science and citizens, and encouraging the consortium between food production and the environment, including the ecosystem services (Johnson et al. 2011 and references herein). Investments are needed in public policies, programs, and actions focused on the perception and knowledge of farmers; this will contribute to the adoption of practices that support the conservation of biodiversity in agricultural areas (Silva-Andrade et al. 2016). The term "gardenification" has been proposed to synthesize the paradigm shift from preserving only large areas of continuous forest to people properly managing their gardens or small pieces of private property to favor biodiversity (Janzen 1999). The expected prospects are reduction of using pesticides, production of healthier foods, and the increase of family income (see Espinel et al. 2018; Fischer et al. 2018; Eissa et al. 2018 in this book).

10.2 Public Policies and Alternatives for Restoration and Conservation

More than half of Brazil's native vegetation is found on private properties, and it is therefore inevitable that this vegetation should be protected for conserving and maintaining biodiversity and the ecosystem services, e.g., stock of billions of tons of carbon (Soares-Filho et al. 2014). Since 1934, when the first version of the National Forest Law, the so-called Forest Code, was introduced, also agricultural owners have been required by law to maintain parts of their lands as legal reserves (RLs) (Santiago et al. 2017) and to protect riparian forest, classified as permanent preservation areas (APPs). However, Article 4 of the current Forest Law 12,651/12 (Brasil 2012) has considerably reduced the protection limits of APPs, without conducting technical studies (Metzger 2010), which can put at risk not only biodiversity but also the future of agricultural production (Develey and Pongiluppi 2010).

The RLs and the APPs were fundamental as a stepping stone for animals and plants (Galetti et al. 2010), capable to connect conservation units through agroecosystems. Based on this, and in order to estimate the real total area of RLs and APPs, the Forest Law established the Rural Environmental Registry (CAR) (Brasil 2016), a declarative electronic system with a georeferenced database containing information of rural properties which is obligatory to use for landowners and a prerequisite to access rural credits (Santiago et al. 2017). The Environmental Regularization Program (PRA) regulated by the Decree 8235 dated May 5, 2014 (Brasil 2014a, b), has obliged rural owners with APPs and RLs in disagreement with the law to carry out restoration projects according to CAR. Although the system of land regularization represented by CAR and PRA can be a procedure that legalizes areas that had been deforested before 2008 (Azevedo and Saito 2013), it is an opportunity to reincorporate vegetation restoration in agroecosystems at the same time.

In addition to reducing the area legally required for RLs, the current Forest Law allows the use of exotic species in 50% of the area (Article 38, §3) to promote commercial use and thus the reduction of restoration costs. Without proper regulation, this situation can lead to the restoration of deforested RLs exclusively with fast-growing exotic species with commercial value such as *Eucalyptus* species or even to a partial recomposition and abandonment of the rest of the RLs. However, if regulated for incorporating exotic species together with endemic, nonexotic commercial, and noncommercial ones, RL restoration has enormous potential for the introduction of productive forest systems for restoring forest fragments and for establishing connectivity between separated areas. Instead, if the debts of RLs and APPs are not offset, the lack of restoration will have implications for the management of agroecosystems with loss of ecosystem services such as carbon sequestration, reduction of pest and disease control, and lack of pollinators (Galetti et al. 2010) even for cultivated species.

There is no consensus on the area to be restored in Brazil. Approximately 21 ± 0.6 Mha (Soares-Filho et al. 2014) to 44.3 Mha (Gouvello 2010) of RLs need to be restored in Brazil. On the one hand, estimates made by *Embrapa Monitoring* indicated figures of the same magnitude, 20 to 40 million hectares, of degraded lands needed to be cultivated and restored. On the other hand, data from the Ministry of the Environment projected in 2016 that 140 million hectares were degraded and need restoration representing 16.4% of the total area of Brazil. This is an enormous potential for tropical silviculture and biodiversity conservation.

Based on the CAR (Brasil 2016), Rio de Janeiro (RJ) state has 36,831 rural properties with RLs, APPs, and remnants of native forests in about 3.2 Mha. From this total, only 20.2% are exempted from restoration measures because they are smaller than four fiscal modules[1] (MF). However, 0.52% of the larger properties (> 1000 ha) occupy 63.2% of the registered area that need to be restored or conserved (Table 10.1).

Even if we estimated that only about 30% of RLs and APPs could be restored at national or state level, the introduction of commercial forest species in RLs can still represent an opportunity for the expansion of biodiversity-friendly agricultural production. This can be possible with the ecological restoration of the RLs, through a mix of native forest species, exotic plants, and plants providing non-timber forest products. Besides restoration, the mixed silvicultural system allows the diversification of production and of income, increases vegetation cover, and reduces restoration costs. In this sense, the National Vegetation Restoration Policy, established by Decree No. 8952/17 (Brasil 2017), provides financial and technical support for a 12.5 Mha land restoration until 2020 to fulfill Brazilian international commitments. Its implementation will be in accordance with the National Plan for Restoration of Native Vegetation (PLANAVEG), which established an initial goal of restoring 390,000 hectares of native forest in 5 years (Brasil 2014a, b). Among the restoration methods suggested by the PLANAVEG are the AFSs. Even the more simplified

[1] Fiscal module is an agrarian unit of measure that represents the minimum area required for rural properties to be considered economically viable (Landau et al. 2012).

Table 10.1 Number and area (ha) of properties registered and classified by social profile, number of fiscal modules[1] (MF), and size (ha) listed in the Rural Environmental Registry (CAR) (Serviço Florestal Brasileiro 2017; Brasil 2016) until November 2016 in the state of Rio de Janeiro

	Description	Number of rural properties	Area of registered rural properties (ha)
Social profile of rural properties	Traditional communities	5	1641
	Settlements of agrarian reform	61	51,235
	Registered rural properties	36,765	3,[illegible]86,284
	Total	**36,831**	**3,239,161**
Fiscal modules (MF)	0 a 4 MF	31,656	648,417
	4 a 15 MF	4004	530,623
	> 15 MF	1171	2,023,720
	Total	**36,831**	**3,202,760**
Size classes of rural properties (há)	0–100	32,889	
	100–500	3247	
	500–1000	326	
	>1000	189	
	Total	**36,651**	

AFSs are efficient stepping stones, increasing landscape connectivity in a more advantageous way than monocultures (Uezu et al. 2008).

The concept of ecosystem services includes provisional, regulating, cultural, and supporting services (de Groot et al. 2010); since publication of the Millennium Ecosystem Assessment (2005), the concept of restoration incorporated the recovery of forest structure, functions, and ecosystem services also at landscape level. In an overall sense, ecosystem services are generated by ecosystem functions and are defined as "the capacity of ecosystems to provide goods and services that satisfy human needs, directly and indirectly" (de Groot et al. 2002). Based on this conceptual framework, the AFSs are an opportunity to aggregate provision and production functions sensu de Groot et al. (2002) by protecting watershed, soil quality, and agricultural and forestry production and by supporting services such as connecting forest fragments, safeguarding wild pollinators, reducing use of pesticides (Power 2010), and thus contributing to biodiversity conservation (Tscharntke et al. 2012).

The restoration of RLs and APPs with complex AFSs incorporating the diversity of species as well as timber and non-timber forest products can reduce land-use pressures and enhance rural livelihoods by the multiple use of the forest (Bhagwat et al. 2008). As an example, the introduction in AFSs of *Euterpe edulis* Mart., a keystone endangered species (Rother et al. 2009), can increase the attraction of fauna (Reis et al. 2000) and, at the same time, add economic value through the sustainable use of its fruits (da Silva 2011), contribute to its *on-farm* conservation (Piña-Rodrigues et al. 2013), and reduce the exploitation pressures in natural areas (Muler et al. 2014). The cultivation of *E. edulis* is particularly feasible in areas of riparian forest (APPs) and in the Serra do Mar and Mantiqueira in the Atlantic Forest of RJ, where climatic conditions favor species establishment and management (Calvi and Piña-Rodrigues 2005).

In general, AFSs occupy small-sized properties embedded in fragmented landscapes, and despite their richness, they use few species of conservation interest (Noble and Dirzo 1997). Even simplified systems such as home gardens, mixed shade coffee, and *Cocus nucifera* systems could also contribute to the conservation of local biodiversity (Bhagwat et al. 2008). Due to its spatial and temporal structure, an AFS provides habitat for disturbance-tolerant species besides connectivity as corridors associated with production functions (Jose 2009, 2012). This was demonstrated in the APP and RL areas of Paraty (RJ) in a 5-year managed AFS with biomass uptake, litter, and nutrient contribution similar to a secondary forest, evidencing the fast recovery of the functions of habitat and production as well as corresponding ecosystem services (Souza et al. 2016).

Both the Forest Law and the states' resolutions such as São Paulo (Secretaria de Meio Ambiente 2008) and RJ (SEA/INEA 2016) defined criteria to enable restoration using agroforestry in APPs and RLs. Nonetheless, at state level questions regarding the presence of exotic cultivated species restrict their use and thus should be reviewed. In RJ, many traditional communities and settlements are small properties. Together they occupy about 1.66% (52,876 ha; see Table 10.1) of the registered areas in the CAR and are exempted from mandatory restoration. However, even for these properties, the AFSs are potentially viable and represent an opportunity to generate income from restoration of RLs and APPs.

Since the 1990s, AFSs have been considered as a carbon sink and a potential offset for greenhouse carbon emission produced by traditional agricultural practices such as pasture, large-scale agriculture, and even the simple ruminant-based silvopastoral AFS (Schroeder 1994; Dixon 1995; Nair 2012). Light and "conventional" AFSs are associated with few or just one forest species planted with agricultural crops, while complex agroforestry systems with many forest species and crops are so-called "forestry" AFSs (Miller 2009). Complex and biodiverse AFSs are common in small to medium properties, while simplified AFSs were planted in large properties or associated with few fruit species such as *Mangifera indica* L. or *Annona muricata* L., largely planted in the north of RJ (Lamônica and Barroso 2008). Therefore, for large properties, different strategies need to be established. As a consequence, the plantation of native forest species with silvicultural potential and environmental compensation via carbon credits seems to be one of the paths pointed out to conciliate biodiversity conservation with forest restoration and agricultural production.

In RJ, only 3.4% of the population live in the rural area dominated by 10 ha properties and 28 "quilombola" communities; until 2010 it was the Brazilian state with most water use conflicts (IBGE 2011). The greatest number of rural properties with AFSs is concentrated in the south of RJ state, encompassing the municipalities of Resende, Piraí, and Japeri, and in the north with the municipalities of Campos dos Goytacazes, São Fidelis, São Sebastião do Alto, Cambuci, and Cardoso Moreira (Fig. 10.3). Together they represent 20.4% of the total rural properties and 37.6% of

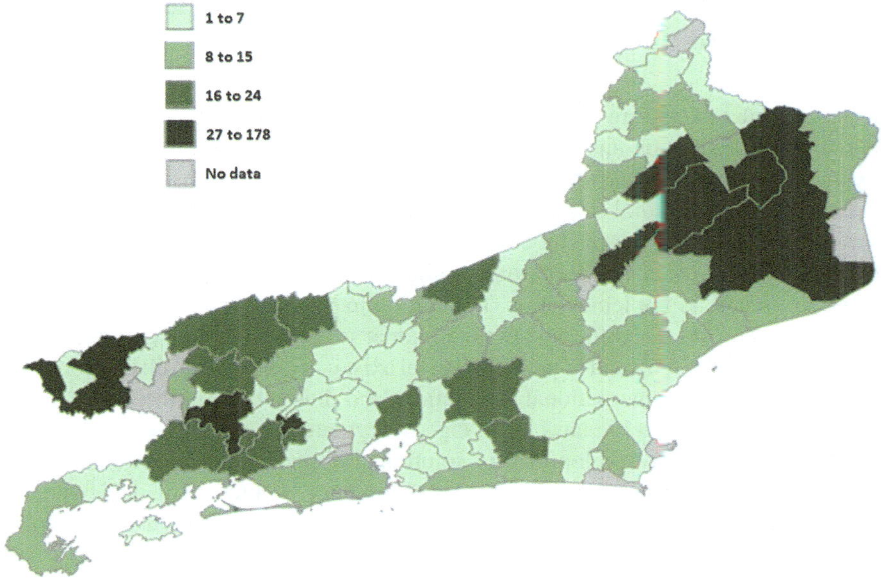

Fig. 10.3 Number of properties with agroforestry systems (AFSs) in the state of Rio de Janeiro, Brazil. Each color of polygon corresponds to the number of rural properties with agroforestry systems, according to the legend in the upper right. (Data from 2010; IBGE 2011)

the rural area of the state (Brasil 2016). The northern mesoregion is dominated by 45.7 to 95% of the properties larger than 15 MF (>1000 ha) (Brasil 2016; Serviço Florestal Brasileiro 2017) the majority with sugarcane and pasture (Santos and Lima 2015). In the mesoregion there are properties of 4–15 fiscal modules (MF) (27.9%) and exceeding 15 MF (47%), except for Japeri, with dominance of areas (73.6%) less than 40 ha (<4 MF) (Landau et al. 2012; Brasil 2016). For these regions, the use of simplified or complex silvicultural systems should have a strategy focused on the agricultural and forestry productive issue associated with the need of restoration and connection between the few remaining secondary forest fragments (SOSMA and INPE 2014). Recommended silviculture practices include species such as the Amazonian native *Hevea brasiliensis* L. (seringueira, Oliveira et al. 2009) and the exotic ones, *Toona ciliata* M. Roem, *Acacia mangium* Willd, *Eucalyptus camaldulensis* Dehnh (SEPLAG 2011) consociated with fruit species (Santos and Lima 2015), and live fences with leguminous species as sources of protein for dairy cattle (Lamônica and Barroso 2008). The restoration actions for the reconnection of forest fragments need to be also implemented in RLs and APPs. Despite the introduction of the CAR (Brasil 2016) by the Forest Law 12,651/12 in order to monitor RL and APP restoration and the forest commercial use of RLs, it is not possible to predict if legal mechanisms will indeed prevent the formation of new environmental liabilities because they still depend on regulation and institutional and financial support (Santiago et al. 2017).

10.3 Biodiversity-Friendly Approaches for Rural Rio de Janeiro

Management options are required to counter the prevailing dichotomy between agricultural production and biodiversity conservation and to bring them together. Such options include measures to maintain and connect areas with near-natural vegetation in an agricultural matrix, measures to monitor ecosystem quality, or measures to allow for more biodiversity-friendly production. Furthermore, the development of management options should consider Brazil's legal environmental jurisdiction and its implementation. In addition to the jurisdiction at national level such as the Forest Law, different laws at regional scale, such as the Law of the Mata Atlântica (Brasil 2006), and at municipal level exist (specific to each city). Moreover, Brazil is member of international conventions and thus subject to their legal specifications. Besides, specific laws exist, e.g., to regulate the cultivation of individual plant species. For the development of solutions for environmental problems, relevant jurisdiction must be integrated. The following chapters will deal with the development of biodiversity-friendly management options in different environmental and different legal context.

In parallel to the top-down regulation for the protection of the remaining forest fragments in Brazil, also a voluntary option to protect private forest area exists. Since between 80 and 90% of the remaining forest fragments of the Atlantic Forest are predicted to be in private hands (Ribeiro et al. 2009; Rodrigues et al. 2011), the establishment of such private protected areas is particularly important for areas in which few alternative larger forest fragments exist, e.g., in Northwestern RJ. As the process to create private protected areas is complex and not well known by land owners, and due to the fact that private forest areas are usually small, Chap. 11 (Guzmán Wolfhard and Raedig 2018) will develop a strategy to systematically identify priority areas for establishing and for linking existing private protected areas.

In the next chapter, an approach to assess hydromorphological quality of waterbodies will be transferred from Germany to RJ. This approach uses physical characteristics of waterbodies as well as the occurrence of region-specific plant species as indicators for habitat quality. The research scrutinizes the applicability of the method in tropical regions as well as required adaptations for standardized implementation in RJ.

For degraded areas, irrespective whether these areas are forest or pasture areas, possible use options are ecological restoration or silvopastoral systems. Whereas there are numerous examples of the use of nonnative species for those measures, which are well known for specific traits such as nitrogen fixation, fast growth, and simple cultivation, only a few studies on the use of native species exist for Brazil. In Chapter 15, the suitability of species native to the Atlantic Forest for restauration and rehabilitation measures will be discussed.

The most famous tree of Brazil, the pau-brasil (*Paubrasilia echinata* Lam.), will be at the heart of Chap. 16. This tree, which occurs exclusively in the Atlantic Forest, has become rare in the wild due to decades of commercial exploitation.

Subject to a law specifically dedicated to this one species, and subject to further jurisdiction at international, national, and regional scale, the potential of pau-brasil as plantation tree in RJ and adjacent states is explored.

All research presented in this section has an applied character with the aim to produce replicable biodiversity-friendly management options suitable for application in RJ state. The unique advantage of research in the frame of INTECRAL project is the partnership with the well-established implementation partner Rio Rural Program, both at regional (RJ state) and local (municipal) scale, and with administrative as well as control functions (Hissa et al. 2018). This allows for the tailor-made development of management options for RJ to be utilized by the Rio Rural Program in the future.

References

Alexandrino ER, Buechley ER, Piratelli AJ et al (2016) Bird sensitivity to disturbance as an indicator of forest patch conditions: an issue in environmental assessments. Ecol Indic 66:369–381. https://doi.org/10.1016/j.ecolind.2016.02.006

Azevedo AA, Saito CH (2013) O perfil do desmatamento em Mato Grosso, após implementação do licenciamento ambiental em propriedades rurais. Cerne 19(1):111–122

Barbier EB (2004) Agricultural expansion, resource booms and growth in Latin America: implications for long-run economic development. World Dev 32(1):137–157

Barrett GW, Ford H, Recher HF (1994) Conservation of woodland birds in a fragmented rural landscape. Pac Conserv Biol 1:245–256

Beecher N, Johnson RJ, Brandle JR et al (2002) Agroecology of birds in organic and nonorganic farmland. Conserv Biol 16(6):1620–1631

Beston JA, Diffendorfer JE, Loss SR et al (2016) Prioritizing avian species for their risk of population-level consequences from wind energy development. Plos One 11(3):e0150813

Bhagwat SA, Willis KJ, Birks HJB, Whittaker RJ (2008) Agroforestry: a refuge for tropical biodiversity? Trends Ecol Evolut 23(5):261–267

Brasil (2006) Lei 11428. http://www.planalto.gov.br/ccivil_03/_ato2004-2006/2006/lei/l11428.htm. Accessed 06 May 2017

Brasil (2012) Lei 12651. Casa civil: Presidência da República http://www.planalto.gov.br/ccivil_03/_ato2011-2014/2012/lei/l12651.htm. Accessed 06 May 2017

Brasil (2014a) Plano Nacional de Recuperação da Vegetação Nativa-PLANAVEG. http://www.mma.gov.br/images/arquivo/80049/Planaveg/PLANAVEG_20-11-14.pdf. Accessed 06 May 2017

Brasil (2014b) Decreto 8235, Brasil: Diário oficial da união, May 5, Ano CLI no 83-A. http://www.planalto.gov.br/ccivil_03/_Ato2011-2014/2014/Decreto/D8235.htm. Accessed 08 May 2017

Brasil (2016) Cadastro ambiental rural. http://www.car.gov.br/#/. Accessed 01 May 2017

Brasil (2017) Lei 8952. http://www.planalto.gov.br/ccivil_03/_ato2015-2018/2017/decreto/D8972.htm. Accessed 01 May 2017

Buechley ER, Şekercioğlu CH, Atickem A et al (2015) Importance of Ethiopian shade coffee farms for forest bird conservation. Biol Conserv 188:50–60

Calvi GP, Piña-Rodrigues FCM (2005) Fenologia e produção de sementes de *Euterpe Edulis* – Mart em trecho de floresta de altitude no município de Miguel Pereira-Rj. Revista Universidade Rural: série Ciências da Vida 25(1):33–40

Caro T, Darwin J, Forrester T et al (2011) Conservation in the Anthropocene. Conserv Biol 26(1):185–188

Cassano CR, Silva RM, Mariano-Neto E et al (2016) Bat and bird exclusion but not shade cover influence arthropod abundance and cocoa leaf consumption in agroforestry landscape in Northeast Brazil. Agric Ecosyst Environ 232:247–253. https://doi.org/10.1016/j.agee.2016.08.013

Classen A, Peters MK, Ferger SW et al (2014) Complementary ecosystem services provided by pest predators and pollinators increase quantity and quality of coffee yields. Proc Biol Sci / The Royal Society 281(1779):20133148

Corlett RT (2015) The Anthropocene concept in ecology and conservation. Trends Ecol Evolut 30(1):36–41. https://doi.org/10.1016/j.tree.2014.10.007

Develey PF, Pongiluppi T (2010) Impactos potenciais na avifauna decorrentes das alterações propostas para o Código Florestal Brasileiro Impactos potenciais na avifauna decorrentes das alterações propostas para o Código Florestal Brasileiro. Biota Neotropica 10(4):43–46

Dixon RK (1995) Agroforestry systems: sources or sinks of greenhouse gases? Agrofor Syst 31(2):99–116

Domínguez-López ME, Ortega-Álvarez R (2014) The importance of riparian habitats for avian communities in a highly human-modified Neotropical landscape. Revista Mexicana de Biodiversidad 85(4):1217–1227

Douglas DJT, Nawanga D, Katebaka R et al (2014) The importance of native trees for forest bird conservation in tropical farmland. Anim Conserv 17(3):256–264

Duelli P, Obrist MK (2003) Regional biodiversity in an agricultural landscape: the contribution of seminatural habitat islands. Basic Appl Ecol 4(2):129–138

Espinel ML, Schlüter S, Resende CM (2018) Towards good agricultural practices (GAP) in smallholders' dairy production systems from an animal welfare perspective. In: Nehren U, Schlüter S, Raedig C, Sattler D, Hissa H (eds) Strategies and tools for a sustainable rural Rio de Janeiro. Springer International Publishing, Cham

FAO (2009) Glossary on organic agriculture. ftp://ftp.fao.org/docrep/fao/012/k4987t/k4987t.pdf. Accessed 08 May 2017

FAO, IFAD, WFP (2015) The State of Food Insecurity in the World 2015. Meeting the 2015 International Hunger Targets: Taking Stock of Uneven Progress. Rome, FAO

Fischer SB, Pedraza Luengas A, Schlüter S, Oliveira Antunes LA (2018) From design to implementation: a participatory appraisal for silvopastoral systems. In: Nehren U, Schlüter S, Raedig C, Sattler D, Hissa H (eds) Strategies and tools for a sustainable rural Rio de Janeiro. Springer International Publishing, Cham

Foley J, Defries R, Asner GP et al (2005) Global consequences of land use. Science 309(5734):570–574

Frishkoff LO, Karp DS, M'Gonigle KL et al (2014) Loss of avian phylogenetic diversity in neotropical agricultural systems. Science 345(6202):1343–1346

Galetti M, Pardini R, Duarte JMB et al (2010) Mudanças no Código Florestal e seu impacto na ecologia e diversidade dos mamíferos no Brasil. Biota Neotropica 10(4):47–52

Galetti M, Guevara R, Côrtes MC et al (2013) Functional extinction of birds drives rapid evolutionary changes in seed size. Science 340:1086–1091

Gebhardt K, Anderson AM, Kirkpatrick KN, Shwiff SA (2011) A review and synthesis of bird and rodent damage estimates to select California crops. Crop Prot 30(9):1109–1116

Gonthier DJ, Ennis KK, Farinas S et al (2014) Biodiversity conservation in agriculture requires a multi-scale approach. Proc R Soc B 281:9–14. https://doi.org/10.1098/rspb.2014.1358

Gouvello C (2010) Estudo de baixo carbono para o Brasil. Banco Internacional para Reconstrução e Desenvolvimento / Banco Mundial.http://siteresources.worldbank.org/BRAZILINPOREXTN/Resources/3817166-1276778791019/Relatorio_Principal_integra_Portugues.pdf. Accessed 12 April 2018

Greenberg R, Bichier P, Angon AC et al (2000) The impact of avian insectivory on arthropods and leaf damage in some Guatemalan coffee plantations. Ecology 81(6):1750–1755

de Groot RS, Wilson MA, Boumans RMJ (2002) A typology for the classification, description and valuation of ecosystem functions, goods and services. Figure 1: framework for integrated assessment and valuation of ecosystem functions, goods and services. Ecol Econ 41:1–20

de Groot RS, Alkemade R, Hein BL, Willemen L (2010) Challenges in integrating the concept of ecosystem services and values in landscape planning, management and decision making. Ecol Complex 7(3):260–272

Guzmán Wolfhard L, Raedig C (2018) Connectivity conservation management: connecting private protected areas. In: Nehren U, Schlüter S, Raedig C, Sattler D, Hissa H (eds) Strategies and tools for a sustainable rural Rio de Janeiro. Springer International Publishing, Cham

Hernandez SM, Mattsson BJ, Peters VE et al (2013) Coffee agroforests remain beneficial for Neotropical bird community conservation across seasons. PLoS ONE 8(9): e65101

Hissa H, Teixeira N, Costa M et al (2018) Sustainable rural development in Rio de Janeiro state: the Rio rural program. In: Nehren U, Schlüter S, Raedig C, Sattler D, Hissa H (eds) Strategies and tools for a sustainable rural Rio de Janeiro. Springer International Publishing, Cham

Hooper E, Legendre P, Condit R (2005) Barriers to forest regeneration of deforested and abandoned land in Panama. J Appl Ecol 42(6):1165–1174

del Hoyo J, Elliott A, Christie DA (2016) Handbook of the birds of the world alive, Barcelona: Lynx Edicions

IBGE – Instituto Brasileiro de Geografia e Estatística, 2011. Estatísticas meio rural- 2010. Ministério do Planejamento

Janzen DH (1999) Gardenification of tropical conserved wildlands: multitasking, multicropping, and multiusers. Proc Natl Acad Sci U S A 96(11):5987–5994

Johnson RJ, Jedlicka JA, Quinn JE, Brandle JR (2011) Global perspectives on birds in agricultural landscapes. In: Campbell W, López Ortíz L (eds) Integrating agriculture, conservation and ecotourism: examples from the field, Issues in agroecology – present status and future prospectus. Springer, Dordrecht/Heidelberg/London/New York, pp 55–140

Jones GA, Sieving KE, Jacobson SK (2005) Avian diversity and functional insectivory on North-Central Florida farmlands. Conserv Biol 19(4):1234–1245

Jose S (2009) Agroforestry for ecosystem services and environmental benefits: an overview. Agrofor Syst 76(1):1–10

Jose S (2012) Agroforestry for conserving and enhancing biodiversity. Agrofor Syst 85(1):1–8

Karp DS, Daily GC (2014) Cascading effects of insectivorous birds and bats in tropical coffee plantations. Ecology 95(4):1065–1074

Klink CA, Machado RB (2005) Conservation of the Brazilian Cerrado. Conserv Biol 19(3):707–713

Lamônica KR, Barroso DG (2008) Sistemas agroflorestais: aspectos básicos e recomendações, 1st edn. Secretaria de Agricultura, Pecuária, Pesca e Abastecimento, Rio de Janeiro

Landau EC, Cruz RK, Hirsch A et al (2012) Variação geográfica dos tamanhos dos módulos fiscais no Brasil Documentos. Embrapa Milho e Sorgo, Sete Lagoas

Lindell CA, Chomentowski WH, Zook JR (2004) Characteristics of bird species using forest and agricultural land covers in southern Costa Rica. Biodivers Conserv 13(13):2419–2441

McKinney ML, Lockwood JL (1999) Biotic homogenization: a few winners replacing many losers in the next mass extinction. Trends Ecol Evolut 14(11):450–453

Metzger JP (2010) O Código Florestal tem base científica? Natureza & Conservação 8(1):1–5

Miller RP (2009) Construindo a complexidade: o encontro de paradigmas agroflorestais. Agrofloresta.net, p.21. http://media0.agrofloresta.net/static/artigos/Construindo_a_complexidade-Robert_Miller.pdf. Accessed 12 April 2018

Millennium Ecosystem Assessment (2005) Ecosystems and human well-being: Current State and Trends. Island Press, Washington

Morrison E, Lindell C (2012) Birds and bats reduce insect biomass and leaf damage in tropical forest restoration sites. Ecol Appl 22(5):1526–1534

Muler AE, Rother DC, Brancalion PS et al (2014) Can overharvesting of a non-timber-forest-product change the regeneration dynamics of a tropical rainforest? The case study of *Euterpe edulis*. For Ecol Manag 324:117–125

Murray DAH, Clarke MB, Ronning DA (2013) Estimating invertebrate pest losses in six major Australian grain crops. Aust J Entomol 52(3):227–241

Nair PKR (2012) Carbon sequestration studies in agroforestry systems: a reality-check. Agrofor Syst 86(2):243–253

Newbold T, Scharlemann JPH, Butchart SHM et al (2013) Ecological traits affect the response of tropical forest bird species to land-use intensity ecological traits affect the response of tropi-

cal forest bird species to land-use intensity. Proc R Soc B Biol Sci 280(1750). https://doi.org/10.1098/rspb.2012.2131
Newbold T, Hudson LN, Hill SL et al (2015) Global effects of land use on local terrestrial biodiversity. Nature 520(7545):45–50
Noble IR, Dirzo R (1997) Forests as human-dominated ecosystems. Science 277(5325):522–525
Nyffeler M, Benz G (1987) Spiders in natural pest control: a review. J Appl Enthomol 103:321–339
Oliveira AB, Cruz RB, Carmo CAFS (2009) A seringueira no estado do Rio de Janeiro: histórico, situação e potencialidade da atividade. Pesagro Rio. http://www.pesagro.rj.gov.br/downloads/infonline/online15.pdf. Accessed 12 April 2018
Peisley RK, Saunders ME, Luck GW (2015) A systematic review of the benefits and costs of bird and insect activity in agroecosystems. Springer Sci Rev 3(2):113–125
de la Peña-Domene M, Martízez-Garza C, Palmas-Pérez S et al (2014) Roles of birds and bats in early tropical-forest restoration. PLoS One 9(8):104656
Petit LJ, Petit DR (2003) Evaluating the importance of human-modified lands for neotropical bird conservation. Conserv Biol 17(3):687–694
Petit LJ, Petit DR, Christian DG, Powell HDW (1999) Bird communities of natural and modified habitats in Panama. Ecography 22(3):292–304
Pimm SL, Raven P (2000) Biodiversity: Extinction by numbers. Nature 403(6772):843–845
Piña-Rodrigues FCM, Franco FS, Capelo FFM et al. (2013). Conservação ex-situ: dos bancos de germoplasma aos sistemas agroflorestais. In Piratelli AJ, Francisco MR (eds) Conservação da biodiversidade: dos conceitos às ações. Rio de Janeiro: Technical Books, pp 274–183
Power AG (2010) Ecosystem services and agriculture: tradeoffs and synergies. Philos Trans R Soc Lond Ser B Biol Sci 365(1554):2959–2971
Reis M, Fantini AC, Nodari RO et al (2000) Sustainable yield management of *Euterpe edulis Martius (Palmae)*. J Sustain For 11(3):1–17
Renwick AR, Vickery JA, Potts SG et al (2014) Achieving production and conservation simultaneously in tropical agricultural landscapes. Agric Ecosyst Environ 192:130–134
Ribeiro MC, Metzger JP, Martensen AC et al (2009) The Brazilian Atlantic Forest: how much is left, and how is the remaining forest distributed? Implications for conservation. Biol Conserv 142(6):1141–1153
Rodrigues RR, Gandolfi S, Nave AG et al (2011) Large-scale ecological restoration of high-diversity tropical forests in SE Brazil. For Ecol Manag 261(10):1605–1613
Rother DC, Rodrigues RR, Pizo MA (2009) Effects of bamboo stands on seed rain and seed limitation in a rainforest. For Ecol Manag 257(3):885–892
Santiago TMO, Rezende JLP, Borges LAC (2017). The legal reserve: historical basis for the understanding and analysis of this instrument. Cienc. Rural 47(2): e20141349
Santos EV, Lima MS (2015) O rural no norte fluminense. In Rio de Janeiro. A diversidade da geografia brasileira: escalas e dimensões da análise e da ação. ENANPEGE, Rio de Janeiro, pp 2828–2839
Sarrazin F, Lecomte J (2016) Evolution in the Anthropocene. Science 351(6276):922–923
Schroeder P (1994) Carbon storage benefits of agroforestry systems. Agrofor Syst 27(1):89–97
SEA/INEA (2016) Diário Oficial de 19 de maio de 2016. https://www.jusbrasil.com.br/diarios/116189547/doerj-poder-executivo-19-05-2016-pg-20. Accessed 12 April 2018
Secretaria do Meio Ambiente (2008) Resolução SMA no 44, São Paulo. http://www.ambiente.sp.gov.br/legislacao/resolucoes-sma/resolucao-sma-44-2008/. Accessed 12 April 2018
Sekercioglu CH (2012) Bird functional diversity and ecosystem services in tropical forests, agroforests and agricultural areas. J Ornithol 153(1):153–161
SEPLAG – Secretaria de Estado de Planejamento e Gestão (2011). Plano básico para o desenvolvimento da silvicultura sustentável para as regiões Norte e Noroeste do estado do Rio de Janeiro: Volume I, Estudo das Florestas Comerciais e Naturais, p 274
Serviço Florestal Brasileiro (2017) CAR - Cadastro Ambiental Rural. http://www.florestal.gov.br/snif/gestao-florestal/cadastro-ambiental-rural. Accessed 12 April 2018
Sick H (1997) Ornitologia Brasileira. Rio de Janeiro: Editora Nova Fronteira

da Silva JZ (2011) Fundamentos da Produção e Consumo de Frutos em Populações Naturais de Euterpe edulis Martius. http://repositorio.ufsc.br/xmlui/bitstream/handle/123456789/96070/298225.pdf?sequence=1. Accessed 12 April 2018

Silva-Andrade HL, de Andrade LP, Muniz LS (2016) Do farmers using conventional and non-conventional systems of agriculture have different perceptions of the diversity of wild birds? Implications for Conservation. PLoS ONE 11(5):e0156307

Soares-Filho B, Rajão R, Macedo M et al (2014) Cracking Brazil's forest code. Science 344:363–364

Somarriba E (1992) Revisiting the past: an essay on agroforestry definition. Agrofor Syst 19(3): 233–240

SOSMA – SOS Mata Atlântica, INPE – Instituto Nacional de Pesquisa Espacial (2014) Atlas dos remanescentes florestais da Mata Atlântica período 2012 a 2013. https://www.sosma.org.br/wp-content/uploads/2014/05/atlas_2012-2013_relatorio_tecnico_20141.pdf. Accessed 12 April 2018

Souza MCS, Piña-Rodrigues FCM, Casagrande JC et al (2016) Funcionalidade ecológica de sistemas agroflorestais biodiversos: Uso da serapilheira como indicador da recuperação de áreas de preservaçãoo permanente. Floresta 46(1):75–82

Tscharntke T, Clough W, Wanger TC et al (2012) Global food security, biodiversity conservation and the future of agricultural intensification. Biol Conserv 151(1):53–59

Uezu A, Beyer DD, Metzger JP (2008) Can agroforest woodlots work as stepping stones for birds in the Atlantic forest region? Biodivers Conserv 17(8):1907–1922

Wenny D, DeVault TL, Johnson MD et al (2011) The need to quantify ecosystem services provided by birds. Auk 128(1):1–14

Chapter 11
Connectivity Conservation Management: Linking Private Protected Areas

Lorena Valeria Guzmán Wolfhard and Claudia Raedig

Abstract Fragmentation has become one of the most pressing threats to Brazil's Atlantic Forest causing habitat loss and the erosion of species diversity, thus compromising ecosystem functioning and the provision of crucial ecosystem services. Beyond the legal obligations of forest owners to protect their forest properties, the creation of private protected areas (Reservas Particulares do Patrimônio Natural, RPPNs) is an important strategy to counteract fragmentation. Such RPPNs allow maintaining or reestablishing connectivity, by acting as stepping stones or as local corridors. However, only few efforts have been carried out to systematically link RPPNs at local implementation level. Therefore, this study aims to explore possible connectivity paths between established and probable future RPPNs in two municipalities. Connectivity routes among RPPNs were identified by using two methods: the "least cost path" method that allowed finding the "cheapest" paths by assigning different weights to land use features and a second method which ranks the different land use features from suitable to non-suitable area for linking RPPNs. Incorporating the willingness of landowners to establish RPPNs into the latter method permitted the identification of suitable linkages between RPPNs and priority areas for future RPPN establishment.

Keywords Atlantic Forest · Connectivity and biodiversity conservation · Corridors · Private natural heritage reserves (RPPNs)

L. V. Guzmán Wolfhard (✉) · C. Raedig
Institute for Technology and Resources Management in the Tropics and Subtropics (ITT), TH Köln – University of Applied Sciences, Köln, Germany
e-mail: claudia.raedig@th-koeln.de

U. Nehren et al. (eds.), *Strategies and Tools for a Sustainable Rural Rio de Janeiro*, Springer Series on Environmental Management,
https://doi.org/10.1007/978-3-319-89644-1_11

11.1 Introduction

The Atlantic Forest has been shown as one of the areas with the highest numbers of plant species worldwide, estimates ranging between more than 5000 vascular plant species per 10,000 km^2 (Barthlott et al. 2005) and approx. 6500 flowering plant species per 10,000 km^2 (Raedig and Lautenbach 2009). Furthermore, the Atlantic Forest represents one of the four centers of flowering plant diversity in the neotropics (Morawetz and Raedig 2007; Raedig et al. 2010). But the Atlantic Forest is also renowned as center for faunal diversity, e.g., for mammals, 229 species (Stotz et al. 1996, Costa et al. 2010), and for avifauna, 682 species (Aleixo and Galetti 2017), and particularly as center of endemic species which only occur along this mostly narrow stripe following the coastline from northern Paraguay to northeast Brazil (Myers et al. 2000; Galindo-Leal and Câmara 2003; Tabarelli et al. 2010; Lamas and Pinto 2011). The largest remaining forest fragments occur southward from Rio de Janeiro (RJ) state (Ribeiro et al. 2009). Further north, many areas only host small forest fragments or are completely deforested. With fragmentation ongoing, often leaving only forest fragments in steep areas which are difficult to use for agricultural, urban, or industrial land uses, the remaining forest fragments represent crucial habitat for many animal and also plant species (Galindo-Leal and Câmara 2003; Raedig and Lautenbach 2009; Fahrig 2010; Prado et al. 2013; Barbosa et al. 2017).

One way to counteract habitat loss and associated biodiversity erosion is to link forest fragments to establish connectivity between fragments (Bennett 2003; Chetkiewicz et al. 2006; Crooks and Sanjayan 2006; Mackey et al. 2010). Such biological corridors or wildlife or habitat corridors are usually linear areas of linked, distinctive patches which differ from the surrounding landscape, allowing species and gene dispersal over its entire range (Sanderson et al. 2006; Lino and Albuquerque 2007). The establishment of corridors to link isolated patches of habitat is the most popular although not uncontroversial approach to counteract habitat fragmentation (Bennett 2003; Baum et al. 2004; Crooks and Sanjayan 2006). On the one hand, proof for the effectiveness of corridors has been furnished experimentally (e.g., Tewksbury et al. 2002; Damschen et al. 2006; Christie and Knowles 2015); on the other hand, corridors have been criticized for ineffectiveness (e.g., Haddad et al. 2014; Resasco et al. 2014). However, conservation biologists and land managers consider wildlife corridors as a strong method capable of protecting habitats and migration routes threatened by fragmentation and climate change (Heller and Zavaleta 2009). RJ forms part of two large-scale biodiversity conservation corridors which are defined as an area determined by both biological and by conservation criteria (Aguiar et al. 2005; Sanderson et al. 2006): the Serra do Mar Corridor and the Central Corridor (Ayres et al. 2005; Fig. 11.1).

Although the importance of the remnant forest fragments for biodiversity conservation is growing, their protection and linkage becomes increasingly difficult in spite of the complex Brazilian legal forest protection system. Essentially three laws are regulating forest remnants and their protection. First is the Forest Law which is the older jurisdiction and in its current version (Law 12.651/2012) governs riparian

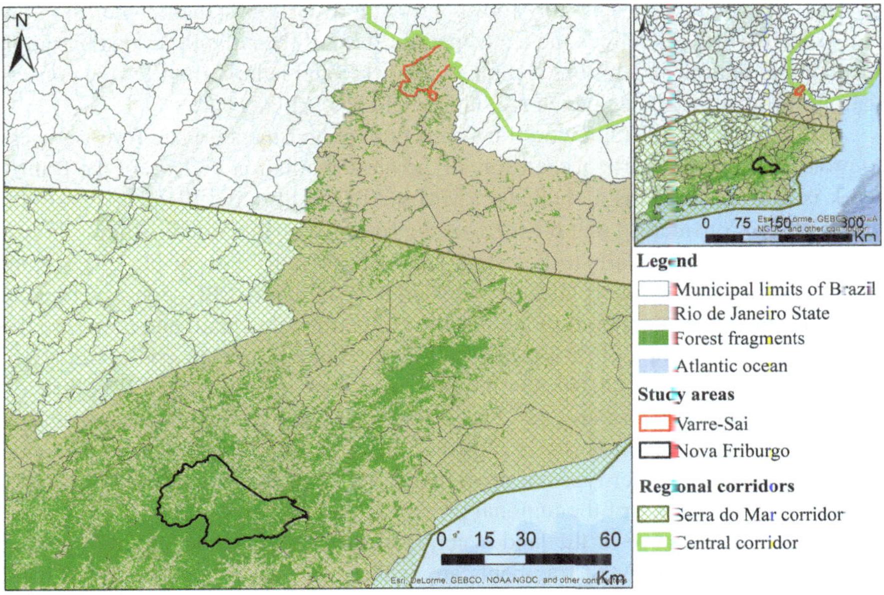

Fig. 11.1 The location of the two target municipalities in the state of Rio de Janeiro and the setting of the two large biodiversity corridors partly covering or situated adjacent to the state. (Sources: Table 11.1; for the corridors: Aliança para a conservação da Mata Atlântica (2009))

permanent preservation areas (APPs),legal forest reserves (RLs), and others (first version enacted in 1934, Decree 23,793/34; Castelo 2006). This law mainly regulates the use and protection of private lands in Brazil (Chiavari and Leme 2015). In all states, all APPs have to be protected, and 20% of the forests on private properties have to be protected as RLs in the Atlantic Forest (May et al. 2015; Piratelli et al. 2018).

Law 9.985/2000, the second law, creates the National System of Units of Conservation (SNUC) and regulates the "Unidades de Conservação" (UCs, conservation units; SNUC 2000). Possible management options are an important aspect for these protected areas. Accordingly, UCs can be divided into two groups: areas with integrated, full protection, such as ecological stations or natural parks, and areas where sustainable use is allowed, such as environmental protected areas (APAs; SNUC 2000). Whereas the Forest Law is a top-down approach, the SNUC also allows the voluntary creation of protected areas. Landowners who want to save their forests can establish private UCs, regardless of the already 20% (RLs) protected under the Forest Law. Such UCs are the private natural heritage reserves, the so-called Reservas Particulares do Patrimônio Natural (RPPNs). RPPNs can be established at federal, state, or municipal level, with different respective authorities (ICMBio Instituto Chico Mendes de Conservação da Biodiversidade, INEA State Institute of the Environment, municipal environmental secretaries), and RPPN owners remain legitimate owners of their lands.

The third law relevant for forest protection is the Atlantic Forest Law (11,428/2006) which defines the legal spatial extent of the Atlantic Forest based on its potential natural vegetation (Nehren et al. 2017) and is important for the conservation of forest fragments which are outside of UCs, APPs, or RLs.

Although this legal framework for forest protection forces private forest owners to protect substantial parts of their forest, the implementation success of the framework is low and remains a challenge (MMA 2010; Ball et al. 2014; Soares-Filho et al. 2014). Considering that 80–90% of the forest remnants in the Atlantic Forest are in private hands (Galindo-Leal and Câmara 2003; Ribeiro et al. 2009; Pinto et al. 2014), RPPNs can contribute substantially to biodiversity conservation, in particular in areas where land properties are comparatively small, and therefore the landowners are exempted from the full APP protection (Chiavari and Leme 2015; Machado and Anderson 2016). On the one hand, linking existing RPPNs directly counteracts fragmentation and thus biodiversity loss, and on the other hand, it contributes to the regional importance of protected areas. When RPPNs are connected to each other and to other types of protected areas, such networks would positively influence the efficiency of the regional protected area network.

From the perspective of the landowner, it is difficult to decide where to protect the forest. Frequently forest fragments are small, and data on biodiversity at local scale are often not available. Some highly detailed publications on specific taxa exist (e.g., Morawetz and Raedig 2007; Raedig and Lautenbach 2009, Scheer and Blum 2011), databases (e.g., Jabot: da Silva et al. 2017) and also several participative data portals are available (e.g., WikiAves), but for fragmented forest in an agricultural matrix, the available data are hard to use at the local level relevant for RPPNs. From the perspective of the local or regional decision-maker, it is also difficult to decide where forest conservation would matter most. At both local and regional levels, decision-makers could better counsel landowners if a scheme for priority areas for RPPN establishment existed.

So far, few studies have been carried out to explore how RPPNs could be systematically linked or where new RPPNs could be established to promote connectivity.

Therefore, the objective of the present study is to explore how existing RPPNs can be systematically linked. More specifically, we addressed the following questions: (1) How are RPPNs geographically distributed in the study area? (2) Is there information that landowners can supply to be integrated into corridor design? (3) How can linkages between RPPNs be designed for data-poor areas guiding landowners and decision makers at implementation level?

11.2 Methods

11.2.1 Study Areas

The study was carried out in the frame of the INTECRAL project (Schlüter et al. 2018) in the two municipalities: Varre-Sai and Nova Friburgo. Varre-Sai is located in the Northwestern Region of the state. This region is characterized by severe degradation and deforestation due to its largest monocultures, coffee and sugarcane, and to cattle grazing that turned the landscape into predominantly pasture lands (Souza et al. 2007; World Bank 2014). In addition, the highest rural poverty indices in the state were registered here with the majority of the inhabitants being small-scale farmers depending on low technological inputs (World Bank 2014). The dominant vegetation is semi-deciduous mountainous forest which is mostly found in small forest fragments in the coffee- and pasture-dominated matrix area. Varre-Sai is located in the gap that separates the two regional corridors, the Serra do Mar Corridor and the Central Corridor (Fig. 11.1).

However, Varre-Sai is considered as one of the municipalities in the Northwestern Region with the highest level of protected areas. From the 3% of the remaining forests found in this region, 21% belong to Varre-Sai (de Martino 2010). RPPNs are the only types of UCs existing in this municipality and are thus considered as a crucial strategy for biodiversity conservation in accordance with sustainable development. Varre-Sai municipality has designated Law 572 (Lei de Repasse) in 2010 which regulates that landowners in this municipality receive an economic incentive for creation of RPPNs (SOS Mata Atlântica 2013) through the direct transfer of the "green tax" on "Transactions on the Circulation of Goods and Service Delivery of Interstate and Inter-municipal Transportation and Communication" (ICMS). The possible use of a proportion of the ICMS for environmental purposes (green ICMS) seeks to compensate opportunity costs of municipalities, mainly due to land given for creation of UCs (Barros and Cenamo 2016).

Nova Friburgo lies in the Serrana region which is named after the coastal Serra do Mar mountain range characterized by steep slopes. It is recognized as one of the four best protected municipalities in the state, providing a large portion of its land covered by forest. It is located inside the Serra do Mar Corridor, considered to be one of the most well-preserved biogeographical regions in the country (Ribeiro et al. 2010). In this municipality, three types of UCs are found, with the state park Três Picos being the best known and also the largest one (587.9 km^2). Many landowners have created RPPNs in Nova Friburgo, although a direct transfer from the green ICMS to owners of RPPNs has not yet been performed. The money the municipality receives from the green ICMS is used for other environmental and sanitation projects (Secretaria do Meio Ambiente 2011).

11.2.2 RPPN Data Collection

For both municipalities, spatial data for RPPNs was collected from respective state authorities and partly from the RPPN owners themselves; subsequently RPPN locations were mapped. The visits to RPPN owners were combined with interviews to pinpoint biodiverse areas or areas planned to become a RPPN. In addition, landowners without RPPNs were interviewed to explore their interest in creating RPPNs. Due to time restrictions, interviews with landowners were only carried out in Varre-Sai.

Table 11.1 Overview on source layers used for identifying landscape elements

Layers	Source	Year(s) of publication
Federal RPPNs	Chico Mendes Institute for Biodiversity Conservation (ICMBio)	2012
State RPPNs	State Institute of the Environment (INEA)	2009–2015
Municipal RPPNs	Technical assistance and rural extension company (EMATER)	2015[a]
Forest fragments	Land use map (layers: Forest fragments, rivers, pastures, agricultural areas; INEA)	2007
Main roads	Ministry of Environment (MMA)	2008
Slope	Brazilian Agricultural Research Corporation (EMBRAPA)	2006

[a]Data were collected in the frame of the field work for this study.

11.2.3 Linking RPPNs

Neither for Varre-Sai nor for Nova Friburgo sufficient species distribution data was available at local scale.

Therefore, we decided against species distribution modeling to support corridor design. For the identification of suitable linkages between RPPNs, the following assumptions were made: RPPNs are situated in a heterogeneous matrix whose different features either support or hinder exchange of species and genes between RPPNs. Since forest fragmentation impacts movement of most species through the matrix, the targeted RPPN linkages correspond to structural corridors. If landowners were interested in biodiversity conservation and willing to create an RPPN on their grounds, their knowledge on species distribution in their forest fragments and their spatial preference for priority areas for RPPN establishment would matter.

Two basic requirements for the selection of the approach existed: easy replicability of the method and usage of commonly existing datasets to allow the targeted user groups, i.e., decision-makers and landowners to approve the results and work with them and, in the best case, to use the methods themselves. Based on these requirements and assumptions, two methods and respective data (Table 11.1) were chosen and will be explained in the following.

11.2.4 Least Cost Path Method: Distance Analysis Method Using ArcGIS (Method 1)

The least cost path method is commonly used for the creation of biological corridors (Calabrese and Fagan 2004; Theobald 2006; Majka et al. 2008; Parks et al. 2013; Suleiman et al. 2015) and also for matrix quality assessment in large-scale corridors (Goulart et al. 2015). This distance analysis method determines "the cheapest" path between two points relative to previously defined weights, i.e., cost units according to the desired objective to be met (Pinto and Keitt 2009). The minimum travel distance for species to cross through the landscape with the lowest exposure to unsuitable habitat by the weights assigned is determined (Worboys et al. 2010; Alves and Fonseca 2012).

The least cost path method allowed calculation of a path between RPPNs and forest fragments, i.e., suitable areas, at the same time avoiding pastures, agricultural areas, urban areas, and rivers, i.e., less suitable areas (Alves and Fonseca 2012). Different weights were assigned based on the suitability of the respective landscape feature to connect RPPNs. Lowest weight was assigned to the areas considered optimal for the path, including RPPNs, forest, and secondary vegetation. Medium weight was assigned to rivers, water bodies, and exposed rock. Highest weight was assigned to the areas that would represent the highest costs for path creation and which therefore must be avoided, like urban areas, pastures, and agricultural areas. Concerning the slopes, the following assumption was made: for mobile and non-flying species, the default element of the least cost path method is relevant. Since its aim is to identify a path which will provide the least space for conflict between land uses, the lowest weight was given to the areas with a slope higher than 20%, since almost all the remaining forest is found in steep, higher mountainous areas (SNUC 2000, Pereira and Neto 2004). As agricultural and construction purposes compete with conservation purposes in areas with lower slope (less than 20%), medium weighting was assigned to such areas (Table 11.2).

11.2.5 Linking RPPNs to Landscape Elements (Method 2)

The second method analyzed landscape elements by using land use features (Bennett 2003; Calabrese and Fagan 2004; Beier et al. 2008; Seoane et al. 2010). Higher resolved satellite images were compared to the broader land use feature dataset (Table 11.1) to allow for the inclusion of more landscape elements relevant at local scale. Thereby, seven connecting elements and five barriers (Table 11.3) were identified. Paths between these elements were manually elaborated in Google Earth Pro and ArcGIS avoiding as many barriers as possible while following a set of rules:

Starting point is an RPPN close to the outer limit of the municipality. The first step is to locate the nearest neighboring RPPN. In the second step, the matrix between the RPPNs is scrutinized for connecting elements and barriers. The shortest possible linkage is chosen considering the strength of connecting elements as well as of barriers (Table 11.3).

Furthermore, considering that RPPNs are privately owned areas, this method was complemented with information obtained through interviews conducted with landowners and experts on the protected area system of Brazil. The interviews were carried out with both landowners with and without RPPNs, and indicated priority areas, where rare species, or flagship species, or endemic species had been observed, or which in the future are likely to become RPPNs. Finally, the paths were elaborated linking the identified connecting landscape elements and relevant locations gathered from the interviews and avoiding the barriers (Fig. 11.2).

Table 11.2 Weighting scheme for least cost path method

Suitability of areas	Landscape elements
Suitable areas (weighting 1)	RPPNs, forest fragments, slope higher than 20%
Medium suitable areas (weighting 6)	Water bodies, exposed rock, slope less than 20%
Non-suitable areas (weighting 10)	Pastures, agricultural areas, urban areas

Table 11.3 Ranking of connecting elements and barriers (in descending order)

Connecting elements		Barriers
RPPNs	Decreasing ↓	Urban areas
Other UCs		Main roads
Forest fragments		Agriculture fields
Riparian vegetation (APP's)		Large pasture lands for cattle grazing
Small forest stands		Plantation of exotic species (mainly Eucalyptus plantations
Hedgerows		Rivers
Fallow lands		
Roadside vegetation		

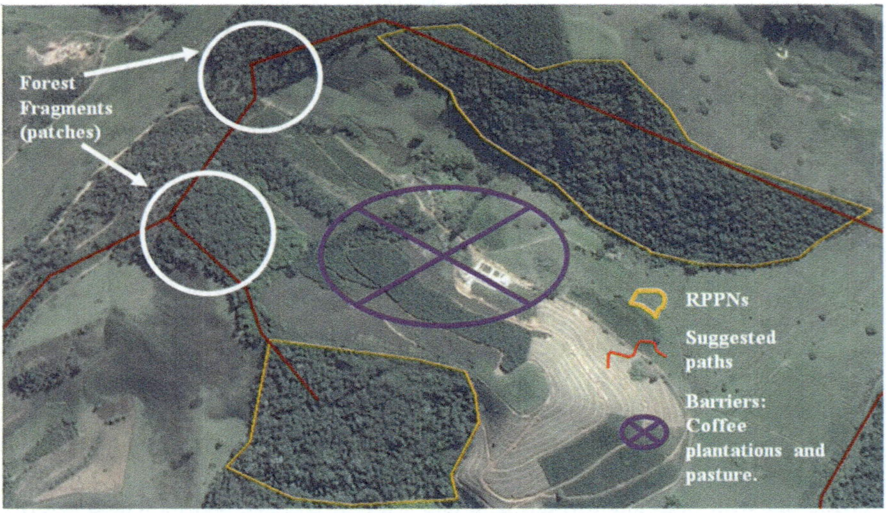

Fig. 11.2 Scheme of path construction (red line) in the second method. In this example, forest fragments (white circles) are used as stepping stones for linking two RPPNs (orange lines). The coffee plantations (purple circle) are considered a barrier for path elaboration. For a ranking of these elements, see Table 11.3. (Source: Google Earth Pro, 2015)

11.3 Results

In Varre-Sai, 11 interviews were carried out with RPPN owners and 18 with landowners without an RPPN. For 8 of the in total 16 RPPNs identified, vector data were provided by INEA. Vector data for the other RPPNs were elaborated based on information from the Rural Environmental Registration (CAR; Piratelli et al. 2018) and from landowners. For some RPPNs, the technicians from Technical Assistance and Rural Extension Company (EMATER) drew the sketches of the RPPNs that were converted into vector data.

In Nova Friburgo, four RPPN owners were interviewed. Seven shapefiles of the in total 20 identified RPPNs were provided by the ICMBio. The shapefiles of the other 13 RPPNs were provided by INEA. With roughly a third of the 60% identified for Nova Friburgo, forest cover was substantially lower in Varre-Sai. Accordingly, the mean size of RPPNs was also smaller. Sizes varied between 1.11 and 101.85 ha (both RPPNs being located in Nova Friburgo; Table 11.4).

Figure 11.3 shows the 16 RPPNs created until 2015 in Varre-Sai, marked in yellow. These RPPNs are linked with each other based on Method 1 (Corridor 1, highlighted in red) and based on Method 2 (Corridor 2, highlighted in orange) For Varre-Sai, landowners' and RPPN owners' knowledge on fauna and flora was considered in the second method, as well as possible and planned future RPPN locations. From 23 locations, 17 are included in the path generated. Consideration of these locations as well as the usage of more elements than in Method 1 yielded a path with a length of approximately 37 km, which is almost double as long as Corridor 1.

In Varre-Sai, no other UCs were found, and thus the connectivity path was elaborated by linking exclusively RPPNs.

In Nova Friburgo, 20 RPPNs were connected using both methods (Fig. 11.4). The length of Corridor 1 is approximately three quarters of the length of Corridor 2 (95 km). A large part of both corridors lies inside of the Tres Picos State Park. In general, Nova Friburgo has a less fragmented landscape showing a larger proportion of forest (Table 11.3). In addition to the Tres Picos State Park, the municipality also has the Macaé de Cima APA that is located in the southeast of Nova Friburgo.

Table 11.4 Number, average size of RPPNs, forest cover, and numbers of interviews performed in Varre-Sai and Nova Friburgo

Municipality	No. of RPPNs	Average size [ha] ± standard error	Forest cover (%)	Interviews with landowners	Interviews with RPPN owners
Varre-Sai	16	14.90 ± 12.19	21	18	11
Nova Friburgo	20	22.38 ± 24.47	60	0	4

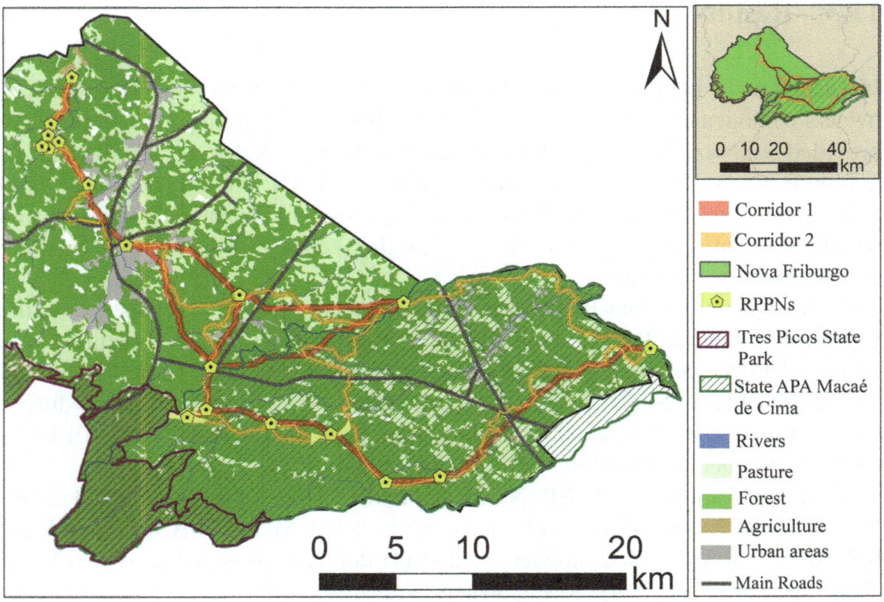

Fig. 11.3 Two suggested corridors in Varre-Sai. Corridor 1: linking RPPNs using the least cost path method with approximately 20 km of total corridor length. Corridor 2: linking RPPNs through landscape elements and incorporating information on priority areas in Varre-Sai with a length of approximately 37 km. Both corridors were buffered, and slope layers are not shown for a better visibility. (Sources: Table 11.1)

11.4 Discussion

Both methods applied can integrate a wide range of ecological as well as social data and yielded systematic linkages between existing RPPNs. The second method yielded a longer and more branched corridor due to the higher resolved spatial data and thus differentiation into more landscape elements, e.g., plantations as barriers (Figs. 11.3 and 11.4), and due to the inclusion of information obtained in interviews with landowners (Fig. 11.3). The integration of data from landowners who had no RPPN yielded a longer path covering areas which otherwise would have been outside the path. Therefore, we consider the inclusion of landowners' knowledge on species occurrences on their grounds as well as their willingness to establish an RPPN important information which should be considered in corridor design at local scale.

The selection of the two methods out of a multitude of possible methods was mainly due to their easy replicability. The first method depends on the availability of different data sources and a GIS software that includes a method for calculating the least cost path for connecting forest patches. The second method is used for satellite images in a higher resolution enabling the identification of elements such

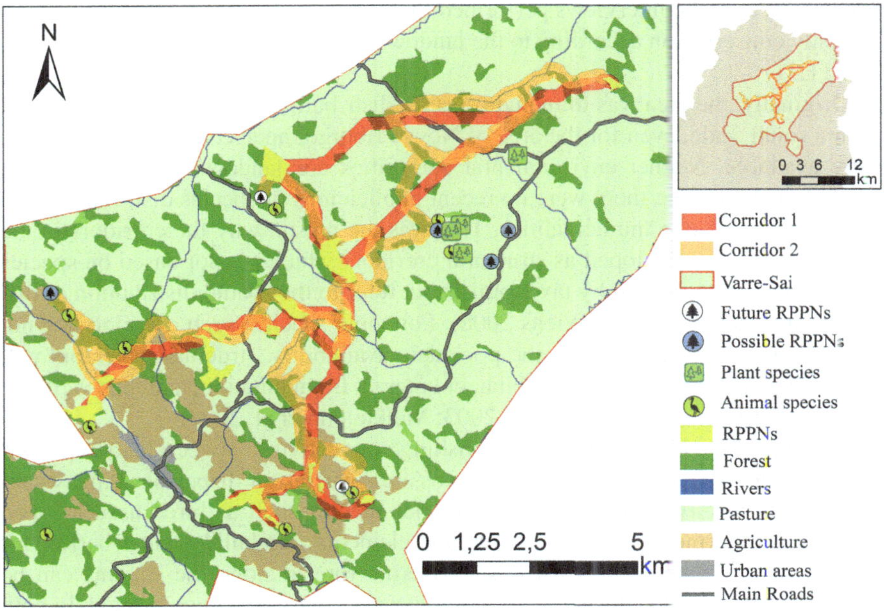

Fig. 11.4 Two suggested corridors in Nova Friburgo. Corridor 1: linking RPPNs using the least cost path Method 1, with a total length of approximately 70 km. Corridor 2: linking RPPNs through landscape elements, with approximately 95 km of corridor length. The corridor was buffered and slope layers are not shown to allow a better visibility. (Sources: Table 11.1)

as small forest fragments or hedgerows. As soon as these elements can be identified, the linkages between RPPNs can be made with any GIS software. The first method will yield replicable results subject to the weighting of the different layers. The second method will yield replicable results subject to the set of rules developed for path generation.

Both methods have proven useful; however, the first method is faster to apply when sufficient data are available, e.g., current land use and land cover data or species distribution data. If this is not the case, however, sufficiently resolved spatial satellite data is available; the second method is more suitable. Since poor availability of georeferenced actual data is a common phenomenon, both methods are presented. Whereas the first method can be also used at large scales, the second method would have to be automated. For the transfer of both methods to other study areas, the more suitable method can be chosen according to data availability and scale consideration.

In the present study, the second method was preferred because of the local scale. In addition, satellite images available from Google Earth Pro were from 2015, whereas the data layers used for the first method were from 2007 (Table 11.1). The results of the first method are shown here to demonstrate that once the data are available in GIS layers, this method yields results that give a clear indication for a feasible connection of RPPNs. The results of the second method implied supplemental results based on the interviews performed including priority areas valuable for biodiversity conservation according to the landowners interviewed (only for Varre-Sai, Fig. 11.3).

Originally, the idea was to include species distribution data. It turned out that at such a small scale, even for the area of Nova Friburgo, no species distribution data were available. Neither environmental authorities nor landowners could provide these data. Moreover, there were no recent publications of species distribution data available for the two municipalities. This was the reason why the corridors in the present study are developed as structural corridors which are not based on species occurrence data collected by taxonomists but refer to the spatial distribution of different landscape elements (Wiens 2006). For some species, such corridors are not promoting dispersal, but given the poor data availability, structural corridors will allow a connection of potential habitat, counteract fragmentation, and thus facilitate dispersal for many species (Bennett 2003; Meiklejohn et al. 2009)

The rules and weightings for establishing connectivity between landscape elements are not easy to set. While, e.g., roads are a barrier for dispersal of many species, this is not true for many flying animal species such as birds used to open habitats. Furthermore, rivers are landscape elements that build a clear barrier for movement for some species. However, riverine forests along the streams can be valuable linkages between forest patches that were otherwise not connected. In both methods, by adequate weighting or by an adequate set of rules, forests along barriers are elements favorable for corridor construction, whereas a river or a road represent barriers which are not crossed when surrounding options are more suitable. For usage of both methods in different areas or when including species distribution data, the weighting of different landscape elements or the ranking for linking them has to be reviewed and adapted. There might be different landscape elements taken into account in the analysis or different weights or rankings necessary.

The set of methods presented here will contribute to find a systematical way of creating and connecting RPPNs. However, one important result from our study was that many landowners have an interest in creating an RPPN, but do not possess the necessary information on how to do that or are discouraged by the partly tedious procedure of RPPN creation. In the frame of the INTECRAL project, a guide was developed providing the information needed completed with practical experiences given by RPPN owners themselves. Particularly in areas where only smaller forest fragments remain – and these are in private hands – the awareness of landowners for the possibility to create RPPNs to protect forest has to be promoted, and the creation of RPPNs to be facilitated.

11.5 Conclusion and Outlook

In the future, corridor design for RPPNs could be complemented with the spatial data from the CAR initiative, resulting in a state-wide strategy for RPPN establishment. A precondition for such a strategy would be the collection of the spatial RPPN data in one common database, irrespective of the type of RPPN, i.e., federal, state, or municipal. So far, the corridors constructed (Figs. 11.3 and 11.4) have stopped at the RPPN or designated RPPN location closest to the limit of the respective municipality. Transboundary corridor design will optimize path creation and allow for a better systematic RPPN planning.

Participation has been shown as crucial for biodiversity conservation in many studies (Kotsakis 2009) and is the key to success for motivating RPPN establishment (Broering and Junior Sandeville 2010). One outcome of the interviews carried out in Nova Friburgo and Varre-Sai was that the landowners had a great interest to learn about biodiversity in their lands. However, there are no checklists or other easily applicable tools available for them to monitor species occurrence on their property. Until now, there has not been process which feedbacks detailed biodiversity information to RPPN landowners. Although the requirement of the establishment of a management plan for RPPNs exists, this has often not been followed up by the respective authorities yet. Therefore, it would be an asset if participative monitoring approaches were developed which landowners themselves could carry out. For such an approach, regional checklists of easily identifiable indicator taxa, e.g., at species, genus, or even family level, should be designed together with taxonomists and distributed to RPPN owners as a standard procedure in regular intervals. In the long run, participative monitoring would allow for monitoring of species loss but also species reestablishment.

In Nova Friburgo, where much more forest areas remain in comparison to Varre-Sai (Table 11.3), RPPNs can be more readily connected with other protected areas. Furthermore, an RPPN with an owner who voluntarily decided for protecting his forest will probably be better protected than "paper" protected areas. Thus, integrating landowners into biodiversity conservation also in municipalities with a higher coverage of protected areas can contribute to biodiversity conservation.

In Varre-Sai there are comparatively few forest remnants left which are often privately owned. Here, RPPNs are the main form of protected area and thus represent central locations for all connectivity routes. This is also true for other municipalities in the Northwestern Region like Porcíuncula or Natividad. For such areas, RPPNs are particularly valuable as they build the backbone of the protected area management. For the future, one challenge is to plan RPPNs considering the location of APPs and RLs on private lands. It is also in this respect that the CAR initiative will be important in providing the spatial data necessary. For decision-making at implementation level, it would be important that RPPN datasets and datasets of other UCs as well as APP and RL datasets are brought together and made available for corridor design.

A long-term goal is the establishment of a connection between the two large-scale biodiversity conservation corridors in the Mata Atlântica. While these corridors are not legally enforced, they have been utilized by NGOs to develop biodiversity conservation strategies, e.g., SOS Mata Atlântica, World Wildlife Found Brazil. Whereas Nova Friburgo is part of the Serra do Mar Corridor, Varre-Sai is adjacent to the Central Corridor (Fig. 11.5). With more data becoming available, the two large-scale corridors can be linked including the large Desengano State Park shared by the municipalities of Santa Maria Madalena, São Fidélis, and Campos dos Goytacazes.

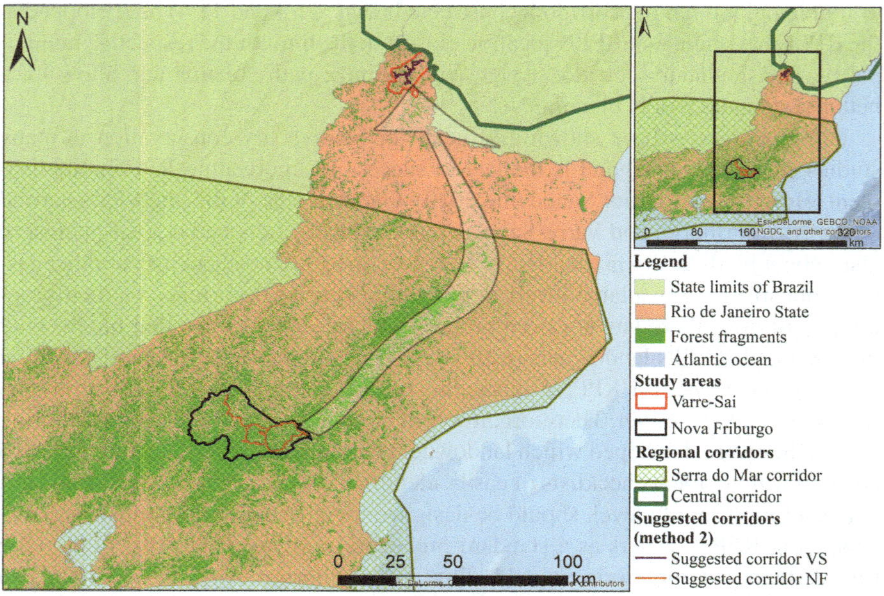

Fig. 11.5 Location of the suggested corridors based on Method 2 in the two target municipalities and missing link between them (visualized by the arrow) that would connect the Serra do Mar Corridor with the Central Corridor. (Sources: Table 11.1; for the corridors: Aliança para a conservação da Mata Atlântica (2009))

Acknowledgments We would like to thank the Rio Rural Project, INEA, EMATER Varre-Sai, and EMATER Nova Friburgo for their support on-field and data provision. Special thanks to Rafael Mussi, Jose Luis Monsores, Manoel Duarte, Deise Morais, Binha, Jonas Daniel, Roberta Guagliardi, Mariana Machado, Aldah de Oliveira, and Rosane Grazioli for their help, their time, and their cooperation. Finally, we would like to give our particular thanks to the landowners, the RPPN owners, and their families in Varre-Sai and Nova Friburgo for their collaboration, interest, enthusiasm, and hospitality.

References

Aguiar AP, Chiarello AG, Mendes SL, Matos SL (2005) Mata Atlântica: Biodiversidade, ameaças e perspectivas: Os corredores central e da Serra do Mar na Mata Atlântica brasileira. In: Galindo-Leal and Câmara (eds) Mata Atlântica Biodiversidade, Ameacas e Perspectivas, Fundacao SOS Mata Atlântica/Conservação Internacional, pp 119–132

Aleixo A, Galetti M (2017) The conservation of the avifauna in a lowland Atlantic forest in south-East Brazil. Bird Conserv Int 7:235–261

Aliança para a conservação da Mata Atlântica (2009) Corredores de biodiversidade da Mata Atlântica. Retrieved from http://www.corredores.org.br/?area=ma (2010). Accessed 14 Mar 2015

Alves DLA, Fonseca BM (2012) Análise espacial para delimitação de áreas potenciais para corredor ecológico na bacia do Rio Santa Bárbara MG 2014 (Apresentação de Trabalho/Congresso)

Ayres JM, Da Fonseca G, Rylands A et al (2005) Os corredores ecológicos das florestas tropicais do Brasil. Sociedade Civil Mamirauá Belém p 256
Ball A, Gouzerh A, Brancalion P (2014) Multi-scalar governance for restoring the Brazilian Atlantic Forest: a case study on small landholdings in protected areas of sustainable development. Forests 5:599–619
Barbosa KVC, Knogge C, Develey PF et al (2017) Use of small Atlantic Forest fragments by birds in Southeast Brazil. Perspect Ecol Conserv 15:42–46
Barros H, Cenamo M (2016) ICMS verde: Incentivo para produção sustentável e redução do desmatamento nos municípios do Amazonas, pp 8–35
Barthlott W, Mutke J, Rafiqpoor D et al (2005) Geographic patterns of vascular plant diversity. Nova Acta Leopoldina 92:61–83
Baum K, Haynes K, Dillemuth F, Cronin J (2004) The matrix enhances the effectiveness of corridors and stepping stones. Ecology 85:671–2676
Beier P, Majka D, Newell S, Garding E (2008) Best management practices for wildlife corridors. America:1–14
Bennett AF (2003) Linkages in the landscape: the role of corridors and connectivity in wildlife conservation. IUCN, Gland, Switzerland and Cambridge, UK, p 254
Broering A, Junior Sandeville E (2010) Mecanismos de envolvimento comunitário na gestão da paisagem de entorno de Reserva Particular Do Patrimônio Natural – RPPN, um estudo de caso, pp 1–4
Calabrese JM, Fagan WF (2004) A comparison-shopper's guide to connectivity metrics. Front Ecol Environ 2:529–536
Castelo TB (2006) Legislação Florestal Brasileira e políticas do governo de combate ao desmatamento na Amazônia Legal. Ambiente e Sociedade 18:221–242
Chetkiewicz CLB, Clair CCS, Boyce MS (2006) Corridors for conservation: integrating pattern and process. Annu Rev Ecol Evol Syst 37(1):317–342
Chiavari J, Leme CL (2015) Policy brief Brazil's new forest code part I: how to navigate the complexity. Climate Policy Initiative and Núcleo de Avaliação de Políticas Climáticas da PUC-Rio de Janeiro, pp 1–8
Christie MR, Knowles LL (2015) Habitat corridors facilitate genetic resilience irrespective of species dispersal abilities or population sizes. Evol Appl 8:454–463
Costa LP, Leite YLR, da Fonseca GAB, da Fonseca MT (2010) Biogeography of south American Forest mammals: endemism and diversity in the Atlantic Forest. Biotropica 32:872–881
Crooks KR, Sanjayan M (2006) Connectivity conservation: maintaining connections for nature. In: Crooks KR, Sanjayan M (eds) Connectivity conservation. Cambridge University Press, Cambridge, pp 1–20
da Silva LAE, Fraga CN, de Almeida TMH et al (2017) Jabot – Sistema de Gerenciamento de Coleções Botânicas: a experiência de uma década de desenvolvimento e avanços. Rodriguésia 68:391–410
Damschen EI, Haddad NM, Orrock JL et al (2006) Corridors increase plant species richness at large scales. Science 313:1284–1286
de Martino MS (2010) Inventário do Patrimônio Ambiental e Urbano do Núcleo Histórico de Varre-Sai. p 396
Fahrig L (2010) Effects of habitat fragmentation on biodiversity. Annu Rev Ecol Evol Syst 34:487–515
Galindo-Leal C, Câmara I (2003) The Atlantic Forest of South America: biodiversity status, threats and outlook. Center for Applied Biodiversity Science at Conservation International and Island Press, Washington, DC. p 488
Goulart FF, Takahashi FSC, Rodrigues M et al (2015) Where matrix quality most matters? Using connectivity models to assess effectiveness of matrix conversion in the Atlantic Forest. Natureza & Conservação 13:47–53
Haddad NM, Brudvig LA, Damschen EI et al (2014) Potential negative ecological effects of corridors. Conserv Biol 28:1178–1187

Heller NE, Zavaleta ES (2009) Biodiversity management in the face of climate change: a review of 22 years of recommendations. Biol Conserv 142:14–32

Kotsakis A (2009) Community participation in biodiversity conservation emerging localities of tension. In: Kessaris AP (ed) Law in the pursuit of development: principles into practice? Routledge, New York, pp 131–145

Lamas IR, Pinto L (2011) CEPF in Atlantic Forest. Final report, Conservation International, Belo Horizonte, p 74

Lino C, Albuquerque J (2007). Mosaicos de Unidades de Conservação no Corredor da Serra do Mar. Conselho Nacional Da Reserva Da Biosfera Da Mata Atlântica, p 97

Machado F, Anderson K (2016) Brazil's new Forest Code: a guide for decision-makers in supply chains and governments. Brasilia, WWF-Brazil

Mackey B, Watson J, Worboys GL (2010) Connectivity conservation and the Great Eastern Ranges corridor, an independent report to the Interstate Agency Working Group (Alps to Atherton Connectivity Conservation Working Group) convened under the Environment Heritage and Protection Council/Natural Resource Management Ministerial Council

Majka D, Beier P, Newell S et al (2008): Best Management Practices for Wildlife Corridors, Northern Arizona University, pp. 1–14

May PH, Bernasconi P, Wunder S, Lubowski R (2015) Environmental reserve quotas in Brazil's new forest legislation: an ex ante appraisal. Center for International Forestry Research (CIFOR), Bogor

Meiklejohn K, Ament R, Tabor G (2009) Habitat corridors & landscape connectivity: clarifying the terminology. Center for Large Landscape Conservation, Bozeman, pp 1–5

MMA – Ministério do Meio Ambiente (2010) Mata Atlântica: Patrimônio Nacional dos Brasileiros. MMA, Brasília, p 408

Morawetz W, Raedig C (2007) Angiosperm biodiversity endemism and conservation in the Neotropics. Taxon 56:1245–1254

Myers N, Mittermeler RA, Mittermeler CG et al (2000) Biodiversity hotspots for conservation priorities. Nature 403:853–858

Nehren U, Raedig C, Sattler D (2017) Der Atlantische Regenwald von Rio de Janeiro: Nutzungsperspektiven im Spannungsfeld von globalen Herausforderungen und lokalen Bedürfnissen. In: Anhuf D (ed) Brasilien – Herausforderungen der neuen Supermacht des Südens. Passauer Kontaktstudium Geographie 14, Selbstverlag Fach Geographie, Universität Passau, pp 51–67

Parks SA, McKelvey KS, Schwartz MK (2013) Effects of weighting schemes on the identification of wildlife corridors generated with least-cost methods. Conserv Biol 27:145–154

Pereira LC, Neto FL (2004) Avaliação da aptidão agrícola das terras: proposta metodológica. Embrapa Meio Ambiente. Documentos 43

Pinto N, Keitt TH (2009) Beyond the least-cost path: evaluating corridor redundancy using a graph-theoretic approach. Landsc Ecol 24:253–266

Pinto SR, Melo F, Tabarelli M, Padovesi A et al (2014) Governing and delivering a biome-wide restoration initiative: the case of Atlantic Forest restoration pact in Brazil. Forests 5(9):2212–2229

Piratelli AJ, Marquez Piña-Rodrigues FC, Raedig C (2018) Integrating biodiversity conservation into agroecosystem management: using birds to bring conservation and agricultural production together. In: Nehren U, Schlüter S, Raedig C, Sattler D, Hissa H (eds) Strategies and tools for a sustainable rural. Cham. Springer International Publishing

Prado FB, Vasconcelos FCW, Chiodi CK (2013) Regime jurídico da Mata Atlântica e o risco à sobrevivência in situ de espécies ameaçadas. Ambiente & Sociedade 17:1–16

Raedig C, Lautenbach S (2009) Broad-scale angiosperm diversity in Brazil's Mata Atlântica: using monographic data to assess prospects for conservation. In: Gaese H, Torrico JC, Wesenberg J, Schlüter S (eds) Biodiversity and land use systems in the fragmented Mata Atlântica of Rio de Janeiro. Cuvillier -Verlag, Göttingen, pp 217–243

Raedig C, Dormann CF, Hildebrandt A, Lautenbach S (2010) Reassessing Neotropical angiosperm distribution patterns based on monographic data: a geometric interpolation approach. Biodivers Conserv 19:1523–1546

Resasco J, Haddad NM, Orrock JL et al (2014) Landscape corridors can increase invasion by an exotic species and reduce diversity of native species. Ecology 95:2033–2039
Ribeiro MC, Metzger JP, Martensen AC et al (2009) The Brazilian Atlantic Forest: how much is left, and how is the remaining forest distributed? Implications for conservation. Biol Conserv 142:1141–1153
Ribeiro MC, Martensen AC, Metzger JP et al (2010) Brazilian Atlantic Forest: a shrinking biodiversity hotspot, In: Biodiversity hotspots, pp 405–343
Sanderson, J, Da Fonseca GAB, Galindo-Leal C et al (2006): Escaping the Minimalist Trap: Design and Implementation of Large-Scale Biodiversity Corridors. In: Crooks K.R. & Sanjayan M. (eds.): Connectivity Conservation. Cambridge University Press, Cambridge 14:620–648
Scheer MCB, Blum CT (2011) Arboreal diversity of the Atlantic Forest of Southern Brazil: from the beach ridges to the Paraná river, The dynamical processes of biodiversity – Case studies of evolution and spatial distribution. In: Grillo O (ed) The dynamical processes of biodiversity – Case studies of evolution and spatial distribution, Sociedade Chauá Brazil, Paraná, pp 109–134
Schlüter S, Nehren U, Sattler D, Raedig C (2018) The INTECRAL project. In: Nehren U, Schlüter S, Raedig C, Sattler D, Hissa H (eds) Strategies and tools for a sustainable rural Rio de Janeiro. Springer International Publishing
Secretaria do Meio Ambiente (2011) Agenda 21 Nova Friburgo, 122. http://www.agenda21comperj.com.br/sites/localhost/files/NovaFriburgo.pdf. Accessed 20 May 2015
Seoane SCE, Sandri DV, Longhi ST, Maranhão FLC (2010) Corredores ecológicos como ferramenta para a desfragmentação de florestas tropicais. EMBRAPA 30:207–216
SNUC – Sistema Nacional de Unidades de Conservação da Natureza (2000) Lei no 9.985, de 18 de julho de 2000; decreto no 4.340, de 22 de agosto de 2002
Soares-Filho B, Rajão R, Macedo M et al (2014) Cracking Brazil's Forest code. Science 344:363–364
SOS Mata Atlântica (2013) XII Edital do programa de incentivo às RPPNs da Mata Atlântcia, p 35
Souza PM, Ponciano NJ, Mata HTC (2007) Strutura fundiária das regiões Norte e Noroeste do Estado do Rio de Janeiro: 1972 a 1998. Rev Econ Sociol Rural 45:71–91
Stotz DF, Fitzpatrick JW, Parker III TA, Moskovits DK (1996) Neotropical birds: ecology and conservation. p 502
Suleiman S, Agarwal VC, Lal D, Sunusi A (2015) Optimal route location by least cost path (LCP) analysis using (GIS) a case study. International Journal of Scientific Engineering and Technology Research 4:9621–9625
Tabarelli M, Aguiar AV, Ribeiro MC et al (2010) Prospects for biodiversity conservation in the Atlantic Forest: lessons from aging human-modified landscapes. Biol Conserv 143:2328–2340
Tewksbury JJ, Levey DJ, Haddad NM (2002) Corridors affect plants, animals, and their interactions in fragmented landscapes. Proc Natl Acad Sci 99(20):12923–12926
Theobald DM (2006) Exploring the functional connectivity of landscapes using landscape networks. In: Crooks KR, Sanjayan M (eds) Connectivity conservation. Cambridge University Press, Cambridge, pp 416–444
Wiens JA (2006) Introduction: connectivity research – what are the issues? In: Crooks KR, Sanjayan M (eds) Connectivity conservation. Cambridge University Press, Cambridge, pp 23–27
Worboys GL, Lockwood M, Francis WL (2010) Challenges and opportunities for connectivity conservation. Connectivity Conservation Management. Earthscan London, London, pp 342–346
World Bank (2014) Project performance assessment report, Brazil. Rio de Janeiro sustainable integrated ecosystem management in production landscapes of North Northwestern fluminese project, p 74

Chapter 12
Applicability of the German Hydromorphological Assessment Approach to Tropical Rivers

Diana Birnbaum and Georg Lamberty

Abstract In addition to water quality, hydromorphological characteristics are crucial for the integrity of river ecosystems: Factors such as flow diversity, substrate variety, or riparian vegetation create diverse habitats for aquatic organisms with their specific demands. These habitats are impaired by channelization, longitudinal river fragmentation, and degradation of the riparian vegetation. In the state of Rio de Janeiro, rivers are exposed to increasing anthropogenic pressures mainly caused by urban growth, intensive agricultural land use, and alterations induced by climate change. While knowledge about the resulting deterioration of the water quality is well established, the knowledge base about consequences for the hydromorphological quality is thinner. Consequently, the hydromorphological potentials and deficits have to be identified and assessed to preserve or revitalize river ecosystems. For this purpose, the applicability of the German on-site method (LAWA-OS) was tested for hydromorphological assessment of streams. This method assesses the hydromorphological state of stream sections against predefined reference conditions. Based on the assessment results, the method identifies stream sections with different ecological development possibilities such as preservation, revitalization, or restriction. This article describes an on-site application test and formulates the applicability of the LAWA-OS method in Brazil by the example of the state of Rio de Janeiro.

Keywords River ecology · Fluvial geomorphology · Basin management · Environmental monitoring · Visual assessment

D. Birnbaum (✉) · G. Lamberty
Institute for Technology and Resources Management in the Tropics and Subtropics (ITT), TH Köln – University of Applied Sciences, Köln, Germany
e-mail: georg.lamberty@th-koeln.de

U. Nehren et al. (eds.), *Strategies and Tools for a Sustainable Rural Rio de Janeiro*, Springer Series on Environmental Management,
https://doi.org/10.1007/978-3-319-89644-1_12

12.1 Introduction

Watercourses are among the ecosystems that have been modified by humans in the most far-reaching manner (Carpenter et al. 2011). In addition to the effects of climate change, excessive water extraction, and the discharge of sewage, fertilizers, and pesticides, in particular hydromorphological interventions have a significant negative impact on ecological functionality of river systems (Vörösmarty et al. 2010). Among the most common human-induced hydromorphological habitat alterations are rectification, channelization, flow regulation, and flood prevention via dikes (Malmqvist and Rundle 2002).

These developments are equally applicable to rivers in the temperate regions as well as in the tropical regions (Dudgeon 2008; Moulton and Wantzen 2006). Several authors point out hydromorphological deficits typical to many tropical rivers in Brazil (Agostinho et al. 2005; Barletta et al. 2010; Sánchez-Argüello et al. 2010; O'Keeffe 2013; Ferreira et al. 2014; Irvine et al. 2016). Increasing deforestation and the lack of riparian vegetation, especially on steep slopes and in areas of extreme precipitation events, increase fine sediment intrusion and its ecological consequences within the water bodies (Wantzen 2006; Couceiro et al. 2007; Couceiro et al. 2009; Lorion and Kennedy 2009; de Souza et al. 2013; Garrastazú et al. 2015). The longitudinal continuity of rivers is impeded through increasing fragmentation by damming for generating electrical energy and for flood control (Callisto et al. 2005; Anderson et al. 2006; Greathouse et al. 2006; Agostinho et al. 2008; de Oliveira Naliato et al. 2009). Straightening and channelization are omnipresent in urban and rural areas (Stevaux et al. 2009; Burford et al. 2011; Mattos et al. 2014).

The devastating consequences of hydromorphological degradation of Brazilian rivers have led to an adaptation of environmental management with the target to return to more natural conditions of rivers (Veiga and Magrini 2011; Elabras Veiga and Magrini 2013; Araújo et al. 2015).

One major goal of Brazilian river management is to mitigate the impact of hydromorphological alterations and to restore natural or seminatural conditions, as formulated in the Brazilian National Water Law ("Lei de Águas," Law 9.433/97) and the Rio de Janeiro (RJ) state Water Resource Policy (State Law 3.239/99) (Araújo et al. 2015). However, to provide river managers with necessary information on hydromorphological potentials and deficits, the hydromorphological status has to be assessed in addition to water quality and quantity.

In contrast to the aforementioned goal, no standardized method addressing hydromorphological assessment is ready for an operationalized use in Brazil and applicable in large scale (Lima et al. 2010). Several Brazilian studies adopted and applied methods for hydromorphological assessment such as the US rapid assessment protocol (Callisto et al. 2001; Rodrigues and Castro 2008; Carvalho et al. 2014; Pereira et al. 2014; Kieling-Rubio et al. 2015; Machado et al. 2015) or the Australian River Styles framework (Nunes de Souza Lima and dos Santos Marçal 2013). Despite numerous strengths such as ease of use (i.e., US rapid assessment protocol) or a process-based, multidimensional assessment approach (i.e., Australian

River Styles framework) and a general potential for being applied in tropical regions, the abovementioned methods have a common weakness: The entire management process from the assessment of current conditions to the processing of assessment results for restoration planning is not covered (Araújo et al. 2015). In contrast, the German LAWA-OS method is characterized by a high level of standardization and detail (LAWA 2000). The central focus of the method is the identification of human-induced morphological pressures and the deviation of appropriate measures for reestablishing of near-natural conditions. The transferability of the LAWA-OS method to areas outside its originally designated region has been confirmed in several studies (Weiß et al. 2008; Sípek et al. 2010; Scheifhacken et al. 2012; Hajdukiewicz et al. 2017). However, the LAWA-OS method has not yet been applied in tropical regions.

The advantages of the LAWA-OS method in combination with its successful application in regions outside of Germany resulted in the selection of this method for the hydromorphological assessment of the study area. The original method was developed in 2000 by the German Working Group on water issues (LAWA). The LAWA-OS method was updated in 2012 by the Ministry for Climate Protection, Environment, Agriculture, Conservation, and Consumer Protection of the state of North Rhine-Westphalia (LANUV-NRW) by adding parameters and increasing assessment accuracy (Gellert et al. 2014).

The cost-efficient and standardized LAWA-OS method assesses morphological characteristics of rivers and creeks through a set of 26 single parameters (SPs) that are aggregated into six main parameters (MPs): MP-1 channel development, MP-2 longitudinal profile, MP-3 bed structure, MP-4 cross profile, MP-5 bank structure, and MP-6 adjacent land zone.

This study examines the transferability of the LAWA-OS method to tropical regions. In order to evaluate its potential applicability, the method was tested at 71 stream sections in RJ. The objectives are (i) to evaluate the applicability of the method to tropical streams, (ii) to quantify assessment bias through statistical analysis, and (iii) to give suggestions for adapting the method to natural and socioeconomic conditions in Brazil.

12.2 Materials and Methods

12.2.1 Study Area

The case study was conducted in RJ in the Mata Atlântica, Brazil's Atlantic Rainforest (Fig. 12.1). The Mata Atlântica stretches along the Atlantic coast of Brazil and extends to Paraguay, Argentina, and Uruguay. It is characterized by a rich biodiversity and endemism, which can be ascribed to a high variation in biotopes due to a wide latitudinal span of 38° and through an elevation that reaches from sea level to 1800 meters (Ribeiro et al. 2009). The studied stream sections are located in the municipal areas of Nova Friburgo and Italva.

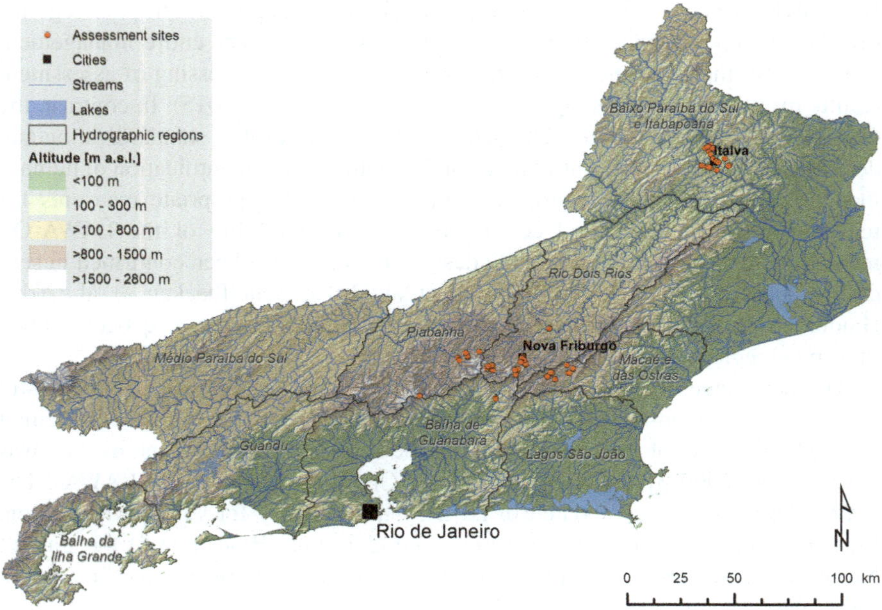

Fig. 12.1 Location of assessed stream sections (n = 71)

In Nova Friburgo, land use is dominated by agriculture, urban areas, and remaining rainforest fragments of the Mata Atlântica. Streams are characterized by coarse substrate and U-shaped or V-shaped valleys.

In Italva, the prevalent land use is pasture. Streams are mainly sand-dominated in flat floodplains, with little or no slopes and slow flow rates well below 0.5 m/s. The Muriaé River, the main river in Italva, was mapped at multiple sites.

12.2.2 Hydromorphological Assessment

The field survey was conducted by an on-site inspection of stream sections and by applying the field protocol according to the guidelines of the method (LANUV-NRW 2012). In total, 71 stream segments were assessed between March and April 2016; 42 sections were located in the municipality Nova Friburgo and 29 sections in Italva. The sections range from small creeks (<1 m channel width) to big streams (> 160 m width) covering all stream widths within RJ. The lengths of the assessed sections are 100 m, 500 m, or 1,000 m depending on the channel width.

The on-site survey consists of 26 single parameters (SPs). Within the assessment procedure, the single parameters aggregate into six main parameters (MPs), three zones (bed, banks, adjacent land), and an overall score for each stream section (Table 12.1).

Table 12.1 Parameter set for the updated version (LANUV-NRW 2012) of the original German LAWA-OS method (LAWA 2000)

Zones	Main parameters	Single parameters	Functional units
Streambed	MP-1 channel development	1-1 curvature/bends	Curvature, movability
		1-2 Erosion at bends	
		1-3 longitudinal bars	
		1-4 special channel structures	
	MP-2 longitudinal profile	2-1 transverse structures	Natural longitudinal structures, migration barriers
		2-2 pumping/culverts	
		2-3 backwater	
		2-4 transverse bars	
		2-5 flow variation	
		2-6 variation in depth	
		2-7 diverted reach	
	MP-3 bed structure	3-1 bed substrate	Type and distribution of substrates, bed fixation
		3-2 substrate diversity	
		3-3 bed protection	
		3-4 special bed features	
Banks	MP-4 cross profile	4-1 profile type	Section type, section depth, variation of width
		4-2 section depth	
		4-3 Erosion of width	
		4-4 variation of width	
		4-5 passage/bridge	
	MP-5 bank structure	5-1 Bank vegetation	Vegetation, bank protection, typical characteristics
		5-2 Bank protection	
		5-3 special bank features	
Adjacent land	MP-6 adjacent land zone	6-1 land use	Riparian surroundings, bank strips
		6-2 riparian zone	
		6-3 harmful land features	

Each single parameter allows for choosing one out of several parameter states. For instance, there are seven states for *SP 1–1 curvature/bends* that range from natural "meandering" to anthropogenically altered "straight" curvature. After assigning one of these states, the parameter is graded through comparing the current state with a predefined, type-specific reference state. A score from 1 to 7 is given to each single parameter, where 1 stands for an unchanged (natural) condition and 7 stands for a completely changed condition (Table 12.2). These scores are then aggregated to MP, zone, and overall scores by mean value calculation.

Two independent evaluation procedures are carried out for validation of the results: index-based and based on functional units. The index-based evaluation

Table 12.2 Assessment scale with scores and corresponding classes (LANUV-NRW 2012)

Class	Scoring index	Degree of hydro-morphological alteration
1	1.0–1.7	No alteration (natural)
2	1.8–2.6	Slight alt.
3	2.7–3.5	Moderate alt.
4	3.6–4.4	Distinct alt.
5	4.5–5.3	Obvious alt.
6	5.4–6.2	Strong alt.
7	6.3–7.0	Complete alteration

makes use of the above-described comparison of current and reference states and mathematically calculated scores. The evaluation based on functional units represents the surveyor's judgment of the situation in the field. Comparing the scores of the two evaluation procedures allows for a plausibility check of scores. In this study, the score differences between the two evaluation approaches indicate which parameters have to be adapted to properly assess tropical streams.

A detailed description of the assessment method is given by Gellert et al. (2014) and the user manual (LANUV-NRW 2012).

12.2.3 Analysis of Assessment Bias

Statistical tests were applied to identify significant score differences between the index-based evaluation and the evaluation based on functional units. The results of these tests served as the basis for adapting the LAWA-OS method to make it suitable to the conditions in the research area.

Differences between the two evaluation approaches were determined by applying the Wilcoxon signed-rank test (Wilcoxon 1945) for ordinal data (in this study scores from 1 to 7) and matched samples (in this study two evaluation procedures for each stream section). We tested for score differences on the main parameter (MP) level by ranking the score differences and calculating positive and negative rank sums. The statistically significant level was set at $p < 0.05$.

In addition to the Wilcoxon test, we applied the robustness indicator (RI) (Lamberty et al. 2016; Meier 2016) for identifying the assessment robustness of main parameters. This indicator depicts the ratio between paired samples with score differences smaller than two classes and the total number of paired samples. Deviations that are smaller than two classes are considered as acceptable (LANUV-NRW 2012). The RI values range from 0 to 1 and are divided into five robustness classes: >0.95 (very high), >0.80–0.095 (high), >0.65–0.80 (moderate), >0.5–0.65 (low), and ≤ 0.5 (very low).

In order to identify potential proxies to replace parameters that are difficult to assess, a Spearman rank correlation analysis was conducted (Caputo et al. 2008). In cases where parameters cannot be assessed (e.g., due to low visibility), a correlated parameter can be used as a proxy to derive the undetectable parameter state. Correlation coefficient values reach from −1 to 1, where the relationship is stronger the closer the coefficient is to ±1 ($p < 0.05$). Values were grouped according to their correlation: 0.00–0.19 (very weak), 0.20–0.39 (weak), 0.40–0.59 (moderate), 0.60–0.79 (strong), and 0.80–1.0 (very strong).

12.3 Results and Discussion

12.3.1 Data Collection and Score Distribution

The field survey delivered assessment scores for single parameters (SPs), main parameters (MPs), river zones, and an overall assessment per river section (Fig. 12.2). Three parameters show high shares of class 7 (*SP 1–3 longitudinal bars*, *SP 1–4 special channel structures*, and *SP 5–3 special bank features*). These parameters are marked with a minus sign (−) in Figure 12.2 to indicate the strong downgrading effect of the single parameters on the respective main parameters. The strong upgrading effect of *SP 2–4 transverse bars*, *SP 3–2 substrate diversity*, *SP 3–4 special bed features*, and *SP 4–1 profile type* is marked with a plus sign (+).

MP-2 longitudinal profile, *MP-3 bed structure*, and *MP-4 cross profile* tend to show qualitatively high scores, whereas *MP-1 channel development*, and especially *MP-5 bank structure*, and *MP-6 adjacent land zone* show qualitatively low scores.

The integrated zone assessment (streambed with MP-1, MP-2, and MP-3, banks with MP-4 and MP-5, adjacent land with MP-6) reflects the findings regarding the respective main parameters. The streambed receives relatively good scores due to good MP-2 and MP-3 scores. The qualitatively high scores for *MP-4 cross profile* and low scores for *MP-5 bank structure* lead to a moderate grading of the riverbank zone.

Generally, the overall assessment of the river sections results in qualitatively moderate or high scores. Almost 60% of the sections were graded with classes 1–4.

12.3.2 Index-Based Evaluation Versus Evaluation Based on Functional Units

As the method was applied outside its designated region (Germany), deviations between index-based evaluation and evaluation based on functional units are likely to result from stream types or landscape characteristics specific to the research area. In order to detect deviations between the two evaluation approaches, the Wilcoxon

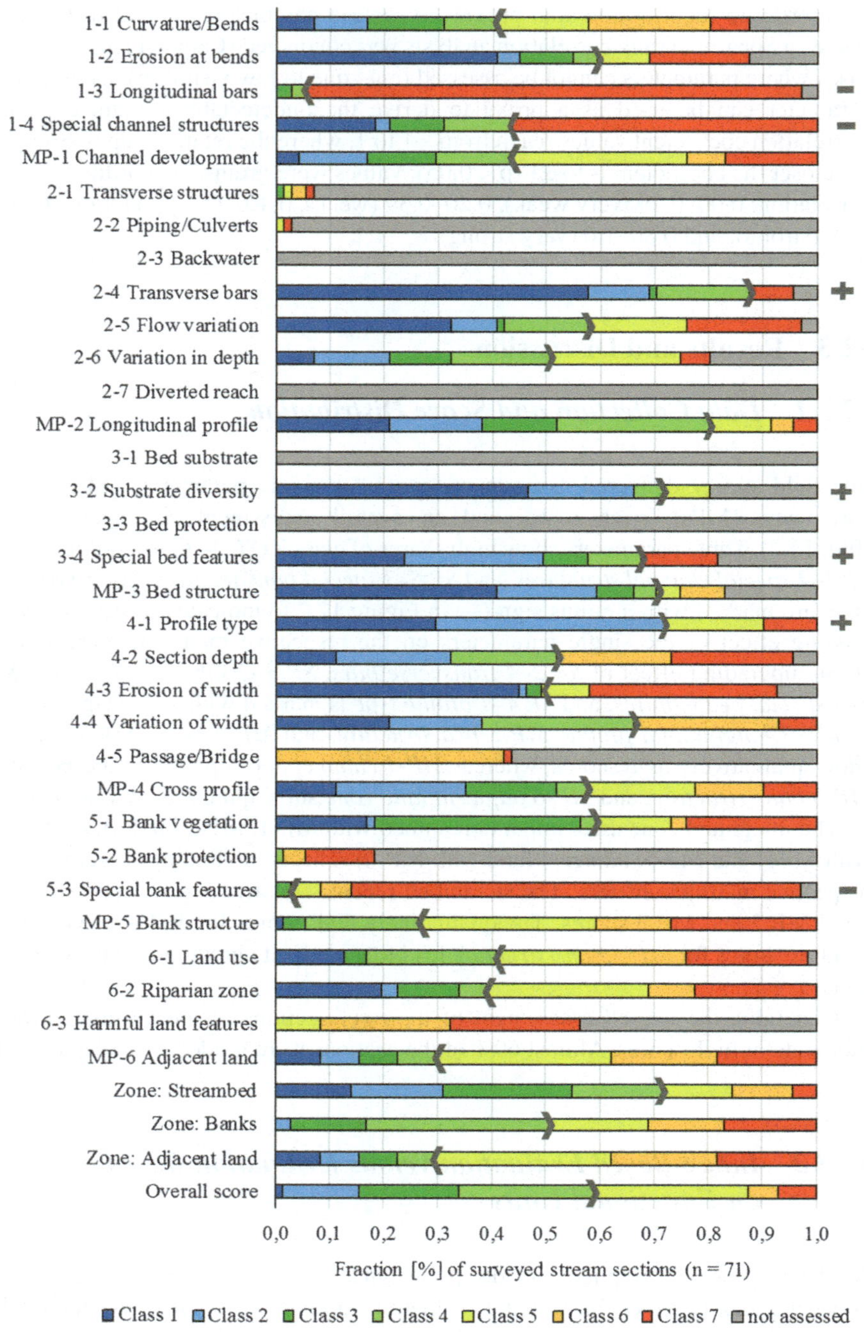

Fig. 12.2 Frequency distribution of assessment classes. The 26 single parameters are graded from 1 to 7 (number of river sections, n = 71). Parameters are not assessed if respective structures are not assessed due to a specific stream type or if they are not assessable

Table 12.3 Comparison of index-based evaluation and functional unit evaluation

Statistics	MP-1	MP-2	MP-3	MP-4	MP-5	MP-6
Index score (mean)	4.75	3.20	2.25	3.76	5.37	4.79
Funct. Unit score (mean)	4.28	3.49	2.24	3.73	5.08	4.90
Z	−3.95	−3.65	−2.43	−4.26	−3.41	−1.53
p	<0.01*	<0.01*	0.81	0.67	<0.01*	0.10
RI	0.87 (−)	0.97 (+)	1.00	1.00	0.99 (−)	1.00

signed-rank test was applied, and the robustness indicator (RI) was calculated (Table 12.3).

Assessment scores were compared on the level of main parameters (MPs) by applying the Wilcoxon signed-rank test (two-tailed; $p < 0.05$). The robustness indicator (RI) represents the fraction acceptable deviations (score differences ≤1). Plus (+) and minus (−) signs indicate positive and negative deviation trends from functional unit evaluation toward index-based evaluation.

The scores of *MP-1 channel development*, *MP-2 longitudinal profile*, and *MP-5 bank structure* differ significantly between the two evaluation approaches ($p < 0.05$). However, score differences of more than one class occur only rarely.

MP-1 and MP-5 show qualitatively lower index scores than functional unit scores. The surveyor judgment (functional unit evaluation) compensates the downgrading effect of *SP 1–3 longitudinal bars*, *SP 1–4 special channel structures*, and *SP 5–3 special bank features* caused by the mathematical calculation of MP scores (index evaluation). According to the surveyor's judgment, the hydromorphological structures specified by SP 1–3, SP 1–4, and SP 5–3 are not type-specific within the research area. Hence, the index-based score (class 7) for nonoccurrence is not appropriate.

For *MP-2 longitudinal profile*, functional unit scores are higher than index scores. Stream sections around Italva were characterized by sandy substrates where transverse bars (SP 2–4) occur rarely in nature (index-based score: class 1). From the surveyor's perspective, this qualitatively high score is not appropriate at several anthropogenically altered stream sections resulting in a qualitatively lower functional unit score for MP-2 compared to the index-based approach.

Scores for *MP-3 bed structure*, *MP-4 cross profile*, and *MP-6 adjacent land zone* do not differ significantly between the two evaluation approaches. All three main parameters show a RI value of 1, which means that only acceptable deviations of ≤1 class occur.

12.3.3 Correlation Analysis

Spearman rank correlations were utilized to examine the relationship between the single parameters (Table 12.4). These correlations can be used as proxies for nonassessable parameters (e.g., invisible bed substrate due to turbidity).

Table 12.4 Correlation between single parameter (SP) scores

SP	1-1	1-2	1-4	2-4	2-5	2-6	3-2	3-4	4-1	4-2	4-3	4-4	5-1	6-1	6-2
1-2	0.42**														
1-4	0.36**	0.33**													
2-4	0.26*	0.38**	0.44***												
2-5			0.49***												
2-6	0.65***	0.38**	0.51***	0.52**	0.65***										
3-2	0.46**		0.36**	0.36**	0.58***	0.51***									
3-4	0.40**	0.37**	0.60***	0.51***	0.65***	0.56***	0.52***								
4-1	0.51***	0.56***	0.68***	0.61***	0.35**	0.67***	0.42**	0.66***							
4-2	0.60***	0.44***	0.55***	0.31*	0.39**	0.68***	0.36**	0.42**	0.77***						
4-3	0.42**	0.49***	0.37**	0.39**	0.31*	0.62***	0.30*		0.69***	0.83***					
4-4	0.58***	0.51***	0.71***	0.66***	0.38**	0.73***	0.41**	0.55***	0.80***	0.56***	0.44***				
5-1		0.29*	0.39**	0.34**	0.25*	0.43**	0.38**	0.46***	0.60***	0.36**	0.40**	0.55***			
5-2			0.57*												
6-1	0.33**	0.52***	0.58***	0.56***		0.53***		0.45***	0.76***	0.44***	0.47***	0.77***	0.62***		
6-2	0.35**	0.30*	0.42***	0.26*		0.48***	0.49***	0.43**	0.54***	0.48***	0.44***	0.47***	0.77***	0.54***	
6-3	0.34*								0.36*			0.36*		0.45**	0.33*

Spearman rank correlation ($p < 0.05$) of single parameters; correlations with $r > 0.6$ are underlined. Significance levels: * $p < 0.05$; ** $p < 0.01$; *** $p < 0.001$ (two-tailed)

Fig. 12.3 Small-sized tributary of the Muriaé River (Corrego Maribondo) with low visibility of riverbed parameters

Concerning all significant correlations (n = 102), only two very strong correlations (r = 0.80–1.0) were examined. Furthermore, 19 strong correlations (r = 0.60–0.79), 48 moderate correlations (r = 0.40–0.59), and 33 weak correlations (r = 0.20–0.39) could be observed.

Regarding the objectives of this study, especially the parameters *SP 4–1 profile type* and *SP 4–4 variation of width* have a high potential for serving as proxies to derive parameters, which regularly are difficult to assess (i.e., due to limited visibility of the riverbed) (Fig. 12.3).

Both parameters, SP 4–1 and SP 4–4, show strong or very strong correlations with nine other parameters. Their good accessibility and their correlations with riverbed parameters provide a good potential for serving as proxies:

- *SP 1–4 special channel structures*; proxy: SP 4–1 (r = 0.68) or SP 4–4 (r = 0.71)
- *SP 2–6 variation in depth*; proxy: SP 4–1 (r = 0.67) or SP 4–4 (r = 0.73)
- *SP 3–4 special bed features*; proxy: SP 4–1 (r = 0.66)
- *SP 4–2 section depth*; proxy: SP 4–1 (r = 0.77)

12.3.4 Applicability of the Main Parameters

MP-1 channel development: For MP-1, scores from index-based evaluation and evaluation based on functional units differ occasionally by more than one class (RI = 0.87). It is likely that the cause of this deviation lies in the natural rare occurrence of longitudinal bars and special channel structures (see also Fig. 12.2). The nonoccurrence of these structures regularly results in disproportionately low scores of MP-1 for natural or near-natural stream sections. In contrast to the index-based evaluation, the functional unit evaluation takes these type-specific conditions into account. The downgrading effect of longitudinal bars is especially inappropriate for streams in V-shaped valleys. According to the method's guidelines, *SP 1–1 curvature/bends* and *SP 1–2 erosion at bends* are not assessed for this valley type; this means that the MP-1 score is only composed of *SP 1–3 longitudinal bars* and *SP 1–4 special channel structures*. As both of these structures are uncommon within the research area, the index-based quality score is unreasonably low for natural streams in V-shaped valleys. As a possible adaption, SP 1–3 and SP 1–4 can be merged, thus including longitudinal bars as special channel structures. Moreover, the parameters *SP 1.3 longitudinal bar, SP 1–4 special channel features, SP 3–4 special bed features*, and *SP 5.3 special bank features* could be merged together into one parameter called "special features."

MP-2 longitudinal profile: For MP-2, the evaluation based on functional units leads to qualitatively lower scores than the index-based evaluation. In case of the stream-type "sand-dominated streams in trough and U-shaped valleys with floodplain," assigned to most of the mapped streams located around Italva (n = 15), *transverse bars (SP 2–4)* do not occur naturally. However, the distinction between natural (class 1) or anthropogenic nonoccurrence (class 7) of transverse bars is sometimes unclear due to limited knowledge about the river's history. In order to simplify the evaluation, SP 2–4 can be omitted completely and replaced by the parameter "riffles and pools" for river types with coarse substrate.

MP-3 bed structure: Especially for the sand-dominated stream sections around Italva, the ability to assess MP-3 was limited due to low visibility. In such cases, MP-3 could be derived by applying the abovementioned proxy parameters.

MP-4 cross profile: On-site surveying confirmed that the *SP 4–1 profile type* is difficult to detect in terms of clearly distinguishing between parameter states such as "approximately natural section" and "decaying technical standard profile" (Lamberty et al. 2016; Meier 2016). As this parameter can serve as a proxy for several other parameters, special emphasis should be put on the description of parameter states in the method's guidelines to omit assessment inconsistency.

MP-5 bank structure: Difficulties with assessing parameter *SP 5–1 bank vegetation* can be attributed to a lack of knowledge regarding the local riparian vegetation, especially regarding the distinction between naturally occurring and invasive riparian trees and shrubs. Assessing *SP 5–3 special bank features* also requires adaption to local conditions as the features specified in the method's guidelines rarely occurred within the research area (see also Fig. 12.2). Many of the natural sections did not show any of the bank features listed in the field protocol. Instead of woody structures, it was rather boulders and rocky walls that characterized the banks of natural sections.

MP-6 adjacent land zone: No particularities were encountered with assessing MP-6 within the research area. Therefore no adjustments of the method are suggested with regard to this main parameter.

12.3.5 General Applicability of the Method for Tropical Streams

In general, the LAWA-OS method is applicable in the research area. However, the following aspects have to be considered with respect to the general applicability of the method for tropical streams. These aspects are concerned with natural conditions as well as socioeconomic considerations.

Segment length: Depending on the width of the river, standardized river sections (100, 500 or 1,000 meters) are assessed. Considering cost-efficiency, several consecutive and homogenous river sections could be combined to assess them as one unit.

Hydrological regime: The LAWA-OS method examines the current hydromorphological state of streams. Seasonal or interannual variability due to the hydrological regime is not considered. However, especially in tropical regions, hydromorphological parameters are influenced by seasonal variance (wet and dry season) (Rodrigues et al. 2012).

Evaluation based on functional units: Cross-checking the results through two independent evaluation procedures is generally judged as a positive asset of the LAWA-OS method (Scheifhacken et al. 2012; Meier et al. 2013; Gellert et al. 2014). Also in our study, comparing the results of the two evaluation approaches for plausibility checks proved to be helpful. However, once the method is adapted to tropical rivers, the evaluation based on functional units might become redundant. This evaluation approach also turned out to be time-consuming and vague.

Access: One of the main problems in applying the hydromorphological assessment method in the research area was the difficult access to river observation sites. Often, there is dense vegetation that cannot be traversed. Additionally, caution is required because of potentially poisonous spiders and snakes. Complementing the field survey with remote sensing tools (i.e., drones) could be a solution to this problem. For large rivers such as the Muriaé, mapping from a boat might be an option.

River types and reference states: The underlying physical processes of erosion, transport, and deposition as a result of the interaction of the two main agents (water flow and substrate) that form hydromorphological structures are generally identical in temperate regions and (sub)-tropical regions (Boulton et al. 2008). We therefore assume that the hydromorphological river types based on dominating substrate and valley form – as used in the LAWA-OS method – are in principle also suitable for tropical rivers. However, in-depth research on tropical ecosystem functioning and its factors, such as geological and geomorphological characteristics, rainfall and discharge, temperature, and riparian vegetation, is required (Gupta 1993; Latrubesse et al. 2005; Boyero et al. 2009), especially with respect to certain hydromorphological parameters and their natural reference states.

12.4 Conclusions

During this study, the German LAWA-OS method proved to correctly assess the hydromorphological status of streams and rivers in the tropics and subtropics. Therefore, we conclude that the method has a great potential in such regions to provide a sound foundation for decision making on preservation and restoration of rivers and adjacent floodplain zones. In general, the study showed that an application of the method is possible in tropical areas, if some adjustments are made. Adjustments include merging redundant parameters in the protocol, focusing merely on the index-based evaluation, dividing segments according to natural homogeneity, and using remote assessment tools (e.g., drones) when access is limited. For the application of the German LAWA-OS method for tropical rivers, the following recommendations are given:

- The protocol should be adapted considering the abovementioned suggestions concerning parameter selection and assessment procedure.
- As a basis for the adjusted method, Brazilian hydromorphological stream types should be identified, and the assessment of the parameters should be adjusted accordingly.
- Field manuals in English, Spanish, and Portuguese should provide descriptions of the regional/local vegetation, valley types, and soil types.
- Software (e.g., mobile apps) should be provided to facilitate data collection, storage, and post-processing.
- Standardized training of staff including certification should be incorporated into field campaigns to ensure assessment robustness and data comparability.

References

Agostinho AA, Thomaz SM, Gomes LC (2005) Conservation of the biodiversity of Brazil's inland waters. Conserv Biol 19(3):646–652. https://doi.org/10.1111/j.1523-1739.2005.00701.x

Agostinho AA, Pelicice FM, Gomes LC (2008) Dams and the fish fauna of the Neotropical region: impacts and management related to diversity and fisheries. Braz J Biol 68(4 SUPPL):1119–1132. https://doi.org/10.1590/s1519-69842008000500019

Anderson EP, Pringle CM, Rojas M (2006) Transforming tropical rivers: an environmental perspective on hydropower development in Costa Rica. Aquat Conserv Mar Freshwat Ecosyst 16(7):679–693. https://doi.org/10.1002/aqc.806

Araújo RS, Alves MDG, Condesso de Melo MT et al (2015) Water resource management: a comparative evaluation of Brazil, Rio de Janeiro, the European Union, and Portugal. Sci Total Environ 511:815–828. https://doi.org/10.1016/j.scitotenv.2014.11.098

Barletta M, Jaureguizar AJ, Baigun C et al (2010) Fish and aquatic habitat conservation in South America: a continental overview with emphasis on neotropical systems. J Fish Biol 76(9):2118–2176. https://doi.org/10.1111/j.1095-8649.2010.02684.x

Boulton AJ, Boyero L, Covich AP et al (2008) Are tropical streams Different from temperate streams? In: Dudgeon D (ed) Tropical stream ecology. Academic, London, pp 257–284

Boyero L, Ramírez A, Dudgeon D, Pearson RG (2009) Are tropical streams really different? J N Am Benthol Soc 28(2):397–403

Burford MA, Revill AT, Palmer DW et al (2011) River regulation alters drivers of primary productivity along a tropical river-estuary system. Mar Freshw Res 62(2):141–151. https://doi.org/10.1071/mf10224

Callisto M, Ferreira WR, Moreno P et al (2001) Aplicação de um protocolo de avaliação rápida da diversidade de habitats em atividades de ensino e pesquisa (MG-RJ). Acta Limnol Bras 14(1):91–98. https://doi.org/10.1590/S2179-975X2012005000024

Callisto M, Goulart M, Barbosa FA, Rocha O (2005) Biodiversity assessment of benthic macroinvertebrates along a reservoir cascade in the lower São Francisco river (northeastern Brazil). Braz J Biol = Rev Bras Biol 65(2):229–240

Caputo A, Fahrmeir L, Künstler R et al (2008) Arbeitsbuch Statistik. Springer, Berlin/Heidelberg

Carpenter SR, Stanley EH, Vander Zanden MJ (2011) State of the world's freshwater ecosystems: physical, chemical, and biological changes. Annu Rev Environ Resour 36:75–99. https://doi.org/10.1146/annurev-environ-021810-094524

Carvalho EM, BA B, Pereira NS (2014) Rapid assessment of habitat diversity in a lotic environment. Interbio 8(1):45–55. https://doi.org/10.1590/s1519-69842008000500007

Couceiro SRM, Hamada N, Luz SLB et al (2007) Deforestation and sewage effects on aquatic macroinvertebrates in urban streams in Manaus, Amazonas, Brazil. Hydrobiologia 575(1):271–284. https://doi.org/10.1007/s10750-006-0373-z

Couceiro SRM, Hamada N, Forsberg BR, Padovesi-Fonseca C (2009) Effects of anthropogenic silt on aquatic macroinvertebrates and abiotic variables in streams in the Brazilian Amazon. J Soils Sediments 10(1):89–103. https://doi.org/10.1007/s11368-009-0148-z

de Oliveira Naliato DA, Nogueira MG, Perbiche-Neves G (2009) Discharge pulses of hydroelectric dams and their effects in the downstream limnological conditions: a case study in a large tropical river (SE Brazil). Lakes Reserv Res Manag 14(4):301–314. https://doi.org/10.1111/j.1440-1770.2009.00414.x

de Souza ALTD, Fonseca DG, Libório RA, Tanaka MO (2013) Influence of riparian vegetation and forest structure on the water quality of rural low-order streams in SE Brazil. For Ecol Manag 298:12–18

Dudgeon D (2008) Tropical stream ecology. Academic, Amsterdam/Boston

Elabras Veiga LB, Magrini A (2013) The Brazilian water resources management policy: fifteen years of success and challenges. Water Resour Manag 27(7):2287–2302. https://doi.org/10.1007/s11269-013-0288-1

Ferreira WR, Ligeiro R, Macedo DR et al (2014) Importance of environmental factors for the richness and distribution of benthic macroinvertebrates in tropical headwater streams. Freshw Sci 33(3):860–871. https://doi.org/10.1086/676951

Garrastazú MC, Mendonça SD, Horokoski TT et al (2015) Carbon sequestration and riparian zones: assessing the impacts of changing regulatory practices in southern Brazil. Land Use Policy 42:329–339. https://doi.org/10.1016/j.landusepol.2014.08.003

Gellert G, Pottgiesser T, Euler T (2014) Assessment of the structural quality of streams in Germany – basic description and current status. Environ Monit Assess 186(6):3365–3378. https://doi.org/10.1007/s10661-014-3623-y

Greathouse EA, Pringle CM, Holmquist JG (2006) Conservation and management of migratory fauna: dams in tropical streams of Puerto Rico. Aquat Conserv Mar Freshwat Ecosyst 16(7):695–712. https://doi.org/10.1002/aqc.804

Gupta A (1993) The changing geomorphology of the humid tropics. Geomorphology 7(1–3):165–186

Hajdukiewicz H, Wyżga B, Zawiejska J et al (2017) Assessment of river hydromorphological quality for restoration purposes: an example of the application of RHQ method to a Polish Carpathian river. Acta Geophys 65:1–18. https://doi.org/10.1007/s11600-017-0044-7

Irvine K, Castello L, Junqueira A, Moulton T (2016) Linking ecology with social development for tropical aquatic conservation. Aquat Conserv Mar Freshwat Ecosyst 26(5):917–941. https://doi.org/10.1002/aqc.2706

Kieling-Rubio MA, Benvenuti T, Costa GM et al (2015) Integrated environmental assessment of streams in the Sinos River basin in the state of Rio Grande do Sul, Brazil. Braz J Biol 75(2):S105–S113. https://doi.org/10.1590/1519-6984.1013

Lamberty G, Zumbroich T, Ribbe L, Souvignet M (2016) Quantifying bias in hydromorphological monitoring: an evaluation of the German LAWA-OS method. Environ Earth Sci 75(22):1–17. https://doi.org/10.1007/s12665-016-6241-x

LANUV-NRW (2012) Gewässerstruktur in Nordrhein-Westfalen: Kartieranleitung für die kleinen bis großen Fließgewässer – LANUV Arbeitsblatt Nr. 18. Pottgiesser T, Müller A (eds) Landesamt für Natur, Umwelt und Verbraucherschutz Nordrhein-Westfalen (Hrsg.). Recklinghausen

Latrubesse EM, Stevaux JC, Sinha R (2005) Tropical rivers. Geomorphology 70(3–4):187–206. https://doi.org/10.1016/j.geomorph.2005.02.005

LAWA (2000) Gewässerstrukturgütekartierung in der Bundesrepublik Deutschland: Verfahren für kleine und mittelgroße Fließgewässer – Empfehlung. Linnenweber, C, Friedrich G, Lacombe J (eds). Länderarbeitsgemeinschaft Wasser LAWA (Hrsg.). Schwerin

Lima FP, Muniz JN, de Marco P Jr (2010) Evaluating Brazilian conservation projects: the weak link between practice and theory. Natureza a Conservacao 8(1):41–45. https://doi.org/10.4322/natcon.00801006

Lorion CM, Kennedy BP (2009) Relationships between deforestation, riparian forest buffers and benthic macroinvertebrates in neotropical headwater streams. Freshw Biol 54(1):165–180

Machado CS, Alves RIS, Fregonesi BM et al (2015) Integrating three tools for the environmental assessment of the Pardo River, Brazil. Environ Monit Assess 187(9). https://doi.org/10.1007/s10661-015-4788-8

Malmqvist B, Rundle S (2002) Threats to the running water ecosystems of the world. Environ Conserv 29(2):134–153. https://doi.org/10.1017/s0376892902000097

Mattos TM, Costa MR, Pinto BCT et al (2014) To what extent are the fish compositions of a regulated river related to physico-chemical variables and habitat structure? Environ Biol Fish 97(6):717–730. https://doi.org/10.1007/s10641-013-0175-x

Meier G (2016) Bewertungsrobustheit der Gewässerstrukturkartierung nach dem Deutschen Vor-Ort-Verfahren (Rating robustness of hydromorphological assessment according to the German on-site method). Bonn University – Department of Geography (Dissertation)

Meier G, Zumbroich T, Roehrig J (2013) Hydromorphological assessment as a tool for river basin management: the German field survey method. J Nat Res Dev 3:14–26. https://doi.org/10.5027/jnrd.v3i0.02

Moulton TP, Wantzen KM (2006) Conservation of tropical streams – special questions or conventional paradigms? Aquat Conserv Mar Freshwat Ecosyst 16(7):659–663. https://doi.org/10.1002/aqc.814

Nunes de Souza Lima R, dos Santos Marçal M (2013) Geomorphic assessment of Macaé catchment – RJ based on River Styles classification framework. Revista Brasileira de Geomorfologia 14(2):171–179

O'Keeffe J (2013) Rivers, time and conservation, especially in developing countries. Aquat Conserv Mar Freshwat Ecosyst 23(2):184–188. https://doi.org/10.1002/aqc.2347

Pereira PS, Fernandes LAC, Dias RJP et al (2014) Ecological water quality assessment in the Guapiaçu-Macacu hydrographic complex (Rio de Janeiro, Brazil) using multiple indicators. Rev Ambiente Agua 9(3):409–423. https://doi.org/10.4136/ambi-agua.1397

Ribeiro MC, Metzger JP, Martensen AC et al (2009) The Brazilian Atlantic Forest: how much is left, and how is the remaining forest distributed? Implications for conservation. Biol Conserv 142(6):1141–1153. https://doi.org/10.1016/j.biocon.2009.02.021

Rodrigues ASL, Castro PTA (2008) Adaption of a rapid assessment protocol for rivers on rocky meadows. Acta Limnol Bras 20(4):291–303. https://doi.org/10.1590/S2179-975X2012005000024

Rodrigues ASL, Malafaia G, Costa AT, Nalini Júnior HA (2012) Adequação e avaliação da aplicabilidade de um Protocolo de Avaliação Rápida na bacia do rio Gualaxo do Norte, Leste-Sudeste do Quadrilátero Ferrífero, MG, Brasil. Revista Ambiente & Água 7:231–244

Sánchez-Argüello R, Cornejo A, Pearson RG, Boyero L (2010) Spatial and temporal variation of stream communities in a human-affected tropical watershed. Ann Limnol 46(3):149–156

Scheifhacken N, Haase U, Gram-Radu L et al (2012) How to assess hydromorphology? A comparison of Ukrainian and German approaches. Environ Earth Sci 65(5):1483–1499. https://doi.org/10.1007/s12665-011-1218-2

Sípek V, Matousková M, Dvorák M (2010) Comparative analysis of selected hydromorphological assessment methods. Environ Monit Assess 169(1–4):309–319

Stevaux JC, Martins DP, Meurer M (2009) Changes in a large regulated tropical river: the Paraná River downstream from the Porto Primavera Dam, Brazil. Geomorphology 113(3–4):230–238. https://doi.org/10.1016/j.geomorph.2009.03.015

Veiga LBE, Magrini A (2011) Water resources management: suggestions to the Brazilian model based on the American experience. WIT Trans Ecol Environ 145:39–50. https://doi.org/10.2495/WRM110041

Vörösmarty CJ, McIntyre PB, Gessner MO et al (2010) Global threats to human water security and river biodiversity. Nature 467(7315):555–561. https://doi.org/10.1038/nature09440

Wantzen KM (2006) Physical pollution: effects of gully erosion on benthic invertebrates in a tropical clear-water stream. Aquat Conserv Mar Freshwat Ecosyst 16(7):733–749. https://doi.org/10.1002/aqc.813

Weiß A, Matouskova M, Matschullat J (2008) Hydromorphological assessment within the EU-water framework directive-trans-boundary cooperation and application to different water basins. Hydrobiologia 603(1):53–72. https://doi.org/10.1007/s10750-007-9247-2

Wilcoxon F (1945) Individual comparisons by ranking methods. Biom Bull 1(6):80–83. https://doi.org/10.2307/3001968

Chapter 13
Use of Native Plant Species for Ecological Restoration and Rehabilitation Measures in Southeast Brazil

Dietmar Sattler, Claudia Raedig, Anja Hebner, and Jens Wesenberg

Abstract Southeast Brazil is the economic backbone of the country where about 70% of the national GDP is generated. The continuously growing pressure from rural and industrial land use is causing progressive land degradation, especially within the unique Atlantic Forest dominion of this region. Restoration and rehabilitation of degraded areas or of impact caused by building activities are still scarce and carried out using mainly technical engineering methods. Natural engineering or soil bioengineering measures using living plant material are very rare, and if applied, they mostly use non-native plants already proven to work in other tropical regions. As the demand for bioengineered rehabilitation and ecological restoration is expected to increase, suitable plant material is urgently needed. The use of non-native plants for engineering carries several ecological risks, in particular the invasive spread of these plants. This chapter provides a brief overview of the actual state of the art regarding the use of native and non-native plant species in rehabilitation and restoration measures applied in the Atlantic Forest dominion. Furthermore, first results are presented from a case study on bioengineered rehabilitation of a degraded pasture where plants native to Southeast Brazil were successfully used.

D. Sattler (✉)
Institute of Geography, Physical Geography and Environmental Research Leipzig University, Leipzig, Germany
e-mail: sattler@uni-leipzig.de

C. Raedig
Institute for Technology and Resources Management in the Tropics and Subtropics (ITT), TH Köln – University of Applied Sciences, Köln, Germany
e-mail: claudia.raedig@th-koeln.de

A. Hebner
Business Unit BioPlanta, Vita 34 AG, Leipzig, Germany
e-mail: anja.hebner@vita34.de

J. Wesenberg
Botany Department, Senckenberg Museum of Natural History Görlitz, Görlitz, Germany
e-mail: jens.wesenberg@senckenberg.de

U. Nehren et al. (eds.), *Strategies and Tools for a Sustainable Rural Rio de Janeiro*, Springer Series on Environmental Management,
https://doi.org/10.1007/978-3-319-89644-1_13

Keywords Soil bioengineering · Atlantic Forest native species · Pasture degradation · Erosion control

13.1 Introduction

In Southeast Brazil, rural and industrial land use is causing progressive land degradation mainly due to unsustainable practices that are being applied. Furthermore, the unique Atlantic Forest (Mata Atlântica), considered as one of the top hotspots of biodiversity worldwide (Eisenlohr et al. 2015), is suffering from deforestation and ecological degradation, depleting its capacity to provide essential ecosystem services (Ribeiro et al. 2011). Hence, overexploitation, land use intensification, and urbanization are additionally degrading the landscape's geo-ecological resilience and increasing the vulnerability to natural hazards and climatic extreme events, which could be counteracted and reduced by, among others, ecosystem conservation and restoration (Alexander et al. 2016; Lange et al. 2018, Sattler et al. 2017a). Land and soil degradation has been widely studied in Brazil, and afforestation and rehabilitation knowledge is well elaborated (e.g., Tavares et al. 2008; Moraes et al. 2013; Muniz et al. 2013; Guerra and Jorge 2014; Santana et al. 2015). Due to the legal requirements of the National Forest Law regulating forest protection on private and public properties, landowners in the Atlantic Forest are obliged to protect 20% of their forest areas (legal reserves) as well as riverine forests and other areas essential for biodiversity and ecosystem services protection, denoted as areas of permanent preservation – APP (Guzmán Wolfhard and Raedig 2018; Piratelli et al. 2018). Based on the most recent version of the Brazilian Forest Code (Brazilian Federal Law 12.727, 2012), a minimum of six million hectares of degraded Atlantic Forest needs to be restored (Pinto et al. 2014; Soares-Filho et al. 2014). Considering this extraordinary challenge for conserving and maintaining biodiversity and ecosystem services provided by the Atlantic Forest, a large-scale nongovernmental initiative, the "Atlantic Forest Restoration Pact," was established in 2009. This diverse set of more than 260 stakeholders (governmental agencies, private sector, NGOs, and research institutions) has the goal to restore 15 million ha in the entire biome of the Atlantic Forest by 2050, explicitly integrating pastures with slopes ≥15% as they are especially prone to soil erosion (Pinto et al. 2014). Hence, there is an increasing awareness of the necessity of restoration of degraded forests, pastures, riverine vegetation, and even urban areas. Moreover, growing efforts in afforesting such degraded areas are evident. On the other hand, implementation of soil bioengineering (Portuguese: *Engenharia natural*) methods for rehabilitation of ecosystem services of farmlands without necessarily restoring the entire native biodiversity by retransforming degraded agricultural areas into forest is still scarce. For afforestation, the paradigm to use plant species native to the Atlantic Forest is widely accepted and put into practice (Rodrigues et al. 2011), as it is an evident and easily recognized concept when taking the unique biodiversity of natural forest remnants into account. In soil bioengineering, if applied, mainly exotic plants are used that have been internationally proven to be effective in tropical environments. This is somehow contradictory to the use of autochthonous species in afforestation measures. In this chapter, we would like to foster the transfer of the native plant use principle to the soil bioengineering sector, exploring the opportunities given by the vast diversity of Atlantic Forest plant species for possible applications in rehabilitation and restoration measures. Furthermore, we present some first results of a case study carried out in the municipality of Itaocara, Rio de Janeiro (RJ) state, where Atlantic Forest native plants were successfully used for the bioengineered rehabilitation of a degraded pasture slope.

13.2 Plants for Rehabilitation of Degraded Areas

Due to excessive and unsustainable land use such as overstocked pasturing on slopes, soil degradation has become a severe and abundant problem in Southeast Brazil and throughout the entire country (Sparovek et al. 2007; Sattler et al. 2017b). The rehabilitation of these degraded areas is still an immense challenge in combatting the environmental, social, and economic effects of deteriorated geo-ecological landscape conditions. A common intervention strategy is afforestation of abandoned pasture land and restoration of riverine forests, which additionally supports ecological connectivity and carbon storage (Pardini et al. 2005; Sattler et al. 2014). The use of native plants for this kind of measures is agreed in practice as there is a broad awareness of ecological quality and biodiversity conservation, especially in the dominion of the Atlantic Forest. In RJ, several guidelines for afforestation measures point out the key role of native plants as catalyzers of natural succession by providing habitat for the native fauna such as insects, birds and small mammals. These animals pollinate flowers, disperse the diaspores of native species planted and bring in new plant seeds from surrounding natural vegetation areas (Moraes et al. 2013; Santana et al. 2015). There is a broad range of native species suggested for afforestation within the dominion of the Atlantic Forest. In a technical manual for the restoration of degraded areas in RJ, Moraes et al. (2013) suggest 143 native tree species: 17 pioneers and 55 species of initial succession (among them 26 Fabaceae), 43 late successional, and 28 climax or mature forest species to be used for afforestation. In 2010, Alonso et al. (2014) evaluated 70 private and public nurseries in RJ with regard to the number of native Atlantic Forest species produced. At this time, 277 native forest species were available, almost a quarter of them (63) from the Fabaceae family with many species known for their ability to improve soil fertility by nitrogen fixation. Even though the authors considered the number of native species produced in RJ nurseries as low when compared to the rich Atlantic Forest flora with about 6,900 endemic vascular plant species (Stehmann et al. 2009), there is at least a reasonable number of species available for afforestation measures.

In Brazil, natural engineering with bioengineered measures represents a comparatively new field of action for restoring degraded soils, river courses and streambanks, and landslides or areas prone to erosion due to road construction or other building activities. Those measures are mainly based on the application of living plants and/or living parts of plants, mostly combined with organic material (wood and fibers) and other building materials such as metal, stones, gravel, and sand (Burri et al. 2009; Evette et al. 2009). In contrast to traditional civil engineering, bioengineered rehabilitation measures use living plants to stabilize the soil structure and increase water infiltration via the plant roots, to protect the soil from water erosion via the vegetation cover, to diminish surface runoff and interflow on the slopes, and to work as a barrier for soil material moving downslope (Zuazo and Pleguezuelo 2008; Preti and Giadrossich 2009; Stokes et al. 2009). For example, in tropical mountain regions such as the Serra do Mar range in Southeast Brazil, some landslides that had resulted from heavy rainfalls, degraded river banks, and road embankments have been stabilized with geotextiles combined with vetiver (*Vetiveria zizanioides* (L.) Nash), a deep rooting grass that is native to tropical India (Balangcod et al. 2015; Machado et al. 2015; Lange et al. 2018). Most of slope stabilizing measures are located in direct vicinity to natural forests or other natural environments. Hence, the plant species used in such implementations should be of native origin to provide ecologically valuable habitat for native insects, reptiles, birds, and mammals and to prevent possible invasive spreading of exotic plants. The vast majority of commercially available material for soil bioengineering in Brazil is of exotic origin, such as the Central American legumes *Gliricidia sepium* (Jacq.) Kunth ex Walp. and *Leucaena leucocephala* (Lam.) de Wit. (Pereira 2006). The latter one is especially known for its invasive potential (Pyšek et al. 2017). Nevertheless, some native plant material has been successfully proven for different bioengineering purposes, such as native *Erythrina* spp., which is used as living poles for fences (Matos et al. 2005).

In Southeast Brazil, fast-growing bamboos have been used for river course stabilization (Durlo and Sutili 2012) or for bioengineered erosion control measures such as the one applied by Clayton-Barbosa (2012) with Chinese *Bambusa multiplex* (Lour.) Raeusch. ex Schult. and Schult. Bamboos are also very commonly used as green demarcation walls between rural properties and shade for cattle, but most of them are as well non-native to the Mata Atlântica region, such as Asian *Phyllostachys* spp., *Dendrocalamus* spp., and *Bambusa* spp. Some exotic bamboo species such as *Bambusa textilis* McClure and *Bambusa vulgaris* Schrad. ex J.C. Wendl. are even considered invasive (Zenni and Ziller 2011). Even though some bamboo species are native to Southeast Brazil (*Chusquea* spp. or *Guadua tagoara* (Nees) Kunth), they are only rarely cultivated, and if so, they can become problematic since these bamboos tend to spread out rapidly in secondary or disturbed mature forests altering the entire forest structure by their overabundance (Manhães 2008; Lima et al. 2012).

13.3 The Itaocara Case Study

Within the framework of the INTECRAL project (Schlüter et al. 2018), a pilot rehabilitation measure (PRM) on a degraded pasture slope was carried out in the municipality of Itaocara, RJ, by implementing hedgerow terraces to control and stop erosion caused by surface runoff and interflow as a result of cattle overstocking. In addition to erosion control, the three hedgerow terraces installed at the foothill, mid-slope, and upper-slope positions should also provide a physical barrier structure for cattle. Furthermore, the planted hedges work as living fences, allowing for the temporary isolation of pasture parts for grass recovery or for the subsequent introduction of alternative and more sustainable pasture systems such as rotational farming. Details on site selection and implementation procedure are described by Seliger et al. (2018) in this volume.

Hedgerow terraces are usually installed using rooted or unrooted live cuttings of shrub or tree species with a high capacity of stump sprouting (Studer and Zeh 2014). In preparation of the PRM, it was difficult to find information on or experience with appropriate native woody plant cuttings or respective source plants for rehabilitation measures in RJ. Taking into account all the limitations regarding plant material usually used for the installation of hedgerow terraces, we decided to modify the established methodology using tree saplings of 0.8 m–1.2 m height instead of live cuttings.

13.3.1 Species Selection and Implementation Procedure

The saplings were donated by a nursery operated by the RJ state water treatment company CEDAE. We decided to use five different species to prevent monoculture and to assure a measure of success in case one or two species failed to grow as expected. We had to use tree species instead of shrubs because such species are rarely available in local nurseries. Furthermore, some natural shrub species of the Mata Atlântica such as *Miconia albicans* (Sw.) Steud and *Baccharis dracunculifolia* DC. tend to spread out and dominate abandoned areas with dense stands (Lima et al. 2006). Species selection criteria for our rehabilitation measure were (i) native to Southeast Brazilian Mata Atlântica, (ii) spiny species to improve the barrier function, (iii) N_2-fixing species to improve soil quality, (iv) tolerance to poor and dry soils, and (v) pioneer or early succession species to cope with site conditions in exposed sunlight. Following these criteria and taking into account the availability of a sufficient number of saplings, we selected four Fabaceae species (*Anadenanthera colubrina* [Vell.] Brenan var. *cebil* [Griseb.] Altschul, *Chloroleucon tortum* [Mart.] Pittier, *Enterolobium contortisiliquum* [Vell.] Morong, *Machaerium nyctitans* [Vell.] Benth.) and the Anacardiaceae species *Schinus terebinthifolius* Raddi. The latter species is neither N_2-fixing nor spiny but matches all other criteria and is of high economic (Brazilian red pepper) and ecologic value (bird nesting and fodder tree). A summary of all relevant species information is given in Table 13.1. We planted the saplings in a very dense, mixed species pattern of 20 cm distance because we were not sure about the performance and survival rate of each species. In total, 2,433 saplings (*A. colubrina* n = 455, *Ch. tortum* n = 870, *E. contortisiliquum* n = 510, *M. nyctitans* n = 88, S. *terebinthifolius* n = 510) were planted horizontally with a slight angle of ca. 10° toward the slope (Fig. 13.1) to stimulate epicormic shoot development and hence a more hedge-like growth.

Table 13.1 Botanical characteristics and functional traits of the species used in the Itaocara PRM

Species	*Anadenanthera colubrina* (Vell.) Brenan var. cebil (Griseb.) Altschul	*Chloroleucon tortum* (Mart.) Pittier	*Enterolobium contortisiliquum* (Vell.) Morong	*Machaerium nyctitans* (Vell.) Benth.	*Schinus terebinthifolius* Raddi
Family	Fabaceae, Mimosoideae	Fabaceae, Mimosoideae	Fabaceae, Mimosoideae	Fabaceae, Faboideae	Anacardiaceae
Common name	Angico-vermelho	Vinhatico-de-espinho	Orelha-de-negro	Jacarandá-de-espinho	Aroeira-pimenteira
Geographic distribution	Brazil, Argentina, Paraguay, Bolivia	Brazil	Brazil, Argentina, Paraguay, Uruguay, Colombia., Bolivia, Peru	Brazil, Argentina	Brazil, Argentina, Uruguay, Paraguay
Native biomes	Mata Atlântica, Cerrado, Caatinga	Mata Atlântica, Cerrado	Mata Atlântica, Cerrado, Caatinga	Mata Atlântica, Cerrado	Cerrado, Mata Atlântica, Pampa
Altitudinal range (m.a.s.l.)	15–1.300	?	5–1.200	60–1.000	0–2.000
Soil requirements	Variable, tolerant to compacted, shallow, acid, and clayey soils	Indifferent toward soil conditions	Very variable, avoids shallow and wet soils	Indifferent toward soil conditions	Variable, sandy to clayey, moist to dry soils, tolerant to short inundation and salt
Drought resistance	Good	Medium	Medium	Medium	Medium
Growth form, height	Tree, 8–20 m	Small tree, 6–12 m	Tree, 10–20 m	Small tree, 8–15 m	Small tree, 5–10 m
Foliage	Semideciduous	Deciduous	Deciduous	Evergreen to semideciduous	Evergreen

Thorns and spines	Yes, trunk	yes, branches	None	Yes, especially young branches	None
Light requirements	Heliophilous	Heliophilous	Helio- to semi-heliophilous	Heliophilous	Heliophilous
Successional stage	Early succession	Mid-succession	Early succession	Early to late succession	Early succession
N_2 fixation	*Rhizobium*	Yes	*Rhizobium*	*Rhizobium*	None
Uses/remarks	Recommended as shade tree at pastures, timber tree, bee pasture	Ornamental, bee pasture	Recommended for riparian afforestation, shade tree at pastures, ornamental, bee pasture	Recommended for degraded areas and riparian afforestation, living fences, ornamental, bee pasture	Recommended for dune stabilization, shade tree at pastures, fruit used as spice (Braz. red pepper)

Sources: Santana et al. 2015; Filardi 2014; Iganci 2014; Silva-Luz and Pirani 2015; Morim 2014; Canosa et al. 2012 Orwa et al. 2009; Carvalho 2008, 2003; Lorenzi 2002

Fig. 13.1 Horizontal planting pattern of five tree saplings per running meter (**a**), covered subsequently with excavated terrace soil (**b**). (Photo: D. Sattler)

Usually, tree saplings are planted at the beginning of the rainy season to take advantage of plentiful water supply in the initial growth phase. To avoid the wash of the terraces and horizontally planted saplings by heavy surface runoff during the rainy season, we decided to plant the hedges at the beginning of the dry season in June 2015. To ensure initial growth under dry conditions, each plant was supplied with 0.5 l of a water-retaining, biodegradable polymer (Forth©, Gel para plantio). Such hydrogels have been successfully applied to improve tree seedling survival in dry environments (Huttermann et al. 1999; Yang et al. 2013). Additionally, we irrigated the terraces until the end of the dry season.

13.3.2 Species Monitoring

We monitored mortality, vigor, and pest infestation of each individual plant 2, 4, 9, and 15 months after planting. After two extraordinarily dry rainy seasons in 2015 and 2016 (Seliger et al. 2018), the average mortality rate of all saplings was 19.2%, increasing with slope position: foothill 16.6%, mid-slope 21.1%, and upper slope 24.5%. Mortality also varied depending on the sapling species (Fig. 13.2). *A. colubrina* was the best performing species with only 6.2 (±0.2) % average mortality,

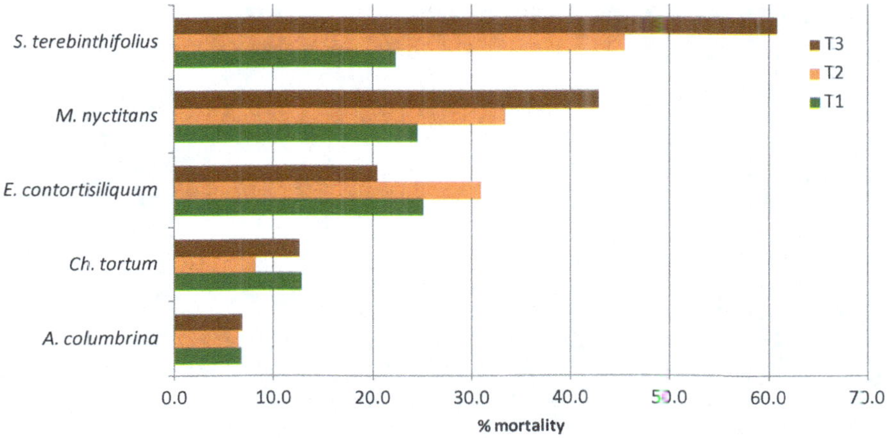

Fig. 13.2 Mortality rate of planted saplings according to species and slope position (*T1* foothill terrace, *T2* mid-slope terrace, *T3* upper-slope terrace)

independent of slope position. The low mortality rate of this species might be closely related to its outstanding drought resistance (see Table 13.1), which facilitated sapling survival during the extraordinarily dry rainy seasons 2015/2016. The second best performing species was *Ch. tortum* with 11.2 (±2.5) % average mortality. The highest loss of saplings was recorded in *S. terebinthifolius* with 42.9 (±19.4) % average mortality, which also strongly increased with slope position. *M. nyctidans* showed the same trend but at a lower mortality rate of 33.6 (±9.2) %. *E. contortisiliquum* was the only species with the highest mortality at the mid-slope position, with an average rate of 25.5 (±5.3) %.

Even though the soil and water supply conditions deteriorate with increasing slope altitude (Seliger et al. 2018), *A. colubrina* and *Ch. tortum* seemed to be able to cope best with local site conditions at the Itaocara PRM. Apart from soil and weather conditions, the growth success of planted saplings was influenced by several pest attacks. Due to ecological considerations, we did not use any chemical or other agents to control pests.

The high mortality rate of *S. terebinthifolius*, especially on the upper-slope terrace, was mainly due to leaf cutter ant attacks. Already 3 months after planting, 25.5% of the saplings died due to complete defoliation by leaf cutter ants. All other planted species only rarely showed signs of damages caused by leaf cutter ants (Fig. 13.3a). The legume species *E. contortisiliquum* and *Ch. tortum* suffered the most due to defoliation caused by the beetle *Epicauta atomaria* Germar 1821, a common pest on Maracuja plants and Solanaceae in agriculture (Boica et al. 2007), in Portuguese called "burrinho da batatinha"(Fig. 13.3b). At the beginning of the rainy season in November 2015, 68.6% of all saplings of *Ch. tortum* and 54.3% of *E. contortisiliquum* had been infested by this beetle. A second but lower wave of infestation was recorded at the beginning of the rainy season of 2016 with 31.8% of

Fig. 13.3 Signs of leaf cutter ant attacks on *M. nyctitans* leaflets (**a**), pest beetles *E. atomaria* feeding on *E. contortisiliquum* (**b**), nest of unknown bird species in *Ch. Tortum* (**c**), epicormic shoot development (red lined) form horizontally planted sapling (yellow lined) of *S. terebinthifolia* 15 months after planting (**d**), foothill terrace with gap for a cattle gate in the front, 16 months after planting (**e**). (Photo **a**–**d**: Dietmar Sattler, **e**: Giuseppe Latorraca)

damaged *E. contortisiliquum* plants, whereas *Ch. tortum* had been almost been ignored (3.6%) by the pest beetle at this time.

Summarizing the overall growth success of the planted species (Fig. 13.3e), *A. colubrina* was the best performing one, even though it is not perfectly suited to grow hedge-like and needs to be cut several times after being planted. *Ch. tortum* was the most appropriate species when combining survival rate and hedge-like growth, which attracted birds to build seven nests exclusively within this species (Fig. 13.3c). Branch sprouting forced by horizontal planting of tree saplings worked out best with *S. terebinthifolius*. Regardless of its high mortality rate, more than 90% of the surviving individuals of Brazilian red pepper showed intense and vigorous branching (Fig. 13.3d). We expect the planted hedgerows to reach an average height of ca. 2 m ca. 24–28 months after planting. At this height, they need to be cut and trimmed to force hedge consolidation. Furthermore, we recommended not cutting/pruning an *A. colubrina* individual every 30 m to let them grow big as pasture shading trees.

13.4 Conclusions

In contrast to considerable experience in afforesting native forests in Brazil, only little information on ecosystem-friendly bioengineering measures is available. Already after 16 months, the PRM consisting of three hedgerow terraces designed for a typical sloped pasture in northwestern RJ made with living plant material of five native species has successfully been implemented, securing the slope and providing habitat for insect and bird species. With increasing hedge size, the separation of three pasture lots at different slope positions will be an additional option allowing extensive rotational grazing. Thus, the success of the Itaocara PRM indicates that (i) bioengineered rehabilitation measures are an additional and valid tool for small-scale land restoration without compromising farmland activities within the dominion of the Atlantic Forest, (ii) the use of species native to the Mata Atlântica biome for soil bioengineering should be considered more often in science and practice, and (iii) the emphasis on soil bioengineering measures using native plant material in needs to be transferred to other, especially tropical, environments, which suffer from degradation. In the long run, species best adapted to the local environmental conditions and only occurring in the Atlantic Forest or other endangered habitats can be integrated into restoration measures to support biodiversity conservation. Native or even endemic live plant material for soil bioengineering would also allow for rehabilitation of river courses, degraded land, or damages caused by building activities within conservation priority areas (Naegeli de Torres et al. 2018) without jeopardizing native vegetation by potential exotic plant invasions. The design of different bioengineered restoration products (different sets of species, different measures) that address different types of degraded areas would also support the national strategies for recovering native vegetation (Política Nacional de Recuperação da Vegetação Nativa, Decree n.8.972/23.01.2017) and combat land degradation and foster soil conservation (TCU 2015).

Acknowledgments We would like to thank the German Federal Ministry of Education and Research (BMBF) for funding the research and development project INTECRAL (033L162F) and the Rio de Janeiro sustainable rural development project (SEAPPA – Rio Rural) for facilitating the presented study. Furthermore, we are grateful to the EMATER-Rio and PESAGRO-Rio Itaocara offices staff, Juliana M. Santos (University of Jena, Germany) and Giuseppe Latorraca, for providing excellent logistic backup support during field work.

References

Alexander S, Aronson J, Whaley O et al (2016) The relationship between ecological restoration and the ecosystem services concept. Ecol Soc 21(1):34

Alonso JM, Leles PSdS, Silveira Filho TB et al (2014) Avaliação da diversidade de espécies nativas produzidas nos viveiros florestais do estado do Rio de Janeiro. Floresta 44(3):369–380

Balangcod KD, Wong FM, Balangcod TD (2015) Chrysopogon zizanioides (vetiver grass) as a potential plant for landslide bioengineering at Atok, Benguet, Philippines. Aust J Bot 63(3/4):216–221

Boica AL, Baptista JZ, de Oliveira JC, de Jesus FG (2007) Attractivity and feeding preference of Epicauta atomaria (Ger.) in some passion fruit species. Rev Bras Frutic 29(3):471–476

Brazilian Federal Law 12.727 (2012). Lei n° 12.727, de 17 de outubro de 2012. http://www.planalto.gov.br/CCIVIL_03/_Ato2011-2014/2012/Lei/L12727.htm. Accessed 30 Mar 2017

Burri K, Graf F, Böll A (2009) Revegetation measures improve soil aggregate stability: a case study of a landslide area in Central Switzerland. For Snow Landsc Res 82(1):45–60

Canosa GA, de Faria SM, de Moraes LFD (2012) Leguminosas florestais da Mata Atlântica brasileira fixadoras de nitrogênio atmosférico. Communicado Técnico, Seropédica, RJ, Doc. Nr. 144

Carvalho PER (2003) Espécies Arbóreas Brasileiras, vol I. Embrapa Informação Tecnológica, Brasilia

Carvalho PER (2008) Espécies Arbóreas Brasileiras, vol III. Embrapa Informação Tecnológica, Brasilia

Clayton-Barbosa A (2012) Bioengenharia utilizando bambus em faixas para o controle de processos erosivos: uma analise qualitativa. Polibotánica 33:223–243

Durlo M, Sutili F (2012) Bioengenharia – Manejo Biotécnico de Cursos de Água, 2nd edn, Santa Maria

Eisenlohr P, de Oliveira-Filho A, Prado J (2015) The Brazilian Atlantic Forest: new findings, challenges and prospects in a shrinking hotspot. Biodivers Conserv 24(9):2129–2133

Evette A, Labonne S, Rey F et al (2009) History of bioengineering techniques for erosion control in rivers in Western Europe. Environ Manag 43(6):972–984

Filardi FLR (2014) Machaerium. In: Lista de Espécies da Flora do Brasil. Jardim Botânico do Rio de Janeiro.http://floradobrasil.jbrj.gov.br/jabot/floradobrasil/FB29774. Accessed 10 Mar 2017

Guerra AJT, Jorge MCO (eds) (2014) Degradação dos solos no Brasil. Bertrand Brasil, Rio de Janeiro

Guzmán Wolfhard L, Raedig C (2018) Connectivity conservation management: connecting private protected areas. In: Nehren U, Schlüter S, Raedig C, Sattler D, Hissa H (eds) Strategies and tools for a sustainable rural Rio de Janeiro. Springer International Publishing, Cham

Huttermann A, Zommorodi M, Reise K (1999) Addition of hydrogels to soil for prolonging the survival of *Pinus halepensis* seedlings subjected to drought. Soil Tillage Res 50(3–4):295–304

Iganci JRV (2014) Chloroleucon. In: Lista de espécies da flora do brasil. Jardim Botânico do Rio de Janeiro. http://floradobrasil.jbrj.gov.br/jabot/floradobrasil/FB18407. Accessed 10 Mar 2017

Lange W, Sandholz S, Viezzer J et al (2018) Ecosystem-based approaches for disaster risk reduction and climate change adaptation in Rio de Janeiro State. In: Nehren U, Schlüter S, Raedig C, Sattler D, Hissa H (eds) Strategies and tools for a sustainable rural Rio de Janeiro. Springer International Publishing, Cham

Lima HC, Pessoa SVA, Guedes-Bruni RR et al (2006) Caracterização fisionômico-florística e mapeamento da vegetação da Reserva Biológica de Poço das Antas, Silva Jardim, Rio de Janeiro, Brasil. Rodriguésia 57(3):369–389

Lima RAF, Rother DC, Muler AE et al (2012) Bamboo overabundance alters forest structure and dynamics in the Atlantic Forest hotspot. Biol Conserv 147(1):32–39

Lorenzi H (2002) Árvores brasileiras, vol I, 4th edn. Instituto Plantarum, Nova Odessa

Machado L, Holanda FSR, da Silva VS et al (2015) Contribution of the root system of vetiver grass towards slope stabilization of the Sao Francisco River. Semina Cienc Agrar 36(4):2453–2463

Manhães AP (2008) Caracterização da cadeia produtiva do bambu no brasil: abordagem preliminar. UFRRJ, Instituto de Florestas, Seropédica
Matos LV, Campello EFC, Resende AS (2005) Plantio de leguminosas arbóreas para produção de moirões vivos e construção de cercas ecológicas. Embrapa Agrobiologia, Sistemas de Produção 3, Seropédica, RJ, Brazil Springer International Publishing, Cham
Moraes LFD, Assumpção JM, Pereira TS, Luchiari C (2013) Manual técnico para a restauração de áreas degradadas no estado do Rio de Janeiro. Instituto de Pesquisas Jardim Botânico do Rio de Janeiro, Rio de Janeiro
Morim MP (2014) Enterolobium. In: Lista de Espécies da Flora do Brasil. Jardim Botânico do Rio de Janeiro. http://floradobrasil.jbrj.gov.br/jabot/floradobrasil/FB83154 Accessed 10 Mar 2017
Muniz CAV, Martins E, Hudson M, França CR (2013) Multirão Reflorestamento – Manual de encarregados. Prefeitura Rio de Janeiro, Rio de Janeiro
Naegeli de Torres F, Richter R, Cardoso Fidalgo EC (2018) Multi-criteria site prioritization for land rehabilitation in the Guapi-Macacu Watershed, Rio de Janeiro. In: Nehren U, Schlüter S, Raedig C, Sattler D, Hissa H (eds) Strategies and tools for a sustainable rural Rio de Janeiro. Springer International Publishing, Cham
Orwa C, Mutua A, Kindt R et al (2009) Agroforest tree database: a tree reference and selection guide version 4.0. http://www.worldagroforestry.org/af/tree. Accessed 18 Apr 2017
Pardini R, de Souza SM, Braga-Neto R, Metzger JP (2005) The role of forest structure, fragment size and corridors in maintaining small mammal abundance and diversity in an Atlantic forest landscape. Biol Conserv 124(2):253–266
Pereira AR (2006) Como selecionar plantas para áreas degradadas e controle de erosão. Editora FAPI, Belo Horizonte
Pinto SR, Melo F, Tabarelli M et al (2014) Governing and Delivering a Biome-Wide Restoration Initiative: The Case of Atlantic Forest Restoration Pact in Brazil. Forests 5(9):2212–2229
Piratelli AJ, Marquez Piña-Rodrigues FC, Raedig C (2018) Integrating biodiversity conservation into agroecosystem management: Using birds to bring conservation and agricultural production together. In: Nehren U, Schlüter S, Raedig C, Sattler D, Hissa H (eds) Strategies and tools for a sustainable rural Rio de Janeiro. Springer International Publishing, Cham
Preti F, Giadrossich F (2009) Root reinforcement and slope bioengineering stabilization by Spanish Broom (*Spartium junceum L.*). Hydrol Earth Syst Sci 13:1713–1726
Pyšek P, Pergl J, Essl F, Lenzner B, Dawson W, Kreft H, Weigelt P, Winter M, Kartesz J, Nishino M et al (2017) Naturalized alien flora of the world: species diversity, taxonomic and phylogenetic patterns, geographic distribution and global hotspots of plant invasion. Preslia 89(3):203–274
Ribeiro MC, Martensen AC, Metzger JP et al (2011) The Brazilian Atlantic Forest: a shrinking biodiversity hotspot. In: Zachos FE, Habel JC (eds) Biodiversity hotspots – distribution and protection of conservation priority areas. Springer, Heidelberg, pp 405–434
Rodrigues RR, Gandolfi S, Nave AG et al (2011) Large-scale ecological restoration of high-diversity tropical forests in SE Brazil. For Ecol Manag 261(10):1605–1613
Santana CAA, Silva VG, Silva AT (2015) Manual de identificação de mudas de espécies florestais, 2nd edn. Secretaria Municipal de Meio Ambiente/CRA, Rio de Janeiro
Sattler D, Murray LT, Kirchner A, Lindner A (2014) Influence of soil and topography on aboveground biomass accumulation and carbon stocks of afforested pastures in South East Brazil. Ecol Eng 73:126–131
Sattler D, Seliger R, Heinrich J (2017a) Landdegradation in Südostbrasilien – Ursachen, Status und Optionen für die Entwicklung einer resilienten Kulturlandschaft. Geographische Rundschau 4(2017):32–38
Sattler D, Seliger R, Nehren U et al (2017b) Pasture degradation in South East Brazil: Status, Drivers and Options for sustainable land use under climate change. In: Leal Filho W, Esteves Freitas L (eds) Climate change adaptation in latin america – managing vulnerability, fostering resilience, Climate change management. Springer International Publishing, Cham. https://doi.org/10.1007/978-3-319-56946-8. (in press)
Schlüter S, Nehren U, Sattler D, Raedig C (2018) The INTECRAL project. In: Nehren U, Schlüter S, Raedig C, Sattler D, Hissa H (eds) Strategies and tools for a sustainable rural Rio de Janeiro. Springer International Publishing, Cham

Seliger R, Sattler D, Soares da Silva A et al (2018) Rehabilitation of degraded sloped pastures – lessons learned in Itaocara. In: Nehren U, Schlüter S, Raedig C, Sattler D, Hissa H (eds) Strategies and tools for a sustainable rural Rio de Janeiro. Springer International Publishing, Cham

Silva-Luz CL, Pirani JR (2015) Anacardiaceae. In: Lista de Espécies da Flora do Brasil. Jardim Botânico do Rio de Janeiro. http://floradobrasil.jbrj.gov.br/jabot/floradobrasil/FB4401. Accessed 10 Mar 2017

Soares-Filho B, Rajão R, Macedo M et al (2014) Cracking Brazil's Forest Code. Science 344(6182):363–364

Sparovek G, Correchel V, Barretto A (2007) The risk of erosion in Brazilian cultivated pastures. Sci Agric 64(1):77–82

Stehmann JR, Forzza RC, Salino A et al (2009) Diversidade Taxonômica na Floresta Atlântica. In: Stehmann JR, Forzza RC, Salino A et al (eds) Plantas da Floresta Atlântica. Jardim Botânico de Rio de Janeiro, Rio de Janeiro, pp 3–12

Stokes A, Sotir R, Chen W, Chestem M (2009) Soil bio- and eco-engineering in China: past experience and future priorities. Ecol Eng 36(3):247–257

Studer R, Zeh H (2014) Soil Bioengineering: Construction Type Manual English – Russian – Portuguese – Chinese – Korean. VDF Hochschulverlag, Zürich

Tavares SRL, Melo AS, Andrade AG et al (2008) Curso de recuperação de áreas degradadas: a visão da ciência do solo no contexto do diagnóstico, manejo, indicadores de monitoramento e estratégias de recuperação. Embrapa Solos, Documentos 103, Rio de Janeiro

TCU – Tribunal de Contas da União (2015) Auditoria operacional governança de solos em áreas não urbanas. Brasíla, TC 011.713/2015-1

Yang LX, Yang Y, Chen Z et al (2013) Influence of super absorbent polymer on soil water retention, seed germination and plant survivals for rocky slopes eco-engineering. Ecol Eng 62:27–32

Zenni RD, Ziller SR (2011) An overview of invasive plants in Brazil. Rev Bras Bot 34:431–446

Zuazo VHD, Pleguezuelo CRR (2008) Soil-erosion and runoff prevention by plant covers. A review. Agron Sustain Dev 28(1):65–86

Chapter 14
Use and Conservation of the Threatened Brazilian National Tree *Paubrasilia echinata* Lam.: A Potential for Rio de Janeiro State?

Silke Lichtenberg, Elisabeth Huber-Sannwald, Udo Nehren, and Juan Antonio Reyes-Agüero

Abstract Several native and endemic tree species of the Atlantic Forest in Brazil, especially precious wood species, are highly endangered. In consequence, the legal extraction from wild forests is completely forbidden, and legal extraction from plantations is strictly regulated. Among them is *Paubrasilia echinata* Lam. – in Portuguese pau-brasil – Brazil's national tree. Its precious wood is used worldwide as raw material for high-quality violin bows. For professional musicians, alternative wood species in violin bows are rarely accepted, even though some of them have been proven to meet the high technical standards of wood characteristics which pau-brasil guarantees. As an alternative solution to cover future wood demand, commercial pau-brasil plantations have been established and supported by violin makers, bow makers, and musicians. While several plantation types with pau-brasil exist,

S. Lichtenberg (✉)
Programa Multidisciplinario de Posgrado en Ciencias Ambientales de la Universidad Autónoma de San Luis Potosí, San Luis Potosí, Mexico

TH Köln - University of Applied Sciences, Institute for Technology and Resources Management in the Tropics and Subtropics (ITT), Köln, Germany

E. Huber-Sannwald
Instituto Potosino de Investigación Científica y Technológica, A.C, San Luis Potosí, Mexico
e-mail: ehs@ipicyt.edu.mx

U. Nehren
TH Köln - University of Applied Sciences, Institute for Technology and Resources Management in the Tropics and Subtropics (ITT), Köln, Germany
e-mail: udo.nehren@th-koeln.de

J. A. Reyes-Agüero
Instituto de Investigación de Zonas Desérticas de la Universidad Autónoma de San Luis Potosí, San Luis Potosí, Mexico
e-mail: reyesaguero@uaslp.mx

U. Nehren et al. (eds.), *Strategies and Tools for a Sustainable Rural Rio de Janeiro*, Springer Series on Environmental Management,
https://doi.org/10.1007/978-3-319-89644-1_14

such as monospecific and mixed tree plantations and agroforestry systems, there is poor understanding concerning their long-term productivity, wood quality, and economic return. In order to guarantee continuous supply of planted pau-brasil for future wood markets and for these plantations to be sustainable, it is essential to identify potential ecological, economic, legal, and social assets and constraints including the potential contribution of wood supply by pau-brasil plantations. The presented data originate from a field study in the Atlantic Forest of Brazil with detailed inventories of nine plantations in Southern Bahia and Espírito Santo. Based on these data we discuss the future pau-brasil wood markets and estimate the conservation and utilization potential of pau-brasil for the state of Rio de Janeiro, the southernmost range of the species.

Keywords *Paubrasilia echinata* · Pau-brasil · Violin bows · Mixed tree plantations · Agroforestry

14.1 Introduction

The Atlantic Forest is ranked fourth in the list of global biodiversity hotspots because of its extraordinary high number of endemic species and for being one of the most threatened biomes and regions in the world (Myers et al. 2000). Around 20,000 plant species are known from the Atlantic Forest, almost 50% of them are endemic (Myers et al. 2000; Murray-Smith et al. 2009). Among them is the endemic tree species pau-brasil (*Paubrasilia echinata* Lam. (Gagnon et al. 2016); formerly *Caesalpinia echinata* Lam.; Fig. 14.1) (Rocha 2010), which, in 1992, was listed in the official list of Brazilian endangered species (IBAMA 1992): in 2014 the conservation status was included, and since then pau-brasil has been categorized as endangered (EN) (MMA 2014). In 1998 the species was listed as endangered (EN) in the list of threatened species of the International Union for Conservation of Nature (Varty 1998; CITES 2007). According to Martinelli and Moraes (2013), pau-brasil can be found in the states of Rio Grande do Norte, Paraíba, Pernambuco, Alagoas, Bahia, Espírito Santo, and Rio de Janeiro (RJ) (from North to South).

With the European discovery of Brazil in the sixteenth century, excessive exploitation of the Atlantic Forest started, in particular of pau-brasil for its red pigment, known as brazilin, as textile dye (Aguiar and Pinho 2007). Pau-brasil gained enormous economic importance, and in 1530 the name of the Portuguese colony in America *Terra do Santa Cruz* was changed to Brazil (Rocha 2004). In 1978, pau-brasil was declared national tree (LEI N° 6.607). However, by the end of the nineteenth century, the color industry had replaced nearly all natural dyes by synthetic products (Michaelson 1993). This historical change temporarily removed the pressure on pau-brasil. However, in the second half of the eighteenth century, F.X. Tourte (1747–1835) introduced the construction of modern violin bows, which were bent

Fig. 14.1 Pau-brasil (*Paubrasilia echinata* Lam.): tree (**a**), bipinnate leaf (**b**), bark with spines of a young tree (**c**), fruit (**d**), seeds (**e**), cross section of wood (reddish heartwood and yellowish sapwood) (**f**), blooming tree (**g**), inflorescence (**h**)

in the opposite direction, what had been common until then, with a concave curve, and for this purpose, pau-brasil was identified as the best and most suitable wood (Fig. 14.2) (Alves et al. 2008). Ever since, this wood has been the most important raw material for high-quality bows (Skeaping 1955; Longui et al. 2010). Today, pau-brasil still is the preferred wood by bow makers and musicians due to its high density, facility to bend, stability in curvature, durability, and beauty (Pierce 2002). For violin bows, the pau-brasil wood needs to dry for a minimum of 12 months; its quality is determined by wood density and speed of sound propagation, this latter with the help of an elasticity tester, i.e., the ultrasonic flow measuring method developed by Giovanni Lucchi (Alves et al. 2008). However, for the Brazilian wood species ipê (*Handroanthus* spp.) and cumaru (*Dipteryx* spp.), there is evidence that

Fig. 14.2 Different stages of sticks in the process of constructing violin bows made of pau-brasil

their wood shows similar key properties and thus can represent an alternative to pau-brasil (Longui et al. 2010, 2014).

Because of the overexploitation of wild populations of pau-brasil, this species has become extremely rare. The largest areas of natural occurrence are found in Bahia, with a total of 73,241 ha; other important remnants are in RJ (50,955 ha), Alagoas (7,288 ha), Paraíba (4,405 ha), Rio Grande do Norte (3,211 ha), and Pernambuco (776 ha) (Rocha and Simabukuro 2008). In Espírito Santo only one small area of 30 ha of pau-brasil was found in 2008 (Zani et al. 2012).

According to the Brazilian Institute of Environment and Renewable Natural Resources (Instituto Brasileiro do Meio Ambiente e dos Recursos Naturais Renováveis – IBAMA), the high demand of pau-brasil wood for violin bows still generates illegal cutting of wild populations. Martinelli and Moraes (2013) estimate a population reduction of minimum 50% within the next 100 years.

Moreover, other uses of pau-brasil might emerge and increase its demand, for example, for pharmaceutical purposes, as it was found that pau-brasil leaves have important anti-carcinoma potentials and could be used for cancer treatment (Parisotto 2011). Furthermore, it was recognized that endophytic fungi from the bark and stem of pau-brasil show leishmanicidal, trypanocidal, antimicrobial, and cytotoxic activities (Siqueira et al. 2014; Fraga Campos et al. 2015).

Due to the alarming situation of pau-brasil and its cultural significance, the Brazilian Ministry of the Environment (Ministério do Meio Ambiente) created the National Conservation Brazilwood Program (Programa Nacional de Conservação do Pau-brasil) to (1) reevaluate the conservation status of pau-brasil, (2) identify protected areas with remnants of pau-brasil, (3) implement the conservation action plan, and (4) promote the sustainable use and commercial plantations of pau-brasil (Ministério do Meio Ambiente 2012).

Current conservation and plantation efforts are mainly enforced and financed by bow makers, violin makers, and musicians, who might be the most affected stakeholders in case of the extinction of pau-brasil or strict law enforcement to use pau-brasil. However, alternative wood species adequate for bow making such as *ipê* and *cumaru* might be one option to minimize the wood demand of pau-brasil in the future and thereby to contribute to the conservation of pau-brasil. The oldest pau-brasil plantations were established in the Northeast by the National Foundation of Pau-

Fig. 14.3 Two types of pau-brasil plantations: monospecific plantation with commercial interest (**a**) and agroforestry system (*cabruca* system) for conservation and potential commercial use (**b**)

Brasil (Fundação Nacional do Pau-Brasil; FUNBRASIL); they date back to 1972 (*personal communication* Ana Cristina Roldão, president of FUNBRASIL). Around 1980, Horst John, owner of a wood trading company and later of a bow making manufacture also started a plantation in Santa Rosa in Espírito Santo (Marques et al. 2012). Important conservation areas outside the natural distribution range are in the state of São Paulo where pau-brasil was already planted in 1930 in the Biological Reserve of Mogi-Guaçu and gained high importance for the genetic species conservation (Rocha and Simabukuro 2008).

Conservation and commerce are the main purposes of pau-brasil plantations (Fig. 14.3). For the latter the ideal diameter for cutting a tree is approximately 50 cm (Dean 1996). For the construction of bows, cultivated trees are allowed to be harvested at the earliest after 30 years when its diameter is adequate (Marques et al. 2012). According to Aguiar and Pinto (2008), the growth in height and width depends on tree density (individuals/ha) and on the geographic location, fertilization, irrigation, and proper pruning. In wild populations, pau-brasil needs much more time until the wood can be harvested (Macedo 2015). Furthermore, good wood quality for bow construction requires trees with straight and regular stems preferably without multiple ramification (West 2014). While local bow makers in Espírito Santo have detected clear differences in wood quality among morphotypes, this correlation remains to be proved scientifically. Pau-brasil develops three different morphotypes known as *arruda* (small leaflets), *café* (medium leaflets), and *laranja* (big leaflets) (Macedo 2015). The widespread concern of bow makers that wood from pau-brasil plantations may not meet the required quality standards compared to trees from wild populations was recently dispelled when two scientific studies (Marques et al. 2012, Taylor et al. 2013) showed that the quality of plantation wood is comparable to that from natural habitats.

Brazil has national and state laws targeted toward the conservation of pau-brasil including specific regulations for nurseries and collections of seeds from native species. The National Forest Law (in the recent version of the Código Florestal from

2012) requires the registration of all areas by their landowners and land users in the environmental registry Cadastro Ambiental Rural (CAR). At the time of this publication, the deadline for registration was postponed to December 2017. Many landowners had not yet registered at the time of the field study (2015); however, registration in CAR is obligatory and of particular importance in case of future commercial use of an area covered with native tree species. Additionally, regional- and state-level legislation control the use of native tree species, e.g., the Law of the Atlantic Forest (Law 11.428) and a law referring explicitly to pau-brasil.

Brazil also needs to comply with international agreements and restrictions regarding species listed in the catalogue of the Convention on International Trade in Endangered Species of Wild Fauna and Flora (CITES). CITES is an international organ to combat illegal trade of endangered species. Since 2007, pau-brasil is listed in Appendix II of CITES within a category of species that are highly vulnerable to extinction, unless strict trade regulations are applied. The listing of species in one of the CITES appendices implies additional expenses for documents and certifications to be issued for legal trade. In the future, for pau-brasil wood extracted from plantations to enter the international market, it will likely be necessary to get phytosanitary certificates ratifying that the specimen were grown in plantations for the purpose of exportation.

Against this background, the objectives of this chapter are (a) to provide an overview of pau-brasil plantations in the Atlantic Forest with respect to their ecological, economic, legal, and social potentials and constraints based on field data and (b) to discuss the use and conservation potential of pau-brasil in RJ. It is assumed that for future conservation efforts of pau-brasil, RJ offers particularly promising opportunities, as the state already covers a large network of conservation areas which could possibly be expanded in particular on degraded pasture lands.

14.2 Methods

For the analysis of ecological, economic, social, and legal potentials and constraints of the cultivation of pau-brasil, the scientific literature was reviewed focusing on studies on the conservation status of pau-brasil and plantation methods in the tropics, especially in the Atlantic Forest of Brazil. In 2015, 15 plantations were visited in Rio Grande do Norte (1), Paraíba (3), Pernambuco (2), Bahia (4), and Espírito Santo (5). During these visits, semi-structured interviews with local farmers and landowners were realized to determine existing practical experiences in planting pau-brasil regarding both the management of trees at different age classes and potential legal constraints for future commercial use of these trees. In order to underpin the analysis of the legal background for cultivating pau-brasil, several expert interviews were conducted with officers of IBAMA. In 2015, empirical data of nine pau-brasil plantations were collected, four of them in Southern Bahia and five in Espírito Santo. The selection of plantations depended on accessibility and was realized with the help of the Executive Committee of the Plan of Cacao Farming

(Comissão Executiva do Plano da Lavoura Cacaueira – CEPLAC) and the Institute of the Living Forest (Instituto Floresta Viva – IFV) in Bahia. Variables of interest included the geographical location and altitude of the plantation (GPS recording), the area of individual plantations (ha), the year of establishment of the plantation, the type of plantation (monospecific, mixed tree plantation, or agroforestry system), the density of planted pau-brasil trees (individuals/ha), and the planted morphotypes. Based on these findings, we discuss to which degree pau-brasil plantations could possibly contribute to sustainable development in RJ taking into account land use and other data from the Rio Rural program and the INTECRAL project.

14.3 Plantation Status of Pau-Brasil in the Atlantic Forest

There are different types of plantations for the cultivation of pau-brasil within the Brazilian Atlantic Forest. The *monoculture or monospecific plantation* is defined as a plantation type with only one species over a large area, also described as "…a uniform or homogeneous planting" (Capinera 2008) (Fig. 14.3). Most forestry plantations worldwide are monospecific (FAO 2001; West 2014). They provide important advantages such as simple nursery practice and easy stand management, the possibility to concentrate on the growth of one specific species, and the uniform harvest of all trees at the same time (Evans and Turnbull 2004). In the case of pau-brasil, in all three major distribution regions (Northeast, Bahia, and Southeast), monospecific plantations are found for conservation purposes in protected areas, as well as in areas for future commercial use for bow making.

Mixed tree plantations consist of two or more tree or shrub species and have several potential advantages (West 2014) as they (a) increase tree growth and production in adequate species combinations, (b) improve the quality of stem form, (c) provide more than one product or ecosystem service and potentially throughout the whole year (whereas mixed species have different rotation lengths), (d) reduce potential damage risk due to biological pest and disease control, (e) regenerate degraded land, (f) increase biological and functional diversity, and (g) improve the aesthetic appearance (cultural service). In case of mixed tree plantations, first experiences were gained by using different plantation densities and/or sequential orders of native tree species including pau-brasil by *Symbiosis Investimentos e Part. Ltda* in Trancoso, Bahia. However, this type of plantation is still not very common throughout Brazil.

Agroforestry systems are agricultural systems with tree components; both high species diversity and biomass production may contribute to environmental sustainability (Franke et al. 2000). The inclusion of pau-brasil and other native tree species in such a system contributes to their conservation as native tree species get (re) planted (Sambuichi et al. 2012). For instance, the *cabruca* system (Fig. 14.3) is a traditional agroforestry system in Southern Bahia, where cacao (*Theobroma cacao* L.) is planted in previously cleared understory of native forests (Sambuichi et al. 2012; Dawson et al. 2013). Trees of native forests are not logged; thus, a significant number of pau-brasil trees from original wild populations remain conserved in this specific agroforestry system (Lobão and Valeri 2009). In existing *cabruca* systems, pau-brasil seedlings are planted to replace fallen or old trees as shade species for cacao plants (Sambuichi et al. 2012). The purpose of planting pau-brasil seedlings within the *cabruca* system is to increase the density of shade trees for cacao plants or to transform a monoculture to the traditional agroforestry system with native tree species. In Southern Bahia, the most important area for cacao production, this *cabruca* system is widely spread. Pau-brasil has been planted in *cabruca* systems for future commercial use ever since 2005 (Table 14.1) as part of the "Programa Pau-brasil" of Southern Bahia.

For aesthetic or cultural purposes, pau-brasil is planted as individual trees in parks and open public spaces, and for conservation purposes, monospecific plantations and mixed plantations are part of areas of permanent preservation (APPs), private reserves of natural heritage (RPPNs), or legal reserves (see Guzmán Wolfhard and Raedig 2018).

Pau-brasil plantations have also been established for commercial purposes; in that case, they are mostly monospecific, but there exist also mixed tree plantations and agroforestry systems with pau-brasil for future commercial use (Fig. 14.3). While pau-brasil was historically used for several purposes, currently the only relevant commercial use is for the construction of violin bows. Yet with re-appreciation of natural dyes, due to the carcinogenic potential of artificial dyes, it is possible that pau-brasil may be valued again for dye.

Geographically, the conservation and plantation efforts of pau-brasil cover three major regions of the Atlantic Forest biome: the Northeast (with Rio Grande do Norte, Paraíba and Pernambuco), Southern Bahia, and Espírito Santo. However, other states within the Atlantic Forest, such as RJ, belong to the natural range of the species and are therefore suitable for plantations as well.

In Northeast Brazil, the NGO Association of Plants from the Northeast (Associação Plantas do Nordeste – APNE) and FUNBRASIL participate in conservation and plantation efforts. According to an interview with Ana Cristina Roldão, the president of FUNBRASIL, the pau-brasil plantations established by FUNBRASIL (from 1972 to 1987) are all monospecific, while the exact number of planted trees was not available. These plantations have not been officially registered until now and thus have not been meant for commercial use; most of them are inside protected areas. Only for scientific studies on the quality of wood grown in plantations, it is currently possible and might also be feasible for the future to remove sample trees from the oldest plantations with a special permission.

For conservation purposes (reintroduction, APPs, RPPNs, or Reserva Legal – see Guzmán Wolfhard and Raedig 2018), APNE planted 2,600 seedlings in 2014 and 2015. For commercial purposes, additionally 5,758 trees were planted in 25 plantations between 2013 and 2015. Of those, 1,631 trees were planted in agroforestry systems, 3,439 trees in monospecific plantations, and 412 trees in an existing coconut plantation as an experiment of a mixed plantation. Another 303 trees were planted together with other native tree species in a mixed plantation in an area of protection.

Table 14.1 Overview of the geographic location, altitude, size, age, and specific characteristics of nine pau-brasil plantations differing in plantation type (Af = agroforestry system, Mono = monospecific plantation, mixed = mixed tree plantation) and morphotype (*arruda* = small leaflets, *café* = medium leaflets and *laranja* = big leaflets)

Localities	State	S lat	W lon	Altitude (m)	Area (ha)	Year of plantation	Density (individuals/ha)	Type	Morpho-type
Area Ceplac	BA	19° 37.393'	40° 11.465'	41	1	2009	277	Af	All
Fazenda Maristela	BA	14° 46.955'	39° 03.031'	101	27.5	2005 (2008 replanted)	1,111	Af	Mainly *arruda*
Fazenda Nova Aurora	BA	14° 44.715'	39° 06.340'	38	3.5	2005	1,111	Af + Mono	*Arruda* + All
Fazenda Transval	BA	14° 44.690'	39° 16.998'	60	17	2005	120	Mono	Mainly *arruda*
Fazenda Guaraná countryside	ES	19° 40.750'	40° 12.871'	37	2	1999 most probably	977	Mono	All
Plantation alto Bergamo	ES	19° 45.695'	40° 25.867'	210	1.1+ 0.7	1997 + 1999	3,265	Mono	Mainly *café*
Plantation Guaraná	ES	19° 40.686'	40° 15.801'	25	1	Before 1997 - today	1,778	Mixed	Mainly *café*
Fazenda Santa Rosa	ES	19° 54.455'	40° 18.123'	7	–	Before 1997 - today	–	Mixed	Mainly *café*
Plantation Domingos Martins	ES	20° 22.577'	40° 36.710'	535	1.2	1999	–	Mono	Mainly *arruda*

In 2015, in Northeast Brazil there was only one officially registered plantation with 625 planted trees for future commercial use, but more plantations with a total number of 5,160 trees were planned and thus had to be registered prior to planting. In order to obtain the legal status of a cultivated area, APNE supports the registration process of plantations for future commercial use. Moreover, plantation owners need to show a management plan including the designated harvest time of trees. The registration process is essential to guarantee future commercial use and to prevent the area from being declared protected area because of the occurrence of native species growing in this area.

In Bahia, it is difficult to evaluate the situation of the plantations of pau-brasil due to the lack of a monitoring program. The challenges are the vast areas of plantations, their scattered distributions, and the partly difficult access. Therefore, the survival rate of seedlings and the mid- and long-term effects of certain management practices of seedlings including the use of fertilizer, pesticides, and pruning to foster straight growth of the seedlings are unknown. The only available information is that between 2004 and 2008, 131,000 trees had been distributed in Southern Bahia during the "Programa Pau-brasil" (CEPLAC 2008). According to the Executive Committee of the Plan of Cacao Farming (Comissão Executiva do Plano da Lavoura Cacaueira – CEPLAC), 21% of the 131,00 seedlings were considered for conservation, 14% for public areas, and 65% for commercial plantations (CEPLAC 2008). However, according to CEPLAC and IFV, only two agroforestry plantations with pau-brasil had been registered for future commercial use at IBAMA. The official registration of these two plantations was also confirmed by its owner. In Southern Bahia all pau-brasil seedlings of the "Programa Pau-brasil" designated for future commercial use had been planted within cacao plantations of the *cabruca* system. Registration of landownership in the CAR has been obligatory since 2012, describing the type of owned land, in the case of the *cabruca* system, as a "cultivation area for cacao". Therefore, it will be necessary to register these areas of the *cabruca system* where pau-brasil was planted for future commercial use as "cultivation areas for cacao and pau-brasil". However, not all areas of the *cabruca* system had been registered by 2015. Due to that fact and the absence of monitoring data of effectively planted trees, the percentage of trees for potential commercial use is unknown.

In Espírito Santo, owners of bow making companies are in charge of most of the pau-brasil plantations. Almost all plantations are monospecific – only the oldest was initially a mixed plantation, where pau-brasil was planted together with other native species and the exotic *Eucalyptus*. However, this fast-growing species was eliminated from that area some years ago. Most of the planted areas are situated within APPs or legal reserves, which implies they cannot be used commercially in the future according to regulations of the National Forest Code, which prohibits wood extraction from protected areas. Some of the existing plantations are in areas that could theoretically be used commercially, as they had been registered at IBAMA as cultivating area. Until 2015, however, none of the plantation areas had been registered in the new CAR. Besides that, management plans had not been completed yet.

A comparison of nine plantations (Table 14.1) shows that their altitude varies from 7 to 535 m and the area between 0.7 and 27.5 ha. The oldest plantations had already been planted before 1997, the youngest in 2009. The density of the planted pau-brasil trees varies between 120 and 3,265 individuals/ha. Three plantations are agroforestry systems (one of them with a small share of monoculture), four are monocultures, and two are mixed plantations.

The experiences of agroforestry systems and the *cabruca* system, respectively, made by local cacao farmers in Bahia are based upon the local traditional cacao production and cannot be transferred to other areas of the Atlantic Forest. The evaluation of the two mixed tree plantations in Espírito Santo does not allow general conclusions for mixed tree plantations of only native tree species, as an exotic *Eucalyptus* species was also included.

A frequently mentioned concern of landowners is that planted pau-brasil trees for future commercial use do not have a guaranteed legal authorization, given by IBAMA, for the harvest of these trees; however, officers of IBAMA had not confirmed this preoccupation. The officers stated that there is no law that prohibits or limits the commercial use of cultivated pau-brasil in mixed plantations or close-to-nature plantations as long as the registration at IBAMA and in the CAR as well as a management plan prove that the specific area is a plantation area where the native species or pau-brasil were planted.

14.4 Conservation and Utilization Potential of Pau-Brasil in the State of Rio de Janeiro

The existing protected areas with natural occurrence of pau-brasil are (1) Área de Proteção Ambiental da Serra da Capoeira, (2) Área de Proteção Ambiental da Massambaba, (3) Área de Proteção Ambiental da Serra de Sapiatiba, (4) Área de Proteção Ambiental do Pau-brasil, and (5) the Parque Municipal da Boca da Barra (Rocha and Simabukuro 2008). There are no commercial plantations in RJ, but pau-brasil is frequently planted in public spaces.

One promising conservation strategy for pau-brasil could be to create a mosaic of protected areas by planting pau-brasil trees in APPs and RPPNs of the same region, as proposed by Rocha et al. (2014). For the genetic conservation, an interconnection between the three protected areas Área de Proteção Ambiental do Pau-brasil, Parque Municipal da Boca da Barra, and Área de Proteção Ambiental da Massambaba through ecological corridors would greatly benefit the conservation. According to Cardoso et al. (2005), there are genetic differences between natural populations. Thus, as it is important to maintain these differences for species reintroduction programs, seeds should be used from populations of that area and seedlings should not be used when their origin is from populations of greater distances, such as from Bahia or Northeast Brazil.

Another conservation strategy is to reintroduce pau-brasil into protected areas and buffer zones where the species existed in the past but is not prevalent anymore. Based on historical reports, pau-brasil was found along the southern coast of RJ, in the municipalities of Mangaratiba and Angra dos Reis (Rocha 2010). This information should be used for the identification of suitable locations within these areas.

For commercial plantations it is important to consider the specific land use situation in RJ. The total farmland of the state covers an area of 2.05 million ha. However, this area includes also 176,659 ha of protected natural forest (APPs, RPPNs, and legal reserves), 102,479 ha of non-protected forest, and 27,960 ha of agriculturally non-usable land (IBGE 2007). As stated in Hissa et al. (2018), 75% of the agricultural production in RJ is family-based. Especially family farming areas offer a great potential to introduce pau-brasil in agroforestry systems including silvopastoral systems. Principal land use of family farming is animal husbandry, especially cattle farming, which is practiced in almost 50% of the family farming establishments (IBGE 2007). The deforested grazing areas for cattle farming are often highly affected by erosion and other types of land degradation (see Soares da Silva et al. 2018; Seliger et al. 2018). The implementation of silvopastoral systems as a specific agroforestry system could represent a sustainable land use strategy contributing to the rehabilitation of these areas improving the water storage capacity of soils and at the same time animal welfare (see Espinel et al. 2018; Fischer et al. 2018). It would allow preserving existing income based on cattle farming and simultaneously generate additional value by planting pau-brasil. Pau-brasil combined with other adequate native tree species (see Sattler et al. 2018) could be planted in these degraded grazed areas and at the moment of plantation be registered for commercial use and legal harvest of the wood in the future.

Important obstacles for commercial plantations of pau-brasil in RJ are (a) the relatively long growing period of minimum of 30 years before the wood can be harvested, (b) the uncertainty of future economic profit (the prize of wood depends on its quality), and (c) legal barriers for commercial use. In order to overcome these hindrances, long-term contracts including guaranteed purchases and price stability with customers (bow makers) could be one promising strategy. For small farmers, associations would help to bridge the time until the wood can be sold. Moreover, especially small farmers require assistance with the registration process and management plans for commercial plantations, as well as with adequate seedling and tree management. In summary, public-private partnerships of farmers, the woodworking companies, and public institutions could facilitate the implementation of commercial plantations.

14.5 Conclusions and Recommendations

Considering the plantation status of pau-brasil, there remain many uncertainties concerning the most adequate plantation type to achieve high wood quality, economic feasibility, and limited amount of labor, but also concerning the highest benefit for nature conservation and sustaining local livelihoods. Lack of general experience and outcome of mixed native tree plantations or agroforestry systems with pau-brasil and native tree species represents a difficulty in the promotion of such plantations and in fully determining their advantages and potentials. The widespread uncertainty and ignorance within the society about the legal framework of wood harvesting of native tree species from plantations limits the willingness and interest of farmers and landowners to commercially plant native tree species, such as pau-brasil. Furthermore, the long growing periods until harvesting impede the motivation especially of smallholder farmers. The collaboration with bow makers, i.e. the future consumer, is required to guarantee the consideration and fulfillment of their interests particularly regarding the wood quality standards, to consequently guarantee as far as possible the future purchase of the wood. But there is also a need of assistance, especially for smallholder farmers when planning to cultivate native or endemic tree species. Agroforestry systems – in particular silvopastoral – could be a sustainable land use strategy for RJ.

Pau-brasil has an exceptional position within the native and endemic trees of the Atlantic Forest due to its extraordinary cultural value for Brazil and its commercial importance for the classical music business. Therefore, it is highly desirable that pau-brasil will be cultivated and conserved together with other endemic tree species of the Atlantic Forest in RJ.

References

Aguiar FFA, Pinho RA (2007) Pau-brasil: *Caesalpinia echinata* Lam., Árvore nacional. Instituto de Botânica, São Paulo

Aguiar FFA, Pinto MM (2008) Manejo silvicultural de um arboreto experimental de pau-brasil em Mogi-Guaçu. In: de CLF RR, Barbedo CJ, Alves ES et al (eds) Pau-brasil da semente à madeira, vol 1. Instituto de Botânica/SMA, São Paulo, pp 67–74

Alves ES, Angyalossy V, Longui EL et al (2008) O arco: arte e ciência. In: Ribeiro R de CLF, Barbedo CJ, Alves ES et al. (eds) Pau-brasil da semente à madeira, 1. Instituto de Botânica/SMA, São Paulo, pp 170–183

Capinera J (2008) Monoculture. Encyclopedia of Entomology 2461

Cardoso SRS, Provan J, Lira CDF et al (2005) High levels of genetic structuring as a result of population fragmentation in the tropical tree species *Caesalpinia echinata* Lam. Biodivers Conserv 14:1047–1057. https://doi.org/10.1007/s10531-004-8409-z

CEPLAC (2008) Programa Pau-Brasil - Program Completion Report (March 2004 to December 2008)

CITES (2007) Examen de las propuestas de enmienda a los apéndices I y II. Convención sobre el comercio internacional de especies amenazadas de fauna y flora silvestres. CITES. 1–11

Dawson IK, Guariguata MR, Loo J et al (2013) What is the relevance of smallholders' agroforestry systems for conserving tropical tree species and genetic diversity in circa situm, in situ and ex situ settings? A review. Biodivers Conserv 22:301–324. https://doi.org/10.1007/s10531-012-0429-5

Dean W (1996) A ferro e fogo: a história e a devastacao de Mata Atlântica brasiliera, São Paolo: Companhia das letras

Espinel ML, Schlüter S, Resende CM (2018) Towards good agricultural practices (GAP) in smallholders' dairy production systems from an animal welfare perspective. In: Nehren U, Schlüter S, Raedig C, Sattler D, Hissa H (eds) Strategies and tools for a sustainable rural Rio de Janeiro. Springer International Publishing, Cham

Evans J, Turnbull JW (2004) Plantation forestry in the tropics, 3rd edn. Oxford University Press, Oxford

FAO (2001) Global Forest Resources Assessment 2000 - Main report. Rome

Fischer SB, Pedraza Luengas A, Schlüter S, Oliveira Antunes LA (2018) From design to implementation: a participatory appraisal for silvopastoral systems. In: Nehren U, Schlüter S, Raedig C, Sattler D, Hissa H (eds) Strategies and tools for a sustainable rural Rio de Janeiro. Springer International Publishing, Cham

Fraga Campos F, Sales Junior PA, Romanha AJ et al (2015) Bioactive endophytic fungi isolated from *Caesalpinia echinata* lam. (Brazilwood) and identification of beauvericin as a trypanocidal metabolite from *Fusarium* sp. Memorias do Instituto Oswaldo Cruz 110:65–74. https://doi.org/10.1590/0074-02760140243

Franke IL, Lunz AMP, Amaral EF do (2000) Metodologia para planejamento, implantação e monitoramento de sistemas agroflorestais: um processo participativo. Rio Branco

Gagnon E, Bruneau A, Hughes CE et al (2016) A new generic system for the pantropical Caesalpinia group (Leguminosae). PhytoKeys 71:1–160. https://doi.org/10.3897/phytokeys.71.9203

Guzmán Wolfhard L, Raedig C (2018) Connectivity conservation management: connecting private protected areas. In: Nehren U, Schlüter S, Raedig C, Sattler D, Hissa H (eds) Strategies and tools for a sustainable rural Rio de Janeiro. Springer International Publishing, Cham

Hissa H, Teixeira N, Costa M et al (2018) Sustainable rural development in Rio de Janeiro state: the Rio rural program. In: Nehren U, Schlüter S, Raedig C, Sattler D, Hissa H (eds) Strategies and tools for a sustainable rural Rio de Janeiro. Springer International Publishing, Cham

IBAMA (1992) Portaria IBAMA no 37-N, de 03 de Abril de 1992

IBGE (2007) Cencso agropecuário 2006 - Agricultura familiar - Primeiros resultados: Brasil, Grandes Regiões e Unidades da Federação. Instituto Brasileiro de Geografia e Estatística - IBGE, Rio de Janeiro

Lobão DE, Valeri SV (2009) Sistema cacau-cabruca: conservaçao de espécies arbóreas da floresta Atlântica. Agrotropica 21:2009

Longui EL, Yojo T, Lombardi DR, Alves ES (2010) The potential of ipê (*Handroanthus* spp.) and Macarandura (*Manilkara* spp.) woods in the manufacture of bows for string instruments. IAWA J 31:149–160

Longui EL, Lima IL de, Lombardi DR et al (2014) Wood with physical, mechanical and acoustic properties similar to those of *Caesalpinia echinata* have high potential as alternative woods for bow makers. CERNE. doi: https://doi.org/10.1590/01047760201420031345

Macedo TM (2015) Variação intraespecífica do lenho e dendrocronologia de *Caesalpinia echinata* Lam. na Floresta Atlântica. Universidade Federal do Rio de Janeiro (UFRJ)

Marques S d S, Oliveira JT d S, Paes JB et al (2012) Estudo comparativo da massa específica aparente e retratibilidade da madeira de pau-brasil (*Caesalpinia echinata* Lam.) nativa e de reflorestamento. Revista Árvore 36:373–380

Martinelli G, Moraes M (2013) Livro vermelho da flora do Brasil. Centro Nacional de Conservação da Flora, Rio de Janeiro 1100. doi: https://doi.org/10.1007/s13398-014-0173-7.2

Michaelson JC (1993) Aniline in history and technology. Endeavour 17:121–126. https://doi.org/10.1016/0160-9327(93)90101-8

Ministério do Meio Ambiente (2012) Portaria No- 320, de 21 de Setembro de 2012. Brazil

MMA M do MA (2014) Lista Nacional Oficial de Espécies da Flora Ameaçadas de Extinção

Murray-Smith C, Brummitt NA, Oliveira-Filho AT et al (2009) Plant diversity hotspots in the Atlantic coastal forests of Brazil. Conserv Biol 23:151–163. https://doi.org/10.1111/j.1523-1739.2008.01075.x

Myers N, Fonseca G a B, Mittermeier R a et al (2000) Biodiversity hotspots for conservation priorities. Nature 403:853–858. https://doi.org/10.1038/35002501

Parisotto EB (2011) Avaliação da atividade antitumoral do extrato das folhas de *Caesalpinia echinata* Lam. In: Dissertação. pp 27–30

Pierce R (2002) The big issue. Strad 8:840–843

Rocha YT (2004) Ibirapitanga: história, distribuição geográfica e conservação do pau-brasil (*Caesalpinia echinata* Lam., Leguminosae) do descobrimento à atualidade. Universidade de São Paulo Faculdade de Filosofia, Letras e Ciências Humanas

Rocha YT (2010) Distribuição geográfica e época de florescimento do pau-brasil (*Caesalpinia echinata* Lam.-Leguminosae). Revista do Departamento de Geografia 20:23–36

Rocha YT, Simabukuro EA (2008) Estratégias de conservação in situ e ex situ do pau-brasil. In de CLF RR, Barbedo CJ, Alves ES et al (eds) Pau-brasil da semente à madeira, vol 1. Instituto de Botânica/SMA, São Paulo, pp 101–113

Rocha YT, Lamarca E, Simabukuro EA et al (2014) Propuesta de una red de áreas protegidas para la protección del pau-brasil (*Caesalpinia echinata* Lam. - Leguminosae), árbol del bosque Atlántico litoral. In: VIII Congreso español de biogeografía: sistemas vegetales y fauna en medios litorales - avances en sus características, dinámica y criterios para la conservación. Sevilla

Sambuichi RHR, Vidal DB, Piasentin FB et al (2012) *Cabruca* agroforests in southern Bahia, Brazil: tree component, management practices and tree species conservation. Biodivers Conserv 21:1055–1077. https://doi.org/10.1007/s10531-012-0240-3

Sattler D, Raedig C, Hebner A, Wesenberg J (2018) Use of native plant species for ecological restoration and rehabilitation measures in Southeast Brazil. In: Nehren U, Schlüter S, Raedig C, Sattler D, Hissa HR (eds) Strategies and tools for a sustainable rural Rio de Janeiro. Springer International Publishing, Cham

Seliger R, Sattler D, Soares da Silva A et al (2018) Rehabilitation of degraded sloped pastures - lessons learned in Itaocara. In: Nehren U, Schlüter S, Raedig C, Sattler D, Hissa H (eds) Strategies and tools for a sustainable rural Rio de Janeiro. Springer International Publishing, Cham

Siqueira EP, Zani CL, Maria T et al (2014) Evaluation of the *In vitro* leishmanicidal and *in vivo* acute oral toxicity of the *Caesalpinia echinata* L. extracts as source of natural products against leishmaniasis. J Nat Prod Plant Resour 4:30–38

Skeaping K (1955) Some speculations on a crisis in the history of the violin. Galpin Soc J 8:3–12. https://doi.org/10.2307/842152

Soares da Silva A, Seliger R, Sattler D, Heinrich J (2018) Soil degradation in Southeast Brazil: a challenge for restoration and rehabilitation. In: Nehren U, Schlüter S, Raedig C, Sattler D, Hissa HR (eds) Strategies and tools for a sustainable rural Rio de Janeiro. Springer International Publishing, Cham

Taylor A, Harper D, Espey C et al (2013) Examination of wood properties of plantation-grown Pernambuco (*Caesalpinia echinata*). IAWA J 34:34–48. https://doi.org/10.1163/22941932-00000004

Varty N (1998) *Caesalpinia echinata*. ICUN Red List Threat Species 1–10. https://doi.org/10.2305/IUCN.UK.1998.RLTS.T33974A9818224.en

West PW (2014) Growing plantation forests, 2nd edn. Springer International Publishing, Cham

Zani LB, Sarnaglia Junior VB, Gomes JML, Thomaz LD (2012) Estrutura de um fragmento de Floresta Atlântica em regeneração com ocorrência de *Caesalpinia echinata* Lam. (pau-brasil). Biotemas 25:75–89. https://doi.org/10.5007/2175-7925.2012v25n4p75

Part III
Integrated Water Management

Chapter 15
Water Security in Rio de Janeiro State

Lars Ribbe, Rosa Maria Formiga-Johnsson, and Juan Luis Ramirez Duval

Abstract Although Rio de Janeiro is considered to be well endowed with renewable water resources, it faces several water challenges. Recurrent floods and landslides and the major drought of 2014–2015 in the Paraíba do Sul river basin, its main source, have given a reality check to the population showing that also water scarcity can be a problem and that water availability is highly dependent on neighboring states. Furthermore, there are still major water quality issues and a relatively high number of people without adequate water supply and sanitation. Overall, it can be stated that the level of water security is considerably far from being at a satisfactory level. The reasons for this situation cannot only be found in the natural variability of natural resources but rather in the prevailing water resources governance system. Even though the Brazilian water laws promote an integrated approach to water management, the practice does not always meet the high aspirations, as observed in Rio de Janeiro state. This chapter analyzes the natural resource base relevant for water resources in the state of Rio de Janeiro and in the interstate Paraíba do Sul basin, its main source. The water resources challenges and the general principles of IWRM and multilevel water governance in Rio de Janeiro are discussed. Furthermore some suggestions on how to assure a higher level of water security in the state are provided.

L. Ribbe (✉) · J. L. Ramirez Duval
Institute for Technology and Resources Management in the Tropics and Subtropics (ITT), TH Köln – University of Applied Sciences, Köln, Germany
e-mail: lars.ribbe@th-koeln.de

R. M. Formiga-Johnsson
State University of Rio de Janeiro (UERJ), Rio de Janeiro, Brazil

U. Nehren et al. (eds.), *Strategies and Tools for a Sustainable Rural Rio de Janeiro*, Springer Series on Environmental Management,
https://doi.org/10.1007/978-3-319-89644-1_15

Keywords Water security · Urbanization · Climate extremes · Rio de Janeiro · Inter-basin transfer

15.1 Introduction

Reaching water security can be considered as the key goal of water resources management in a given region. In recent years, the interest in water security increased in particular with the discussion of the linkage of the water sector to other sectors, resulting in concepts of the water-energy-food security nexus (Smajgl et al. 2016). Furthermore, it is widely recognized that water security is a precondition for several other elements of social and economic development (Sadoff et al. 2015). In that same sense, water security is considered as a pivotal element to reach several of the sustainable development goals, i.e., not only that on water (6) but also those related to it. In addition, food security (2), health (3), energy (7), cities (11), climate (13), and land (15) have obvious close linkages to water security of the SDGs (http://www.un.org/sustainabledevelopment/).

While there is no internationally agreed upon definition on water security, the intergovernmental panel on water (UN 2013: 1) defines it as "The capacity of a population to safeguard sustainable access to adequate quantities of and acceptable quality water for sustaining livelihoods, human well-being, and socio-economic development, for ensuring protection against water-borne pollution and water-related disasters, and for preserving ecosystems in a climate of peace and political stability." This broad definition emphasizes that water security not only refers to access of humans to sufficient water of good quality but also to different sectors, including the environment, and also to protection from water-related natural hazards. Even though the analysis of household water supply coverage or the relation of water availability and demand has been conducted since long time from local to global scale (see, e.g., Alcamo et al. 2003), the water security aspects related to ecosystems or water-related hazards deserve more attention in research and management.

Considering that the assessment of water security can be conducted at different spatial levels, the choice of the analysis unit is quite relevant for the further interpretation. Taking Brazil as a country, for example, it may be difficult to interpret indicators of water security. While Brazil is the most water-rich country in the world, water availability and demand are distributed quite unevenly. Analyzing water security at the state or basin level seems to be more appropriate. The state level is also the level where adequate responses can be taken in terms of water governance or investments.

In this chapter we analyze the different aspects of water security in the state of Rio de Janeiro (RJ) – a state facing relevant challenges regarding water resources development and water-related disasters in recent years.

15.2 Water Resources in Rio de Janeiro State

RJ is the second smallest state of Brazil, nonetheless with 16.7 million inhabitants, a high urbanization rate, and a very strong industry. The state accounts for 10.4% of the Brazilian GNP, which was the second of the country in 2016 (R$ 651,597 million). The water availability in the region is 1279 m^3/inhabitant/year (Agência Nacional de Águas – ANA 2013), which according to the "Falkenmark indicator" is a water-stressed region (1700–1000 m^3/inhabitant/year), where periodic or limited water shortages can be expected. It should be noted that availability per capita is only a rough estimate for water scarcity. Many different metrics exist; however, no standard or "the" best indicators for water scarcity or stress are available (Damkjaer and Taylor 2017).

The population is concentrated in the coastal region where water availability is less. That is why the metropolis of Rio de Janeiro, which concentrates more than 70% of the state population, depends on the diverted waters of Paraíba do Sul river to Guandu basin. Around 75% of the state population (including 9.4 million people from the metropolis) depend on the water from Paraíba do Sul basin which also meets large parts of the water demand from industrial, agricultural, and hydropower sectors of the entire state (INEA; Fundação Coppetec 2014). As the headwaters are located in the state of São Paulo (SP) and major tributaries are located in the state of Minas Gerais (MG), water security in RJ is subject to a high level of vulnerability (Fig. 15.1).

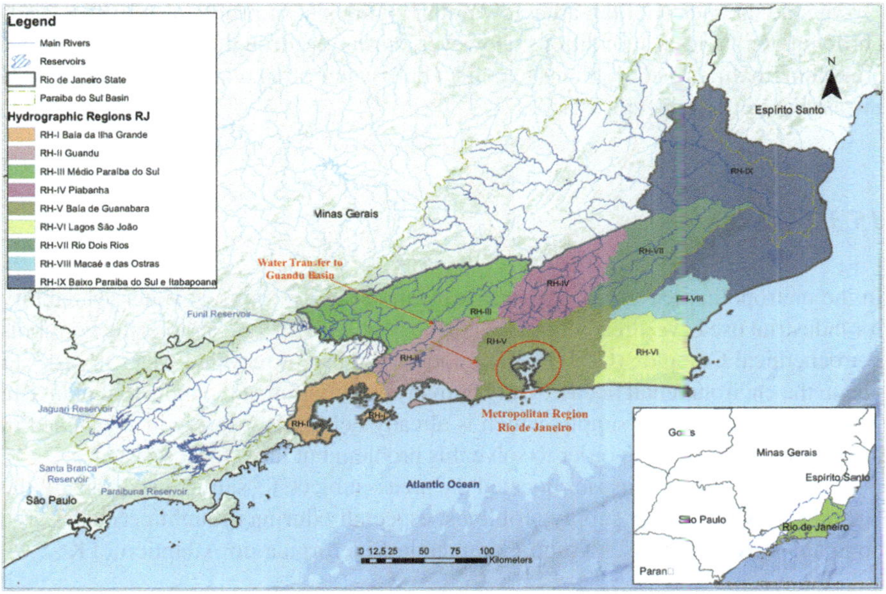

Fig. 15.1 State of Rio de Janeiro and its location in the Paraíba do Sul basin

In 2014 the Paraíba do Sul river basin suffered the worst drought in the last 85 years of historical record, which lasted until the rainy season of 2015. For this reason, the Paraíba do Sul river basin became a point of conflict with SP (Formiga-Johnsson et al. 2015).

15.3 Water-Related Challenges in Four Dimensions: Households, Economy, Environment, and Extreme Events

15.3.1 Households and Urban Demand

In the last 50–60 years, the state's population has become essentially urban with today over 96% of the population living in urban areas, 70% of which live in Rio de Janeiro metropolitan area (IBGE 2010). Currently, around 89% of the population have access to water supply, 60% of sewage is collected, and 35% of sewage is treated (IBGE 2010).

In the eastern metropolitan region, there is a current deficit of water supply (2.2 m^3/s) due to a lack of available bulk water and low water security because of the absence of regulation in its main rivers, Guapiaçu and Macacu (INEA; Fundação Coppetec 2014). On the contrary, the Guandu system has adequate water availability and legal water reserves for future expansions to the Western Metropolitan supply system.

Outside the metropolitan area, few municipalities have problems of water availability under "normal" conditions. However, during the drought of 2014–2015, several municipalities had problems to collect bulk water often related to inadequacy of infrastructure (Costa et al. 2015).

15.3.2 Economy

In the metropolitan east area, which faces water scarcity, limits of water availability for industrial use have already been set (Britto et al. 2016). For example, the Petrobrás Petrochemical Complex (COMPERJ) imposed the use of recycled water in order to obtain the environmental license. Only human consumption of the industrial plant can be met with local water supply, which is already deficient. A reservoir is planned in the Macacu river basin in order to solve this problem but faces local resistance.

Similarly, some industries located in areas subject to saline intrusion of the Guandu river face water supply problems, especially during drought periods. Large companies as the ThyssenKrupp Companhia Siderúrgica do Atlântico (TKCSA), one of the largest steel companies in Latin America, were heavily impacted by the 2014–2015 water crises (see Costa et al. 2015).

In contrast to the large industrial plants that take water from regulated rivers, irrigation demand is concentrated in rivers that have no regulation in the Paraíba do Sul basin (Serrana and Baixo Paraíba do Sul regions). The Serrana region is expected to face strong growth in irrigated areas in the coming decades.

So far, industrial and agricultural demands have been generally met, but the drought of 2014–2015 had important impacts, i.e., in agriculture, losses were as follows: 60% in fruit growing, 30% in coffee production, 40% in tomato, 60% in sugarcane, and 60% in cattle farming (EMATER-Rio; PESAGRO-Rio 2016). A survey of the Federation of Industries of RJ state found that the drought had impacts on 16% of 517 industries, whether temporary interruption, reduction of production, or dismissal (FIRJAN 2016). The droughts also have an impact on the hydropower sector, which is concentrated in the Paraíba do Sul and Guandu rivers; Costa et al. (2017) concluded that the energy generation decreased by 19.4% in 2013–2014 and by 30.9% in 2014–2015 when compared to the previous years (2010–2013).

15.3.3 Environment

RJ also faces important challenges concerning water resources and the environment. For example, the Paraíba do Sul and Guandu rivers and its tributaries are heavily polluted due to irregular industrial activities along its banks and the nonexistence of adequate wastewater collection and treatment systems in some municipalities. Other immediate environmental challenges that need to be addressed are the source water protection closely related to the loss of vegetation cover and deforestation.

15.3.4 Extreme Events

RJ is highly vulnerable to natural hazards and climatic extreme events (Nehren et al. 2018). Floods and landslides are the most frequent events in RJ (Frank et al. 2018). In 2014 and 2015, the Paraíba do Sul river basin suffered its worst recorded drought in the past 85 years (see Fig. 15.2). As a result, the regulated flow of the Paraíba do Sul and Guandu rivers had to be progressively reduced facing users of both rivers with unprecedented challenges regarding water security (Costa et al. 2015).

There was consensus between representatives of public authorities and civil society that the current operating rules pertaining to reservoirs in the Paraíba do Sul river basin were no longer appropriate in the light of the new situation of potential low flow in combination with increasing water demands. In December 2015, after a long negotiation process involving the managers of the Paraíba do Sul waters, new

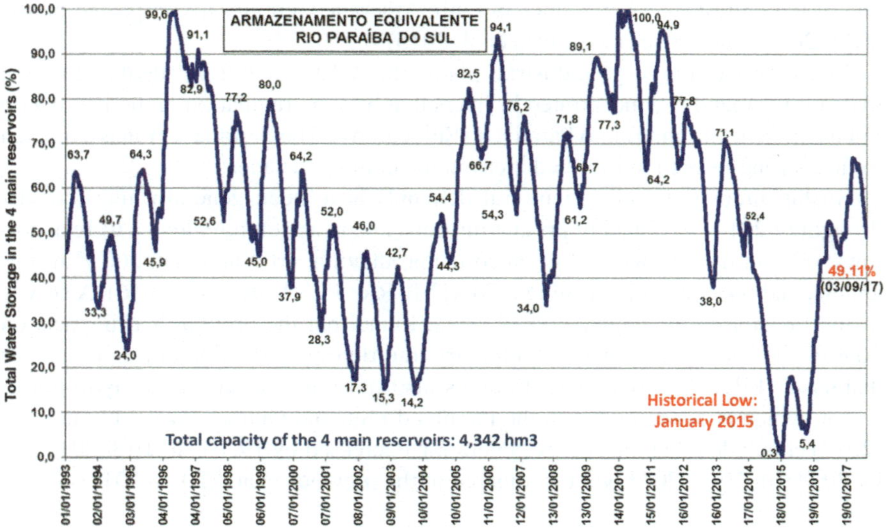

Fig. 15.2 Evolution of total reservoir storage in the **Paraíba** do Sul basin, 1993–2017

operating rules were adopted, but only a year later, they were implemented, when the water crisis was over. It remains to be seen if the new rules suffice to deal with potential future droughts. In fact, the variability of the natural and human systems is not yet sufficiently well understood.

15.4 IWRM Principles in Brazil and Implementation of Water Governance in Rio de Janeiro

All waters in Brazil are in the public domain, federal or state[1]. For this reason, the water management requires separate institutional frameworks at both levels of government. New water legislation was first approved in SP in 1991 and at national level in 1997; only in 1999 RJ has approved its water law (Law 3239/99). All of them embraced the new approach taken by the pioneering states, in consonance with the principles recommended by major international charters such as the Dublin Statement of 1991 (Box 15.1).

[1] The federal government has jurisdiction over waters that cross state or international boundaries. The water located entirely within the territory of a single state, as well groundwater resources, are in state domain, except when they are used by federal infrastructure projects.

Box 15.1: Main Elements of Federal Water Law 9433/97 and Complementary Legislation

Objectives of the National Water Resources Policy:

- Reliability of water availability for current and future generations, at the desired level of water quality
- Rational and integrated use of the resource for sustainable development
- Prevention and relief of critical hydrologic events resulting from natural causes or inadequate natural resource use

Principles:

- Integrated water management, with the river basin as the planning unit
- Water as a finite and fragile resource
- Water as an economic good
- Decentralized management
- Participatory management

Organizations:

- National Water Resource Council (CNRH): deliberative body, with stakeholder participation, in charge of supervising and regulating the National Water Resources Management System
- Water Resources Secretariat (SRH): in charge of elaborating the National Water Policy
- National Water Agency (ANA): in charge of implementing the water resources management system (*)
- State Water Resources Councils: similar to the National Water Resources Council
- State Water Management Agencies: similar to the SRH and ANA (SEA and INEA, in RJ).
- River Basin Committees: deliberative stakeholder body with decision-making and regulatory powers
- River Basin Agencies: their technical and administrative arms

Instruments:

- Basin water resources plans/state water resources plans/national water resources plan (**)
- Bulk water use permits

(continued)

Box 15.1 (continued)

- Bulk water charges (**)
- Classification of water bodies according to predominant use and water quality standards[2]
- National and state water resources information system (**)

(*)ANA was created in 2000 (Federal Law 9.984/00)

(**)These instruments were first introduced by the federal and state laws. The remaining tools already existed but either were not functioning as planned or their structure and functioning were modified significantly by the law.

By defining the river basin as the new territorial scale for planning and management, the proposal is to go beyond sectoral management, promoting territorial integration, through river basin plans. The design of the management system was largely based on the French system, centering on the committees and the basin agencies, with the committee as the deliberative institution where negotiations and participatory decision-making take place and the agency as the committee's executive arm for technical support and implementation. The laws also created a new water management instrument: bulk water charges, which would give financial support to the new basin institutions. The concession and control of water user rights remained under the jurisdiction of either federal or state governments.

The development of a new, modern, and complete water resources information system is one of the basic needs for the implementation of the water resources management system considered under Law 9433/97. A complete and comprehensive dataset on water availability; users; water quality monitoring, with the ample use of current technologies like databases, geographic information systems (GIS); and a comprehensive reporting system is certainly the way to produce an efficient decision support framework. This information is also needed to adequately cope with environmental and natural disasters, such as flash floods and landslides, as well as natural seasonal fluctuations in precipitation that cause pronounced rainy and dry periods.

At national level, ANA is responsible for maintaining a national monitoring network in every hydrographical region. At state level, most federal states operate monitoring networks that have adopted different approaches, i.e., regarding number of parameter, sampling frequency, etc. In the case of RJ, the State Institute of Environment (INEA) is responsible for monitoring water resources. This raises acute problems in the Paraíba do Sul basin since there is insufficient harmonization of monitoring and water allocation criteria of shared waters between states (RJ, SP, and MG). The existing information system is fragmented and underutilized and

[2] National Environmental Council (CONAMA) Resolution 20/86 determines that river reaches and other water bodies have to be assigned to one of four water classes according to their prevalent use.

needs to be expanded and consolidated to improve the decision-making process (OECD 2015). The water agency of the Paraíba do Sul river (AGEVAP) recently launched the information system SIGA-CEIVAP to try to solve these issues for the Paraíba do Sul basin (sigaceivap.org.br/).

Nonetheless, despite the increasing abstraction and the increasing pollution of water resources in RJ, there is a lack of systematic monitoring networks, especially on water quality. Local decision-makers lack the resources and information to tackle such problems; models to interpret the data and to develop scenarios or other decision tools are only used exceptionally. Thus, planners have to make decisions based upon insufficient information.

In spite of the before-mentioned shortcomings, IWRM in the territory of RJ is considered one of the most advanced in the country in politico-institutional terms as well as regarding technical tools (OECD 2015). Water governance in the state is extremely complex when it comes to the river basins of the Paraíba do Sul and Guandu because it involves federal waters shared with neighboring states. In other hydrographic regions, management is simpler since all rivers are "state waters" and remain under the governance of state institutions.

At state level, the implementation of the water resources management system was strongly accelerated with the creation of the State Environmental Institute (INEA) in 2009. At national level, management related to the Paraíba do Sul basin has also made many advances, including the pioneering experience of implementing the IWRM in federal river basins, by the National Water Agency (ANA) in conjunction with the states: the Committee (CEIVAP), which covers the whole of the basin in SP, MG and RJ, was created in 1996; its basin agency (AGEVAP), established in 2004, has been substantially strengthened over the years.

In sum, a remarkable change occurred in the political and institutional context with the creation of ANA at federal level and INEA at state level, resulting in significant technical and financial capacity building. Similarly, an extraordinary mobilization around water issues, problems, and management occurred. The reform process has allowed groups that had been traditionally excluded from decision-making in the water resources sector i.e. municipalities, users, and civil society, to enter into the negotiation process. These groups are now represented in the State Water Resources Council and basin committees.

However, these great advances are clearly not sufficient to face the magnitude of problems related to water security and management. Many had high expectations that committees and their agencies, counting on basin plans and water charges, would have a great transforming power to trigger changes in practice. But, with few exceptions, the results of IWRM implementation cannot yet be measured in terms of water quality improvements, rational water use, integration with other policy arenas, etc.

15.5 Recommendations Towards Improved Water Security

We presented the major water resources challenges in RJ as well as the current structures and instruments installed. Based on this we are now elaborating some suggestions on how to assure a higher level of water security in the state.

15.5.1 Strengthening the Management and Governance System

Although technical issues remain significant, the greatest challenges for water security in RJ are related to governance. Mechanisms through which the federal and state management and policy are to interact at the Paraíba do Sul basin still need to be developed. This is crucial, since RJ's water supply is largely dependent on the neighboring states through the Paraíba do Sul river and main tributaries. Perhaps the major challenge for RJ is to maintain, in the long run, the minimum flow of 190 m^3/s in the Paraíba do Sul river at the point of the *basin transfer* and the minimum flow of transfer to the Guandu (119 m^3/s). This was the point that RJ defended most in negotiations regarding SP's proposal to divert waters from Paraíba do Sul river basin to reinforce greater SP's water supply: reducing the flow of the Paraíba do Sul river was nonnegotiable for RJ (Formiga-Johnsson et al. 2015). Similarly, there is a need to establish a drought risk management strategy, since the state is also facing water quantity issues as observed in the Paraíba do Sul basin in 2001–2003 and 2014–2015.

Also river basin plans are in general too traditional and conservative, consisting of a "wish list" with no indication on the allocation of responsibilities, timeline, and funding for implementation (OECD 2015).

15.5.2 Technical Base for Management (Monitoring, Information Management, Modeling)

As mentioned before, the lack of information is one of the most critical points regarding the development and implementation of the new management system in the country and at the state/basin level. In the Paraíba do Sul basin, the water agency AGEVAP attempts to integrate all the information available from federal and states agencies into a single information system (SIGA-CEIVAP); however, there are still problems in obtaining relevant datasets for this information system. Moreover, the amount of monitoring stations/monitoring data for water quality and quantity on the Paraíba do Sul and its tributaries is still not sufficient, even though current projects of ANA and AGEVAP address this topic.

At state level, INEA has invested a lot of resources in monitoring extreme events and water quantity, especially after the floods and landslides in the Serrana Region of 2011 (compare Nehren et al. 2018), but still the monitoring of water quality has been left behind. In many hydrographic regions of RJ, the monitoring of water quality only takes part a couple of times a year, with a combination of in-situ measurements and collection of sample for laboratory analysis. There is no monitoring and information system financed by the state with continuous and real time data on water quality for the main rivers.

In general, there is a need for modern water quality monitoring strategies and equipment, as well as information systems, to store, analyze, and share the information. Similarly, the technical expertise level on these issues in many of the institutions is not sufficient. There is a real need for more training and hiring of technical staff for these areas.

Within the research project INTECRAL (working package 4; see Schlüter et al. 2018), some of the key issues addressed were the construction of monitoring concepts and measurement networks and also the building of information and model systems in order to create an appropriate database for decision-making. INTECRAL has also focused on conducting training for local stakeholders and project partners to increase the knowledge base.

15.5.3 Improve Water Infrastructure

Inadequate sanitation access and low level of sewage treatment are massive problems in the metropolitan area of Rio de Janeiro. The investment in sanitation infrastructure has been low in all parts of the city and even worse in poorer areas (favelas). According to the Ministry of Cities, 35% of the population in RJ is not connected to a formal sanitation system, and even in areas with formal connections, only about half of sewage waste is treated before entering into waterways and eventually the ocean.

The majority of residents without access to sewage collection belong to municipalities located around Guanabara Bay (Rio de Janeiro, Duque de Caxias, Niterói, and São Gonçalo) or have rivers and streams that flow into it. Of the 750,000 homes without sewerage, it is estimated that 630,000 homes belong to this area. Consequently, 84% of uncollected, untreated sewage in the metropolitan area of RJ flows directly into Guanabara Bay. As a result, the condition of the Guanabara Bay is deplorable.

In order to improve water quality and thus water availability in RJ, authorities need to reverse their prevailing vision of sanitation as a cost, rather than an investment. The government has chosen to invest in other areas such as industry, urban projects, etc. Moreover, a new approach is needed for sanitation systems in poor areas or favelas. In those areas it is more suitable to build many small systems, i.e., decentralized wastewater treatment, instead of one large system, in order to adapt to differing conditions, such as topography and spatial constraints, and ensure access for all residents. Other factors such as education on how to use each system and proper maintenance are crucial. Up to now, the government has yet to devise a widespread sanitation plan for favelas that adequately addresses their unique conditions.

15.5.4 Protection of Springs and Headwaters

Headwaters and springs should be given special attention because they are responsible for the regular flow of water of good quality. Inadequate land uses in the upstream or riverbank region devastate the riverbed or recharge areas responsible for replenishing aquifers and springs.

For the recovery and preservation of springs and headwaters in rural properties, some soil and vegetation protection measures can be adopted, ranging from the elimination of burning practices to the enrichment of native forests. Furthermore, strategies of preservation of springs should include control of soil erosion through physical structures and buffer strips while also minimizing chemical and biological contamination.

In the case of RJ, the Rio Rural program and the payment for environmental services (PES) programs are a good example that encourages the protection of the springs and seeks the support of the various local actors, companies, and public institutions for the realization of projects and investments, aiming at the conservation of water resources. The new Forest Code could play an important role for source water and riverbank protection through the promotion of buffer strips and forest protection on mountain tops and slopes (Buriti 2016).

15.5.5 Strategies for Water Demand Management and Water Use Efficiency

Demand management plays a key role toward the sustainable use and management of water and is a fundamental part of the Integrated Water Resources Management (IWRM) concept. According to Kampragou et al. (2010), demand management should be based on five core principles: (i) enforce water conservation, (ii) encourage water saving, (iii) invest in water saving, (iv) set the price of water, and (v) educate water users.

In RJ there are many opportunities for the implementation of water demand management strategies. Additionally, many institutions and stakeholders recognize the importance of such programs. However, water managers may lack the skills and knowledge to undertake demand management programs. Some recommended strategies for water demand management are:

- The tariff structure of water and wastewater services need to be redesigned in order to generate sufficient revenues for the utilities to expand the wastewater infrastructure.

- While the Paraíba do Sul basin generates 10% of national GDP, federal and state bulk water charges collect less than 20 million Real, i.e., around 7 Mio US$ yearly.
- Rio de Janeiro is by far the metropolis in Brazil with the highest per capita water consumption with 251 L per capita per day (São Paulo 179, Brasilia 181, Belo Horizonte 154 L). Its distribution network has losses of 30.6% (Stepping 2016). Therefore, there is a high potential for the reduction of water consumption and waste by using different strategies that should be analyzed: improvement of infrastructure, new water-saving devices (retrofitting), water reuse, rainwater harvesting, awareness campaigns among the public, etc.

Water resources in RJ are under growing pressure due to increasing human demands and impacts in combination with climate extremes. In order to cope with the current and future challenges, water governance and infrastructure need to be adapted. Generation of and access to adequate information seems to be a crucial condition for better decision-making and water management in the state.

References

Agência Nacional de Águas – ANA (2013) Conjuntura dos Recursos Hídricos no Brasil. Relatório 2013, Brasília

Alcamo J, Döll P, Henrichs T et al (2003) Development and testing of the WaterGAP 2 global model of water use and availability. Hydrol Sci J 48(3):317–338

Britto AL, Formiga-Johnsson RM, Carneiro PRF (2016) Water supply and hydrosocial scarcity in the Rio de Janeiro metropolitan area. Ambient Sociedade 19:183–206

Buriti RdS (2016) Policy analysis for source water protection in Rio de Janeiro state: assessing institutions and strategies. Master of Science Natural Resources Management and Development Degree, Cologne University of Applied Sciences/ITT – Institute for Technology and Resources Management in the Tropics and Subtropics

Costa LF, Farias Junior JEF, Formiga-Johnsson RM, Acserald MV (2015) Crise hídrica na Bacia do rio Paraíba do Sul: enfrentando a pior estiagem dos últimos 85 anos. Revista Ineana (Revista técnica do Instituto Estadual do Ambiente, RJ) 3:26–47

Costa LF, Farias Junior JEF, Formiga-Johnsson RM (2017) Impactos da estiagem 2014–2015 sobre os principais setores usuários de água no Estado do Rio de Janeiro. Paper accepted for oral presentation, Simpósio Brasileiro de Recursos Hídricos, Florianópolis

Damkjaer S, Taylor R (2017) The measurement of water scarcity: defining a meaningful indicator; Ambio n.d.46: 513

Emater-Rio; Pesagro-Rio (2016) Relatório de Perdas no Setor Agropecuário no Estado do Rio de Janeiro com foco no crédito rural – Estiagem 2014/2015. Rio de Janeiro, 2016

FIRJAN (2016) Apresentação sobre os impactos da escassez de água na indústria 2014–2015. Rio de Janeiro

Formiga-Johnsson RM, Farias Junior JEF, Costa LF, Acserald MV (2015) Segurança hídrica do Estado do Rio de Janeiro face à transposição paulista de águas da Bacia Paraíba do Sul: relato de um acordo federativo. Revista Ineana (Revista técnica do Instituto Estadual do Ambiente, RJ) 3:48–69

Frank B, Formiga-Johnsson RM, Viana VJ, Ikemoto SM (2018) Disaster risk management in the state of Rio de Janeiro. In: Nehren U, Schlüter S, Raedig C, Sattler D, Hissa H (eds) Strategies and tools for a sustainable rural Rio de Janeiro. Springer International Publishing, Cham

IBGE (2010) Censo demográfico 2010. Características da população e dos domicílios: resultados do universo. IBGE, Rio de Janeiro, p 2011
INEA; Fundação Coppetec (2014) Plano estadual de recursos hídricos do estado do Rio de Janeiro (PERHI-RJ). Final report
Kampragou E, Lekkas DF, Assimacopoulos D (2010) Water demand management: implementation principles and indicative success stories. Environmental & Energy Management Research Unit, School of Chemical Engineering, National Technical University of Athens
Nehren U, Kirchner A, Lange W et al (2018) Natural hazards and climate change impacts in the state of Rio de Janeiro: a landscape historical analysis. In: Nehren U, Schlüter S, Raedig C, Sattler D, Hissa H (eds) Strategies and tools for a sustainable rural Rio de Janeiro. Springer International Publishing, Cham
OECD (2015) Water resources governance in Brazil, OECD studies on water. OECD Publishing, Paris
Sadoff CW, Hall JW, Grey D et al (2015) Securing water, sustaining growth: report of the GWP/OECD task force on water security and sustainable growth. University of Oxford, pp 180. ISBN: 978-1-874370-55-0
Schlüter S, Nehren U, Sattler D, Raedig C (2018) The INTECRAL project. In: Nehren U, Schlüter S, Raedig C, Sattler D, Hissa H (eds) Strategies and tools for a sustainable rural Rio de Janeiro. Springer International Publishing, Cham
Smajgl A, Ward JR, Pluschke L (2016) The water-food-energy Nexus – realising a new paradigm. J Hydrol 533(02/2016):533–540
Stepping K (2016) Urban sewage in Brazil: drivers of and obstacles to wastewater treatment and reuse. Governing the water-energy-food nexus series. Discussion paper 26/2016, German Development Institute
UN (2013) Water security & the Global Water Agenda. A UN-Water analytical brief, United Nations University, 2013. Revised version, October 2013

Chapter 16
Analyzing Hydro-Climatic Data to Improve Hydrological Understanding in Rural Rio de Janeiro, Southeast Brazil

Juliana M. Santos, Annika Künne, Sven Kralisch, Manfred Fink, and Alexander Brenning

Abstract The rural area of Rio de Janeiro (RJ) state has experienced increased pressure on water resources, due to an increasing population linked with the growth of the industrial and agricultural sectors. High interannual variability of rainfall causes frequent extreme events leading to droughts, floods, and landslides. Therefore, it is crucial to understand how climate affects the interaction between the timing of extreme rainfall events, hydrological processes, vegetation growth, soil cover, and soil erosion. Ecohydrological modeling can contribute to a better understanding of spatial–temporal process dynamics to develop adaptation strategies. However, prior to modeling, it is crucial to evaluate the reliability of the climate and hydrological data. This study aims to homogenize the climatic data and to analyze the hydro-climatic time series needed for further hydrological studies (e.g., ecohydrological modeling) and to contribute to a better understanding of long-term hydro-climatic patterns in a mesoscale watershed, the Muriaé River Basin. The analyses include homogeneity assessment, statistical analyses, and trend detection for a time period of over 50 years. The assessment provides important insights into long-term hydro-climatic patterns, such as an increase of the annual mean temperature, a decrease of the annual relative humidity, and an increase of the frequency of intense rainfall events.

Keywords Data analysis · Homogeneity assessment · Statistical analyses · Trend detection · Ecohydrological modeling

J. M. Santos (✉) · A. Künne · S. Kralisch · M. Fink · A. Brenning
Geographic Information Science Group, Institute of Geography, Friedrich Schiller University of Jena, Jena, Germany
e-mail: juliana.santos@uni-jena.de; annika.kuenne@uni-jena.de; sven.kralisch@uni-jena.de; manfred.fink@uni-jena.de; alexander.brenning@uni-jena.de

U. Nehren et al. (eds.), *Strategies and Tools for a Sustainable Rural Rio de Janeiro*, Springer Series on Environmental Management,
https://doi.org/10.1007/978-3-319-89644-1_16

16.1 Introduction

The rural parts of Rio de Janeiro (RJ) state in Southeast Brazil have experienced increasing pressure on water resources, due to the population growth of the city of Rio de Janeiro and the associated growth of the industrial and agricultural sectors (Zeng et al. 2016). This has led to a number of negative environmental impacts, including water scarcity, riverine forest degradation, soil erosion, land degradation, and water quality problems (Merten and Minella 2013). Additionally, this region experiences high seasonal rainfall variations, which can lead to droughts, floods, and landslides (Martirani and Peres 2016). The area is part of the Brazilian Atlantic Forest biome, and only approximately 11.4–16.0% of the historic cover remains after drastic deforestation (Ribeiro et al. 2009). Besides historical land-use changes, climate variations have also had a major impact on landscape development in the Brazilian Atlantic Forest (Mata Atlântica; Nehren et al. 2013).

Southeastern Brazil has climatically distinct dry and wet seasons (ranging from 2 to 5 dry months), with average annual temperatures ranging from 22°C to 25°C, and mostly semi-deciduous and deciduous tree species (Colombo and Joly 2010). These environmental conditions may be subject to change due to climate variability, however, resulting in different temperature and rainfall patterns (Pachauri et al. 2014). Some tree species that are part of the Atlantic Forest biome are at particular risk of extinction due to the interaction between climate (i.e., temperature, rainfall, and humidity) and vegetation growth dynamics (Colombo and Joly 2010). Pachauri et al. (2014) also pointed out that increased rainfall amounts and intensities could lead to higher rates of erosion, partly due to changes in plant canopy caused by shifts in plant biomass production associated with changed soil moisture regimes.

Recent rainfall trend studies in South America have indicated that trend analysis of extreme rainfall patterns is challenging due to the lack of homogeneous long-term climate observations (i.e., data collected using equivalent methods; Vincent et al. 2005; Haylock et al. 2006; Carvalho et al. 2014). In Brazil, the Law on Access to Information from 18 November 2011 (Lei 12527) regulates the constitutional right of access to public information (Brasil 2011). As a result, historical hydro-climatic time series from governmental sources were made publicly available in 2012 (Martirani and Peres 2016). The largest online platform in the country is the National Water Agency (Agência Nacional de Águas, ANA), which distributes hydro-climatic data from different local sources, such as the Environmental Agency of RJ (Instituto Estadual do Meio Ambiente, INEA). Other platforms that distribute climatic data are the Integrated System of Environmental Data (Sistema Integrado de Dados Ambientais, SINDA) and the National Meteorological Institute (Instituto Nacional de Meteorologia, INMET). Hence, different platforms are available, all with the aim of data collection and distribution. However, metadata are frequently difficult to obtain, and different equipment and measurement approaches (e.g., measurement time, calculation of daily averages and sums) are used, making it a challenge to evaluate the usability of the data. Although much of these data have been checked and corrected, the procedure is not always transparent, can be difficult to access, and has only been made available for older time series (e.g., streamflow time series from ANA are only available until 2005).

The objective of this study is to develop a consistent procedure for homogenizing climatic data, to analyze the hydro-climatic time series needed for hydrological studies (e.g., trend analyzes and ecohydrological modeling), and to contribute to a better understanding of the long-term hydro-climatic patterns in this region. These steps should help to address the data shortcomings in RJ. The selected study area is the Muriaé River Basin, a mesoscale watershed located in the rural part of RJ. The Muriaé River Basin was selected because it is representative of the characteristics of the rural countryside in RJ, and it is an important site for addressing the problems of degraded pasturelands in this region (Seliger et al. 2018; Sattler et al. 2018, in this book). In addition, the prolonged dry season and intense rainfall events, which are predicted to be induced by climate change, may lead to increased land degradation (Carvalho et al. 2014). Hence, a trend analysis was conducted to prove a marked long-term (50–70 years) monotonic increase or decrease of climate variables in recent years.

16.2 Study Area

The Muriaé River Basin has an area of 8161 km^2, 55% of which is located in RJ (downstream part) and 45% in the state of Minas Gerais (MG; headwater part). The total population within the basin is 430,000 inhabitants, who live in 36 municipalities. The origin of the Muriaé River is in the Miraí municipality in Serra da Mantiqueira mountain range, with an elevation of 900 m.a.s.l., and its main tributaries are the Glória and the Carangola rivers. The Muriaé River flows from west to east, terminating at the Paraíba do Sul River, with a total river channel length of approximately 300 km (Ávila et al. 2016). The Köppen climate classification of the study area is "humid subtropical with dry winter and hot summer" (Cwa) along the upper part of the basin and "tropical with dry winter" (Aw) for the remaining part (Alvares et al. 2014).

During the last 100 years, the Muriaé River Basin has experienced major floods in 1925, 1945, 1979, 2008, and 2011 due to intense rainfall events at the headwater of the basin. The most recent major flood was in 2011, during which the state government of Rio de Janeiro declared a state of emergency in seven municipalities in the north and northwest parts of the state (Lange et al. 2018; Nehren et al. 2018). In addition to naturally occurring floods, the Muriaé River was also affected in 2006 and 2007 by breaking mining waste dams in the municipality of Miraí (MG), which released a volume of 2 million m^3 of mud, causing floods and destruction in all downstream cities (SACE 2016).

16.3 Data and Methods

The selected datasets for the Muriaé River Basin used in this study are listed in Table 16.1; a workflow diagram of the data analysis steps is shown in Fig. 16.1. The location of the study area and the stations used within this study are presented in Fig. 16.2. Table 16.2 presents details of the stations.

The ANA provides rainfall and streamflow datasets, available through the "Hydrological Information System" (Hidroweb), with two levels of consistency. Level 1 corresponds to the raw data, and level 2 corresponds to revised data (ANA 2015). However, for the revised daily data, the time series stopped in 2005. Therefore, the original non-homogenized time series were used for this study. In the case of rainfall from the CRS, the readings were conducted once per day at 7 a.m. In the case of the water-level data from the CHS from ANA, the stage readings were conducted twice per day (7 a.m. and 5 p.m.) and/or by a linigraph. When no linigraph was available, the daily value provided by ANA is the average of the two readings (ANA 2012).

The climate datasets consist of rainfall, minimum and maximum temperatures, relative humidity, wind speed, and sunshine duration (CWS) or solar radiation (AWS). Regarding CWS, meteorological instruments record a measurement up to three times per day, depending on the parameter. For instance, the rainfall record (mm) was recorded for the previous 24 h at 12UTC; the minimum temperature (°C) at 12UTC; the maximum temperature (°C) at 00UTC; the relative humidity (%) and the wind speed (m/s), measured at 10 m, at 12UTC, 18UTC, and 00UTC; and the

Table 16.1 Data collected and used within the study

Datasets	Station – abbreviation	Responsible organization – Portuguese acronym	Source	Operating since; time step
18 Long-term rainfall series	Conventional rainfall station – CRS	Brazilian National Water Agency – ANA	(ANA 2015)	1939; daily
2 Long-term climate series	Conventional weather station – CWS	National Meteorological Institute/Meteorological Database for Teaching and Research – INMET/BDMEP	(INMET 2014)	1961; daily
5 Climate series	Automatic weather station – AWS	National Meteorological Institute/Brazilian Environmental Data System – INMET/SADMET	(Requested by email, INMET 2016)	2005; 1 h
3 Climate series	Automatic weather station – AWS	Integrated Data Environmental System/ National Institute for Spatial Research – SINDA/INPE	(SINDA 2015)	2005; 3 h
8 Long-term streamflow series	Conventional hydrometric station – CHS	Brazilian National Water Agency	(ANA 2015)	1939; daily

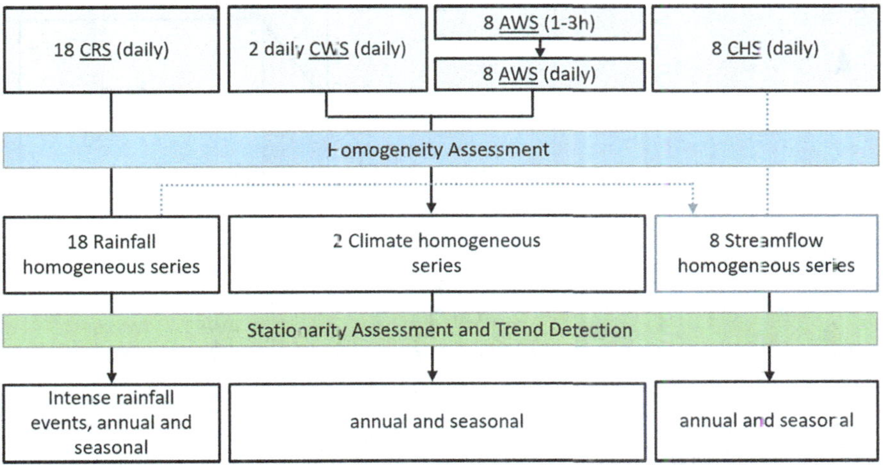

Fig. 16.1 Workflow diagram of the data analysis work. The underlined labeling refers to the stations used within the study

sunshine duration (h) is the daily sum, read at 00UTC. The daily values from the CWS (e.g., mean air temperature and relative humidity) are then calculated by INMET, using specific equations (INMET 2017). This collection time information is crucial for comparing daily climate datasets from CWS with the AWS datasets.

16.3.1 Data Management and Homogeneity Assessment

In homogeneous climate records, variations are caused only by deviations in climate, while inhomogeneities may occur due to many factors, such as changes in gauge locations, surroundings, collecting agency/individual, instrumentation, etc. (WMO 2011).

A large number of available stations (18 CRS, 2 CWS, 8 AWS, and 8 CHS) were selected in this study, to provide the most comprehensive spatial representation possible (Fig. 16.1). All daily corrected data were added to the internally used database, RBIS-INTECRAL (web-based River Basin Information System of the INTECRAL project), to be used by the partners of the project (RBIS-INTECRAL 2014).

16.3.1.1 Long-Term Daily Data Analysis: Rainfall

During the homogeneity assessment, the following steps were conducted: (1) metadata and quality control; (2) identification of homogeneous rainfall zones, considered as "reference time series" (RTS); (3) breakpoint detection; and (4) the deletion

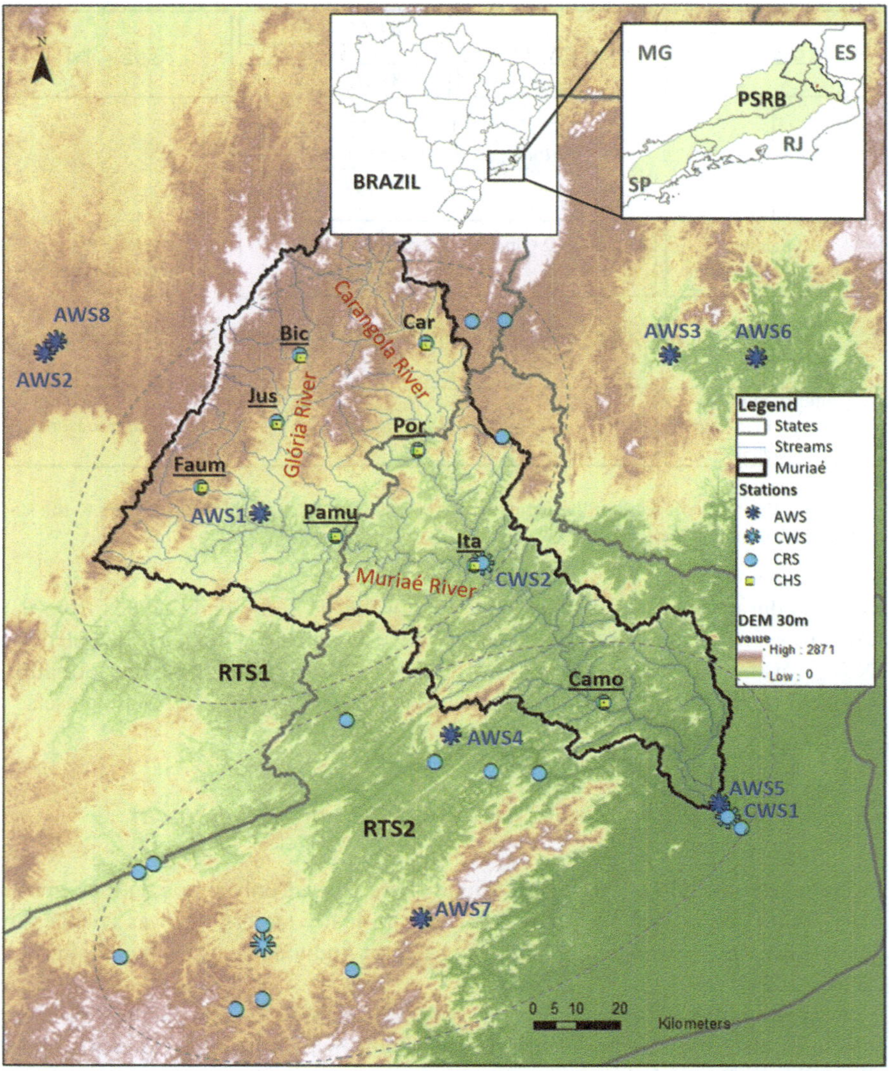

Fig. 16.2 Map of the study area with hydro-climatic stations. ES, Espírito Santo state; MG, Minas Gerais state; PSRB, Paraíba do Sul River Basin; RJ, Rio de Janeiro state; RTS1, reference time series 1; RTS2, reference time series 2; SP, São Paulo state. The underlined labeling refers to the CHSs

Table 16.2 Details of the weather stations (CWS and AWS) and the hydrometric stations (CHSs) in the region

Stations (name – legend)	R.O. – identifier	X 23S; Y 23S	*E* (m)	DA (km^2)	Data period	% of missing data/(missing years)
Climate data						
Campos – CWS1	INMET – 83698	879600.38; 7592638	11.2	–	1961–2014	4.22–18.53 (1983–1991)
Itaperuna – CWS2	INMET – 83695	822883.54; 7652508.0	124	–	1961–2014	(1962–1967, 1984–1989)
Muriaé – AWS1	INMET – 86852	771836.0; 7664549.3	243	–	2006–2015	2.11–6.48
Viçosa – AWS2	INMET – 86824	722096.2; 7702209.2	698	–	2006–2015	3.24–16.52
Alegre – AWS3	INMET – 86828	866261.6; 7701534.4	129	–	2006–2015	1.34–3.85
Cambuci – AWS4	INMET – 86854	815880.7; 7612043.3	60	–	2002–2015	14.03–47.59 (2004)
Campos – AWS5	INMET – 86855	877730.3; 7595925.4	17	–	2006–2015	1.74–15.45
Jerónimo Monteiro – AWS6	SINDA – 31958	886222.05; 7701010.9	146	–	2005–2013	30.87–3.38 (2008–2010)
Santa Maria Madalena – AWS7	SINDA – 31956	809029.6; 7569104.5	615	–	2005–2014	0.42–15.50
Viçosa – AWS8	SINDA – 32509	724820.3; 7704695.6	686	–	1999–2014	3.24–16.52
Streamflow data						
Bicuiba (R. Glória) – Bic	ANA – 58916000	781046.7; 7700680.8	616	393	1963–2014	0.95
Carangola (R. Carangola) – Car	ANA – 58930000	809940.1; 7703774.5	399	773	1936–2014	8.13
Cardoso Moreira (R. Muriaé) – Camo	ANA – 58960000	850929; 7619608.3	28	7210	1955–2014	4.37
Fazenda Umbaúbas (R. Preto) – Faum	ANA – 58910000	758351.8; 7670304.6	506	151	1938–2014	6.25
Itaperuna (R. Muriaé) – Ita[a]	ANA – 58940000	821157.3; 7651993.1	110	5800	1936–2014	31.91 (1954–1978)
Jussara (R. Glória) – Jus[a]	ANA – 58917000	775706.6; 7685192.2	383	744	1979–2014	0.32
Patrocínio do Muriaé (R. Muriaé) – Pamu	ANA – 58920000	789176.8; 7658869.8	243	2660	1943–2014	2.50
Porciúncula (R. Carangola) – Por	ANA – 58934000	808098.35; 7679062.3	188	1340	1942–2014	1.09

DA, drainage area (km^2); *E*, elevation (m.a.s.l.); R., river; R.O., responsible organization
[a]Presence of linigraph

of inhomogeneous data. Regarding step 1, it is best to know how the data were collected, processed, and compiled (WMO 2011). However, only a limited amount of metadata is available (ANA 2017). Therefore, the raw data was plotted to visually detect inhomogeneities, by comparing the data with other data from a nearby station with similar conditions. Data that appeared to be erroneous (e.g., outliers, rain gauges' changing behavior due to mechanical problem) were then removed. The next step was to create an RTS using several close stations with the same climate patterns, and to compare this with the station to be homogenized, the "candidate station series" (CSS), according to Aguilar et al. (2003). In this study, the creation of the RTS was based on a previous study in the region that had identified nine RTS within the Paraíba do Sul River Basin and Ribeirão das Lajes (A = 56,700 km^2), using the same rainfall datasets from ANA (ONS 2008) and applying the method proposed by Hosking and Wallis (1997). Moreover, the proximity, elevation range, and correlation coefficient were also considered in order to include a few additional CRS (Aguilar et al. 2003). Two tests for homogeneity assessment were performed using the RTS to compare with the CSS, the double-mass curve method (Searcy and Hardison 1960) and the standard normal homogeneity test (SNHT) (Alexandersson 1986). The SNHT is one of the most widely used homogeneity tests (Hanssen-Bauer and Førland 1994; González-Rouco et al. 2001) and is a parametric test (normal distribution is required). Thus, it was used here for homogenization of annual time series. The SNHT is a likelihood ratio test performed on a ratio (in the case of rainfall) in series between the CSS and a RTS. Using this approach, heterogeneity in the data due to non-climatic factors was detected by comparing the CSS with the RTS, according to Aguilar et al. (2003). This was anticipated to be the most critical step in the homogenization procedure, so the selection of suitable nearby stations as input for a RTS had to be undertaken carefully (Begert et al. 2005).

16.3.1.2 Climate Data Analysis

In order to make a comparison between the daily climate datasets from the CWS and AWS datasets, the sub-daily data from AWS (1–3 h time steps) were first converted to daily data, according to the standards of the CWS, as presented above in the data section. The average daily temperature was set as the mean between the maximum and minimum daily temperatures, following the "Guide to Climatological Practices – World Meteorological Organization" (WMO 2011).

To prepare the climate datasets for use in modeling applications, additional calculations were needed. For example, for CWS, global solar radiation was estimated from the sunshine duration using the Ångström–Prescott equation (Ångström 1924; Prescott 1940). The daily average wind speed at 2 m above ground was determined from the wind speed at 10 m above ground, according to Allen et al. (1998), as wind speed at 2 m above ground is required for the estimation of evapotranspiration. Regarding the data homogeneity assessment, in the case of the CWS, no metadata were available, and not enough stations were available to create a RTS, as was done for CRS. A SNHT was carried out on differences (TMAX, TMIN, RHUM) in series between the two CWSs. The monthly correlation coefficients between the two CWSs were 0.94, 0.98, and 0.76 for TMAX, TMIN, and RHUM, respectively. Missing values in the climate datasets were patched using values derived from either a monthly linear regression with the other station or from the average of observations in the series. For longer periods with gaps, the series were not completed.

16.3.1.3 Long-Term Daily Data Analysis: Streamflow

Mean daily streamflow data measured at a gauging station represent the surface outflow of the drainage basin above a specific point on the stream, here considered as sub-basins of the Muriaé River Basin. These data are summarized as a flow duration curve (FDC), which shows the percentage of time that streamflow exceeds specific different magnitudes. According to Searcy (1959), the shape of the FDC is determined by the hydrologic and geologic characteristics of the drainage area, and therefore FDCs may be used to compare the characteristics of one basin with those of another. After checking streamflow data homogeneity and filling data gaps, the datasets from the CHSs were compared with each other based on the proximity and similarities between the sub-basins (e.g., size, monthly FDC, and daily correlation coefficient). For each sub-basin, the daily streamflow data were converted from m^3/s to mm/day.

16.3.2 Trend Detection

The Mann–Kendall nonparametric statistical test was used to assess climate and streamflow trends (Mann 1945; Kendall 1975). The Mann–Kendall test is one of the most widely used tests for trend analysis in climatic variables (Subash et al. 2011a, 2011b; de la Casa and Nasello 2012; Shahid et al. 2012; Carvalho et al. 2014; Rosmann et al. 2016) and streamflow datasets (Helsel and Hirsch 2002; Helsel et al. 2006; Hawtree et al. 2014). The confidence level adopted in the Mann–Kendall statistical test was 95% for all data series. The nonparametric method called the Sen curvature was used to estimate the magnitude of the trends found in the data series (Sen 1968). For this analysis, the long-term daily series (see Sect. 16.3.1) were used, and only stations with more than 30 years of homogeneous data were considered (WMO 2011).

Before carrying out the Mann–Kendall (MK) test, the augmented Dickey–Fuller test (Dickey and Fuller, 1979, 1981) was first applied to the data to assess stationarity. This test examines temporal lags in relation to the variable being analyzed, to check for autocorrelation in the residuals (Carvalho et al. 2014). To eliminate the effect of autocorrelation on the MK test, effective sample size (ESS) has been proposed by Yue and Wang (2004) to modify the MK test. This procedure was applied during the MK test using the XLSTAT statistical software.

Regarding the detection of intense rainfall events, the method of Groisman et al. (2004) was applied. Following this approach, all events with rainfall greater than 1 mm are evaluated and are classified as "heavy" when they fall into the upper 5.0% of all rainfall events, as "very heavy" when they fall into the upper 0.3% of rainfall events, and as "extreme" when they falls into the upper 0.1% of all rainfall events. Therefore, for each station, the upper threshold values of 95.0, 99.7, and 99.9 percentiles of daily rainfall events were determined for daily rainfall during the reference period (1939–2013, January–December). Based on these thresholds, the number of very heavy rainfall events or above was calculated for each year and then averaged over the previously identified RTS, as recommended by Groisman et al. (2004). For the time period of 1939–2013, the following variables were tested: historical changes in annual and seasonal rainfall and streamflow and annual frequency of intense rainfall events above the 99.7 percentiles of daily rainfall events. In Brazil, hydrological years are defined as starting in October and ending in September, with the period from October to March corresponding to the summer rainy season in the southeast region of Brazil (Carvalho et al. 2014). However, in the study area, April was also considered part of the rainy season due to the long-term monthly rainfall exceeding 60 mm (Climpag 2017). Regarding the detection of trends in the climate data, the long-term climate data from the CWSs were used for the period from 1962 to 2013. Additionally, historical seasonal changes in maximum temperature (TMAX), minimum temperature (TMIN), and relative humidity (RHUM) were tested, as well as seasonal rainfall and streamflow changes. The seasons are defined as DJF (December, January, and February), MAM (March, April, and May), JJA (June, July, and August), and SON (September, October, and November).

16.4 Results and Discussion

16.4.1 Homogeneity Assessment

Inhomogeneities were detected in 20 out of 29 CRS rainfall series (69%). Some errors and outliers were identified by plotting the raw daily data, for example, an apparent error in the rainfall gauge (at 10 mm steps) was detected in the Dois Rios CRS, resulting in 17.4% of the data from 1989 to 2001 being removed (Fig. 16.3). Unfortunately, the cause of this error is unknown. According to the station metadata, the rainfall gauge was replaced in 2000 (ANA 2017). The preliminary data quality control was crucial for checking the data, as it allowed for the detection of some errors that would not had been detected using monthly values during the homogeneity assessment. After the preliminary data, quality control check was completed; two RTS in the study area were identified, which have nine and ten stations, respectively. The artificial boundary between the two RTS can be seen in Fig. 16.2. The corrected long-term mean annual rainfall over RTS1 and RTS2 was 1340 mm (mean elevation = 436 m.a.s.l.) and 1153 mm (mean elevation = 245 m.a.s.l.), respectively. The creation of an RTS, or cluster analysis, from

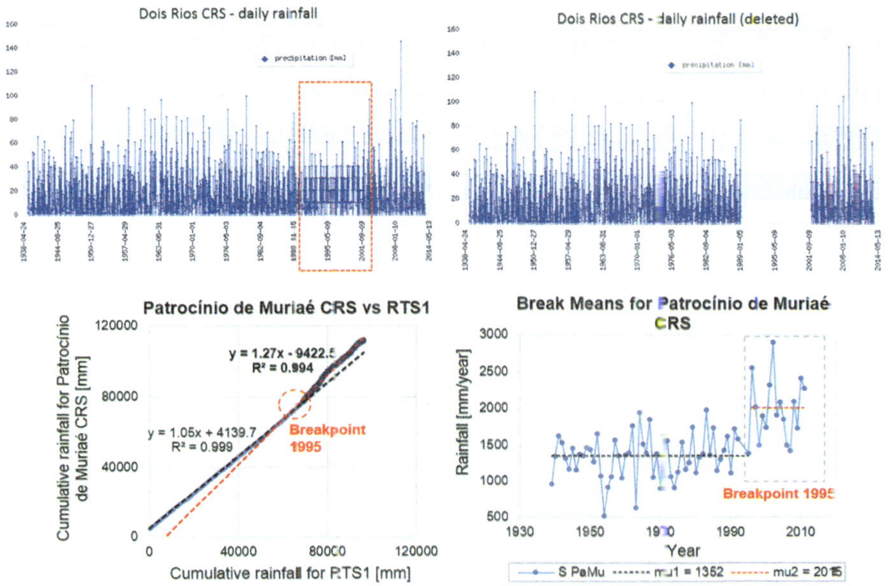

Fig. 16.3 Homogeneity assessment: example for the Dois Rios CRS: Left: raw daily data with location of uncorrected rainfall marked as red rectangle (**a**); Right: rainfall data deleted from the dataset (RBIS-INTECRAL 2014); Example for the Patrocínio de Muriaé CRS: Left: double-mass curve for Patrocínio de Muriaé CRS against RTS1 (**b**); Right: SNHT – uncorrected rainfall from Patrocínio de Muriaé CRS, with location of breakpoint (year 1995) marked as a gray rectangle

monthly rainfall series was applied in previous studies in this study region. For example, André et al. (2008) identified six RTS in RJ (A = 43 700 km^2), and ONS (2008) identified nine RTS within the Paraíba do Sul River Basin and Ribeirão das Lajes (A = 56,700 km^2), corresponding the RTS1 and RTS2 of the present study to the RTS VIII and RTS IX, respectively. Figure 16.3 shows the results of the homogeneity assessment performed for the Patrocínio de Muriaé CRS. Both graphs show a sudden change in 1995, with an increase of the mean annual rainfall. The SNHT annual homogeneity test detected the year of 1995 as a breakpoint with a 99.9% level of confidence (SNHT test: $T_0 = 29.6$; $t = 1995$; p-value <0.001). Based on this, the data after 1995 were excluded from the dataset. Table 16.3 shows the percentage of the deleted data for all of the CRS that were considered. After the homogeneity assessment, the stations presented here show a monthly correlation coefficient that is higher than 0.9 with the corresponding RTS. Besides the above-referenced errors, gaps were observed in almost all of the station datasets, consisting of both large gaps (e.g., entire years missing) and small gaps distributed throughout the datasets.

Inhomogeneities and gaps in the data from the HidroWeb System (ANA 2015) have been previously reported in studies conducted in Brazil, both for rainfall data (Oliveira et al. 2005; Silva et al. 2006; de Oliveira et al. 2008a, 2008b; ONS 2008) and for streamflow data (Kuwajima 2013). Oliveira et al. (2010) pointed out that the main reasons for the noncontinuous rainfall datasets (i.e., gaps) are the absence of

Table 16.3 Available data and corrections for RTS1 and RTS2

Station	Station code	Obs. raw data sets	Obs. hom. data sets	Deleted data (%)	Number of years	Mean annual rainfall (1939–2012) (mm)
RTS1						
Carangola	02042000	26,687	26,625	0.20	59	1169.9
Porciúncula	02042027	27,140	27,109	0.10	74	1268.3
Itaperuna	02141004	23,390	23,270	0.50	70	1148.1
Itaperuna – climate station*		13,953	12,122	13.10	70	1172.8
Patrocínio de Muriaé	02142002	28,441	21,643	23.90	56	1352.1
Fazenda Umbaúbas	02142004	25,162	23,701	5.80	69	1420.3
Bicuiba	02042014	18,646	13,697	26.50	36	1502.9
Jussara	02142009	17,145	15,597	9.00	41	1316.6
Caiana	02041005	26,285	22,330	15.00	63	1306.5
Dores do Rio Preto	02141006	23,574	23,543	0.10	64	1432.3
Varre-Sai*	02041046	13,596	12,507	8.00	38	1403.7
RTS2						
São Fidélis	02141005	22,919	22,919	0.00	59	962.8
Dois Rios	02141006	27,209	22,462	17.40	60	982.9
Aldeia	02142022	27,137	27,137	0.00	69	1212.9
Bom Jardim	02242021	26,337	26,337	0.00	68	1380.2
Cardoso Moreira	02141003	27,174	27,174	0.00	72	941.8
Três Irmãos	02141007	25,629	25,629	0.00	67	1083.9
Uhe Ilha dos Pombos Barramento*	02142065	4253	4200	1.20	8	
Santo António de Padua I	02142058	16,947	16,532	2.40	44	1181.8
Santo António de Padua II*	02142067	4406	4406	0.00	8	
Sumidouro	02242029	22,522	22,095	1.90	59	1373.8
Cantagalo I*	02142078	2274	2230	1.90	5	
Cordeiro – Climate station*		10,322	8607	16.60	15	
Ponte Bersot*	02242033	3566	3566	0.00	8	
Barra Alegre	02242018	17,195	17,164	0.20	45	1343.2
Visconde de Imbe	02242017	17,379	17,379	0.00	46	1222.4
Campos – Ponte municipal*	02141002	13,961	9406	32.60	28	
Macabuzinho*	02241003	24,239	23,143	4.50	60	
Campos – climate station*		15,456	15,456	0.00	24	

Stations with asterisks (*) present less than 30 years of homogeneous data and therefore were excluded for RTS creation and trend detections rainfall data for some stations, due to equipment problems, or the absence of operator during certain periods. Also, large gaps in the data from weather stations in Brazil were reported by Rosmann et al. (2016). For instance, for the CWS2, data from 1962 to 1967 and from 1984 to 1989 are missing, and, for the AWS, gaps are also present, but it is mostly in the wind speed and solar radiation datasets.

Regarding the homogeneity assessment of the streamflow data, it was found to be homogeneous when compared to similar sub-basins. For the revised streamflow datasets from the CHSs, the best daily correlation coefficients were found for Car and Por ($R = 0.90$), Bic and Jus ($R = 0.95$), Camo and Ita ($R = 0.95$), Faum and Jus ($R = 0.83$), and Ita and Pamu ($R = 0.93$). The daily linear regression of the Bic CHS versus Jus CHS and also for four out of the eight FDC of the sub-basins can be seen in Fig. 16.4. Similarities between the Bic–Jus CHS and the Camo–Ita CHS were noticeable. Some gaps were filled by linear regression according to the similarities between the sub-basins. However, some exceptions occurred. For example, a change was detected in the Glória River in the streamflow regime of the Jussara sub-basin and its tributary, the Bicuíba sub-basin (headwater of the basin) from 2011 onward. Land-use change may be responsible for this change, for example, due to an increase in agriculture and irrigated areas. The streamflow data from CHSs were less prone to errors than the rainfall data, as staff gauges typically have less operational problems than more complex technical equipment. Water levels can be easily read by an observer and need no additional technical equipment such as laptops to read the data. However, the major source of error in estimating streamflow is the inadequacy of the stage-discharge rating curves used to convert water levels into streamflow. Moreover, the level of accuracy and reliability is particularly low for high flow and flood estimations (Fenton and Keller 2001).

It is important to highlight that given the high vulnerability of the Muriaé River Basin to flooding, it has belonged to the 11 river basins in Brazil that constitute the National Hydrological Alert System – Critical Events Warning System (Sistema de Alerta de Eventos Críticos, SACE) since 2014, which was developed and is

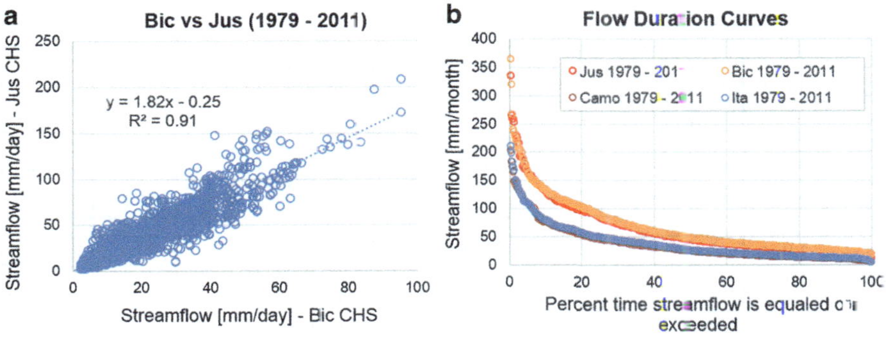

Fig. 16.4 Homogeneity assessment: scatter plot of the daily streamflow for the Bic versus Jus CHS (**a**); flow duration curves (FDC) of the monthly streamflow for four CHSs (**b**)

operated by Mineral Resources Exploration Company, Brazilian Geologic Survey (Companhia de Pesquisa de Recursos Minerais, Serviço Geológico do Brasil, CPRM). The SACE–Muriaé has five automatic stations that have been operating since 2014 (SACE 2016). However, due to the short dataset, the data were not used in the present study.

16.4.2 Historical Trends

The number of years used for trend detection varied from 30 to 73 (Table 16.4). For RTS2, a statistically significant decrease in winter rainfall (JJA; dry season) was detected (p = 0.001). A statistically significant increase in MAM was detected (p = 0.05) for the two RTS.

With respect to intense rainfall events for the two regions (RTS1 and RTS2), the average annual thresholds at 5.0%, 0.3%, and 0.1% of daily rainfall events were 47.4 mm (σ = 1.9), 92.9 mm (σ = 5.7), and 104.8 mm (σ = 7.8) for RTS1 and 44.1 mm (σ = 2.7), 89.0 mm (σ = 6.0), and 106.6 mm (σ = 9.8) for RTS2. Groisman et al. (2004) obtained a value of 95.0 mm (99.7 percentiles of daily rainfall events) for a location within the same study area. Based on these thresholds, a statistically

Table 16.4 Results of the Mann–Kendall Test. * is the trend at 0.05 level of significance; ** is the trend at 0.01 level of significance; *** is the trend at 0.001 level of significance; n is the number of years; Q is the Sen's slope estimate; gray cells represent not stationarity; stationarity confidence level is 95%; beige cells indicate "not applicable"

		Period 1962 - 2013												Period 1939 - 2013														
		DJF			MAM			JJA			SON			ANNUAL - HY			Rainy Season			Dry Season			ANNUAL (Jan - Dec)			Intense Rainfall Events / Year (Jan - Dec)		
STATIONS	Parameter	n	Sig.	Q	n	Sig.	Q	n	Sig.	Q	n	Sig.	Q	n	Sig.	Q	n	Sig.	Q	n	Sig.	Q	n	Sig.	Q	n	Sig.	Q
Rainfall Data (CRS)																												
RTS1	RAIN	51	*	2.73	51	*	2.42	51		-0.57	50		0.88	73	***	3.14							75	***	2.82	58	***	0.00
RTS2	RAIN	51		1.66	51	*	0.98	51	***	-0.70	50		-0.23	73	*	2.03							75		1.81	59	***	0.00
Discharge (CHS)																												
Bic		49	*	-0.84	50	**	0.71	51		0.14	51		-0.34	48		0.53	48		-0.18	50	**	0.49						
Car		49		0.76	48		0.28	47		-0.01	49		-0.28	66		0.88	67		0.70	67		0.17						
Camo		48		-0.08	50		0.46	50		-0.08	49		-0.43	50	**	1.22	51	*	1.02	55	*	0.24						
Faum		47	***	1.37	49	*	1.37	49		0.00	49		0.05	68		1.55	68		1.49	70		-0.08						
Ita		34		1.33	35		0.54	35		-0.15	35		0.41	47		-1.04	47		-0.49	48	**	-0.47						
Jus		33		0.36	35		0.38	35		0.13	34	*	0.45	32	*	4.09	32		3.33	33		0.73						
Pamu		50		-0.08	51		0.26	51		-0.08	51	***	-0.36	67		-0.07	67		0.09	69	*	-0.29						
Por		52		0.16	51		0.31	51	*	0.15	52		-0.16	70		-0.46	70		-0.36	71		-0.11						
Climate Data (CWS)																												
CWS1	TMAX	51	***	0.02	52	*	0.02	51	***	0.02	50	***	0.03										53	***	0.02			
	TMIN	51	***	0.03	51	***	0.03	51	***	0.03	50	***	0.03										53	***	0.03			
	RHUM	43	***	-0.09	44	***	-0.07	42	***	-0.11	43	***	-0.10										45	***	-0.09			
CWS2	TMAX	39	***	0.02	41		0.01	40	**	0.03	40	*	0.05										40	***	0.02			
	TMIN	39	***	0.04	41	***	0.03	40	***	0.02	40	***	0.03										40	***	0.03			
	RHUM	37	***	-0.10	40	***	-0.06	39	***	-0.14	39	***	-0.18										40	***	-0.12			

significant increase ($p = 0.001$) was found for the annual frequency of very heavy and extreme rainfall events for both RTS (from 1939 to 2013). Considering total precipitation, a statistically significant increase ($p = 0.001$) was found only for RTS1, as presented in Table 16.3.

Similar results have been found in other regions, where changes in mean rainfall have been found to be less significant (or insignificant), while changes in heavier rainfall events have been found to be statistically significant. Examples of similar trends include Stone et al. (2000) for Canada for the latter half of the twentieth century, Roy and Balling (2004) for India from 1910 to 2000, and Zhai et al. (1999, 2005) for China from 1956 to 2009. For the period from 1941 to 2001, Groisman et al. (2004) and Marengo et al. (2009) studied trends in intense rainfall events in southeastern Brazil and have detected an increase in frequency since the 1950s. A similar analysis on extreme values was conducted by Carvalho et al. (2014). They have observed an increase in maximum daily rainfall for the southeast region of Brazil since 1940. Lombardi Neto and Moldenhauer (1992), Colodro et al. (2002), and Mello et al. (2007) also have highlighted the importance of higher maximum daily rainfall, which in turn increases the erosivity potential. Large-scale meteorological phenomena, such as the El Niño–La Niña cycle, can also have a large effect on rainfall distribution (Bergamaschi 1999). However, this effect tends to be more pronounced in the southern and northeastern regions of Brazil (Brunini and Pinto 1997). CPTEC (2012) observed a moderate increase in average temperatures in the southeastern region of Brazil due to El Niño, but no characteristic pattern of changes in rainfall.

For CWS1, in all seasons from 1962 to 2013, a statistically significant increase in TMIN and a statistically significant decrease in the RHUM were found ($p = 0.001$; Table 16.3). Regarding TMAX, a statistically significant increase was also detected ($p = 0.001$) but less pronounced in MAM ($p = 0.05$). Similar results were also observed in the CWS2. Therefore, the temperature trends since the 1960s show stronger increasing of the minimum temperatures than the maximum, indicating that there has been a decrease in the diurnal temperature range. Similar results were found in southern Brazil for the period of 1960–2002 by Marengo and Camargo (2008). The warming identified by several previous studies in various parts of Brazil has been attributed in some cases to changes in land use, in particular the development of large cities such as São Paulo and Rio de Janeiro (Marengo 2001, 2003; Sansigolo et al. 1992). Since the 1960s, the annual temperature has risen by 0.20–0.25°C (TMAX) and 0.27–0.30°C (TMIN) per decade for the two CWSs.

With respect to streamflow trend-testing, a statistically significant increase was detected for two CHSs for the complete hydrological year. However, for seasonal periods, a statistically significant increase streamflow during DJF was detected for the Faum CHS, and a statistically significant decrease streamflow during SON was detected for the Pamu CHS ($p = 0.001$). Additionally, when the year was split between the rainy season (i.e., October to April) and dry season (i.e., May to September), a statistically significant decrease in streamflow was found during the dry season for two of the eight CHSs in the study area (Pamu and Ita; Fig. 16.4b).

16.5 Conclusions

Inhomogeneities were observed and corrected in almost all datasets examined in this study, which reinforces the need to perform a homogeneity assessment prior to their use for applications such as trend-testing or modeling. In addition, the process of assessing and revising these datasets provided insight into the long-term hydro-climatic patterns in southeastern Brazil over the last 50–70 years. All the corrected data have been made available to project partners (INTECRAL) through a project database (RBIS-INTECRAL).

This study provides a firm foundation for further hydrological analysis in this region, such as ecohydrological modeling studies, which are presented in Chap. 17 of this book (see Künne et al. 2018). The study also provides insight into the climate and streamflow trends in the historic time series, namely, that the climate has become warmer in all seasons, dryer in JJA (dry season), and that intense rainfall events have become more frequent during the last 70 years. Further, the pronounced dry season as well as higher frequency of intense rainfall events, i.e., higher interannual variability induced by climate change, can further trigger soil erosion and land degradation because of the changes in soil moisture regimes, vegetation growth dynamics, and soil cover.

It is important to consider that changes in land use, such as deforestation of the Atlantic Forest, increasing agricultural areas, and increasing irrigated crops in winter, may provide additional justification to the trends observed in this study, which highlights the need for further detailed ecohydrological assessment in this region.

Acknowledgments This work was supported by the German Ministry of Education and Research under the project INTECRAL, grant number033L162K.

References

Aguilar E, Auer I, Brunet M et al (2003) Guidelines on climate metadata and homogenization. WMO-TD World Meteorological Organization, WMO/TD No. 1186

Alexandersson H (1986) A homogeneity test applied to precipitation data. J Climatol 6:661–675

Allen RG, Pereira LS, Raes D et al (1998) Crop evapotranspiration: guidelines for computing crop water requirements. Irrigation and Drainage Paper No. 56, Food and Agriculture Organization of the United Nations, Rome, Italy

Alvares CA, Stape JL, Sentelhas PC et al (2014) Köppen's climate classification map for Brazil. Meteorol Z 22(6):711–728

ANA (2012) Orientações para consistência de dados fluviométricos/Agência Nacional de Águas; Superintendência de Gestão da Rede Hidrometeorológica. ANA, SGH, Brasília. 19 p

ANA (2015) Agência Nacional de Águas, HidroWeb, Sistema de Informações Hidrológicas. http://hidroweb.ana.gov.br/. Accessed 14 Nov 2015

ANA (2017) Software Hidro 1.3 (Sistema de informações hidrológicas) – 3. Inventário pluviométrico/fluviométrico atualizado. http://hidroweb.ana.gov.br/HidroWeb.asp?TocItem=6010. Accessed 14 Jan 2017

André RGB, Marques VS, Pinheiro FMA et al (2008) Identificação de regiões pluviometricamente homogêneas no estado do Rio de Janeiro, utilizando-se valores mensais. Rev Bras Meteorol 23(4):501–509

Ångström A (1924) Solar and terrestrial radiation. Q J R Meteorol Soc 50:121–125
Ávila MW, Hora MAGM, Ávila CR et al (2016) Gestão qualitativa dos recursos hídricos. Proposta metodológica para o planejamento de uma rede de estações para monitoramento da qualidade de águas superficiais. Estudo de caso: bacia hidrográfica do Rio Muriaé RBRH 21(2):401–415
Bergamaschi H (1999) Estratégias para reduzir riscos por estiagens. Seminário PFA/UFRGS
Begert M, Schlegel T, Kirchhofer W (2005) Homogeneous temperature and precipitation series of Switzerland from 1864 to 2000. Int J Climatol 25:65–80
Brasil (2011) Law n. 12.527, 18 November 2011. Regulates the constitutional right of access to public information. Law on access to information. Presidência da República, Brasília. http://www.planalto.gov.br/ccivil_03/_Ato2011-2014/2011/Lei/L12527.htm. Accessed 06 Jun 2015
Brunini O, Pinto HS (1997) As oscilações climáticas e agricultura. Laranja & Cia 47:6–7
Carvalho JRP, Assad ED, Fortes de Oliveira A et al (2014) Annual maximum daily rainfall trends in the Midwest, southeast and southern Brazil in the last 71 years. Weather Clim Extrem 5–6:7–15
Climpag (2017) Climate impact on agriculture, FAO – Food and Agriculture Organization of the United Nations. http://www.fao.org/nr/climpag/climate/index_en.asp/. Accessed 29 May 2017
Colodro G, Carvalho MP, Roque CG et al (2002) Erosividade da chuva: distribuição e correlação com a precipitação pluviométrica de Teodoro Sampaio (SP). Rev Bras Ciênc Solo 26:809–818
Colombo AF, Joly CA (2010) Brazilian Atlantic Forest lato sensu: the most ancient Brazilian forest, and a biodiversity hotspot, is highly threatened by climate change. Braz J Biol 70:697–708
CPTEC (2012) Centro de previsão de tempo e estudos climáticos. El Niño e La Niña. http://enos.cptec.inpe.br. Accessed 01 Jun 2016
de la Casa A, Nasello OB (2012) Low frequency oscillation of rainfall in Córdoba, Argentina, and its relation with solar cycles and cosmic rays. Atmos Res 113:140–146
Dickey DA, Fuller WA (1979) Distribution of the estimators for autoregressive time series with unit root. J Am Stat Assoc 74:427–431
Dickey DA, Fuller WA (1981) Likelihood ratio statistics for autoregressive time series with a unit root. Econometrica 49:1057–1072
Fenton JD, Keller RJ (2001) The calculation of streamflow from measurements of stage. Technical Report 01/6. Cooperative Research Centre for Catchment Hydrology, Melbourne, Australia
González-Rouco JF, Jiménez JL, Quesada V et al (2001) Quality control and homogeneity of precipitation data in the Southwest of Europe. J Clim 14:964–978
Groisman PY, Knight RW, Easterling DR et al (2004) Trends in intense precipitation in the climate record. J Clim 18:1326–1350
Hanssen-Bauer I, Førland E (1994) Homogenizing long Norwegian precipitation series. J Clim 7:1001–1013
Hawtree D, Nunes JP, Keizer JJ et al (2014) Time-series analysis of the long-term hydrologic impacts of afforestation in the Águeda watershed of North-Central Portugal. Hydrol Earth Syst Sci Discuss 11:12223–12256
Haylock MR, Peterson TC, Alves LM et al (2006) Trends in total and extreme South American rainfall 1960-2000 and links with sea surface temperature. J Clim 19:1490–1512
Helsel DR, Hirsch RM (eds) (2002) Statistical methods in water resources. Elsevier, New York
Helsel DR, Mueller DK, Slack JR (2006) Computer program for the Kendall family of trend tests. U.S. Geological Survey Scientific Investigations Report 2005–5275, p 4
Hosking JRM, Wallis JR (1997) Regional frequency analysis – an approach based on L-moments. Cambridge University Press, New York
INMET (2014) BDMEP – Banco de Dados Meteorológicos para Ensino e Pesquisa. INMET – Instituto Nacional de Meteorologia. http://www.inmet.gov.br/portal/index.php?r=bdmep/bdmep. Accessed 16 Jun 2014
INMET (2016) Instituto Nacional de Meteorologia (National Meteorological Institute)/Seção de Armazenamento de Dados Meteorológicos (Brazilian Environmental Data System) – INMET/SADMET. http://www.inmet.gov.br/portal/index.php?r=home/page&page=central_servicos
INMET (2017) Normais Climatológicas do Brasil/1961–1990. http://www.inmet.gov.br/portal/index.php?r=clima/normaisclimatologicas. Accessed 29 May 2017
Kendall MG (1975) Rank correlation methods, 4th edn. Charles Griffin, London

Kuwajima JI (2013) Application of SWAT in a Brazilian watershed with inconsistent hydrological data. SWAT Conference 2013, Toulouse

Künne A, Kralisch S, Santos JM, Flügel W-A (2018) Ecohydrological modeling and scenario impact assessment in Rural Rio de Janeiro. In: Nehren U, Schlüter S, Raedig C, Sattler D, Hissa H (eds) Strategies and tools for a sustainable rural Rio de Janeiro. Springer International Publishing, Cham

Lange W, Sandholz S, Viezzer J et al (2018) Ecosystem-based approaches for disaster risk reduction and climate change adaptation in Rio de Janeiro state. In: Nehren U, Schlüter S, Raedig C, Sattler D, Hissa H (eds) Strategies and tools for a sustainable rural Rio de Janeiro. Springer International Publishing, Cham

Lombardi Neto F, Moldenhauer WC (1992) Erosividade da chuva: sua distribuição e relação com as perdas de solo em Campinas. Bragrantia 51:189–196

Mann HB (1945) Non-parametric tests against trend. Econometrica 13:163–171

Marengo J (2001) Mudanças climáticas globais e regionais: Avaliação do clima atual do Brasil e projeções de cenários climáticos do futuro. Rev Bras Meteor 16:1–18

Marengo J (2003) Condições climáticas e recursos hídricos no Norte Brasileiro. In: Tucciand CE, Braga B (eds) Clima e recursos hídricos no Brasil. Associação Brasileira de Recursos Hídricos FBMC/ANA, Brazil, p 117–161

Marengo JA, Camargo CC (2008) Surface air temperature trends in southern Brazil for 1960–2002. Int J Climatol 28:893–904

Marengo JA, Jones R, Alves LM et al (2009) Future change of temperature and precipitation extremes in South America as derived from the PRECIS regional climate modeling system. Int J Climatol 29:2241–2255

Martirani LA, Peres IK (2016) Water crisis in São Paulo: news coverage, public perception and the right to information. Ambiente Soc 1:1–20

Mello CR, de Sá MAC, Curi N et al (2007) Erosividade mensal e anual da chuva no Estado de Minas Gerais. Pesq Agropec Bras 42:537–545

Merten GH, Minella JPG (2013) The expansion of Brazilian agriculture: soil erosion scenarios. Int Soil Water Conserv Res 1:37–48

Nehren U, Kirchner A, Sattler D et al (2013) Impact of natural climate change and historical land use on landscape development in the Atlantic Forest of Rio de Janeiro, Brazil. Ann Braz Acad Sci 85(2):497–518

Nehren U, Kirchner A, Lange W et al (2018) Natural hazards and climate change impacts in the state of Rio de Janeiro: a landscape historical analysis. In: Nehren U, Schlüter S, Raedig C, Sattler D, Hissa H (eds) Strategies and tools for a sustainable rural Rio de Janeiro. Springer International Publishing, Cham

Oliveira VPS, Zanetti SS, Pruski FF (2005) Climabr parte I: Modelo para a geração de series sintéticas de precipitação. Rev Bras Eng Agríc Ambient 9(3):348–355

Oliveira LFC, Antonini JCA, Griebeler NP (2008a) Estimativas de chuvas intensas para o estado de Góias. Eng Agríc 28(1):22–33

Oliveira LFC, Antonini JCA, Fioreze AP et al (2008b) Métodos de estimative de precipitação máxima para o Estado de Góias. Rev Bras Eng Agríc Ambient 12(6):620–625

Oliveira LFC, Fioreze AP, Medeiros AMM et al (2010) Comparação de metodologias de preenchimento de falhas históricas de precipitação pluvial annual. Rev Bras Eng Agríc Ambient 14(11):1186–1192

ONS – Operador Nacional do Sistema Eléctrico (2008) Relatório de Análise dos Dados Pluviométricos da Bacia do rio Paraíba do Sul e Ribeirão das Lajes. Contrato n° GPD-CT-185/06-2, Consórcio Enerconsult-Hidrosistem-Internave

Pachauri RK, Allen MR, Barros VR et al (2014) Climate Change 2014: Synthesis Report. Contribution of Working Groups I, II and III to the Fifth Assessment Report of the Intergovernmental Panel on Climate Change / R. Pachauri and L. Meyer (editors), Geneva, Switzerland, IPCC, 151 p., ISBN: 978-92-9169-143-2

Prescott J (1940) Evaporation from a water surface in relation to solar radiation. Trans R Soc South Aust 64:114–125

RBIS–INTECRAL (2014) River Basin Information System – Integrated Eco Technologies and Services for a Sustainable Rural Rio de Janeiro. http://leutra.geogr.uni-jena.de/intecraRBIS. Accessed 14 Jul 2014
Ribeiro MC, Metzger JP, Martensen AC et al (2009) The Brazilian Atlantic Forest: how much is left, and how is the remaining forest distributed? Implications for conservation. Biol Conserv 142:1141–1153
Rosmann T, Dominguez E, Chavarro J (2016) Comparing trends in hydrometeorological average and extreme data sets around the world at different time scales. J Hydrol Reg Stud 5:200–212
Roy SS, Balling RC (2004) Trends in extreme daily rainfall indices in India. Int J Climatol 24:457–466
SACE-Muriaé (2016) Sistema de Alerta de Eventos Críticos – Bacia do Rio Muriaé. http://www.cprm.gov.br/sace/index_bacias_monitoradas.php?getbacia=bmuriae#. Accessed 15 Feb 2016
Sansigolo C, Rodriguez R, Etchichury P (1992) Tendências nas temperaturas médias do Brasil, vol 1. Congresso Brasileiro de Meteorologia, São Paulo, Brazil, pp 367–371
Sattler D, Raedig C, Hebner A, Wesenberg J (2018) Use of native plant species for ecological restoration and rehabilitation measures in Southeast Brazil. In: Nehren U, Schlüter S, Raedig C, Sattler D, Hissa H (eds) Strategies and tools for a sustainable rural Rio de Janeiro. Springer International Publishing, Cham
Searcy JK, Hardison CH (1960) Double-mass curves. In: Manual of hydrology: part 1. General surface water techniques. United States Government Printing Offices, Washington, DC
Searcy JK (1959) Flow-duration curves. In: Manual of hydrology. Water supply paper 1542-A. United States Geological Survey, Washington, DC
Seliger R, Sattler D, Soares da Silva A et al (2018) Rehabilitation of degraded sloped pastures – lessons learned in Itaocara. In: Nehren U, Schlüter S, Raedig C, Sattler D, Hissa H (eds) Strategies and tools for a sustainable rural Rio de Janeiro. Springer International Publishing, Cham
Sen PK (1968) Estimates of the regression coefficient based on Kendall's tau. J Am Stat Assoc 63:1379–1389
Shahid S, Harun SB, Katimon A (2012) Changes in diurnal temperature range in Bangladesh during the time period 1961–2008. Atmos Res 118:260–270
Silva AM, Oliveira PM, Mello CR et al (2006) Vazões mínimas e de referência para outorga na região do Alto Rio Grande, Minas Gerais. Rev Bras Eng Agríc Ambient 10:374–380
SINDA (2015) Sistema Integrado de Dados Ambientais (Integrated Data Environmental System)/Instituto Nacional de Pesquisas Espaciais (National Institute for Spatial Research) – SINDA/INPE. http://sinda.crn2.inpe.br/PCD/SITE/novo/site/historico/index.php. Accessed 15 Mar 2015
Stone DA, Weaver AJ, Zwiers FW (2000) Trends in Canadian precipitation intensity. Atmos Ocean 38:321–347
Subash N, Singha SS, Priya N (2011a) Extreme rainfall indices and its impact on rice productivity — a case study over sub-humid climatic environment. Atmos Res 98:1373–1387
Subash N, Singha SS, Priya N (2011b) Variability of rainfall and effective onset and length of the monsoon season over a sub-humid climatic environment. Atmos Res 99:479–487
Vincent LA, Peterson TC, Barros VR et al (2005) Observed trends in indices of daily temperature extremes in South America 1960–2000. J Clim 18:5011–5023q
WMO (2011) Guide to climatological practices, 3rd edn. WMO-N°. 100, Geneva
Yue S, Wang CY (2004) The Mann-Kendall test modified by effective sample size to detect trend in serially correlated hydrological series. Water Resour Manag 18:201–218
Zeng C, Deng X, Dong J et al (2016) Urbanization and sustainability: comparison of the processes in “BIC” countries. Sustainability 8:400
Zhai PM, Sun A, Ren F et al (1999) Changes of climate extremes in China. Clim Chang 42:203–218
Zhai PM, Zhang XB, Wan H et al (2005) Trends in total precipitation and frequency of daily precipitation extremes over China. J Clim 18:1096–1108

Chapter 17
Ecohydrological Modeling and Scenario Impact Assessment in Rural Rio de Janeiro

Annika Künne, Sven Kralisch, Juliana M. Santos, and Wolfgang-Albert Flügel

Abstract Understanding hydrological process dynamics is a crucial requirement for river basin management and environmental planning. Possible future climate changes raise questions about their impact on human livelihoods, which strongly depend on water availability and quality, soil fertility, and other ecosystem services. This chapter presents a physically based, spatially distributed ecohydrological model that was applied within three meso- to macroscale watersheds in the hinterland of Rio de Janeiro. While an increasing population and a fast-growing industrial sector create a high demand for water supply, the study region faces serious problems of forest fragmentation, overexploitation, and soil degradation, which create increasing pressures on water resources. This situation is further intensified by the climate conditions with distinct wet and dry periods that can cause floods and landslides in the rainy season and water shortages during dry periods, especially affecting the agricultural and domestic supply sectors. Recent water shortages raise questions how future climate changes will impact the hydrological dynamics and if river basin management needs to take appropriate counteractions. The results show that the developed models allow simulating hydrological processes at a high spatio-temporal resolution. Given the fact that their process representation is physically based, these models can help answer questions about hydrological dynamics under changing environmental conditions.

Keywords Ecohydrological modeling · System analysis · Scenario assessment · Climate change · Atlantic Forest

A. Künne (✉) · S. Kralisch · J. M. Santos
Geographic Information Science Group, Institute of Geography, Friedrich Schiller University of Jena, Jena, Germany
e-mail: annika.kuenne@uni-jena.de; sven.kralisch@uni-jena.de; juliana.santos@uni-jena.de

W.-A. Flügel
Professor (ret.) and former Head of the Department of Geoinformatics, Hydrology and Modelling (DGHM), Friedrich Schiller University of Jena, Jena, Germany

U. Nehren et al. (eds.), *Strategies and Tools for a Sustainable Rural Rio de Janeiro*, Springer Series on Environmental Management,
https://doi.org/10.1007/978-3-319-89644-1_17

17.1 Introduction

Brazil is one of the countries with the largest freshwater resources in the world (WBCSD 2006). However, during the years 2014 and 2015, the southeastern area of the country experienced a severe water shortage, also described as water collapse (Dobrovolski and Rattis 2015) in two of the biggest metropolitan regions in the country, Sao Paulo and Rio de Janeiro, with altogether around 32 million inhabitants.

The metropolitan region of Rio de Janeiro with approx. 12 million inhabitants is still growing. This growth attracts more and more industry and agricultural corporations within the hinterland of the city to meet the increasing economic demands. Thus, the pressure on water resources is becoming crucial and is visible, e.g., by reservoirs and hydropower plants or increasing agricultural production within the region. Despite rising water withdrawals and pollution, the region still lacks water provision plans, water system infrastructure, and adequate political management, also described as hydrosocial scarcity (Britto et al. 2016).

The challenges related to these conditions are further intensified by the hydro-climatological factors. The region possesses distinct wet and dry periods that can cause floods and landslides in the rainy season as well as water shortages during dry periods, especially affecting the agricultural and domestic supply sectors. A sound river basin management that would safeguard the various water-related ecosystem functions and services requires an evaluation of the ecohydrological system to understand the extent of the related process alteration. Due to their explicit representation of involved processes and related interactions, physically based computer models can provide valuable help to understand complex, interlinked systems. Because of their predictive capabilities and the option to evaluate the impacts of future development scenarios, e.g., changing land use patterns or climate conditions, they are often used to support decision-making processes (Bossel 2004). Process-based ecohydrological models are therefore adequate tools to verify the current understanding of the observed system as well as to quantify and predict hydrological dynamics on a specific spatiotemporal scale.

The overall objective of the study presented here was to understand the ecohydrological system and develop a transferrable ecohydrological modeling platform that can be applied in the whole Rio de Janeiro state (RJ). More specifically, one objective was to develop physically based, spatially distributed ecohydrological models with high spatiotemporal resolutions within different meso- to macroscale pilot watersheds that are representative for the rural parts of the state. Based on these models, another objective was to simulate and evaluate the impact of climate change scenarios on the hydrological dynamics within river basins to answer questions about water availability and occurrence of extreme events in the future. Different Earth system models (ESM) and scenarios were used to set the boundary conditions for this objective.

17.2 Data and Methods

The choice of data and methods was based on the following additional objectives: (i) identify study areas that represent the environmental conditions and land management that are dominant in RJ, (ii) select a hydrological simulation model that represents the controlling hydrological and management processes on a high spatio-temporal resolution, and (iii) hydrological system understanding to properly parameterize the selected models. This section describes the chosen data and methods.

17.2.1 Study Area

To address the objectives listed above, three meso- to macroscale basins were selected for the model applications to represent the ecohydrological conditions throughout the rural parts of RJ (see Fig. 17.1):

1. The Guapi-Macacu River basin (GMRB) is located close to the metropolitan region of Rio de Janeiro and Niterói and covers an area of 1,265 km^2. The land use is dominated by forest and pasture. The climate varies from a maritime

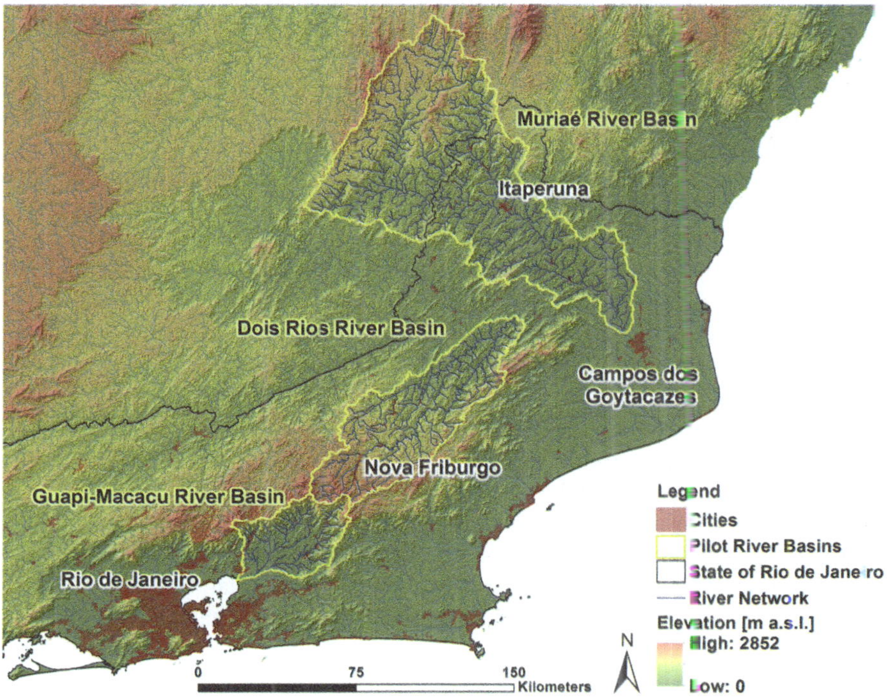

Fig. 17.1 Location of the study area and pilot river basins: GMRB, DRRB, and MRB

influence at the coast (Aw after Köppen and Geiger; Kottek et al. 2006) to a mild temperate, fully humid climate with warm summers in the northeast (Cfb after Köppen and Geiger; Kottek et al. 2006). In the area where the mountain range of the Serra dos Órgãos dominates the weather regimes, there is a more distinct wet and dry season as well as lower temperatures.

2. The second study area – the Dois Rios River basin (DRRB; 3122 km^2), in the northeast of the GMRB – shows similar conditions. The water flows from the Serra dos Órgãos mountain range to the northeast until the Dois Rios River flows into the Paraíba do Sul River. The land use is mainly forest and pasture. Both adjacent river basins have high relief energy, emanating in the Serra dos Órgãos at ca. 2,300 m.a.s.l. to altitudes of just a few m.a.s.l. at the outlets.
3. The Muriaé River basin (MRB) is located in the north of RJ state, with its upstream part in the state of Minas Gerais. With 8,165 km^2, the MRB is the largest of the pilot river basins within this study. It is characterized by a warmer and drier climate compared to the other basins. However, the dominant land use is also forest and pasture, whereby the latter one is more intensively used here than in the other two basins due to high milk production in this area (Zoccal et al. 2006). Geology and soils are more calcareous, which together with the climate and farming causes high degradation, especially on pasture (see Santos et al. 2018; Sattler et al. 2018; Seliger et al. 2018, in this book).

17.2.2 System Analysis, Data Acquisition, and Data Processing

Table 17.1 gives an overview of the different types of collected data, which were analyzed and preprocessed for the modeling purpose.

Primary data was collected for the GMRB from 2010 to 2012. The hydrological and meteorological measurements were carried out by Künne and Penedo (Penedo et al. 2011). Soil sampling and physical and chemical soil measurements were conducted by Leipzig University (Nehren et al. 2013; Kirchner 2014). Vegetation studies were also conducted by Leipzig University (Wesenberg and Sattler 2011; Lindner and Sattler 2012; Sattler et al. 2014). As a result of these measurements, the analysis of the leaf area index and root depth could be used for the parameterization of the model.

All time series data were analyzed, tested for homogeneity (e.g., double-mass curve method and in case of errors corrected (see Santos et al. 2018). The spatial data (e.g., soil, land use, geology) were checked for validity and consistency, by using measured physical information compiled out of different sources.

With regard to the modeling process, the data analyzed and processed were either prepared as model input time series or as data for parameterization. Table 17.2 provides an overview of the data preparation, explaining which data was used for what purpose.

Table 17.1 Data collected and used within the study

Data	Responsible organization/person	Resolution and date
Digital elevation model to delineate topographical information (e.g., slope, aspect)	Shuttle radar topography mission (SRTM non-void filled) (Data available from the U.S. Geological Survey 2015)	1 arc-sec, 30 m
Soil maps	Carvalho Filho et al. (2000)	1:250,000
Physical soil information (e.g., texture, bulk density)	Soil Department of Empresa Brasileira de Pesquisa Agropecuária (EMBRAPA 1980), Kirchner (2014)	2010–2012
Land use maps	Serviço Geológico do Brasil – CPRM (Duringer and Shinzato 2000)	1:250,000
Physical vegetation information (e.g., LAI, root depth)	Wesenberg and Sattler (2011), Lindner and Sattler (2012), Sattler et al. (2014), Künne and Penedo (2012)	2010–2012
Geology	Departamento de Geologia da CPRM (Silva and Silva Cunha 2001)	1:500,000
Physical geological information (e.g., type of aquifer, conductivity)	Menezes Raposo de Almeida et al. (2010)	
Discharge data	ANA Agência Nacional de Águas (2015)	Daily (1933–2015)
	Penedo et al. (2011), Künne and Penedo (2012), Künne et al. (2016)	Daily (2010–2012)
Precipitation data	INMET Instituto Nacional de Meteorológia (2015), ANA Agência Nacional de Águas (2015)	Daily (1935–2015)
	Penedo et al. (2011), Künne and Penedo (2012), Künne et al. (2016)	Daily (2010–2012)
Meteorological data (temperature, wind speed, humidity, solar radiation, or sunshine duration)	INMET Instituto Nacional de Meteorológia (2015)	Daily (1931–2015)
	Penedo et al. (2011), Künne and Penedo (2012), Künne et al. (2016)	Daily (2010–2012)

17.2.3 Ecohydrological Modeling

This section describes the development and adjustment of the models to the specific conditions of the pilot river basins. Furthermore, the delineation of the spatial model entities and the definition of the parameters for describing soils, geology, and land use are presented. Additionally, a short overview is given on model calibration and validation, followed by a description of the model optimization, sensitivity, and uncertainty analysis.

Table 17.2 Overview of the model parameters needed as well as data and sources used for the reclassification and the final parameterization

Maps state of RJ	Source	Reclassification	Sources for parameterization (physical soil analysis)	Parameters for J2K per each class
Soil map	Carvalho Filho et al. (2000)	Yes. FAO classification used (IUSS Working Group WRB 2006)	Kirchner (2014), (EMBRAPA 1980), Menezes Raposo de Almeida et al. (2010)	Soil depth, field capacity (sum and per dm depth), sum of air capacity, capillary rise
Land use/land cover map	Duringer and Shinzato (2000)	No	Lindner and Sattler (2012), Wesenberg and Sattler (2011), Sattler et al. (2014), Fidalgo et al. (2008), Xavier and Vettorazzi (2003), Freitas et al. (2005), Campoe et al. (2010)	Albedo, stomata resistance for each month of the year, LAI and height of plants every quarter of the year, grade of sealing, root depth
Geology map	Departamento de Geologia da CPRM (Silva and Silva Cunha 2001)	Yes, depending on aquifer conductivity	Menezes Raposo de Almeida et al. (2010)	Storage capacity and conductivity of two components RG1 (weathering belt) and RG2 (base rock)

17.2.3.1 Modeling Software

The software JAMS/J2K was used for the ecohydrological modeling within the studied river basins and their sub-catchments. J2K is a process-oriented, fully distributed ecohydrological model that was designed to simulate hydrological processes on the catchment scale (Krause 2001). J2K is built within the JAMS software, which is an object-oriented modular modeling framework to design custom-tailored models. The modeling framework allows the analysis of data and supports the parametrization as well as the manual and automatic calibration of the models. Moreover, JAMS enables an evaluation of results and sensitivity and uncertainty analyses (Kralisch and Krause 2006). J2K uses the concept of hydrological response units (HRU) – as suggested by Flügel (1996) and Pfennig et al. (2009) – to represent the spatial heterogeneity of the basins. The concept considers topographical, pedo-geological, and land use information as well as a routing scheme incorporating different flow topologies.

17.2.3.2 Model Design and Development

The modular approach of the JAMS framework allows designing custom-tailored models (Kralisch and Fischer 2012). The user can select different simulation components from a component library or use full-fledged sub-models in order to read inputs, simulate processes, and generate outputs, depending on the specific research questions. The extensible JAMS component library provides a plethora of different process simulation routines, for example, to simulate evapotranspiration, soil water processes, or stream routing. The component libraries offer routines to simulate processes in different ways that require varying amounts of input data, e.g., the algorithms after Penman-Monteith (Penman 1948; Monteith 1965) and Hargraves-Samani (Samani 2000) in the case of evapotranspiration.

With regard to the research questions of this study, the model design allows simulating the following processes: interception, evapotranspiration, infiltration, percolation, runoff generation, and stream routing. Depending on the data availability and in regard to the physical-based approach, the model was designed using specific components and equations. The snow module, which is a standard component of the J2K model, was removed as snow does not play a role within the study area. Since J2K was applied for the first time in the region, other minor adaptations had to be carried out, e.g., for the calculation of the precipitation correction and the calculation of solar radiation. The austral summer also had to be taken into consideration in some components.

17.2.3.3 Parameter Identification

As mentioned before, the J2K model was developed and first tested under temperate climate conditions (Krause 2001). Despite the fact that the model has been used all over the world in tropical as well as low-temperate regions influenced by glaciers, this is the first application in the Mata Atlântica. Hence, the existing parameter libraries of soil classes, land use classes, and geological classes needed to be adapted and updated to describe the ecohydrological conditions of this region. This was done by the analysis of collected primary and secondary data sources.

The parameterization was done for the entire state based on the maps of geology, soil, and land use/land cover, which were provided by the state environmental agency INEA (Instituto Estadual do Ambiente), from different sources (see Table 17.2).

For the parameterization of the soil classes, the data that was analyzed by Leipzig University (Kirchner 2014), Embrapa Solos (EMBRAPA 1980), and the Projeto Macacu (Sin Hwa et al. 2010) was used to parameterize all soil classes for the whole RJ and the upper part of the Muriaé river basin located in Minas Gerais (Table 17.2).

17.2.3.4 Model Calibration, Validation, and Sensitivity Analysis

Sensitivity analysis was done with the OPTAS tool, which is part of the JAMS software (Fischer et al. 2009). In doing so, we were able to identify the most sensitive parameters that have a high impact during calibration. In addition, understanding about the river basin systems within the region could be enhanced by studying the parameter sensitivity.

The main differentiation of calibration methods for hydrological models is made between automatic and manual calibration (Boyle et al. 2000). We decided to use a semiautomatic model calibration using both calibration approaches. Hereby, 11 parameters and their specific boundaries were manually selected out of 48 calibration parameters in the model. The selected parameters were calibrated using the multi-objective genetic algorithm NSGA-2 (Deb et al. 2002). Choosing this semiautomatic calibration procedure allowed us to exploit the understanding and representation of ecohydrological processes gained in field surveys and measurements, which played a major role in our study.

17.2.4 Modeling Future Climate Impacts

The selection of the regional climate models for impact assessment was mainly guided by their spatiotemporal resolutions. To simulate flood events or low flow situations, daily input data were needed. Moreover, regional climate models where used to make sure that the regional orography was well represented. The evaluation of different models and platforms resulted in the following six models of the CORDEX (COordinated Regional climate Downscaling Experiment) initiative (Evans 2011) for South America:

(i) EC-Earth-hist
(ii) EC-Earth-RCP4.5
(iii) EC-Earth-RCP8.5
(iv) MPI-ESM-hist
(v) MPI-ESM-RCP4.5
(vi) MPI-ESM-RCP8.5

Each of these models represents a combination of an Earth system model (EC-EARTH, MPI-ESM), a regional model (RCA4), and a development scenario (hist, RCP4.5, RCP8.5). These are in detail:

- EC-Earth: The European Earth system model (EC-Earth) (Hazeleger and Bintanja 2014).
- MPI-ESM: This model was developed by the Max Planck Institute (MPI) for Meteorology in Hamburg, Germany, and consists of different components for simulating dynamics of the atmosphere (ECHAM6), the oceans (MPIOM), land surfaces (JSBACH), and their interactions (Hagemann et al. 2013).

- hist: This hindcast scenario assumes the historic conditions with radiative forcing between the years 1951 and 2005.
- RCP4.5: Representative Concentration Pathways (RCPs) represent an assumed radiative forcing until the year 2100. This scenario assumes a stabilization of radiative forcing by different greenhouse gases at 4.5 W/m^2 (ca. 650 ppm CO2 equivalent) until the year 2100 (Thomson et al. 2011).
- RCP8.5: This RCP scenario assumes an increase of radiative forcing leading to 8.5 W/m^2 (rd.1370 ppm CO2 equivalent) stabilization until the year 2100 (Vuuren et al. 2011).

Both Earth system models were downscaled using the Rossby Centre regional atmospheric model (RCA4), which resulted in a spatial resolution of 0.44°, equivalent to approx. 49 km grid size in the project region. In order to create input data for the J2K model, the grid cells that intersect the catchments plus one additional layer of surrounding cells were selected, which resulted in 22 grid cells. For each of the six models and the selected cells, J2K input datasets for precipitation, temperature (minimum, maximum, and mean), relative humidity, sunshine duration, and wind speed at 2 m above ground were extracted.

Using these climate data as a driver, the catchment's hydrology was simulated with J2K for the following time slices:

(i) 1976–2005 (based on EC-Earth-hist/MPI-ESM-hist)
(ii) 2011–2040 (based on RCP models)
(iii) 2041–2070 (based on RCP models)
(iv) 2071–2100 (based on RCP models)

For each of the resulting 14 time slice/model combinations, the outputs of the hydrological simulation were analyzed in two ways, (i) temporally, by aggregating the spatial outputs of J2K and by analyzing the resulting daily time series, and (ii) spatially, by aggregating the temporal outputs on a quarterly basis and by creating long-yearly averages of the quarterly data.

For the analysis of the daily time series, the software IHA (indicators of hydrologic alteration) (The Nature Conservancy 2009) was used. The software was developed to analyze the impacts of hydrological alterations (e.g., dam constructions), with respect to both short-term and long-term changes. For this study, the simulated streamflow for the historical period was compared to the results of the three future time slices. This comparison was repeated for each of the four RCP-based models.

For each of the 14 time slice/model combinations, overall 67 statistical parameters were calculated with the IHA software based on daily, monthly, and yearly values (e.g., 7-day minimum, 25th/50th/75th percentiles, and extremes). Moreover, by using IHA, the following five flow components of ecological importance – also denoted as environmental flow components (EFCs) – were analyzed: extreme low flows, low flows, high flow pulses, small floods, and large floods. These were in detail:

- *Extreme low flows*: occur during drought periods and restrict water availability. Defined here as all low flow days with a flow value below 10% of daily flows for the related 30-year period.

- *Low flows*: main flow condition, especially during the dry season; the low flow determines water availability within a region. Defined here as all flows below 75% of daily flows for the related 30-year period.
- *High flow pulses*: are related to rainy events or the rain season, where the flow is above low flow levels but not more than bank full. They ensure water availability. Defined here as flows that exceed 75% of daily flows for the assigned 30-year period.
- *Small floods*: The river exceeds and crosses the main channel but is not defined as a large flood. Defined here as an initial high flow pulse with a peak flow greater than a 2-year return interval within the related 30-year period.
- *Large floods*: occur more rarely but with higher impacts by destroying habitats and changing river morphology to a greater or lesser extent. Besides negative anthropogenic impacts, large floods also bring nutrients to the flood plain and surrounding area and can recharge groundwater storage. However, within the study region, floods show a destructive character, demolishing lives as well as urban and arable land, and hence are connected to major expenses for human communities. They are defined here as an initial high flow pulse with a peak flow greater than an event with a 10-year return interval within the related 30-year period.

In addition to the ECF analysis, different statistical analyses were carried out, both spatially and temporally, e.g., the distribution of all parameters modeled for all time slices or the climatic water balance (CWB), which takes precipitation and actual evapotranspiration into account. The rationale is to highlight changes of future water availability for a basin.

17.3 Findings

Within this section, the results of this study will be presented in the same order as presented within the Data and Methods section (Sect. 17.2).

17.3.1 System Understanding/Data Acquisition and Processing

Throughout this study, more than 200 time series and more than 80 geo data sets and 200 soil samples from RJ were collected from different project partners. The data was processed, verified and corrected, and made available within the river basin information system (INTECRAL-RBIS). Using the web-based user interfaces of INTECRAL-RBIS, all required J2K input data were downloaded for the study regions, in the case of time series data already formatted according to the J2K requirements. In order to delineate the required spatial modeling units (HRUs), spatial input data on elevation, soil, geology, and land use were processed using the web-based *HRUweb* geodata processing software. The parameterization was done for whole RJ and the upper part of MRB in order to ensure a fast and transparent model setup in any other basin within the region.

17.3.2 Ecohydrological Modeling

The J2K model was applied in three study areas, namely, DRRB, GMRB, and MRB (see Sect. 17.2.1). However, only DRRB will be used to exemplify the modeling results. Figure 17.2 shows the results of the simulated and observed runoff at gauge Dois Rios and the precipitation for the hydrological years 2009 until 2012. The comparison between measured and simulated hydrographs illustrates that the model simulates the hydrological dynamic very well. This is especially true for the fast response to precipitation events. Moreover, the low flows during the dry season are represented with high accuracy. This is reflected in the statistical efficiency criteria used to evaluate model performance. Table 17.3 shows the efficiency criteria chosen and the rationale behind their selection. The base flow was simulated very well, which is displayed by the logarithmic Nash-Sutcliffe efficiency of 0.79. Looking at the graph, it becomes evident that peak flows were predicted precisely, which is also represented by the Nash-Sutcliffe efficiency (Nash and Sutcliffe 1970) of 0.78 as well as by the relative volume error (pBIAS), which shows that the model overpredicts discharge by 5.9%. The Kling-Gupta efficiency (KGE; Gupta et al. 2009) of 0.88 emphasizes a very good match between simulated and observed runoff with regard to correlation, bias, and variability. However, both the visual data plot and the statistical evaluation show that the model was able to represent the hydrological dynamic and processes within the basin well.

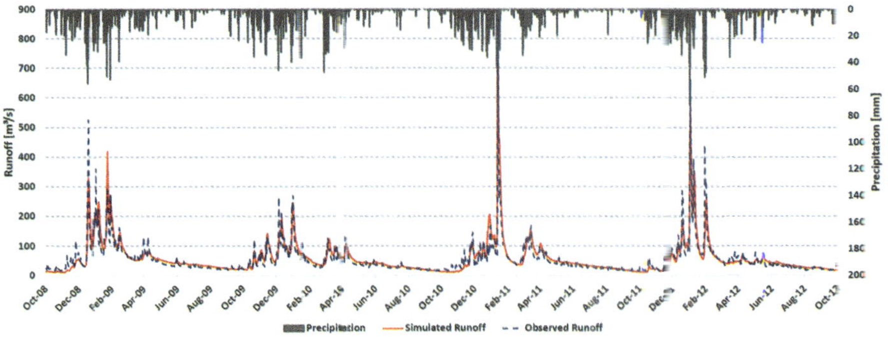

Fig. 17.2 Observed and simulated runoff at gauge Dois Rios

Table 17.3 Efficiency criteria used to evaluate model performance of the model (Gupta et al. 2009). Values here for the J2K model for DRRB and hydrological years 2009–2012

Efficiency criteria	Reason for selection	Range	Value
Nash-Sutcliffe efficiency (NSE)	Multi-objective function, strong focus on simulation of peak flows	$(-\infty, 1)$	0.78
Logarithmic Nash-Sutcliffe efficiency (LogNSE)	Like NSE, but logarithm focuses on representation of simulation of base flow	$(-\infty, 1)$	0.79
Kling-Gupta efficiency (KGE)	Aggregated multi-objective function, representing bias, correlation, and flow variability	$(-\infty, 1)$	0.88
Relative volume error (pBias)	Representing overall under- or overestimation	$(-\infty,\infty)$	5.9%

17.3.3 Modeling Future Climate Impacts

Within the DRRB, the EC-Earth-based models predict an increase of the mean temperature by 1.8 °C until the end of the century in scenario RCP4.5 and by 3.1 °C in scenario RCP8.5 (Fig. 17.2). These differences are less pronounced for the scenario RCP4.5 but show the same pattern. To highlight the extreme development scenarios, we focus on the RCP8.5 emission scenario, the EC-Earth model, and the time slice from 2071 to 2100 (Fig. 17.3).

Besides the overall increasing temperature, precipitation shows a more diverse picture. Considering scenario RCP8.5, Fig. 17.4 emphasizes that until the year 2100, less rainfall (−15%) will occur during the dry season, here represented by the quarter September, October, and November (SON), and more rain (6%) will occur during the rainy season, represented by the quarter December, January, and February (DJF). However, the analysis of the time series shows that the increase of rain until the year 2100 is mainly driven by extreme events within the rainy season.

The changes in temperature and rainfall have a direct effect on the hydrological dynamics (including extreme events) and water availability in the region. Figure 17.5 shows the modeled quarterly sum of total runoff of each HRU within DRRB for SON, between 1976–2005 (EC-Earth-hist) and 2071–2100 (EC-Earth-RCP8.5 scenario). A comparison of these two time slices of SON clearly indicates that less runoff is predicted for the future scenario (a 34% reduction for the entire basin). As expected, the actual evapotranspiration shows an increase of 30% resulting in less available water. With our spatially distributed simulation of the total runoff, we identified hot spot areas with pronounced changes (Fig. 17.5). For instance, the scenarios show that areas with a total runoff of less than 100 mm will increase – not only close to the outlet, but also in the headwater areas where these patterns rarely occurred in historic times.

Actual evapotranspiration is predicted to increase throughout the whole year until the end of the century, whereby the overall rise is higher in SON (15%) than in DJF (6%).

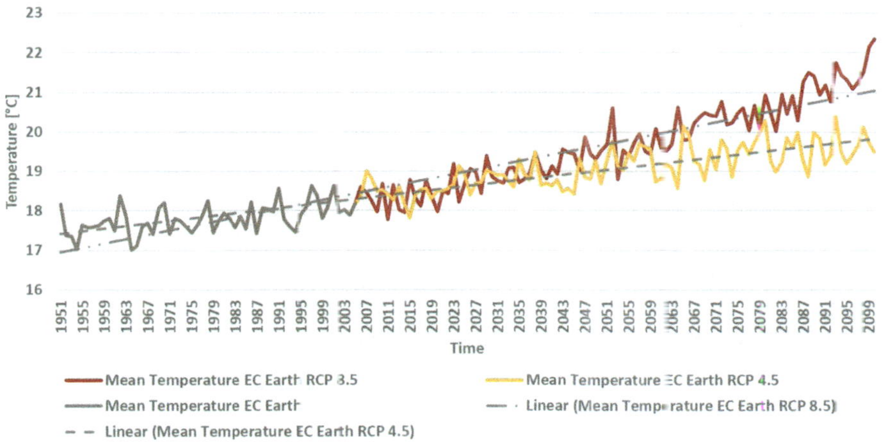

Fig. 17.3 Mean temperature derived from EC-Earth historical data and future climate scenarios RCP4.5 and RCP8.5 for the DRRB

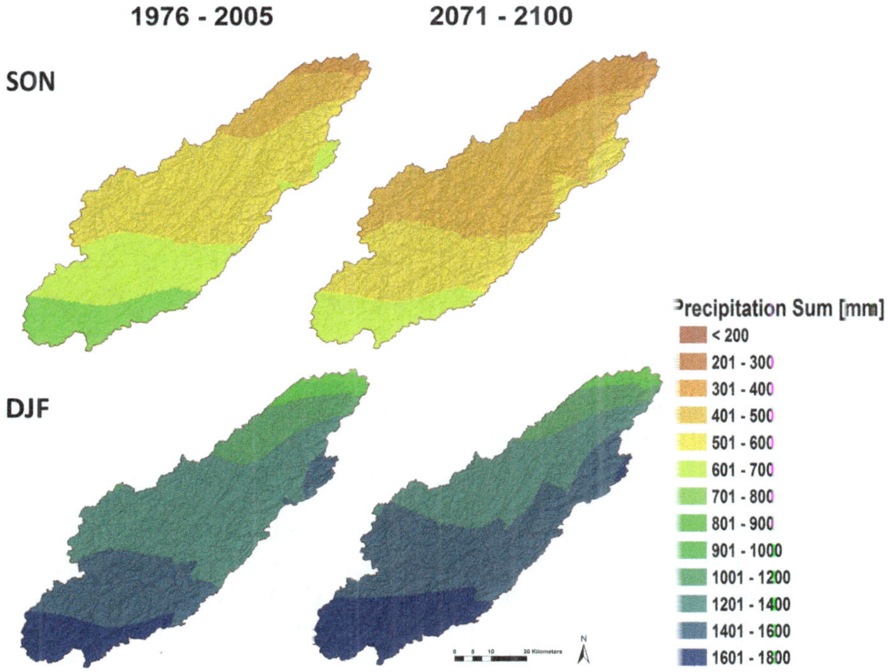

Fig. 17.4 Spatial distribution of mean sum of quarterly precipitation for DRRB between historic data and the time slice 2071–2100 of EC-Earth-RCP8.5 scenario

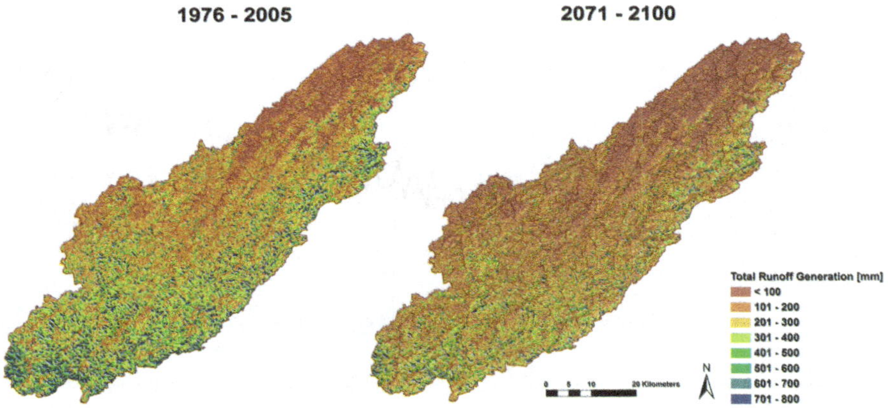

Fig. 17.5 Spatial distribution of mean sum of total runoff per HRU in SON between historic data (1976–2005) based on EC-Earth-hist and the time slice 2071–2100 based on EC-Earth-RCP8.5

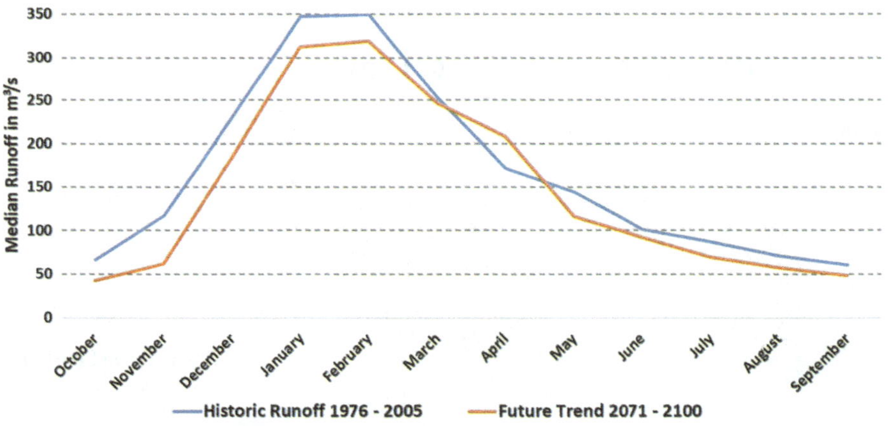

Fig. 17.6 Median runoff for the 30-year period of historic data based on EC-Earth-hist and the time slice 2071–2100 based on EC-Earth-RCP8.5

The hydrological predictions show less water availability in the future, which is also indicated by a lower median value throughout almost the entire year (cf. Fig. 17.6). However, the number and intensity of single flood events increase considerably, especially in the time slice 2071–2100 (cf. Fig. 17.7). Figure 17.7 shows that especially large floods, defined here as peak flow greater than an event with a 10-year return interval, will increase in number and intensity. The J2K model predicts for the DRRB that 0.6% of all daily flows will be large floods (based on the EC-Earth-hist model), while it simulates an enormous increase (3.4% of total flow as large floods) for the EC-Earth-RCP8.5 model (time slice 2071–2100).

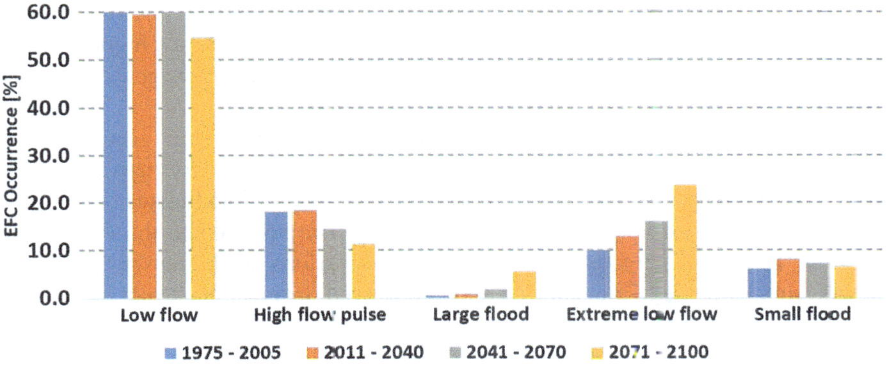

Fig. 17.7 Occurrence of environmental flow components (EFCs) for the different time slices as simulated with J2K based on the EC-Earth-hist and EC-Earth-RCP8.5 scenarios. The definition of the EFCs is described in Sect. 17.2.4

In contrast, low flows would decrease by 11%, but also high flow pulses are projected to decrease by around 7% (while small floods will increase by 0.4%). These projections are all for the time slice 2071–2100 and based on EC-Earth-RCP8.5. However, extreme low flows are predicted to show an increase of 14% until the end of the century (cf. Fig. 17.7).

17.4 Discussion and Conclusion

Sound river basin management requires an understanding of the impacts of future changes on dominant hydrological processes and their dynamics, e.g., with regard to climate and land use changes. Ecohydrological models as applied in the presented study are valuable tools to address this problem. Events like the water crises in the southeast of Brazil during 2014 and 2015 illustrate the need for integrated river basin management, which is based on a thorough social-ecological system understanding that takes the specific conditions of the region into account. Besides human impact of, e.g., agricultural, industrial, and service-oriented water use, this management has to consider how future climate or changing land use can impact water availability in general and hydrological dynamics in particular.

Our study of future climate impacts on water resources for representative basins within RJ reveals that less water will be available in the future. While our models projected this effect throughout the entire year, it was especially pronounced during the dry season. As an example, our models simulated 34% less runoff for the dry season within the time period 2071–2100 (EC-Earth-RCP8.5 scenario). On the other hand, extreme precipitation events will increase during the wet season, which will lead to a higher intensity and frequency of flood events. An analysis of the model results using the indicators of hydrologic alteration (IHA) showed that the number of days with occurring floods will be more than five times higher for the mentioned scenario. In order to prevent severe socioeconomic damages, counteractions therefore need to be established sooner than later.

Modeling is always linked to uncertainties that stem from data, general model assumptions, and the underlying process representations. Every step of model development, from the observation of data to the generation and linkage of complex models, comes with its own uncertainties, which can accumulate throughout the process. Different methods were applied in the various work steps to minimize errors, e.g., by using homogeneity tests on observed data (see Santos et al. 2018 in this book). Furthermore, selection of model parameters focused both on model performance criteria and overall robustness and transferability (Künne et al. 2016).

This chapter presents only the results of the ecohydrological modeling and climate impact assessment for the DRRB. However, ecohydrological models were also applied in two other study areas, i.e., in GMRB and MRB. Additionally, the calibrated and validated models were applied in different sub-basins within the region. The results could not all be presented here but showed similar or even better performances with regard to statistics as well as visual model interpretation. The model achieved best results for the headwater areas, which are characterized by steep slopes and less soil depth. The three river basins (GMRB, DRRB, and MRB) all feature a great variety in elevation and relief energy and show differences in soil types and land use. Global model parameters, which were adapted during the calibration process, make it hard to account for these differences. However, our overall objective was the development of a transferable ecohydrological modeling platform that can be used for change impact assessment in the entire rural RJ region. The results show that we met this objective, which can be seen in the following achievements:

(i) The improvement of the understanding of the relevant processes and their interactions
(ii) The design of computer models to analyze and predict integrated ecohydrological systems
(iii) The evaluation of future development scenarios and the quantification of their impacts on hydrology

The work and results presented in this chapter were made available for all project partners and stakeholders and can be used for knowledge-based decision-making in river basin management of the pilot river basins and all their sub-basins. Moreover, the solid data basis and parameter definition for all types of hydro-geology, soil, and land use occurring in RJ and the Muriaé river basin together with the flexible JAMS/J2K modeling platform build the basis for ecohydrological modeling studies in the future. Further, the presented methods to assess climate impacts can be transferred and reproduced in other regions and can thus build the basis for a science-based search for related adaptation measures.

Acknowledgments The authors would like to thank all Brazilian partners and friends from the Projeto Rio Rural (SEAPPA) and Embrapa Solos for their help throughout the project and during the fieldwork. We also thank our German partners from the TH Köln in Cologne and from Leipzig University who supported our work with important soil and vegetation measurements plus data and helped us with field measurements and equipment. We finally thank the German Ministry of Education and Research for funding this research via the DINARIO and INTECRAL projects (grant numbers 01LB0801A/C & 033L162K).

References

ANA Agência Nacional de Águas (2015) Sistema nacional de informações sobre recursos hídricos. Banco de Dados Hidrometeorólicos – Módulo de Dados Quáli-Quantitativos – SNIRH. http://www.ana.gov.br/PortalSuporte. Accessed 25 Nov 2015

Bossel H (2004) Systeme, dynamik, simulation. Modellbildung, analyse und simulation komplexer systeme. Books on Demand GmbH, Norderstedt

Boyle DP, Gupta HV, Sorooshian S (2000) Toward improved calibration of hydrologic models: combining the strengths of manual and automatic methods. Water Resour Res 36:3663–3674. https://doi.org/10.1029/2000WR900207

Britto AL, Formiga Johnsson RM, Ferreira Carneiro PR (2016) Water supply and hydrosocial scarcity in the Rio de Janeiro metropolitan area. Ambient Soc 19:183. https://doi.org/10.1590/1809-4422ASOC150159R1V1912016

Campoe OC, Stape JL, Mendes JCT (2010) Can intensive management accelerate the restoration of Brazil's Atlantic forests? For Ecol Manag 259:1808–1814. https://doi.org/10.1016/j.foreco.2009.06.026

Carvalho Filho A, Lumbreras JF, Santos RD (2000) Os solos do estado do Rio de Janeiro. Estudo geoambiental do Estado do Rio de Janeiro. Executed by Empresa Brasileira de Pesquisa Agropecuária – Embrapa Solos. Ministério de Agricultura e Abastecimento, Centro Nacional de Pesquisa de Solos, Serviço Geológico do Brasil. Brasilia

Deb K, Member A, Pratap A et al (2002) A fast and elitist multiobjective genetic algorithm: NSGA-II. IEEE Trans Evol Comput 6:182–197

Dobrovolski R, Rattis L (2015) Water collapse in Brazil: the danger of relying on what you neglect. Nat Conserv 13:80–83. https://doi.org/10.1016/j.ncon.2015.03.006

Duringer Jacques P, Shinzato E (2000) Uso e cobertura do solo. Estudo geoambiental do estado do Rio de Janeiro. Executed by CPRM – Serviço Geológico do Brasil. Departamento de Informações Institucionais e Departamento de Gestão Territorial. Brasília

Evans JP (2011) CORDEX – An international climate downscaling initiative. In: 19th international congress on modelling and simulation, Perth. 12–16

Embrapa (1980) Estudo expedito do solos no Estado do Rio de Janeiro para fins de classificação, correlação e legenda preliminar. EMBRAPASNLCS. Boletim Tecnico 62. Rio de Janeiro

Fidalgo EC, Pedreira BCC, Abreu MB et al (2008) Uso e cobertura da terra na bacia hidrográfica do rio Guapi-Macacu. Empresa Brasileira de Pesquisa Agropecuária Embrapa Solos Ministério da Agricultura, Pecuária e Abastecimento. 31 p. Série Documentos No. 105, Rio de Janeiro, Brazil

Fischer C, Kralisch S, Krause P et al (2009) Calibration of hydrological model parameters with the JAMS framework. In: Anderssen RS, Braddock RD, Newham LTH (eds) Proceedings of the 18th world IMACS congress and MODSIM09 international congress on modelling and simulation, Cairns, pp 866–872

Freitas SR, Mello MCS, Cruz CBM (2005) Relationships between forest structure and vegetation indices in Atlantic rainforest. For Ecol Manag 218:353–362. https://doi.org/10.1016/j.foreco.2005.08.036

Flügel W-A (1996) Hydrological Response Units (HRU) as modelling entities for 28 hydrological river basin simulation and their methodological potential for modelling 29 complex environmental process systems. – Results from the Sieg catchment. p 42–62
Gupta HV, Kling H, Yilmaz KK, Martinez GF (2009) Decomposition of the mean squared error and NSE performance criteria: implications for improving hydrological modelling. J Hydrol 377:80–91. https://doi.org/10.1016/j.jhydrol.2009.08.003
Hagemann S, Loew A, Andersson A (2013) Combined evaluation of MPI-ESM land surface water and energy fluxes. J Adv Model Earth Syst 5:259–286. https://doi.org/10.1029/2012MS000173
Hazeleger W, Bintanja R (2014) Accomplishments, current status and future plans of EC-EARTH: a European Earth System Model. 24
INMET Instituto Nacional de Meteorológia (2015) BDMEP – Banco de dados meteorológicos para ensino e pesquisa. http://www.inmet.gov.br. Accessed 25 Nov 2015
IUSS Working Group WRB (2006) World reference base for soil resources 2006, World soil resources report, vol 103. FAO, Rome
Kirchner A (2014) Rekonstruktion der spätpleistozänen und holozänen Landschaftsgenese im Guapi-Macacu Einzugsgebiet, Rio de Janeiro, Südostbrasilien. 1–171
Kottek M, Grieser J, Beck C et al (2006) World map of the Köppen-Geiger climate classification updated. Meteorol Z 15:259–263. https://doi.org/10.1127/0941-2948/2006/0130
Kralisch S, Krause P (2006) JAMS – a framework for natural resource model development and application. In: Voinov A, Jakeman A, Rizzoli AE (eds) Proceedings of the iEMSs third biannual meeting, Burlington
Kralisch S, Fischer C (2012) Model representation, parameter calibration and parallel computing – The jams approach. In: Seppelt R, Voinov AA, Lange S, Bankamp D (eds) iEMSs 2012 – Managing Resources of a Limited Planet: Proceedings of the 6th Biennial meeting of the international environmental modelling and software society. Leipzig, Germany, pp 1177–1184
Krause P (2001) Das hydrologische Modellsystem J2000. Ein Modellsystem zur physikalisch basierten Nachbildung der hydrologischen Prozesse in großen Flusseinzugsgebieten. Jülich
Künne A, Penedo S (2012) Field survey. Unpublished raw data
Künne A, Penedo S, Kralisch S, Schuler AE, Bardy Prado R, Santos JM, Flügel W-A (2016) Modelagem de processos hidrológicos em bacias de meso-escala escassamente monitoradas na mata atlântica, rj, brasil. In: Rodrigues LN, Moura MF, Junio RC de O (eds) Água e Agricultura: incertezas e desafios para a sustentabilidade frente às mudanças do clima e do uso da terra: anais do IV Seminário da Rede Agrohidro. Embrapa, Brasília, pp 7–10
Lindner A, Sattler D (2012) Biomass estimations in forests of different disturbance history in the Atlantic Forest of Rio de Janeiro. Brazil. New Forests 43(3):287–301
Menezes Raposo de Almeida R, Neto Mi de M, Tubbs D, et al (2010) GEOTECNIA E HIDROGEOLOGIA. In: Sin Hwa C, Aquinio G. Massera da Hora M de, Fróes V (eds) Projeto Macacu. Planejamento Estratégico da Região Hidrográfica dos Rios Guapi - Macacu e Caceribu - Macacu. FEC/UFF, PETROBRÁS AMBIENTAL, Niterói,
Monteith JL (1965) Evaporation and environment. Symp Soc Exp Biol 19:205–234
Nash E, Sutcliffe V (1970) Part I- a discussion of principles * The problem of determining river flows from rainfall, evaporation, and other factors, occupies a central place in the technology of applied hydrology. It is not only the essential problem of flood forecasting but a. 10:282–290
Nehren U, Kirchner A, Sattler D et al (2013) Impact of natural climate change and historical land use on landscape development in the Atlantic Forest of Rio de Janeiro, Brazil. An Acad Bras Cienc 85:497–518. https://doi.org/10.1590/S0001-37652013000200004
Penedo S, Künne A, Bardy Prado R, Schuler AE, Roehrig J, Ribbe L (2011) Implementation of a Hydro-climatic Monitoring Network in the Guapi-Macacu River Basin in Rio de Janeiro. Proceedings of the XIV world water Congress, 25–29 Sept 2011, Pernambuco, Brazil
Penman HL (1948) Natural evaporation from open water, bare soil and grass. Proc R Soc Lond A Math Phys Sci 193:25. https://doi.org/10.1098/rspa.1948.0037
Pfennig B, Kipka H, Wolf M, Fink M, Krause P, Flügel WA (2009) Development of an extended routing scheme in reference to consideration of multidimensional flow relations between

hydrological model entities. 1972–1978. Proceedings of the 18th World IMACS / MODSIM Congress, Cairns, Australia 13–17 July 2009

Samani Z (2000) Estimating solar radiation and evapotranspiration using minimum climatological data. J Irrig Drain Eng 126:265–267. https://doi.org/10.1061/(ASCE)0733-9437(2000)126:4(265)

Santos JM, Künne A, Kralisch S et al (2018) Analyzing hydro-climatic data to improve hydrological understanding in rural Rio de Janeiro, southeast Brazil. In: Nehren, U, Schlüter S, Raedig C, Sattler D, Hissa H (eds) Strategies and tools for a sustainable rural Rio de Janeiro. (in press)

Sattler D, Raedig C, Hebner A, Wesenberg J (2018) Use of native plant species for ecological restoration and rehabilitation measures in Southeast Brazil. In: Nehren, U, Schlüter S, Raedig C, Sattler D, Hissa H (eds) Strategies and tools for a sustainable rural Rio de Janeiro. (in press)

Sattler D, Murray LT, Kirchner A, Lindner A (2014) Influence of soil and topography on aboveground biomass accumulation and carbon stocks of afforested pastures in South East Brazil. Ecol Eng 73:126–131. https://doi.org/10.1016/j.ecoleng.2014.09.003

Seliger R, Sattler D, Soares da Silva A et al (2018) Rehabilitation of degraded sloped pastures – lessons learned in Itaocara. In: Nehren U, Schlüter S, Raedig C, Sattler D, Hissa H (eds) Strategies and tools for a sustainable rural Rio de Janeiro, Springer International Publishing

Silva LC da, Silva Cunha HC (2001) Geologia do Estado do Rio de Janeiro: texto explicativo do mapa geológico do Estado do Rio de Janeiro, 2nd edn. CPRM, Brasília

Sin Hwa C, Aquinio G. Massera da Hora M de Fróes V (2010) Projeto Macacu. Planejamento Estratégico da Região Hidrográfica dos Rios Guapi – Macacu e Caceribu – Macacu. FEC/UFF, PETROBRÁS AMBIENTAL, Niterói

The Nature Conservancy (2009) Indicators of hydrologic alteration version 7.1 User's manual. Nat Conserv 76. https://doi.org/10.1002/0470848944

Thomson AM, Calvin KV, Smith SJ et al (2011) RCP4.5: a pathway for stabilization of radiative forcing by 2100. Clim Chang 109:77–94. https://doi.org/10.1007/s10584-011-0151-4

U.S. Geological Survey (2015) SRTM 1 arc-second for global coverage (~30 meters) non-void filled. downloaded 2015 from https://earthexplorer.usgs.gov/

Vuuren DP, Edmonds J, Kainuma M et al (2011) The representative concentration pathways: an overview. Clim Chang 109:5–31. https://doi.org/10.1007/s10584-011-0148-z

WBCSD (2006) Facts and trends water. World Bus Counc Sustain Dev 16. https://doi.org/10.1080/0379772780030302

Wesenberg J, Sattler D (2011) Field survey. Unpublished data

Xavier AC, Vettorazzi CA (2003) Leaf area index of ground covers in a subtropical watershed. Sci Agric 60:425–431. https://doi.org/10.1590/S0103-90162003000300002

Zoccal R, Assis AG and Medeiros Evangelista SR (2006) Distribuição geográfica da pecuária leiteira no Brasil 88. Embrapa Gado de Leite, Juiz de Fora, MG

Chapter 18
Reuse-Oriented Decentralized Wastewater and Sewage Sludge Treatment for Small Urbanized Rural Settlements in Brazil: An Environmental Cost-Benefit Analysis

Jaime Cardona, Alena Lepilova, Daniel Gieseler, and Kristina Kreter

Abstract Latin-American countries lack efficient solutions for wastewater and sewage sludge treatment. In particular, small urbanized rural settlements (SURUS) in many of these countries face significant challenges with respect to the selection and operation of sustainable sewage treatment facilities. Decentralized sanitation and reuse (DESAR) solutions can significantly contribute toward the improvement of wastewater sanitation coverage in SURUS in Latin-American regions. The major advantages of DESAR for SURUS are a reduction in final treatment costs because these systems allow for water reclamation and sewage sludge reuse for agriculture. To reflect the applicability of DESAR on a regional scale, we present here an integrative assessment, including a cost-benefit analysis (CBA) and geographic information systems (GIS) surveying, as a "decision support methodology" for conducting environmental-economic analyses. As a case study, this methodology was applied to six SURUS located at the Rio Dois Rios basin of Rio de Janeiro state. The CBA shows that DESAR could recover between 15% and 34% of total operational and maintenance costs for SURUS populations between 222 and 1,585 inhabitants. The findings suggest that DESAR systems can respond to the need to reduce

J. Cardona (✉)
Centre for Environmental Biotechnology (UBZ), Helmholtz Centre for Environmental Research (UFZ), Leipzig, Germany

Training and Demonstration Centre for Decentralized Sewage Treatment – BDZ e.V., Leipzig, Germany
e-mail: cardona@bdz-abwasser.de

A. Lepilova · D. Gieseler
Training and Demonstration Centre for Decentralized Sewage Treatment – BDZ e.V., Leipzig, Germany
e-mail: lepilova@bdz-abwasser.de; gieseler@bdz-abwasser.de

K. Kreter
Institute for Technology and Resources Management in the Tropics and Subtropics, TH Köln – University of Applied Sciences, Cologne, Germany

U. Nehren et al. (eds.), *Strategies and Tools for a Sustainable Rural Rio de Janeiro*, Springer Series on Environmental Management,
https://doi.org/10.1007/978-3-319-89644-1_18

costs and improve nutrient recovery capabilities of sanitation interventions in rural communities.

Keywords Decentralized sanitation · Rural development · Sludge reuse Nutrient recovery · INTECRAL project

18.1 Introduction

Integrative and sustainable wastewater and sewage sludge treatment solutions (WASTES) for rural areas in low- to middle-income countries are keys for achieving the Sustainable Development Goals (SDGs). However, implementing WASTES is a challenge for both planners and decision-makers, especially in small urbanized rural areas (SURUS) with population densities similar to those found in urban areas. WASTES are normally not implemented in SURUS, in part, due to the absence of economies of scale and the high costs per capita compared to urbanized areas (UN-Habitat 2006; Hophmayer-Tokich 2010). Yet, SURUS generally have population densities for which conventional on-site sanitation facilities prove less cost-effective than collecting and treating wastewater via sewer networks and treatment plants (Bakir 2001). Therefore, SURUS are becoming "gray areas" in which implementation of WASTES is commonly postponed. This produces significant asymmetries for investments between rural and urban settlements, especially in middle to less developed countries (Cardona et al. 2016).

The past decade has seen increased interest in the potential of decentralized wastewater management (DWM) systems to address sanitation challenges in SURUS. In contrast to their centralized counterparts, decentralized approaches are designed to treat the effluent close to the point of generation (Crites and Tchobanoglous 1998). They require less funds to be allocated to the collection and transportation of effluent (Otis and Mara 1995; Bakir 2001) while supporting local reuse of treated wastewater and partial recovery of treatment costs (Massoud et al. 2009; Lienhoop et al. 2014; van Afferden et al. 2015). As a result, they balance cost and effectiveness more successfully in areas with low to medium population densities (Otterpohl et al. 1997; Wilderer and Shereff 2000).

Unfortunately, the selection and implementation of DWM systems in a number of low- and middle-income countries have presented many challenges. In particular, the methodology for the design of DWM systems should ensure that local conditions are well understood and, furthermore, that local reuse practices are appropriately integrated for maximum economic benefits. In this context, local conditions encompass technical, social, economic, and environmental aspects. Provided that sound techno-economic feasibility studies are conducted, decentralized sanitation and reuse (DESAR) systems can assist rural communities in the development and implementation of low-cost, simple to operate effluent treatment systems (Lens et al. 2001).

This chapter summarizes the results of the "Integrated Eco technologies and Services for a Sustainable Rural Rio de Janeiro (INTECRAL)" project, funded by the German Ministry of Education and Research. The project aims to address issues pertaining to sustainable sanitation solutions in rural areas of Rio de Janeiro state (RJ). One key objective is to perform an environmental CBA of decentralized wastewater and sewage sludge treatment solutions in small urbanized rural areas within the Rio Dois Rios Basin. This study summarizes the methodology developed and applied and highlights the necessary steps for conducting CBA within SURUS at a regional level.

18.2 Problem Setting and Case Study Region

In Brazil, 75% of the total rural population receives inadequate sanitation services (Costa and Guilhoto 2012). 73% of the 5,564 Brazilian municipalities are small and medium towns of maximum 20,000 inhabitants. Only 55% of these municipalities have sewer systems, and 52% of them use septic tanks for primary treatment (IBGE 2010b). Some authors have further emphasized that treated wastewater in many rural areas does not meet the minimum Brazilian legal requirements for effluent release (Gallotti 2008). This situation is likely to instigate wider environmental and social problems, such as groundwater pollution and outbreaks of waterborne diseases (Segovia 2014; Cardona et al. 2016).

Although the municipal basic sanitation plan is the first mandate under the Brazilian national sanitation policy, the provision of sanitation services in rural areas remains largely unattended. The lack of environmental awareness and technical expertise means that sanitation plans are poorly formulated and access to state and federal resources is frequently denied. Therefore, planning instruments for WASTES in Brazilian SURUS are important from both an environmental and an economic perspective in order to improve the decision-making process in the basin.

18.2.1 Selected Basin, Rio Dois Rios

As a part of the INTECRAL project, the catchment of the Rio Dois Rios (Fig. 18.1) was selected based on its great lack of wastewater treatment infrastructure. The conditions within the Rio Dois Rios Basin are representative of the situation in many SURUS in Brazil, where the wastewater infrastructure investment has been very poor in comparison with urban areas. For instance, according to the 2010 IBGE census, only 11.3% of the total Rio Dois Rios population (51,332 people) were connected to a sewer network, 23.9% relied on septic tanks, and 64.7% had no access to wastewater treatment facilities (IBGE 2010b). The basin drains 3,200 km^2 and covers nine municipalities that are home to 220 thousand inhabitants. The Rio Dois Rios Basin includes 82 micro-basins with decentralized administrative structures

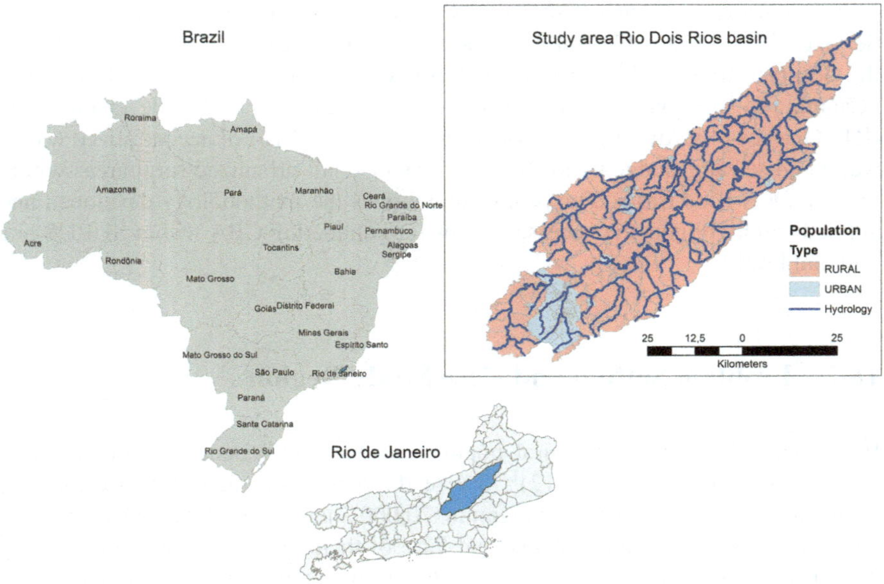

Fig. 18.1 Study area basin Rio Dois Rios, Rio de Janeiro, Brazil (own elaboration)

and community-based integrations to facilitate water resources management and encourage sustainable agricultural practices. Moreover, we chose the basin due to the significant amount of existing information related to socioeconomic and water/wastewater infrastructure surveys.

In this context, identification of additional water resources for irrigation and methods of nutrient recovery for fertilizer substitutes provided opportunity for the application of DESAR.

18.3 Materials and Methods

The DESAR environmental and economic valuation methodology, developed by Cardona et al. (2016) within the framework of the INTECRAL project, includes the following steps:

1. Identification of SURUS priority areas:
   (a) Regional assessment, including a socioeconomic survey of the selected community.
   (b) Population density analysis, using geographic information systems (GIS) surveying (based on the socioeconomic survey and satellite imagery).
   (c) The densities of building and distances between them can be used to determine the requirements placed on the sewer network.
   (d) Connection degree.
   (e) Future population.

2. Definition of wastewater quality standards and sludge quality parameters
3. Selection of most suitable SURUS communities for WASTES
4. CBA of selected WASTES applied to identified SURUS
   (a) Cost estimates for construction, land acquisition, as well as operating and maintenance costs.
   (b) Estimation of economic benefits associated with the treatment solutions. In order to calculate the economic benefits of DESAR solutions in monetary terms, we followed a methodology proposed by Chen and Wang (2009). In this approach, a net benefit value (NBV) model quantifies the benefits of costs avoided. For our case, the following benefits were chosen because of local environmental legislation:
      (i) Avoided BOD_5 (5-day biochemical oxygen demand) discharge penalties
      (ii) Avoided costs associated with sludge management (transportation, drying, and final disposal)
      (iii) Avoided costs related to uptake of water for irrigation
      (iv) The benefits of sludge reuse as fertilizer
   (c) Estimation of net present value (NPV) for selected SURUS communities.
5. Calculation of cost-recovery capabilities, especially in terms of operating and maintenance costs

18.3.1 Identification of SURUS Priority Areas

18.3.1.1 Population Density Analysis and Future Population Estimation

Population density estimates were obtained based on a socioeconomic survey (IBGE Census 2000 and 2010a) and a rural census (IBGE 2007) and were supported by ArcGIS software. The 2007 IBGE census guided the identification of rural buildings, and necessary adjustments were performed manually using ArcGIS satellite imagery. The point density tool (ArcGIS), with a person per square kilometer resolution, determined a cell size of 25 m and a rectangular neighborhood setting of 125 m to identify the densest communities (exceeding 2,000 inhabitants per km^2) (Fig. 18.2b–c). The population density value was assumed based on the fact that the maximum urban population density in RJ by the year 2010 rose to 5,265 inhabitants per km^2 (IBGE 2010a). In our case, we chose a lower population density value because remote sensing techniques identified that most of the SURUS had densities under 2,000 inhabitants per km^2.

In the six selected communities, the number of residential houses was evaluated using satellite imagery. Assuming that 3.5 inhabitants occupy each house, the present community size was evaluated for the year 2017. Using arithmetical projection methods based on previous censuses (IBGE 2000, 2010a), population growth in the communities of interest was estimated in the range of 0.95–1.72. This result was used to forecast population expansion.

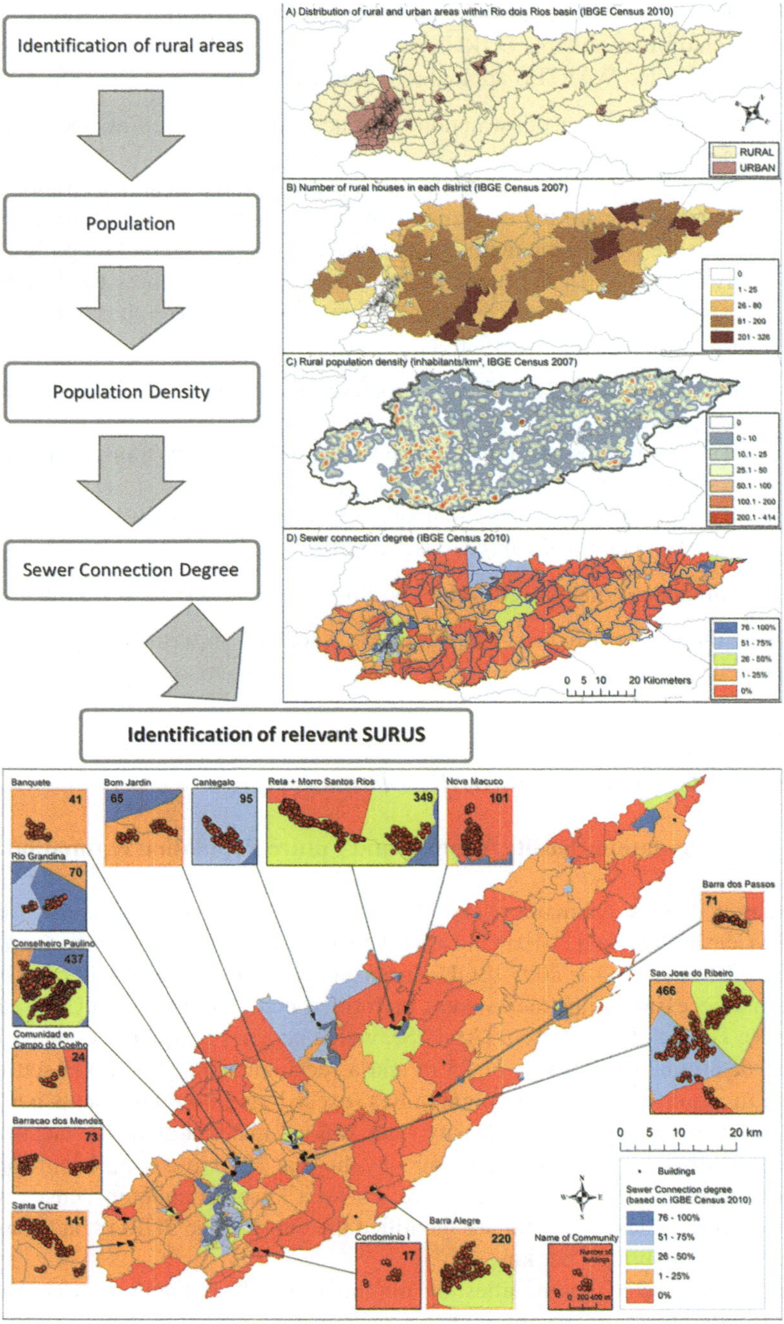

Fig. 18.2 GIS-based analysis of population density and degree of sewer connection within the Rio Dois Rios basin (own elaboration)

18.3.1.2 Connection Degree Estimation

The connection degree is the percentage of the settlement that is connected to a sewer network. This, however, does not necessarily mean that it is linked to a treatment facility. Based on the 2010 IBGE census (2010a), six communities were selected based on a degree of sewer connection (below 30%) and community size (minimum 40 houses). A detailed analysis of Barracão dos Mendes had already been performed (Segovia 2014; Cardona et al. 2016). For this reason, the community was not evaluated further but rather used as an exemplary case for the adaptation of various indicators, such as inhabitants per household or sewer line requirements per capita (Segovia 2014; Cardona et al. 2016). The locations of the remaining six settlements can be seen in Fig. 18.2c.

As a result of the GIS-based assessment, six SURUS (Table 18.1) were identified as priority settlements for conducting a CBA analysis on decentralized WASTES.

Table 18.1 Population, infrastructural, and effluent quality characteristics of six Rio Dois Rios rural communities

SURUS	Current buildings	Pop 2010[a]	Pop change rate in 10 years[b]	Pop in 2017[c]	Pop in 2037[c]	Connection degree[d] (%)	Pop without wastewater treatment system[e]	Equivalent daily BOD_5 load (kg)[f]
C1	101	354	1.72	533	1585	0	1585	63.42
C2	71	249	1.54	342	808	1.12	799	31.98
C3	220	770	0.95	745	679	15	577	23.09
C4	141	494	1.07	517	588	3.2	570	22.78
C5	41	144	1.17	161	222	16.23	186	7.43
C6	65	228	1.14	249	323	20.5	257	10.27

Nova Macuco (C1), Barra dos Passos (C2), Barra Alegre (C3), Santa Cruz (C4), Banquete (C5), Bom Jardim (C6)

[a]Based on IBGE population census, 2010a

[b]Population change rate in 10 years is calculated based on the geometrical progression method, $IG = (P_2 - P_1) / P_1$, where P_1 is the population by year 2000 (IBGE 2000) and P_2 is the population by year 2010 (IBGE 2010a)

[c]Population between 2017 and 2037; the latter is based on the geometrical progression method, $P_n = P \times (1 + IG/100)^n$, where IG = geometric mean (%), P = present population, and n = number of decades

[d]Connection degree calculated based on the data extracted from the water and sanitation infrastructure survey (IBGE 2010b)

[e]Population without wastewater treatment system is calculated based on the formula, $P_{WWT} = P_n \times (100 - CD)$, where P_n is the current total population and CD is the connection degree

[f]Equivalent daily BOD_5 load calculated based on the formula $BOD_5 = (P_{WWT} \times BOD_{Ca}) / 1000$, where P_{WWT} is current population without wastewater treatment system and BOD_{Ca} is BOD load per inhabitant per day ($g/c.d^{-1}$). BOD load was assumed to be equal to 45 ($g/c.d^{-1}$) (Mara 2013)

18.3.2 Selection of WASTES Technology and Quality Standards

The following criteria were considered for the selection of suitable WASTES:

- Low operation and maintenance costs
- The Brazilian wastewater effluent standards (Brazilian National Environmental Council (CONAMA) Resolution 357 of 2005), BOD discharge limits (CONAMA DZ 215.R4 of 2007), and sewage sludge standards (CONAMA Resolution 375 of 2006)
- The possibility of treating sludge locally

As a result, upflow anaerobic sludge blanket (UASB) technology combined with a vertical flow constructed wetland (VFCW) was proposed and is considered to suit the socioeconomic climate in rural areas (de Sousa et al. 2001; El-Khateeb and El-Gohary 2003; Wendland et al. 2006; Halalsheh et al. 2008). In addition, sludge treatment wetlands (STWs) (Uggetti et al. 2010), which are shallow tanks filled with gravel and planted with emergent vegetation, such as *Phragmites australis*, were incorporated into the design (Cole 1998; Uggetti et al. 2011). The treatment train utilized for the present study is shown in Fig. 18.3.

Other technologies such as sequential batch reactors (SBRs) and membrane bio reactors (MBRs) can be also considered as decentralized treatment units for the community. However we did not consider them due to their higher running costs and the qualifications required for the staff for operation.

18.3.3 Cost-Benefit Analysis

The CBA model presented in Eq. (18.1) was developed for evaluating DESAR in the selected SURUS and simulates the most relevant parameters for an environmental-economic analysis of the communities considered:

$$\mathrm{NPV} = \sum_{n}^{t=0} \frac{(Bi - Ci)_t}{(1+r)^t} \tag{18.1}$$

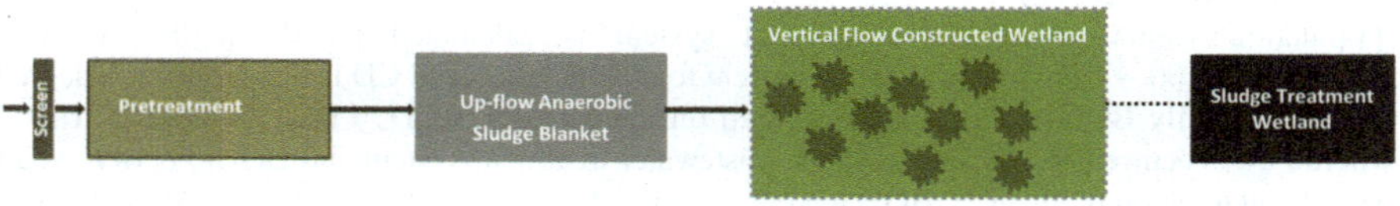

Fig. 18.3 DESAR treatment train implemented in the present case study (Source: Cardona et al. 2014)

where NPV is the net present value, B_i is the value of the benefit item i, C_i is the value of the cost of item i, r is the discount rate, t is year, and r is analytical horizon in number of years. The NPV measures the economic value of a project. The CBA takes NPV as the main financial indicator decision rule. A project with a positive NPV value (NPV > 0) is economically viable, and if the NPV is negative (NPV < 0), the project should be rejected. The best option will offer the highest CBA (Pearce et al. 2006).

18.3.3.1 Economic Benefit Estimation

Using the CBA approach, various models were generated to quantify the economic benefits that are associated with WASTES and are determinant to improving the recovery of operating and maintenance costs. In the present study, benefits associated with the following were evaluated:

- Reuse of water and sewage sludge for agriculture
- Avoidance of BOD_5 discharge penalties
- Avoidance of conventional sludge transport, treatment, and disposal costs

The economic benefits were calculated according to the methodology and equations proposed by Cardona et al. (2016) and are summarized in Table 18.2.

18.3.3.2 Cost Estimations

The costs associated with sanitation projects are assigned a monetary value that, in most cases, is based on a literature review and the empirical values that were acquired from wastewater engineering companies. Using the available literature and the empirical cost data obtained from the analysis of Barracão dos Mendes, (i) the capital investment requirements for the sewage network and (ii) operation and maintenance costs for the associated facilities were calculated. All costs were calculated in Brazilian reals.[1]

Equation (18.2) shows the quantification of aggregated costs:

$$\Sigma C_I = C_1 + C_L + C_{O\&M} \tag{18.2}$$

where C_I represents the initial investment cost, C_L the land cost, and the $C_{O\&M}$ the operating and maintenance costs.

Table 18.3 summarizes the assumptions made and parameters defined for the cost estimation of the WASTES investigated in the present work.

[1]Values in Brazilian reals (R$) October 2014 (1R<Footnote ID="Fn1"><Para ID="Par47">Values in Brazilian reals (R$) October 2014 (1R$ = 0.4073 USD).</Para></Footnote>#x00A0;= 0.4073 USD).

Table 18.2 Estimation of the economic benefits of the applied DESARs (Cardona et al. 2016)

Benefits	Equation	Description of variables
Avoided sludge transportation The evaluation of costs for sludge transportation and disposal ($C_{ST\,\&\,D}$)	$C_{ST\,\&\,D} = Q_{Si} \times (C_T + C_D)$	Q_{Si} is annual sludge production (T/year), C_t is costs for sludge transportation, and C_D is costs for sludge disposal
Benefits of nutrient compounds[a] Nutrients presented in treated wastewater by DESAR solutions can be applied as fertilizer substances for soil and crops in agriculture. Produced sludge can contain nitrogen, phosphorus, and potassium. The production of nutrients was estimated for UASB	$BN = \sum (Q_{Ni} \times P_{Ni})$	Q_{Ni} is produced nutrients from sludge (kg/year), and P_{Ni} is the market price of nutrients in Brazil (Marcon et al. 2015)
Avoided costs of uptake water The uptake water payments for irrigation in the community were calculated based on water directives CEIVAP (2014)	$AC_{UWI} = 365 \sum_{n}^{i=1} Q_{WW}\, P_{WC}$	AC_{UWI} represents the avoidance of costs of uptake water for irrigation purposes, Q_{ww} the quantity of wastewater treated every day, and P_{wc} the price of water catchment, as established for the region (CEIVAP 2014). This value is discounted for each year
Avoided costs of BOD_5 discharges Kg amount of BOD_5 generated for the system on a yearly basis multiplied by the value of BOD_5 discharge in Brazilian reals (0.0763 R$/kg), as established by CEIVAP directives (CEIVAP 2014)	$AC_{BOD5} = 365 \sum_{n}^{i=1} Q_{BOD5}\, P_{BOD5}$	AC_{BOD5} represents the avoided cost of BOD_5, Q_{BOD5} the quantity of BOD_5 and P_{BOD5} the public unit prices of BOD_5 organic discharge

[a]The total sludge produced by the UASB reactor under aerobic conditions was calculated using the methodology of Andreoli et al. 2007. The applied COD load (kg COD/d) of the sludge generated in the UASB reactor was computed according to $COD_{load\ applied} = Q \times COD_{typical}$, where Q is the estimated flow rate (m^3/d) and $COD_{typical}$ is the influent COD concentration of typical wastewater (947 mg/L) (Andreoli et al. 2007). The quantity of sludge production is $= Y \times COD_{load\ applied}$, which is a function of the applied COD load and the sludge mass production, $Y = 0.18$ kg suspended solids/kgCOD (Andreoli et al. (2007).

18.4 Results and Discussion

A CBA was performed based on a simulation model after 20 years of operation, and the results are presented in terms of NPV (Table 18.4) and specific treatment costs (Table 18.5). DESAR projects in selected SURUS in Rio Dois Rios turn out to be clearly beneficial for recovery operation and maintenance costs. CBA models for DESAR were generated using two different biosolid market prices, namely, a worst case of R$ 30.83 and a best case of R$ 167.32 per ton (Marcon et al. 2015). For

Table 18.3 Assumptions and parameters applicable to the WASTES cost estimates

Parameter units	Value	Source
Design parameters		
Water consumption (l/(c d))	250	(Andreoli et al. 2007)
Biological oxygen demand (mg/l)	463	(Andreoli et al. 2007)
Chemical oxygen demand (mg/l)	947	(Andreoli et al. 2007)
Inhabitants per household	3.5	(Cardona et al. 2016)
Costs parameters UASB + VFCW		
Sewer line requirements per capita (m)	2.5	(Cardona et al. 2016)
Land prices ($R/m)	16	(Segovia 2014)
Construction costs in USD for VFCW = $1650.4*Q^{0.697}$		(Cardona 2005)
O&M in EUR UASB+VFCW = $402.55*(PE)^{0.454}$		(van Afferden et al. 2015)
Costs parameter STW		
O&M costs for STW STW = $16.19*(PE)^{0.8589}$		(Uggetti et al. 2011)
Costs parameter sludge transportation and disposition		
Transport costs = (pf/cap) *2D		(Quintana et al. 2012)
Where pf (assumed price per km) = R$3.79/km, cap = cargo capacity of the truck assumed in 16 t. Two times distance for collecting and disposition (2D)		
Landfill disposal (R$/T)	120	(Cardona et al. 2016)
Benefits of nutrient recovery Best case (R$/T)	154.32	(Marcon et al. 2015)
Benefits of nutrient recovery Worst case (R$/T)	30.83	(Marcon et al. 2015)
Financial parameters		
Evaluation period in years	20	
Discount rate	12%	(Ministerio da Cidades 2010)

instance, community C1 with a projected population of 1585 inhabitants by year 2037 showed the highest cost recovery potential (34% in the best case and 24% in the worst case), while the smallest community C5 with an estimated population of 222 inhabitants by year 2037 showed only 12% in the best case and 8% in the worst case (Fig. 18.4). From an economic (cost efficiency) perspective, the model results suggest that on-site treatment of wastewater and sewage sludge should be given first priority.

A clear correlation between the number of beneficiaries (community inhabitants) and the economic benefits obtained was observed. Communities with a greater

Table 18.4 Costs and benefits associated with DESARs presented in net present value for a project life span of 20 years and a discount rate of 12%. Values in thousand Brazilian reals R$ October 2014 (1R$ = 0.4073 USD)

Item	Units	*c1*	*c2*	*c3*	*c4*	*c5*	*c6*
Costs							
Total capital costs	R$	730	420	370	330	150	210
Total O&M costs in 20 years	R$	303	219	202	189	120	143
NPV of project costs	R$	1,060	657	583	528	277	354
Benefits *(After 20 years of operation)*							
Avoided costs BOD_5 discharges	R$	26	13	11	10	4	5
Avoiding costs uptake water for irrigation	R$	11	5	5	4	1	2
Avoided costs sludge transportation and disposal	R$	22	11	9	8	3	4
Benefits for nutrient contents as fertilizer	R$	45	23	19	17	6	9
NPV of project benefits	R$	104	53	45	39	15	21
Recovery of O&M costs	%	34%	24%	22%	21%	12%	15%

Table 18.5 Specific treatment costs and benefits based on net present value for a project life span of 20 years. Values in thousand Brazilian reals R$ October 2014 (1R$ = 0.4073 USD)

Item	Units	*c1*	*c2*	*c3*	*c4*	*c5*	*c6*
Flow rate	m^3/d	317	162	136	118	44	65
Specific treatment costs STC							
STC (including investments and O&M)	R$/$m^3$	0.44	0.53	0.56	0.59	0.81	0.71
STC (O&M)	R$/$m^3$	0.12	0.18	0.19	0.21	0.35	0.29
STC (O&M – Benefits)	R$/$m^3$	0.08	0.13	0.15	0.17	0.31	0.25

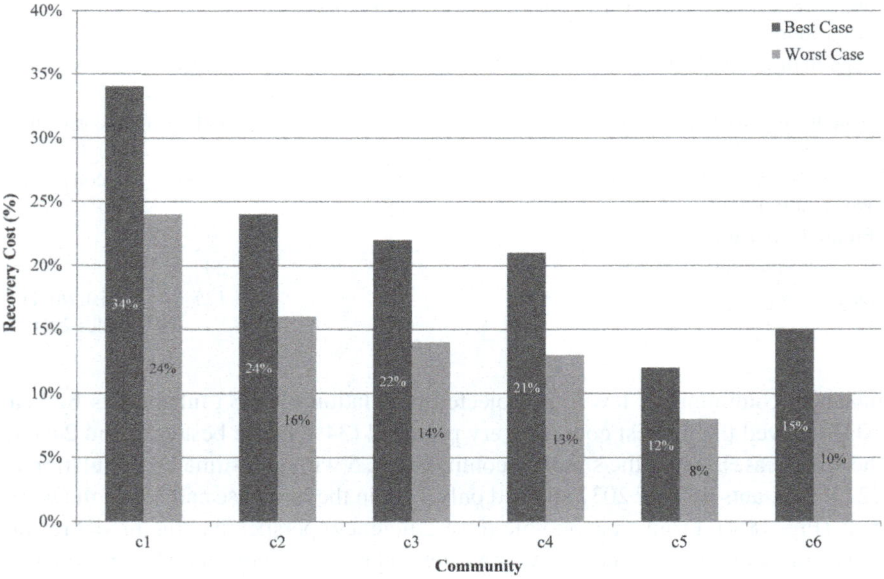

Fig. 18.4 Comparison between recovery of operation and maintenance costs considering best and worst cases scenarios

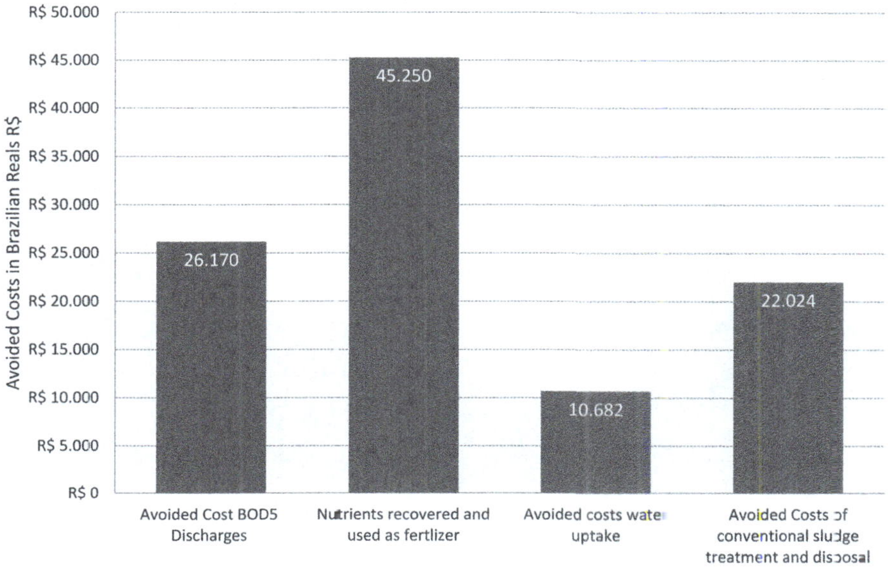

Fig. 18.5 Comparison of economic benefits of DESARs implemented in community C1 (1,585 inhabitants) considering the highest market price for biosolids as chemical fertilizer substitutes (164.32 R$/T)

number of inhabitants can recover a larger portion of the operation and maintenance costs. This can be explained by the larger quantities of biosolids and water for reclamation and irrigation generated, as well as the proportional costs of BOD discharge avoided by implementing decentralized WASTES.

Under the best case biosolid's market price scenario, it was observed that the benefits associated with nutrient reutilization made the biggest contribution to cost recovery (Fig. 18.5). In the worst case scenario, considering the lowest biosolid market price, costs avoided from BOD_5 discharge contributed the least to cost recovery (Fig. 18.6). Cost reduction due to avoidance of sludge transportation and disposal was the second most sensitive parameter (Figs. 18.4 and 18.5).

The STCs related to O&M vary from 0.12 to 0.35 R$/m^3. Integration of the operational and maintenance benefits associated with a decentralized system yielded a noticeable STC reduction (Table 18.5) and reflects potential for a full recovery of O&M costs using DESARs. This trend has been also reported in previous studies conducted in Jordan by van Afferden et al. (2015).

The dry session in 2015 drastically affected the availability of water for consumption and irrigation. This highlighted the fact that alternative water sources for irrigation of agricultural land can play a significant role in reducing the effects associated with drought and climate change. Therefore, water management programs in the region are currently reviewing reuse practices and investigating alternative water resources.

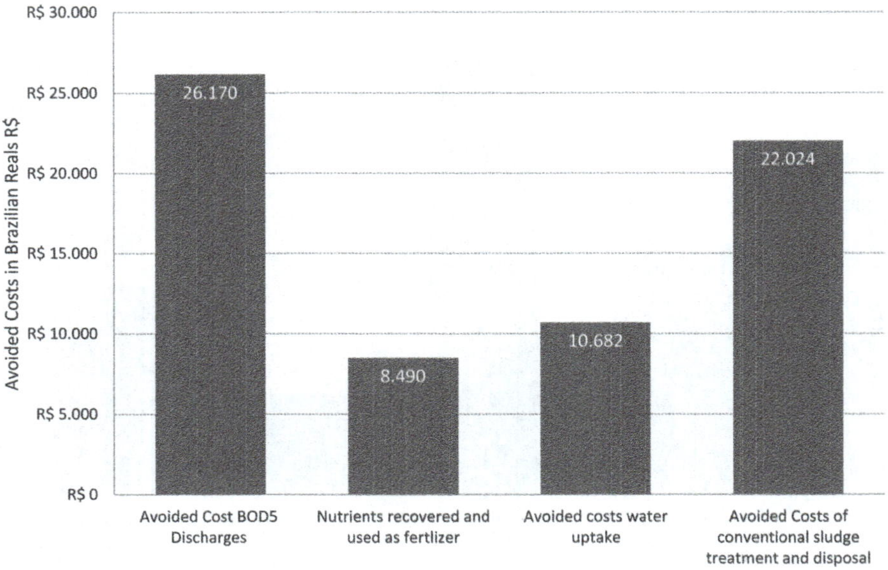

Fig. 18.6 Comparison of economic benefits of DESARs implemented in community C1 (1,585 inhabitants) considering the lowest market price for biosolids as chemical fertilizer substitutes (30.83 R$/T)

Over the past few years, Brazil has been faced with an economic crisis, which resulted in an increase in the price of chemical fertilizers. As a result, alternative nutrient sources, such as treated sewage sludge, are gaining attention from farmers as a fertilizer substitute (Rigo et al. 2014). Additionally, a large portion of the land in the selected study area has been degraded due to poor agricultural practices and over farming (Sattler et al. 2014). Consequently, farmers are looking for affordable nutrient sources to rehabilitate the unproductive areas. In this context, DESARs are well suited to respond to both sanitation needs and the improvement of agricultural practices.

18.5 Conclusions

The efficacy of DESARs is highly dependent on local social, environmental, geographical, economic, and technological conditions. Combining CBA with GIS surveying can guide the decision-making process for implementation of decentralized wastewater and sewage sludge treatment. This approach is also a powerful tool for determining the economic benefits associated with DESARs considering the relevant socioeconomic, geographical, and environmental conditions. The present work has shown that geographical and spatial dimensions are critical in identifying suitable SURUS communities at a regional level.

SURUS communities can benefit from DESARs in terms of provision of improved sanitation services. In addition, treatment of sewage sludge locally using DESAR provides famers in rural areas with an alternative source of chemical fertilizers. Hence, eco technologies for wastewater treatment and sludge stabilization have great potential in rural areas and, more specifically, in SURUS, due to low operational costs, ease of integration into such communities, and nutrient recovery capabilities. These advantages facilitate decentralized system implementation in Brazilian hydrological basins while promoting green markets and agroecological practices.

Additionally, DESAR makes wastewater treatment more affordable by maximizing the benefits associated with O&M cost recovery. This increases the opportunities for initiating wastewater treatment projects in rural areas with low payment capacities.

References

Andreoli CV, von Sperling M, Fernandes F (2007) Sludge treatment and disposal, vol 6. IWA publishing, London

Bakir HA (2001) Sustainable wastewater management for small communities in the Middle East and North Africa. J Environ Manag 61(4):319–328

Cardona JA (2005) Análsis económico de sistemas de tratamiento de aguas residuales en Colombia. In: Conferencia Internacional: Ecotecnologías para el Tratamiento de Aguas Residuales, Pereira, July 2005 2005. Universidad Tecnológica de Pereira (UTP)

Cardona J, Saraiva J, Boettger S, Prata Filho D (2014) Contribution to the sustainable managment of water resources through decentralized wastewater treatment and reuse solutions. Kick off the workshop in March, 19–21 at Terésopolis, Brazil. In: Kick off conference of the project: integrated eco technologies and services for a Rural Rio de Janeiro, Teresópilos

Cardona J, Segovia O, Böttger S et al (2016) Reuse-oriented decentralized wastewater and sewage sludge treatment for rural settlements in Brazil: a cost-benefit analysis. In: Paper presented at the 13th IWA specialized conference on small water and wastewater systems &. 5th IWA specialized conference on resources-oriented sanitation, Athens, 14–16 September 2016

CEIVAP (2014) Deliberação CEIVAP N° 218/2014. "Estabelece mecanismos e propõe valores para a cobrança pelo uso de recursos hídricos na bacia hidrográfica do rio Paraíba do Sul, a partir de 2015". vol Deliberação CEIVAP N° 218/2014. Committee for the Integration of the Hydrographic Basin of Paraíba do Sul River, CEIVAP, Resende

Chen R, Wang XC (2009) Cost-benefit evaluation of a decentralized water system for wastewater reuse and environmental protection. Water Sci Technol 59(8):1515–1522. https://doi.org/10.2166/Wst.2009.156

Cole S (1998) The emergence of treatment wetlands. Environ Sci Technol 32(9):218A–223A

CONAMA (2005) National Environmental Council (2005). Resolução no 357/2005

CONAMA (2006) National Environmental Council (2006). Resolução no 375/2006

CONAMA (2007) National Environmental Council (2007). Resolução no DZ 215.R4/2007

Costa C, Guilhoto J (2012) Importância de uma política de saneamento rural no Brasil

Crites R, Tchobanoglous G (1998) Small and decentralized wastewater management systems. McGraw-Hill series in water rescources and environmental engineering. WCB McGraw-Hill, Boston

de Sousa JT, van Haandel AC, Guimaraes AAV (2001) Post-treatment of anaerobic effluents in constructed wetland systems. Water Sci Technol 44(4):213–219

El-Khateeb M, El-Gohary F (2003) Combining UASB technology and constructed wetland for domestic wastewater reclamation and reuse. Water Recycl Mediterr Reg 3(4):201–208
Gallotti R (2008) Tratamento Decentralizado de efluentes como alternativa a despoluição dos recursos hídricos da region metropolitana de Aracaju/SE Universidad Federal de Sergipe
Halalsheh M, Dalahmeh S, Sayed M et al (2008) Grey water characteristics and treatment options for rural areas in Jordan. Bioresour Technol 99(14):6635–6641. https://doi.org/10.1016/j.biortech.2007.12.029
Hophmayer-Tokich S (2010) Wastewater management strategy: centralized v. Decentralized technologies for small communities. University of Twente at the Cartesius Institute, Leewarden
IBGE – Instituto Brasileiro de Geografia e Estatística (2000) Population Census 2000. Base de informações municipais. 2 ed. (Brazilian Institute of Geography and Statistics. Municipal information base). Rio de Janeiro
IBGE – Instituto Brasileiro de Geografia e Estatística (2007) Contagem Populacional de 2007. Brazilian Institute of Geography and Statistics. Population Census 2007. Rio de Janeiro, Brazil
IBGE – Instituto Brasileiro de Geografia e Estatística (2010a) Population Census 2010. Instituto Brasileiro de Geografia e Estatística, IBGE, Pesquisa Nacional de Saneamento Básico 2008 (Brazilian Institute of Geography and Statistics. National Research of Basic Sanitation 2008). Rio de Janeiro
IBGE – Instituto Brasileiro de Geografia e Estatística (2010b) Pesquisa Nacional de Saneamento Básico 2008. National Research of Basic Sanitation 2008. IBGE, Rio de Janeiro
Lens P, Lens PNL, Zeeman G, Lettinga G (2001) Decentralised sanitation and reuse: concepts, systems and implementation. IWA Publishing, London
Lienhoop N, Al-Karablieh EK, Salman AZ, Cardona JA (2014) Environmental cost–benefit analysis of decentralised wastewater treatment and re-use: a case study of rural Jordan. Water Policy 16(2):323. https://doi.org/10.2166/wp.2013.026
Mara D (2013) Domestic wastewater treatment in developing countries. Routledge, London
Marcon MKF, Frigo EP, Nogueira CEC et al (2015) Economic viability of the agricultural recycling of sewage sludge in Brazil. Afr J Agric Res 10(20):2159–2164
Massoud MA, Tarhini A, Nasr JA (2009) Decentralized approaches to wastewater treatment and management: applicability in developing countries. J Environ Manag 90(1):652–659
Ministerio da Cidades (2010) Diretrizes para Elaboraçãode Projetos de Engenharia. Fonte de Recursos: Financiamento. Ministério das Cidades, Secretaria Nacional de Saneamento Ambiental, Brasilia
Otis R, Mara D (1995) The design of small bore sewer systems, TAG technical note, vol 14, Washington DC: World Bank
Otterpohl R, Grottker M, Lange J (1997) Sustainable water and waste management in urban areas. Water Sci Technol 35(9):121–133. https://doi.org/10.1016/S0273-1223(97)00190-X
Pearce D, Atkinson G, Mourato S (2006) Cost-benefit analysis and the environment: recent developments. Organisation for Economic Co-operation and Development, Paris
Quintana NRG, de Carvalho Bueno O, de Melo WJ (2012) Custo de transporte do esgoto para viabilidade no uso agricola de transporte de lodo de esgoto para viabilidad no uso agrícola energia na agricultura 27(3):90–96
Rigo MM, Ramos RR, Cerqueira AA et al (2014) Destinação e reuso na agricultura do lodo de esgoto derivado do tratamento de águas residuária domésticas no Brasil. Gaia Scientia 8(1):174–186
Sattler D, Murray LT, Kirchner A, Lindner A (2014) Influence of soil and topography on aboveground biomass accumulation and carbon stocks of afforested pastures in South East Brazil. Ecol Eng 73:126–131
Segovia O (2014) Environmental costs-benefit analysis of decentralized wastewater and sanitation technologies in the microbasin of Barracão dos Mendes, Brazil. Universidad Autónoma de San Luis de Potosí, Cologne University
Uggetti E, Ferrer I, Llorens E, Garcia J (2010) Sludge treatment wetlands: a review on the state of the art. Bioresour Technol 101(9):2905–2912. https://doi.org/10.1016/j.biortech.2009.11.102

Uggetti E, Ferrer I, Molist J, García J (2011) Technical, economic and environmental assessment of sludge treatment wetlands. Water Res 45(2):573–582. https://doi.org/10.1016/j.watres.2010.09.019

UN-Habitat (2006) Meeting development goals in small urban centres: water and sanitation in the wold's cities 2006. United Nations Human Settlements Programme (UN-HABITAT), Nairobi

van Afferden M, Cardona JA, Lee MY et al (2015) A new approach to implementing decentralized wastewater treatment concepts. Water Sci Technol 72(11):1923–1930. https://doi.org/10.2166/wst.2015.393

Wendland C, Behrendt J, Elmitwalli TA et al (2006) UASB reactor followed by constructed wetland and UV radiation as an appropriate technology for municipal wastewater treatment in Mediterranean countries. In: Proceedings of the 7th specialised conference on small water and wastewater systems in Mexico, 2006

Wilderer P, Shereff D (2000) Decentralised and centralised wastewater management: a challenge for technology developers. Water Sci Technol 41(1):1–8

Chapter 19
A Tool to Assess Land Use Impacts on Surface Water Quality: Case Study from the Guapi-Macacu River Basin in Rio de Janeiro

Santiago Penedo-Julien, Annika Künne, Rachel Bardy Prado, and Lars Ribbe

Abstract Impacts of human activities, mainly land use and land cover (LULC) changes, on hydrology and water quality are manifold and need to be ascertained. In order to understand and assess such impacts, water quality monitoring programs are crucial for collecting the required data. Transforming data into information is an important step, and water quality indices (WQIs) can be a useful and concise method for aggregating several parameters and measurements into a single number to facilitate problem identification and decision-making. For this study, a monitoring program was designed and implemented to measure water quality in a rural river basin in Rio de Janeiro (Guapi-Macacu). Three sub-watersheds, together with other relevant sampling points, were selected to assess the influence of the most relevant LULC classes on water quality. The monitored parameters were used to calculate the Canadian water quality index (CCME-WQI). This index was able to capture the impacts of water quality impairments such as untreated sewage and agricultural activities. The index calculation and resultant map, used to depict the spatial distribution, aim at becoming tools for practitioners and decision-makers in the basin.

Keywords Water quality index (WQI) · Land use and land cover (LULC) · Monitoring · Brazil

---

S. Penedo-Julien (✉) · L. Ribbe
Institute for Technology and Resources Management in the Tropics and Subtropics (ITT), TH Köln – University of Applied Sciences, Köln, Germany
e-mail: santiago.penedo@th-koeln.de; lars.ribbe@th-koeln.de

A. Künne
Geographic Information Science Group, Institute of Geography, Friedrich Schiller University of Jena, Jena, Germany
e-mail: annika.kuenne@uni-jena.de

R. Bardy Prado
Brazilian Enterprise for Agricultural Research – Embrapa Solos, Rio de Janeiro, Brazil
e-mail: rachel.prado@embrapa.br

U. Nehren et al. (eds.), *Strategies and Tools for a Sustainable Rural Rio de Janeiro*, Springer Series on Environmental Management,
https://doi.org/10.1007/978-3-319-89644-1_19

19.1 Introduction

Economic development and population growth have led to changes in land use and land cover (LULC) worldwide. These changes have been identified as the main drivers affecting biodiversity and water quality (Horn et al. 2004). However, their relative impacts on water quality have yet to be quantitatively ascertained (Tong and Chen 2002).

Agriculture and urban areas are two of the most relevant LULC affecting water quality and aquatic ecosystems (McCoy et al. 2015). For instance, crop production exports chemical compounds such as nutrients and pesticides, thus increasing their concentrations in both surface and groundwater. The main negative impacts of high nutrient concentrations in surface water are well-known and include algal blooms, eutrophication, and hypoxia (Nelson 2005; Ouyang et al. 2009). Furthermore, urban development is creating tremendous stress on water resources at local, regional, and global scales (Duh et al. 2008). In many developing countries, wastewater is not properly treated, or simply not treated at all, and can be contaminated with a myriad of different components such as pathogens, nutrients, organic matter, and heavy metals, among others (Corcoran et al. 2010).

Consequently, a large body of research has been devoted to understanding the interlinkages between LULC and surface water quality at different spatiotemporal scales (Johnes and Heathwaite 1997; Coats et al. 2008; Walton and Hunter 2009; Tu 2013; Wilson 2015). However, fundamentally understanding these interlinkages remains a challenge since water quality in a watershed is highly variable both in space and time because the driving processes (e.g., precipitation, pollution sources, discharge, LULC) also vary spatially and temporally (Walton and Hunter 2009). Thus, in the last decades sound water quality monitoring concepts at the watershed level have gained importance (Telci et al. 2009). The resultant monitoring networks are paramount for collecting relevant and representative data to understand causes of and impacts on water quality and to design ad hoc management strategies to improve it.

In order to avoid the so-called "data-rich but information-poor" syndrome (Ward et al. 1986), large data sets that are generated by monitoring programs need to be summarized, analyzed, and visualized to simplify their interpretation. Environmental indices are widely used in environmental policymaking to describe the state of the environment. Such indices aim at aggregating different environmental parameters into one number, which can offer a general diagnosis of the ecosystem being investigated (Esty et al. 2006).

In the context of water quality, indices are single numbers used to integrate the measurements of diverse water quality parameters. They provide a simple and concise method to facilitate problem identification and decision-making (Lumb et al. 2011).

For this study, a water quality monitoring network was designed and implemented in a tropical river basin in the Brazilian Mata Atlântica (Atlantic Forest). The monitored data was used to calculate a WQI to assess the impacts of different LULC classes on surface water quality.

19.2 Research Area

The Guapi-Macacu River Basin (GMRB) is located within the state of Rio de Janeiro (RJ), northeast from the Guanabara Bay and the city of Rio de Janeiro (Fig. 19.1). The basin has a drainage area of 1,265 km^2 and is situated between 42°32′–43°3′W and 22°21′–22°41′S. The elevation varies from the coastal plain at the bay up to the mountain range Serra dos Órgãos at 2,250 m.a.s.l. (Fidalgo et al. 2008). The Serra dos Órgãos is covered with tropical forest and hosts a great variety of biological species making it a biodiversity hotspot (Webb et al. 2005) as described by Myers et al. (2000). The major rivers, the Macacu and Guapiaçu, drain the basin on a NE-SW axis and originally had separate outlets into the bay. However, in the 1950s, the main rivers were rectified and channelized to avoid flooding and to reduce the habitat for mosquitoes and, therefore, water-related diseases such as malaria and yellow fever (Benavides et al. 2009).

According to the Köppen classification, climate in the basin belongs to the Aw group: tropical humid with dry winters (Pinheiro et al. 2012) Climate is strongly influenced by orography and by Atlantic tropical air masses. The most important climatic system influencing precipitation patterns and temperatures is the South Atlantic convergence zone (Barreiro et al. 2002). The coastal area in the southwest of the watershed shows higher temperatures and lower precipitation rates compared to the mountain range in the northeast (UFF/FEC 2010). The maximum temperatures are reached in February and the lowest in July. Moreover, according to the

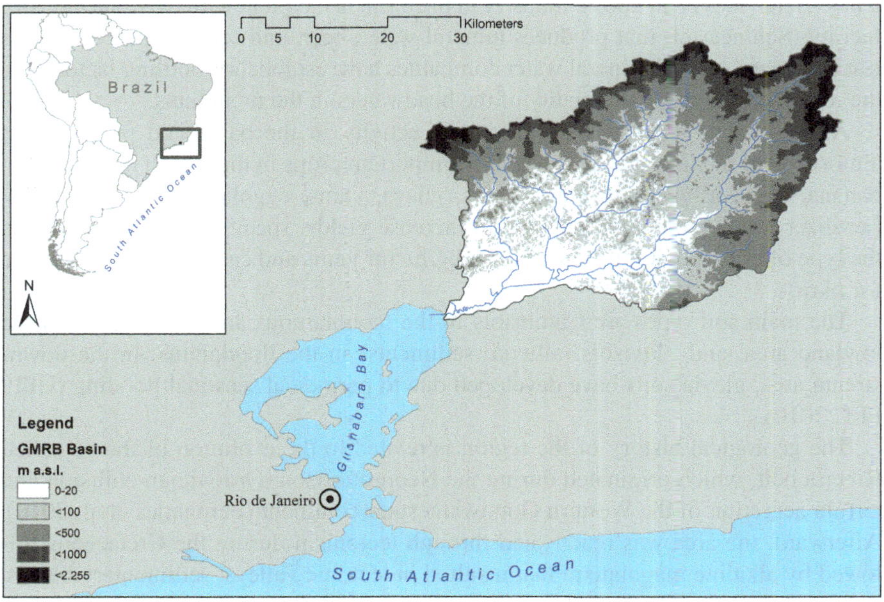

Fig. 19.1 Location and elevation of the GMRB (own figure). (Source data Fidalgo et al. (2008))

available precipitation data, the mean annual precipitation (1996–2010) within the watershed varies between 1,200 mm and 2,800 mm (ANA 2012). The yearly distribution presents an interannual variability with a maximum in the summer months from November to March and a minimum in the winter season from June to August. According to the available streamflow data (1996–2013), January is the wettest month with an average streamflow of 14.0 $m^3 \cdot s^{-1}$, while August is the driest month with 4.2 $m^3 \cdot s^{-1}$.

Although not densely populated (~90 inhabitants•km^{-2}), the basin plays an important role in the eastern metropolitan region of Rio de Janeiro (Niterói, São Gonçalo, Paquetá, parts of Itaboraí, and other municipalities) since it supplies around 2.5 million people with drinking water (Costa Dantas et al. 2007; UFF/FEC 2010). The pumping station is located downstream after the two main rivers – Macacu and Guapiaçu – converge. Due to its importance as a drinking water source, river quality is a major issue of concern for society in general but also for the research community and decision-makers.

The three main urban settlements in the basin are Cachoeiras de Macacu, the capital city of the municipality, Guapimirim, and Papucaia with populations fluctuating between 20,000 and 30,000 inhabitants. Over 80% of the population live in urban areas, while the rest live spread in rural settlements across the basin. Both urban and rural settlements generate wastewater that is a pressure to the fluvial system since treatment is very low. 61% of the total wastewater do not receive treatment whatsoever (47% of the households are connected to the wastewater collection system without treatment, 7% discharge it into open channels and another 7% directly into rivers); 17% treat their wastewater using septic tanks and 20% with rudimentary tanks (Benavides et al. 2009). Industry in the basin is incipient with only one industry facility (Schincariol) that produces mineral water, beer, and sodas. However, in the last few years, several mineral water companies have established bottling factories in the region due to the good quality of the headwaters in the mountains.

Agriculture is an important economic activity in the basin and practiced on almost 5% of the total surface. The most important crops in the region include corn, banana, cassava, coconut, lemon trees, guava, yams, eggplant, okra, and beans. Fertilizer use is a common practice to increase yields; specific dosing depends on the type of crop, but it ranges from 250 kg/ha for yams and cassava up to 900 kg/ha for okra.

The main soil types are Cambisols in the mountainous areas, Ferralsols in the lowland area, and Fluvisols (alluvial sediments) in the floodplains. In the downstream area, gleyic soils have developed due to periodical seasonal flooding (UFF/FEC 2010).

The geological history of the region is related to the evolution of the so-called Ribeira belt, which originated during the Neoproterozoic-Ordovician collision and terrain accretion of the Western Gondwana supercontinent (Fernandes et al. 2010). Afterward, the area was reactivated through tectonism during the Cretaceous followed by alkaline magmatism that resulted in tectonic valleys, sedimentary basins, and mountains known as the Continental Rift of Southeastern Brazil (Ferrari 2001; Riccomini et al. 2004; Fernandes et al. 2010). The main geological layers found in the basin are crystalline gneiss in the mountainous part, alluvial deposits in the floodplains, and deposits from the Macacu Formation in the southern part.

19.3 Methodology

For this study, a monitoring network was designed to gain understanding about the processes that affect water quality in the basin and how it is spatially distributed. The GMRB is spatially heterogeneous with several LULC classes and a high level of fragmentation. The most important anthropogenic pressures on water quality include agriculture (e.g., nutrients, manure, and erosion) and untreated sewage that is generated in urban and rural settlements. Due to their diffuse nature, these pollution sources are difficult to delimit and quantify.

First, the most relevant LULC in the basin were identified and related to their potential impacts on water quality. Land use data for the region was available from a study conducted by the Brazilian research institute Embrapa Solos (Pedreira et al. 2007; Fidalgo et al. 2008). The resultant map (Fig. 19.2) depicts the spatial distribution of LULC in the basin with forest (in its different succession stages) covering the mountainous areas in the north and northeast, while pasture and arable land can be found in the floodplains and flatter areas located in the center and south.

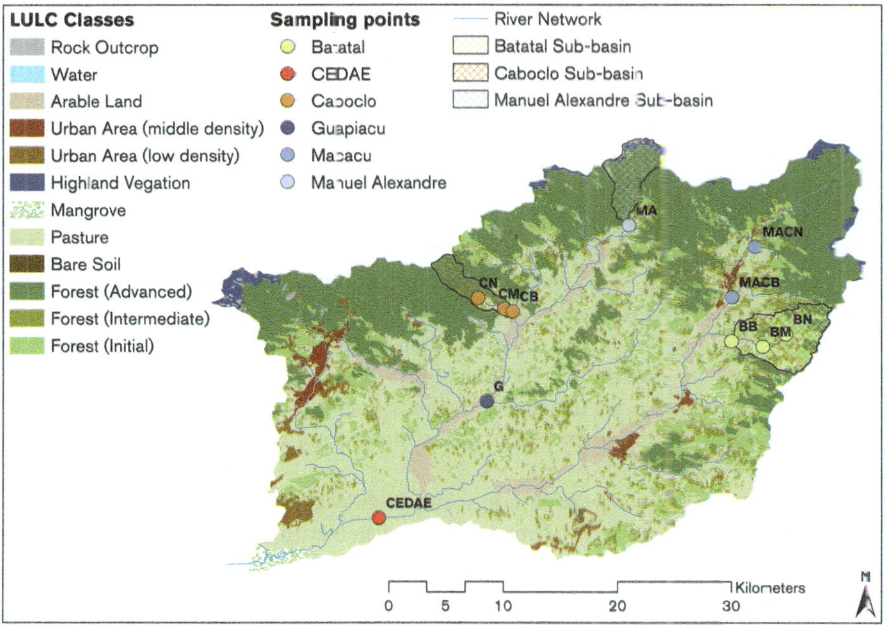

Fig. 19.2 LULC map showing the location of the three monitored sub-basins and water quality sampling points (own figure). (Source data: Fidalgo et al. (2008), Pedreira et al. (2007))

Considering the main pressures affecting water resources, a monitoring concept was developed to try to capture the influence of the different LULC classes on surface water quality. Consequently, three sub-basins were selected according to the most expressive LULC classes: Manuel Alexandre (MA), Batatal, and Caboclo. The rationale behind this concept was to monitor small sub-basins to try to isolate the impact of different LULC classes on water quality.

The MA river was selected as "background area" (Arreghini et al. 2007) due to the predominance of forest and the low impact of human activities. Although the Caboclo sub-basin also has forest as its dominant use, according to the information provided by the director of the extension services in the region (EMATER), the agriculture practiced is more intensive, and the percentage of cultivated area is also higher than the average for the GMRB. Moreover, a swine farm (~2,000 pigs) is located downstream, which contributes to impairing water quality through sewage and untreated sludge. Lastly, the Batatal sub-basin presents a mixture of the most relevant LULC with forest and banana as agroforestry systems in the high parts followed by pasture and horticulture in the lower parts. The size of the sub-basins fluctuates from 12.5 km^2 (Caboclo) to 32 km^2 (Batatal). Furthermore, the sub-basins were also selected based on information collected during field trips conducted in 2010 as well as expert information provided by agricultural advisory services and local planners. Logistical constraints such as "ease of access" were also taken into consideration.

Although the sub-basins are not small enough to be homogeneous in relationship to soil type, land use, climate, etc., their size allows an increased understanding of the relationship between LULC and water quality. Each sub-basin was sampled up-, mid-, and downstream to assess the water quality evolution in relation to the river course and the changes in LULC pattern distribution. However, in MA, due to the difficult accessibility and more homogeneous land use (forest), only one sampling site at the exit of the watershed was chosen.

In addition, to estimate the influence of untreated wastewater on water quality, two sampling points were located up- and downstream from Cachoeiras de Macacu, the biggest urban settlement in the basin. A site was selected at the Guapiaçu River shortly after an intensively cultivated agricultural area and where an existing discharge measuring station is located. Last, a sampling site was chosen where the Drinking Water and Wastewater State Company (CEDAE) constructed a small weir to divert water to the pumping station that provides drinking water. This point was selected due to its importance for the region but also as an integration site that aggregates all human activities and influences that occur upstream. Table 19.1 and Fig. 19.2 present the sampling points of the monitoring network together with their names and codes.

The sampling frequency of a monitoring network should be based on the variability of the parameters to be measured that needs to be assessed through surveys. Unfortunately, ideal sampling frequency is often limited by financial and logistical constraints. In this case, the available human and economic resources limited the sampling to a bimonthly frequency from August 2010 to February 2012. However, logistical problems, sensor malfunctions and measuring errors led to a decrease in the number of monitored data. In some cases, such as total

Table 19.1 Water sampling sites including codes and names

Site code	Name	Description
MA	Manuel Alexandre	Background area in headwaters
MACN	Macacu Nascente (upstream)	Upstream urban settlement
MACB	Macacu Baixo (downstream)	Downstream urban settlement
CN	Caboclo Nascente (upstream)	Headwater sub-basin
CM	Caboclo Meio (midstream)	Middle "intensive" agriculture
CB	Caboclo Baixo (downstream)	Downstream swine farm
BN	Batatal Nascente (upstream)	Headwater sub-basin
BM	Batatal Meio (midstream)	Next to rural settlement midstream
BB	Batatal Baixo (downstream)	Next to horse riding school downstream
G	Guapiaçu	Downstream agricultural area
CEDAE	CEDAE	At weir for drinking water station

nitrogen, the sampling population was reduced to a total of five values. Nevertheless, the monitored water quality data was statistically analyzed and used to compute the CCME-WQI for each sampling point. Although a bimonthly sampling is not sufficient to properly represent the variability of the basin including seasonal behavior, it does provide the required data to calculate the water quality index. The increased spatial coverage of the monitoring network aimed at compensating the sparse temporal series.

The parameters monitored were selected based on the main LULC classes that affect water quality. Salinization caused by agricultural activities or urban runoff can be estimated through electrical conductivity (EC). Moreover, the impact of fertilizer use and discharge of untreated sewage can be assessed by measuring in-stream nutrient (both nitrogen and phosphorus) concentrations. Higher nutrient concentrations are an important cause of algal blooms, eutrophication, and hypoxia (Nelson 2005; Ouyang et al. 2009). These impacts can be identified by a decrease in dissolved oxygen (DO) concentration, which negatively affects aquatic ecosystems. Furthermore, DO concentration determines the type of predominant nitrogen species in the water, which determines their toxicity. Finally, pH greatly influences acid/base reactions and oxidation states and is a good indicator to assess acidification processes. Consequently, the monitored parameters were:

- Electrical conductivity (EC, 1 μS/cm), pH (0.01 pH), dissolved oxygen (DO, 0.01 mg/L), and temperature (0.01 °C) using a multiparameter probe in situ
- Total nitrogen (TN, 0.1 mg/L) and total phosphorous (TP, 0.05 mg/L) using a spectrophotometer in the laboratory

As mentioned earlier, water quality indices (WQIs) are a useful tool to integrate river quality information into a single number to facilitate decision-making. There are numerous WQIs available that focus on different parameters depending on the characteristics of the specific water body. Based on a literature review and the available data, the WQI from the Canadian Council of Ministers of the

Environment (CCME) was selected for this study. One of the main advantages of this index is its versatility since the variables, objectives, and time period used in the index are not specified and since these may vary from region to region, depending on local conditions and information available (CCME 2001). The CCME-WQI has been endorsed by the United Nations Environment Programme (UNEP) as a suitable model for grading the quality of drinking water bodies worldwide (Lumb et al. 2012). By applying the World Health Organization (WHO) guidelines as targets for the calculations, the index reflects the quality of water for both health and acceptability (UNEP/GEMS 2007). Furthermore, the index can measure ecosystem health, and it addresses issues which are globally – and also in the Brazilian context – relevant such as (i) eutrophication, (ii) nutrient pollution, (iii) acidification, and (iv) salinization (Carr and Rickwood 2008). Therefore, the parameters considered by the index to address the abovementioned water quality issues are pH, total nitrogen, total phosphorous, EC, and DO. The index calculation is based on the comparison between proximity-to-target (PTT) values (see Table 19.2 for the specific targets for each water quality parameter) and measured data.

The CCME index has been applied in different regions and water bodies. For instance, Venkatramanan et al. (2016) used this index to assess groundwater quality in India; Rachedi and Amarchi (2015) and Haldar et al. (2016) applied it to assess surface water quality in Algeria and India, respectively. In Brazil, this index has been used to monitor and manage shrimp farms (Ferreira et al. 2011) but has not yet been applied to assess the impacts of different types of LULC on water quality.

CCME-WQI values vary from 0 to 100, with 0 being the poorest and 100 the best water quality (see Table 19.3 for qualitative designations). CCME (2001) recommends that water quality indices that are based on this algorithm be calculated with at least four measurements per year for at least four parameters, known as "four by four" rule. This condition is met in the GMRB monitoring network implemented for this study.

The calculation of this index is based on three different functions called F1, F2, and F3 that are calculated as follows. F_1 – or scope – represents the percentage of variables not meeting the objectives at least once during the period under consideration:

$$F_1 = \left(\frac{\text{Number of failed variables}}{\text{Total number of variables}} \right) \cdot 100 \tag{19.1}$$

F_2 – or frequency – represents the percentage of individual tests not meeting the objectives:

$$F_2 = \left(\frac{\text{Number of failed tests}}{\text{Total number of tests}} \right) \cdot 100 \tag{19.2}$$

F_3 – or amplitude – representing the amount by which the failed tests do not meet the objectives:

Table 19.2 Target values for each water quality parameter

Parameter	Target
EC	500 μS·cm^{-1}
DO	6 mg·L^{-1}, for temperature > 20 °C 9.5 mg·L^{-1}, for temperature ≤ 20 °C
pH	6.5–8.5
TN	1 mg·L^{-1}
TP	0.05 mg·L^{-1}

Source: UNEP/GEMS (2007)

Table 19.3 Water quality index designations

Designation	Index value	Description
Excellent	95–100	All measurements are within objectives virtually all of the time
Good	80–94	Conditions rarely depart from the desirable levels
Fair	65–79	Conditions sometimes depart form the desirable levels
Marginal	45–64	Conditions often depart from the desirable levels
Poor	0–44	Conditions usually depart from the desirable levels

Source: UNEP/GEMS (2007)

$$F_3 = \left(\frac{nse}{0.01nse + 0.01} \right) \quad (19.3)$$

where

$$nse = \frac{\sum_{i=1}^{n} \text{excursion}_i}{\#\,\text{of tests}} \quad (19.4)$$

and

$$\text{excursion}_i = \left(\frac{\text{Objective}_j}{\text{Failed Test Value}_i} \right) - 1 \text{ or } \left(\frac{\text{Failed Test Value}_i}{\text{Objective}_j} \right) - 1 \quad (19.5)$$

Finally, the total index is calculated by adding up the three factors that are treated as if they were vectors. The index represents a three-dimensional space defined by each factor along one axis. Therefore, the index varies in direct proportion to changes in all three factors (CCME 2001):

$$WQI = 100 - \left(\frac{\sqrt{F_1^2 + F_2^2 + F_3^2}}{1.732} \right) \quad (19.6)$$

The divisor 1.732 normalizes the resultant value to a range between 0 and 100. Based on this formula, a WQI was calculated for each sampling point of the GMRB monitoring network.

19.4 Results and Discussion

In percentage, forest (initial, intermediary, and advanced) is the most important LULC class with 48% of the total surface of the basin and is followed by pasture with 41% (see Table 19.4). Although arable land accounts for only 4.40%, the use of fertilizers makes it a nonpoint source pollution that increases the in-stream nutrient concentration. If compared to natural LULC classes such as forests, arable lands also have higher erosion coefficients that contribute sediments to rivers (Alam et al. 2013; McCoy et al. 2015). Finally, urban areas represent only 2.6% of the total surface, but since sewage is discharged untreated into the main rivers, it represents an important pressure into the system. The following table depicts the percentages of the LULC classes for the whole basin and for the three selected sub-basins.

Table 19.5 shows the values of the individual functions and the total WQI for each sampling point together with the qualitative designation to facilitate interpretation. Furthermore, the WQI values were used to prepare a map (Fig. 19.3) to show the spatial distribution of the results and to allow for a quick interpretation of the water quality in the basin.

Logically, the highest scores can be found at the headwaters of the respective sub-basins (BN, CN, and MA), where forest is the dominant use upstream. The influence of untreated sewage can be easily visualized through the deterioration of the WQI from 76.4 (at MACN) to 57.4 (at MACB), up- and downstream from an urban settlement, respectively. The impact of intensive agricultural practices and untreated sewage from the swine farm in Caboclo can be observed in an abrupt decrease (CB) of the WQI after a relative constant quality in CN and CM. Other sampling sites with an influence of mixed land uses such as MACN, BM, and BB have fair values due to the lower intensity of human activities practiced upstream. Finally, as expected both downstream points (GUAPI and CEDAE), which are impaired by the activities occurring upstream, have noticeably low marginal values, especially CEDAE. The lowest WQI score can be found at CEDAE (the most downstream point of the monitoring network) because it aggregates all the impairments affecting water quality within the basin including agriculture, rural, and urban settlements, as well as animal husbandry.

Nutrients – both nitrogen and phosphorus – were the most sensitive parameters affecting the WQI calculation. These parameters are mainly responsible for the marginal values obtained for CB, CEDAE, GUAPI, and MACB. In fact, the maximum values for nitrogen (3.98 mg/L) and phosphorus (0.98 mg/L) were measured in CEDAE and MACB, respectively, showing how untreated sewage is the main source of phosphorus in the basin. Moreover, pH and DO values fell out of range only in isolated cases. In general, oxygen concentrations were relatively high to moderate

Table 19.4 Percentages of most relevant LULC classes for the GMRB and the three monitored sub-basins

Basin	Surface [km^2]	LULC classes [%]					
		Pasture	Arable land	Forest (Ini)	Forest (Int)	Forest (Adv)	Others
Batatal	31.8	23.7	3.1	23.9	28.8	20.5	–
Caboclo	12.5	7.5	8.9	1.2	6.2	76.2	0.1
MA	18.5	0.5	–	2.9	2.3	83.8	10.5
GMRB	1265	41.4	4.4	8.0	11.9	28.9	5.4

Table 19.5 WQI functions, results, and status for each sampling point

Sampling point	F1	F2	F3	WQI	Status
MA	5.12	20	6.11	88.0	Good
MACN	7.69	40	4.05	76.4	Fair
MACB	20.51	60	37.57	57.4	Marginal
CN	2.56	20	1.36	88.3	Good
CM	2.56	20	0.25	88.4	Good
CB	15.38	50	6.21	64.1	Marginal
BN	2.56	20	1.76	88.3	Good
BM	5.12	40	0.87	75.7	Fair
BB	7.69	40	4.59	75.3	Fair
GUAPI	17.94	60	7.10	63.6	Marginal
CEDAE	41.02	80	19.52	46.9	Marginal

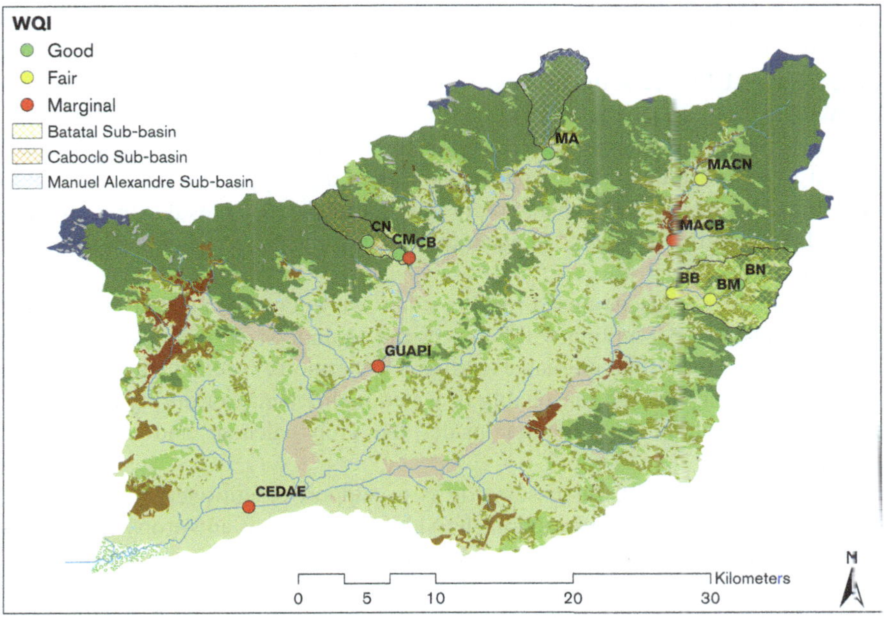

Fig. 19.3 LULC map showing WQI results for each sampling point. (Source data: Fidalgo et al. (2008), Pedreira et al. (2007))

(no hypoxic conditions) and pH values slightly alkaline with values ranging from 7.1 to 8.0 in the headwaters while slightly acidic (6.3–6.9) in the sampling points downstream. EC never exceeded the target value.

While high nutrient concentrations do not dramatically increase water drinking treatment costs, they can have serious environmental impacts such as eutrophication. Since these high concentrations are a consequence of fertilizer and manure use, as well as untreated sewage, ad hoc land management strategies have to be developed for this basin in order to tackle the problem.

19.5 Conclusions and Outlook

Monitoring is a paramount activity to assess the state of the environment and generate the necessary information to make good decisions. However, monitoring water quality is a very challenging task, especially in developing countries, due to (i) spatial and temporal heterogeneity, (ii) limited financial resources, (iii) time requirements, and (iv) lack of technical know-how.

Therefore, the design of monitoring networks has to be based on clear management objectives to representatively measure the most appropriate parameters while optimizing available resources and minimizing costs. A deep understanding of the system and the pressures influencing water quality – such as LULC – is required for this kind of tailor-made design.

The monitoring network implemented in this rural watershed in Brazil is an example of a useful approach to collect relevant information to understand the influences of LULC on surface water quality. Although it was impossible to completely isolate each LULC class, the monitoring concept was able to capture the impact of the main anthropogenic uses.

The resultant monitoring data was summarized using the CCME-WQI which proved to be a useful tool to simplify monitored data into one single number to facilitate interpretation. The spatial representation of the index results through maps can help to identify "hotspots" of water quality impacts, but a closer look is needed to identify the root causes.

The sampling frequency used for this study was enough to calculate the index. However, increased sampling will be necessary to understand the specific impacts of fertilizers, manure, and untreated sewage. Furthermore, special campaigns could be designed to capture the seasonality of water quality behavior during the dry and wet seasons and to be able to consider the crop calendar and demographic development.

The main purpose of this case study is to support practitioners and decision-makers with a sound scientific approach to identify problems and propose solutions to ensure better water management, especially since this basin is so relevant for the drinking water supply of the region. For instance, these results are relevant for regional stakeholders such as the Drinking Water and Wastewater State Company (CEDAE), the State Environmental Institute (INEA), or the extension services

(EMATER). Furthermore, the measured data can be useful to set a baseline and be used for later comparisons if measures are undertaken to improve the water quality in the basin.

Future research and monitoring should include parameters such as agrochemicals and pathogens (e.g., pesticides, fungicides) since they are a by-product of agricultural production. If urbanization and population growth continues, then pharmaceutical products (e.g., antibiotics, hormones) should also be included in the analysis since their impact is still uncertain. None of these were considered in this study due to limited equipment and lab availability.

Finally, in order to isolate and quantify the impacts of different LULC classes on water quality, other approaches such as modeling and remote sensing can offer new directions to simulate and validate the hydro-chemical processes that occur in a basin. Since LULC changes are still a major driver of global environmental degradation, research in this field of study is very much needed.

References

Alam RQ, Dufreche S, Hayatdavoudi A, Gang DD (2013) Nonpoint source pollution. Water Environ Res 85:1715–1733. https://doi.org/10.2175/106143013X13698672322822

ANA – Agência Nacional de Águas (2012) Sistema Nacional de Informações sobre Recursos Hídricos (SNIRH). http://www.snirh.gov.br/portal/snirh. Accessed 01 May 2017

Arreghini S, de Cabo L, Seoane R (2007) A methodological approach to water quality assessment in an ungauged basin, Buenos Aires, Argentina. GeoJournal 70:281–288. https://doi.org/10.1007/s10708-008-9134-z

Barreiro M, Chang P, Saravanan R (2002) Variability of the South Atlantic convergence zone simulated by an atmospheric general circulation model. J Clim 15:745–763. https://doi.org/10.1175/1520-0442(2002)015<0745:VOTSAC>2.0.CO;2

Benavides Z, Pezza Cintrão R, Fidalgo EC et al (2009) Consumo e abastecimento de água nas bacias hidrográficas dos rios Guapi-Macacu e Caceribu – RJ. Embrapa Solos Report. Embrapa Solos - the Brazilian research institute, Documentos 115, Rio de Janeiro

Carr GM, Rickwood CJ (2008) Water quality index for biodiversity technical development document. Prepared for Biodiversity Indicators Partnership, World Conservation Monitoring Center, Cambridge, UK, p. 64

CCME (2001) Canadian water quality guidelines for the protection of aquatic life: CCME Water Quality Index 1.0, User's Manual. Winnipeg

Coats R, Larsen M, Heyvaert A et al (2008) Nutrient and sediment production, watershed characteristics, and land use in the Tahoe basin, California-Nevada. J Am Water Resour Assoc 44:754–770. https://doi.org/10.1111/j.1752-1688.2008.00203.x

Corcoran E, Nellemann C, Baker E et al (2010) Sick water? The central role of wastewater management in sustainable development. A Rapid Response Assessment. United Nations Environment Programme (UNEP), UN-Habitat, GRID-Arendal

da Costa Dantas JR, de Almeida JR, Lins GA (2007) Impactos ambientais na bacia hidrográfica de Guapi/Macacu e suas conseqüências para o abastecimento de água nos municípios do leste da Baia de Guanabara. Série gestão e planejamento ambiental, coleção artigo técnicos no 7. CETEM/MCT, Rio de Janeiro

Duh JD, Shandas V, Chang H, George LA (2008) Rates of urbanisation and the resiliency of air and water quality. Sci Total Environ 400:238–256. https://doi.org/10.1016/j.scitotenv.2008.05.002

Esty DC, Levy MA, Srebotnjak T et al (2006) Pilot 2006 environmental performance index. Yale Center for Environmental Law & Policy, New Haven

Fernandes NF, Tupinambá M, Mello CL, de Peixoto MNO (2010) Rio de Janeiro: a metropolis between granite-gneiss massifs. In: Migon P (ed) Geomorphological landscapes of the world. Springer, p 387

Ferrari AL (2001) Evolução tectônica do graben da Guanabara. Doctoral Dissertation, Biblioteca Digital de Teses e Dissertações da USP

Ferreira NC, Bonetti C, Seiffert WQ (2011) Hydrological and water quality indices as management tools in marine shrimp culture. Aquaculture 318:425–433. https://doi.org/10.1016/j.aquaculture.2011.05.045

Fidalgo EC, da BCC Gomes Pedreira, Bueno de Abreu M, et al (2008) Uso e cobertura da terra na bacia hidrográfica do rio Guapi-Macacu. Embrapa Solos Report, Documentos 105, Rio de Janeiro

Haldar D, Halder S, Das (Saha) P, Halder G (2016) Assessment of water quality of Damodar River in south Bengal region of India by Canadian Council of Ministers of Environment (CCME) Water Quality Index: a case study. Desalin Water Treat 57:3489–3502. https://doi.org/10.1080/19443994.2014.987168

Horn AL, Rueda FJ, Hörmann G, Fohrer N (2004) Implementing river water quality modelling issues in mesoscale watershed models for water policy demands—an overview on current concepts, deficits, and future tasks. Phys Chem Earth Parts A/B/C 29:725–737. https://doi.org/10.1016/j.pce.2004.05.001

Johnes PJ, Heathwaite AL (1997) Modelling the impact of land use change on water quality in agricultural catchments. Hydrol Process 11:269–286

Lumb A, Sharma TC, Bibeault J-F (2011) A review of genesis and evolution of Water Quality Index (WQI) and some future directions. Water Qual Expo Heal 3:11–24. https://doi.org/10.1007/s12403-011-0040-0

Lumb A, Sharma TC, Bibeault J-F, Klawunn P (2012) A comparative study of USA and Canadian Water Quality Index models. Water Qual Expo Heal 3:203–216. https://doi.org/10.1007/s12403-011-0056-5

McCoy N, Chao B, Gang DD (2015) Nonpoint source pollution. Water Environ Res 87:1576–1594. https://doi.org/10.2175/106143015X14338845156263

Myers N, Mittermeier RA, Mittermeier CG et al (2000) Biodiversity hotspots for conservation priorities. Nature 403:853–858. https://doi.org/10.1038/35002501

Nelson GC (2005) Drivers of ecosystem change: summary chapter. In Ecosystems and Human Well-Being: Volume I: Current State and Trends, R Hassan, R Scholes, and N Ash, Eds., Island Press, Washington, DC, 73–76

Ouyang W, Wang X, Hao F, Srinivasan R (2009) Temporal-spatial dynamics of vegetation variation on non-point source nutrient pollution. Ecol Model 220:2702–2713. https://doi.org/10.1016/j.ecolmodel.2009.06.039

Pedreira BCCG, Abreu MB, Fidalgo ECC (2007) Proposta de legenda para o mapeamento do uso e cobertura da terra na bacia hidrográfica do Rio Macacu, RJ. Embrapa Solos Report, Documentos 91, Rio de Janeiro

Pinheiro HSK, da Chagas CS, de Carvalho W Jr, dos Anjos LHC (2012) Modelos de elevação para obtenção de atributos topográficos utilizados em mapeamento digital de solos. Pesqui Agropecuária Bras 47:1384–1394

Rachedi LH, Amarchi H (2015) Assessment of the water quality of the Seybouse River (North-East Algeria) using the CCME WQI model. Water Sci Tech-W Sup 15(4):793–801. https://doi.org/10.2166/ws.2015.033

Riccomini C, Sant'Anna LG, Ferrari AL (2004) Evolução geológica do Rift Continental do sudeste do Brasil. In: Mantesso-Neto V, Bartorelli A, Carneiro CDR, de Brito Neves BB (eds) Geologia do Continente Sul-Americano: Evoluçao da obra de Fernando Flávio Marques de Almeida. Beca, Sao Paulo, pp 383–405

Telci IT, Nam K, Guan J, Aral MM (2009) Optimal water quality monitoring network design for river systems. J Environ Manag 90:2987–2998. https://doi.org/10.1016/j.jenvman.2009.04.011

Tong STY, Chen W (2002) Modeling the relationship between land use and surface water quality. J Environ Manag 66:377–393. https://doi.org/10.1006/jema.2002.0593

Tu J (2013) Spatial variations in the relationships between land use and water quality across an urbanization gradient in the watersheds of northern Georgia, USA. Environ Manag 51:1–17. https://doi.org/10.1007/s00267-011-9738-9

UFF/FEC (2010) Planejamento estratégico da região hidrográfica dos rios Guapi-Macacu e Caceribu-Macacu, Niterói. Published by Universidade Federal Fluminense and Fundação Euclides da Cunha

UNEP/GEMS (2007) Global Drinking Water Quality Index: development and sensitivity analysis. Prepared and published by the United Nations Environment Programme Global Environment Monitoring System (GEMS)/Water Programme

Venkatramanan S, Chung SY, Ramkumar T et al (2016) Assessment of groundwater quality using GIS and CCME WQI techniques: a case study of Thiruthuraipoondi city in Cauvery deltaic region, Tamil Nadu, India. Desalin Water Treat 57:12058–12073. https://doi.org/10.1080/19443994.2015.1048740

Walton RS, Hunter HM (2009) Isolating the water quality responses of multiple land uses from stream monitoring data through model calibration. J Hydrol 378:29–45. https://doi.org/10.1016/j.jhydrol.2009.09.004

Ward RC, Loftis JIMC, Collins F, Graham B (1986) The " data-rich but information-poor " syndrome in water quality monitoring. Environ Manag 10:291–297

Webb TJ, Woodward FI, Hannah L, Gaston KJ (2005) Forest cover-rainfall relationships in a biodiversity hotspot: the Atlantic Forest of Brazil. Ecol Appl 15:1968–1983. https://doi.org/10.1890/04-1675

Wilson CO (2015) Land use/land cover water quality nexus: quantifying anthropogenic influences on surface water quality. Environ Monit Assess 187:424. https://doi.org/10.1007/s10661-015-4666-4

Part IV
Land Restoration, Disaster Risk Reduction and Climate Change Adaptation

Chapter 20
Natural Hazards and Climate Change Impacts in the State of Rio de Janeiro: A Landscape Historical Analysis

Udo Nehren, André Kirchner, Wolfram Lange, Marco Follador, and Dieter Anhuf

Abstract Floods, landslides, and mudslides are frequent phenomena in Rio de Janeiro state (RJ). In the past decades, several catastrophic events have occurred and caused severe damages to people and infrastructure. In contrast, the persistent droughts that affected Southeast Brazil between 2014 and 2017 are phenomena that were not known earlier – at least in such frequency and intensity. Climate scenarios predict that extreme events will further increase in the future leading to increased heavy rainfall events on the one hand and longer dry spells on the other. In this chapter, we provide an overview on the different types of natural hazards, their occurrence (frequency) and intensity, and historical disasters caused by these hazards in RJ. Furthermore, we reconstruct in how far climate variability and human impact (in particular deforestation) affected the occurrence of hydrometeorological hazards in the Holocene. Based on the analysis of historical trends and modeling outcomes under different climate scenarios, we discuss potential future hazards.

U. Nehren (✉)
TH Köln – University of Applied Sciences, Institute for Technology and Resources Management in the Tropics and Subtropics (ITT), Köln, Germany
e-mail: udo.nehren@th-koeln.de

A. Kirchner
University of Hildesheim, Department of Geography, Hildesheim, Germany
e-mail: andre.kirchner@uni-hildesheim.de

W. Lange
University of Cologne, Department of Geography, Köln, Germany

M. Follador
ALSO Consulting, Belo Horizonte, Brazil

Institute BioAtlantic (IBIO), Rio de Janeiro, Brazil
e-mail: follador@alsocabra.com

D. Anhuf
University of Passau, Department of Geography, Passau, Germany
e-mail: dieter.anhuf@uni-passau.de

U. Nehren et al. (eds.), *Strategies and Tools for a Sustainable Rural Rio de Janeiro*, Springer Series on Environmental Management,
https://doi.org/10.1007/978-3-319-89644-1_20

Keywords Natural hazards · Disasters · Climate change impacts · Rio de Janeiro state · Land use history

20.1 Introduction

In recent years, Southeast Brazil has been hit by numerous severe disasters, such as the event in January 2011 in Rio de Janeiro state (RJ), when floods, landslides, and mudslides caused 900 fatalities in the Serrana region (EM-DAT; www.emdat.be), or the droughts in 2014–2017 that severely affected the states of São Paulo (SP), RJ, Minas Gerais (MG), and Espírito Santo (ES). Among scientists, it is controversially discussed whether there has been a random bunching of disaster events caused by hydrometeorological hazards within the last few years or whether there are certain causes for the cumulative occurrence of incidents. Regardless of the long-term disaster trend, the past events have revealed the high vulnerability of society and key economic sectors in particular toward persistent droughts (Escobar 2015) and disclosed the need for policy intervention to increase disaster resilience (Nehren et al. 2017).

To obtain a clearer picture of disaster risk related to hydrometeorological hazards in RJ, we reviewed the scientific literature and added own research results and observations to complete the following objectives:

(a) Reconstruct the Holocene landscape dynamics, and analyze in how far climatic variations and land degradation affected geomorphic stability and triggered catastrophic events.
(b) Identify causes and impacts of historical disaster events, and determine risk factors for different types of natural hazards.
(c) Assess future disaster risk based on existing climate scenarios.

The chapter is organized in five sections. After the introduction (Section 20.1), Section 20.2 provides an overview on climate variations and landscape dynamics in Southeast Brazil during the Holocene. This section aims at reconstructing in how far climate variations caused environmental changes that may have triggered catastrophic events. Further, it explores to which extent human impact in different time periods caused land degradation with possible impact on susceptibility to natural hazards. Section 20.3 is dedicated to historical disasters related to hydrometeorological hazards in RJ. In this section, we analyze historical disaster events with respect to causes and impacts and try to identify distinct risk factors. In Section 20.4 we use existing climate scenarios to estimate future disaster risk. The results provide a basis for policy interventions and the development of suitable adaptation measures (Lange et al. 2018). Section 20.5 synthesizes the previous results and concludes this chapter.

20.2 Holocene Climate Variations and Landscape Dynamics in Southeast Brazil

Over the last few decades, multidisciplinary research on different environmental archives (e.g., peat bogs, speleothems, or lakes) has led to enormous progress in reconstructing Holocene climate conditions in Southeast Brazil (e.g., Behling 1995; Wang et al. 2007; Behling and Safford 2010). Nowadays it is widely accepted that Holocene climate experienced significant variations, especially in moisture regimes (e.g., Behling 2002; Prado et al. 2013). As a result, multiple changes in vegetation cover (see, e.g., de Oliveira et al. 2005; Rodrigues et al. 2016) and consequently in geomorphic dynamics appeared throughout the Holocene (e.g., Hiruma et al. 2013; Kirchner et al. 2015; Nehren et al. 2016).

Regional pollen records from the Serra dos Órgãos (Behling and Safford 2010) and the Serra da Bocaina (Behling et al. 2007) reconstruct warm and wet climate conditions for the Younger Dryas (12,700 to 11,600 cal BP) followed by an onset of drier climate during the early Holocene. Geomorphological studies from Southeast Brazil reported on a significant erosional-depositional cycle during that time indicating geomorphic activity (Coelho Netto 1999; Hiruma et al. 2013; Kirchner 2014), obviously related to this Pleistocene-Holocene climatic transition. Especially steep slopes with a weaker-resistant soil mantle and valley bottoms experienced a widespread erosion-depositional cycle most likely caused by heavy rainfall events.

The early to mid-Holocene is considered to be substantially drier compared to the latest Pleistocene (Younger Dryas) and late Holocene which is explained by a weakened South American monsoon circulation and reduced precipitation over the South Atlantic Convergence Zone (SACZ) (e.g., Behling and Safford 2010; Prado et al. 2013). Contemporaneously, geomorphological studies indicate reduced hillslope/fluvial dynamics and overall geomorphic stability with soil formation (Bigarella and Becker 1975; Semmel and Rohdenburg 1979; Coelho Netto 1999; Modenesi-Gauttieri 2000; Kirchner et al. 2015), which seems to be a direct consequence of the drier climate conditions during the first half of the Holocene. According to ^{14}C-dated shell mounds from the southern coast of RJ, early tribes of Sambaqui mound builders had already settled along the coastline of RJ under these relatively dry climatic conditions from 8700 cal BP onward (Lima et al. 2002; compare Ribeiro de Oliveira et al. 2018). It is, however, assumed that these indigenous tribes preferably used the resources of their estuarine environment and did not practice agriculture. Therefore, deforestation that would have led to increased erosion and landslide dynamics can be largely excluded (Kirchner and Nehren 2017).

Toward the late Holocene, slowly increasing climatic moisture led to a dense forest cover and an onset of quasi-modern climatic conditions after around 5640 cal BP (Behling and Safford 2010). These wetter conditions were most likely triggered by a southward migration of the mean annual position of the Intertropical Convergence Zone (ITCZ) (Haug et al. 2001; Peterson and Haug 2006; Wanner et al. 2008) and the reinforcement of the South Atlantic Convergence Zone (SACZ) as suggested by Prado et al. (2013).

Kirchner et al. (2015) presented evidence that this climatic change from drier to wetter conditions was obviously the main driver for considerably enhanced fluvial dynamics in the upper Guapi-Macacu River system (RJ) after approx. 4700 cal BP. Large quantities of coarser-grained deposits (high-energy fluvial environment) were deposited during the entire Late Holocene, whereas poorly sorted, fine-grained sediments had been deposited in a medium- to low-energy fluvial environment during the drier mid-Holocene (Fig. 20.1). For the steep parts of the catchment, these deposits postulate increased frequency of mass movements and sediment supply driven by intense and frequent heavy rainfall events. In this context, the regional environmental setting of the catchment with a steep relief, heavy biological load, loamy to clayey soil texture seems to be of great importance. However, these vulnerable natural conditions are notably fulfilled in large areas of the windward side of the Serra do Mar, which makes them prone to extensive erosion on hillsides and steep slopes during heavy rainfall events (de Ploey and Cruz 1979; Furian et al. 1999; Dykes 2002). In contrast, the lee of the mountain range is less affected by landslides due to lower rainfall and less steep relief.

The most recent period of geomorphic activity started with large-scale deforestation during the European colonization, even though tribes of the Tupi-Guarani had already inhabited the coastal zone of RJ since at least 1800 BP (Silva Noelli 2008). However, Nehren et al. (2013) argue that even though these tribes have been practicing shifting cultivation, deforestation rates and related erosion processes were predominately low. The situation changed dramatically with the prominent colonial exploitation cycles: extraction of brazilwood [*Paubrasilia echinata*] starting in the early sixteenth century, establishment of sugar cane plantations from the mid-sixteenth century onward, mining activities [the so-called gold routes] starting in the early eighteenth century, and coffee plantations from about 1790 to 1860 (compare Dantas and Coelho Netto 1995; Nehren et al. 2013; Ribeiro de Oliveira et al. 2018).

For the Guapi-Macacu catchment, Kirchner et al. (2015) demonstrated a markedly increased sediment deposition in the floodplains after 290 cal BP, which can be explained with large-scale deforestation in erosion prone areas (Fig. 20.2). Additionally, young colluvial soils in foot slope positions with embedded modern pottery and high amounts of charcoal document high erosion rates triggered by land use intensification (Nehren 2008; Kirchner 2014). Finally, land use intensification is also documented by significantly higher modern sedimentation rates in the Guanabara Bay, as proved by various studies (Amador 1980; Wilken et al. 1986; Godoy et al. 2012).

In postcolonial times, large areas were converted into agricultural and pasture land, which accelerated soil erosion and enhanced landslide risk (Nehren et al. 2016). This modern exploitation cycle is characterized by extensive grazing systems, often in steep relief positions, which caused severe gully erosion (Nehren et al. 2013, 2016; Fig. 20.3). With increasing urbanization, landslides and mudslides also occurred in densely populated areas, particularly on the steep slopes of the marginal settlements (*favelas*), and led to human casualties and high material dam-

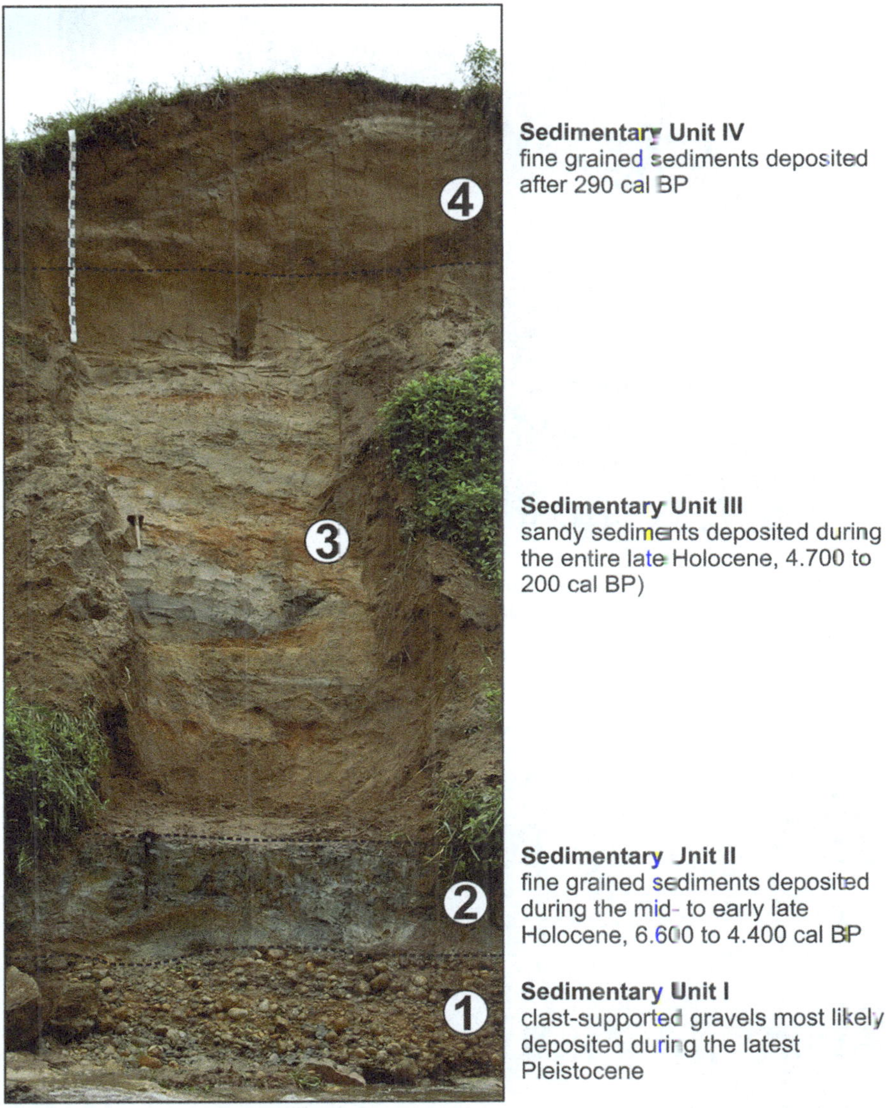

Fig. 20.1 Sedimentary sequence and sediment units of the profile *Guapiaçu 3* in the upper Guapi-Macacu River system (RJ). (Credit: A. Kirchner)

age (Sandholz et al. 2018). Moreover, urban dwellers were increasingly affected by floods. The growing population of RJ faced severe water scarcity due to water over-use already in the mid-nineteenth century. Subsequently, many coffee plantations were expropriated, and a large-scale reforestation program was initiated (Dantas and Coelho Netto 1995).

Fig. 20.2 The historical photo taken around 1920–1925 in the upper Guapi-Macacu catchment (RJ) shows intense deforestation during that period. (With permission of Nicholas Locke, Reserva Ecológica de Guapiaçu (REGUA))

Fig. 20.3 Gully erosion on a degraded pasture in the municipality of Itaperuna (RJ). (Photo credit: U. Nehren)

20.3 Causes and Impacts of Historical Disasters in the State of Rio de Janeiro

Landslides, mudslides, and floods are the most frequent natural hazards in RJ that lead to severe disasters (UFSC 2013). The 2011 event in the Serrana region, which is among the deadliest hydrometeorological disasters in Brazilian history (EM-DAT; www.emdat.be), was triggered by persistent and heavy rainfall. According to Coelho Netto et al. (2011), in less than 10 h, the region received a volume of rainfall close to that of an average month. However, the authors also point out that high-intensity rainfall does not always show close correlation with landslide occurrences so that other triggering and controlling factors must also play an important role. These factors include lightning bolts that might have caused vibrations and strong sounds, fractures in the predominating granite and magmatite rocks that favor superficial or subsuperficial flows, and the stability of saturated or near-saturated soils, as well as the slope gradient.

Coelho Netto et al. (2011) also stressed the importance of a well-preserved forest cover that can mitigate the propagation of landslides. However, the 2011 event and other landslides events show that landslides also occurred in well-developed forests with low human impact, in particular in steep slope positions (own observations). This means that a closed forest cover can reduce, but not completely eliminate landslide risk (Fig. 20.4), which supports the assumption that landslides are a natural process of the steep Serra do Mar landscapes, even under closed forest cover.

Fig. 20.4 Landslides within a formerly closed secondary forest in the municipality of Nova Friburgo (RJ), January 2011. (Photo credit: A. Kirchner)

Finally, the 2011 event showed that the urban environments were the most vulnerable, as the highest concentration of landslides with the most losses and damages occurred in the urbanized areas of Nova Friburgo, Teresópolis, and Petropolis, even under less proportions of rain (Coelho Netto et al. 2011). This leads to the question to which extent the natural risk is increased by human intervention.

Taking a look at other past landslide, mudslide, and flood events in RJ, the most severe events, apart from the 2011 event, occurred in 1966 and 1967 (Barata 1969; da Costa Nunes 1969), 1988 (Brandão 1997; Schuster et al. 2002), 1996 (Coelho Netto 1999; Coelho Netto et al. 2007; Fernandes et al. 1999), and 2010 (Avelar et al. 2011; Machado de Mello et al. 2014). All events were triggered by heavy and persistent rainfall, and most human losses and material damages were recorded in urbanized areas. According to Nehren et al. (2014) and Sandholz et al. (2018), this is due to the following reasons: (a) increasing exposure of humans and their assets to natural hazards as a consequence of population growth and urbanization processes; (b) high vulnerability of exposed neighborhoods, in particular informal settlements in steep slopes positions and in floodplains; (c) road and settlement infrastructure as well as garbage dumps in endangered slope positions that trigger landslides by heavy load, undercutting slopes, and destabilizing soils and weathering mantles; and (d) degradation of protecting ecosystems, in particular old-growth Mata Atlântica forests.

Historical documents and photos show that landslides have occurred in urban areas in historical times when forests were cleared for human habitation. However, landslides and severe erosion problems were also already described several decades ago in rural areas associated with land clearing in particular for pasture use (Duarte de Barros 1956). Highly eroded slopes with deep gullies testify to the land degradation process that is characterized by a combination of different geomorphic processes triggered or at least intensified by deforestation and overexploitation (Valentin et al. 2005).

With respect to floods, it must be considered that regular flooding was a natural process in the fluvial and alluvial plains of the lowlands before many streams were rectified and channelized and floodplains meliorated for cultivation. The narrowing of the floodplains in combination with deforestation resulted in higher and faster runoff with increased erosion, river incision, and more frequent and higher flood peaks (Pereira et al. 2016). At the same time, more people settled along rivers and exposed themselves to flood hazards (Fig. 20.5).

In comparison to geomorphic processes triggered by huge amounts of precipitation, drought events as slow onset-hazards are less frequent in RJ and in the literature only explicitly described for the recent past. Even the water scarcity in the mid-nineteenth century cannot be clearly assigned to a meteorological drought event; it was rather caused by massive overuse of water resources particularly in coffee plantations (Dantas and Coelho Netto 1995). However, for the last few decades, there have been some indications that the regional climate is becoming drier, even though there are no sufficient climate data available yet to statistically verify this assumption. Hence, the causes for the severe water scarcity in Southeast Brazil in 2014–2017 are controversially discussed among scientists. Some see the

Fig. 20.5 Flood damages in the municipality of Teresópolis (RJ). (Photo credit: U. Nehren)

main causes in large-scale deforestation in the Amazon basin as well as land use changes in the Cerrado and Mata Atlântica biomes that reduce the transport of wet air masses to the south (Marengo 2014). In addition, these scientists assume that urban heat island effects coupled with heat waves reduce the formation of rain clouds. Other scientists, however, argue that at least in the case of SP, poor water management and old infrastructure led to high water losses and were the main reasons for the water scarcity (Getirana 2015; Nobre et al. 2016). Consequently, they conclude that the disaster was caused by a combination of climatic factors and mismanagement.

20.4 Potential Future Climate Change Impacts

Although the abovementioned disasters are not direct results of global climate change, climate models suggest that similar extreme weather events might become more frequent and intense in the future (Marengo 2014). The effects of climate change are already visible in some regions. In the southeast of the country, a downward trend in annual rainfall has been recognizable since 1990, with an average rate of 3 mm per year (Marengo et al. 2015). The water balance in the region is already showing a negative trend (Fundação COPPETEC and Laboratório de Hidrologia e Estudos de Meio Ambiente 2014).

The most detailed climate projections used to support recent studies about climate change impacts in Brazil by the National Institute for Space Research (CPTEC/INPE) have been based on the regionalization of two global climate models, namely, Eta-HadGEM-ES and Eta-MIROC5 (Chou et al. 2014). Two representative concentration pathway (RCP) scenarios (Moss et al. 2008), RCP 4.5 (optimistic one that estimates a peak of greenhouse gas emissions shortly after 2040 and a stabilization of CO_2 concentration around 2100) and RCP 8.5 (pessimistic scenario characterized by increasing greenhouse gas emissions over time), have been considered for climate forecasting for the periods 2011–2040, 2041–2070, and 2070–2100. The models cover the entire area of Brazil with a resolution of 20 × 20 km^2 and the southeast with a finer resolution of 5 × 5 km^2 (available only for the RCP 8.5 scenario).

A recent study by the Brazilian Ministry of Environment, in collaboration with the World Wildlife Fund (WWF), shows that the climate in Southeast Brazil will be drier in the winters of 2011–2040, while in the summers of this period, more concentrated rainfall is expected (Camarinha et al. 2015). Taking into account a continuing population growth and a rising demand for water, this will further increase the vulnerability of the region, especially against droughts (Camarinha et al. 2015). Motta (2016) supports this prognosis and assumes a temperature rise of up to 4 °C by the year 2100. Less precipitation and higher temperatures could therefore further dry up semiarid and subhumid regions of RJ. However, it should be noted that the effects of climate change are distributed regionally, which is illustrated by the distribution of the expected change in days with severe precipitation (Fig. 20.6a) and water availability in the soil (Fig. 20.6b) (Follador et al. 2016). Both maps are based on the RCP 4.5 scenario.

Follador et al. (2016) also presented a climate projection for Southeast Brazil based on the pessimistic scenario RCP 8.5 in a higher resolution of 5 × 5 km^2 that focuses on the specific risks of floods, landslides, and variation of agroclimatic zoning (Fig. 20.7a–c). According to this scenario, within the state boundaries of RJ, the mountain regions of the Serra do Mar register the highest values of the flood indicator (Fig. 20.7a), which is expected to increase by about 10–15% in the rainy season until 2040. A few hotspots along the coast, including the city of Rio de Janeiro, will also be more susceptible to flooding but to a lesser extent (up to +5%), while in other parts of the state, in particular the north, flood risk is likely to decrease.

An increase of landslide events is expected to be observed in the mountainous areas of RJ (Fig. 20.7b), due to a synergy between the augmented rain erosivity and the biophysical characteristics of the region, i.e., steep slopes and friable Cambisol soils that are highly eroded. Within the urban perimeter, informal settlements in vulnerable areas with high declivity, soil erodibility, and a thin vegetation cover are projected to favor landslides when subjected to heavy precipitations (Bittencourt et al. 2017).

The assessment of climate change impacts on agriculture susceptibility (Fig. 20.7c) is based on the water requirement satisfaction index (WRSI) that tends to follow the trends in the distribution of water availability in the soil. In the RCP

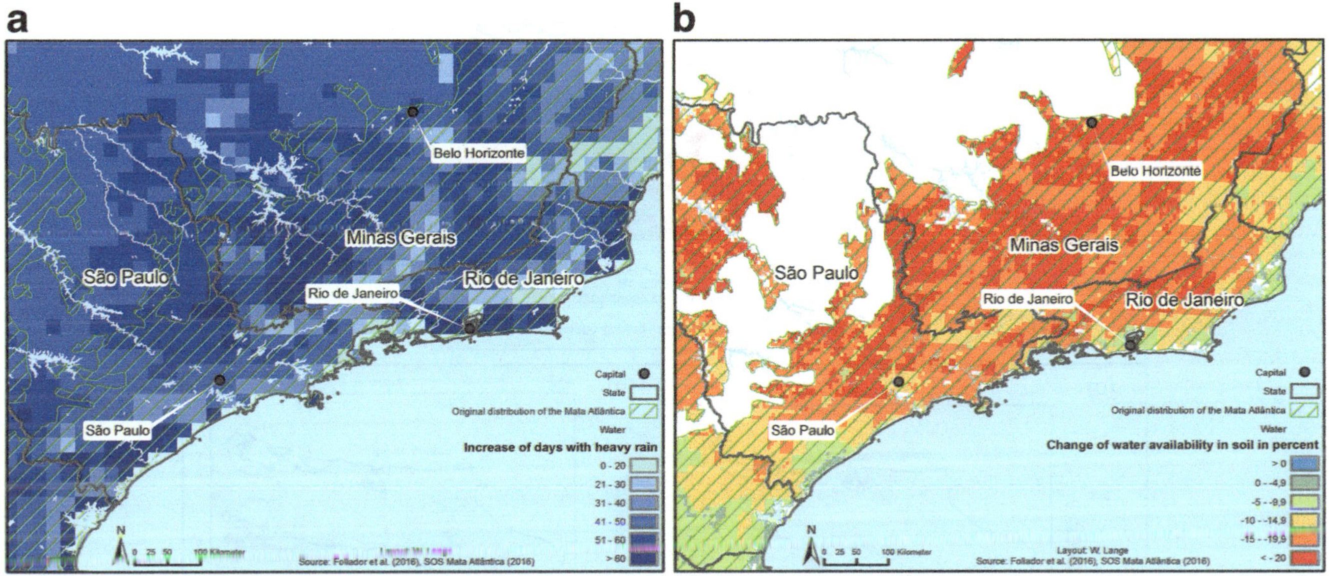

Fig. 20.6 Predicted change of days with heavy rainfall (**a**) and water availability in soil (**b**) between the periods 2011–2040 and 1960–2005. (RCP 4.5, June to August)

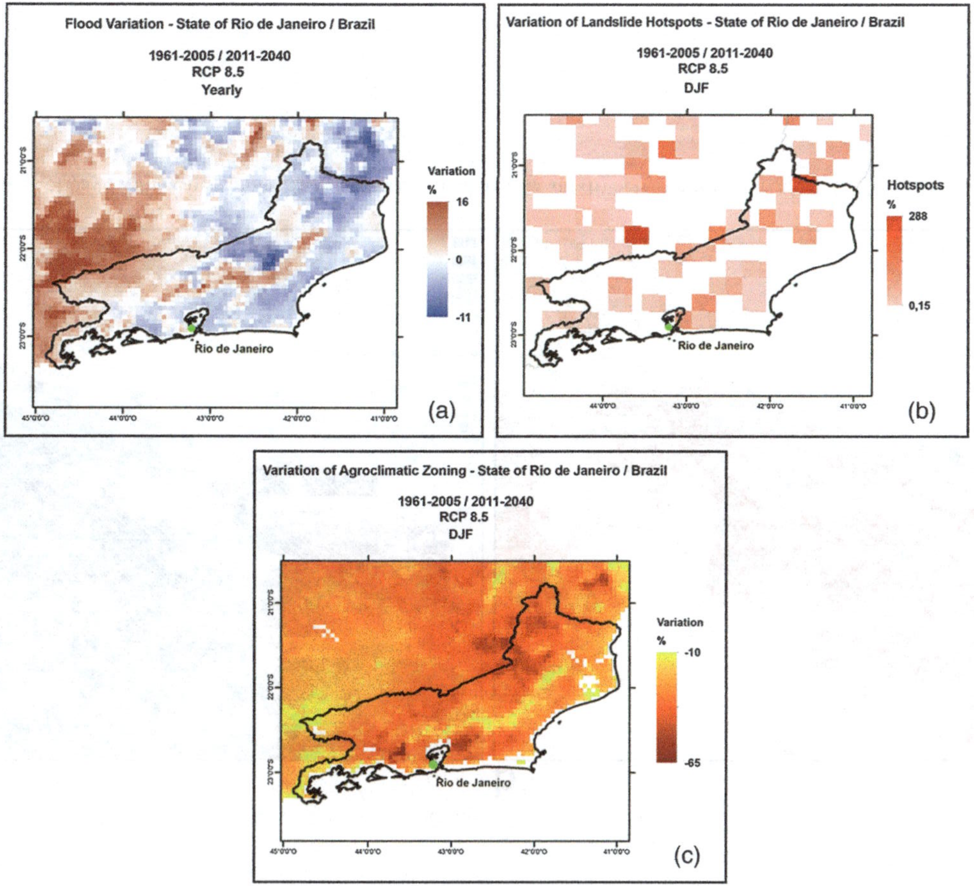

Fig. 20.7 Potential impacts of climate change in RJ until 2040 in scenario RCP 8.5. Variation (%) between time-window 2011–2040 and historical baseline 1961–2005. Floods (**a**); hotspots of landslides (**b**); agro-zoning and loss of viable agricultural lands (**c**). (Source: Follador et al. 2016)

8.5 scenario, RJ registers relevant losses in the agriculture potential during the rainy season (planting period for most crops), with a predicted decrease of up to 45% of the WRSI and a consequent probable negative impact on agriculture. Even though the expected impacts on the agricultural sector seem not to greatly affect the overall economic indicators of RJ (Follador et al. 2015; Ferreira Filho and Moraes 2015), they are unequally distributed within its territory, and they could severely damage family-run agriculture and poor smallholders who are highly vulnerable to climate change and with a low adaptive capacity.

20.5 Synthesis and Conclusions

The landscape historical analysis shows that with a shift from warm and dry climate conditions to a wet-tropical moisture regime in the late Holocene, the Mata Atlântica forests spread throughout Southeast Brazil and widely replaced the open grassland vegetation that had developed under drier climate conditions (Fig. 20.8). Increasing rainfall led to enhanced fluvial and landslide dynamics during the transition phase, but with an increasingly dense forest cover at least the gently to medium inclined slopes became more and more stabilized. However, the steep parts of the Serra do Mar mountain range, in particular those facing the Atlantic Ocean, received high amounts of precipitation, triggering landslides, even without human interference.

With the arrival of the Tupi-Guarani in the coastal zone of Southeast Brazil, forests were locally burned and cut to establish slash and burn agriculture. This might have led to increased erosion and landslide dynamics in hilly parts of the coastal landscape, but overall the landscape can be considered as stable from a geomorphological and ecological perceptive. This situation changed dramatically with the arrival of the European colonizers in the early sixteenth century.

The European colonization with the previously described exploitation cycles led to a progressive forest loss that in turn caused a destabilization of slopes and change of the river dynamics (Fig. 20.8). In this context, Coelho Netto (1999) use the term "catastrophic landscapes" to describe the geomorphic response to regional deforestation that is accompanied by increased erosion rates, landslides, mudslides, flooding, and river incision.

Modern development in RJ is characterized by rapid urbanization and a shift from colonial exploitation cycles to modern land use systems that are dominated by livestock production and peasant agriculture (Fig. 20.8). Land reclamation and infrastructural development in the rural areas of the state led to different forms of land degradation, such as forest fragmentation, loss of valuable habitats, soil erosion and compaction, and losses of retention room due to melioration of wetlands and channelization and rectification of rivers. These interventions made the landscape system more vulnerable and thereby enhanced the risk of floods, landslides, and mudslides. In urban areas, forest loss and degradation increase the risk of landslides particularly – but not exclusively – in *favelas* that are often located on very

System components and effects	Transition mid-to late Holocene (approx. 5,600 cal BP)	Late Holocene to colonialization	Colonial exploitation cycles (early 16th century to early 20th century)	Modern development (starting early 20th century)	Scenario 2050
Climate	Transition from warm-dry, to humid-tropical conditions (temperate in higher altitudes)	Humid-tropical conditions (temperate in higher altitudes)		Humid-tropical conditions (temperate in higher altitudes); impacts of global climate change and regional climate effects; extreme climate events more likely	
Vegetation	Prevalence of a closed forest cover	Tupi-Guarani occupation leads locally to deforestation in coastal zone	Progressive deforestation; coffee plantations penetrate steeper slope positions	High forest fragmentation rates higher altitudes affected	Relatively stable forest cover: only small scale gains and losses; potential for reforestation
Human impact	Sambaqui mound builders settled in coastal zone	Tupi-Guarani occupation from about 1,800 cal BP onwards; slash and burn agriculture	Deforestation and land degradation due to land exploitation: Brazilwood, sugarcane, mining, coffee	Expansion of pasture and agricultural systems; urbanization; river rectification, melioration	Ongoing urbanization processes; degradation processes in rural areas continue, but less intense
Geomorphic process dynamics	Enhanced fluvial and landslide dynamics after 4,700 cal BP	Locally increased erosion and landslide dynamics possible (deforested and landslide prone areas)	High erosion and landslide dynamics	High erosion and landslide dynamics, river incision, river bank erosion	High erosion and landslide dynamics, ongoing river incision in lowlands, river bank erosion
Disaster risk	Sambaqui mound builders exposed to coastal hazards	Tupi-Guarani particularly exposed to coastal hazards and floods	Increasing landslide and flood risk; water scarcity due to overuse and poor infrastructure	High risk of landslides, mudslides, floods; urban poor most vulnerable	High risk of landslides, mudslides, floods; urban poor most vulnerable; increased drought risk, water shortages possible

Today — Time scale

Fig. 20.8 Historical change of key landscape system components and related effects on landscape dynamics in RJ from the mid- to Late Holocene to 2050 (scenario) steep slopes. Besides this, poor urban dwellers also settle along rivers where they are highly exposed to floods.

For the next few decades, climate models predict an increased occurrence of climatic extreme events on both, the wet and the dry sides. Droughts are a rather new phenomenon in RJ, and therefore mitigation and adaptation strategies are still in a rudimentary stage. The droughts of 2014–2017 in Southeast Brazil have shown that the metropolitan areas are particularly vulnerable, and the water supply infrastructure is in an inadequate condition. Droughts endanger both agriculture and livelihood security in rural areas, particularly in the drier northwest of the state. To avoid reaching a tipping point in the social-ecological system, the resilience of the landscape needs to be enhanced, which requires restoration, rehabilitation, and risk reduction measures (Seliger et al. 2018; Soares da Silva et al. 2018), in addition to sound agricultural practices and nature conservations measures that are described in previous chapters of this book. Future research should therefore put particular emphasis on potential drought impacts on the landscape system and its consequences for land use and water management.

References

Amador ES (1980) Assoreamento da Baía de Guanabara: taxas de sedimentação. Anais Acad Brasil Ciénc 52(4):723–742

Avelar AS, Coelho Netto AL, Lacerda WA et al (2011) Mechanisms of the recent catastrophic landslides in the mountainous range of Rio de Janeiro, Brazil. In: Proceedings of the second world landslide forum, Rome, 3–7 Oct 2011

Barata FE (1969) Landslides in the tropical region of Rio de Janeiro. In: Proceedings of the 7th international conference on soil mechanics and foundation engineering, Mexico City, vol 2, pp 507–516

Behling H (1995) A high resolution Holocene pollen record from Lago do Pires, SE Brazil: vegetation, climate and fire history. J Paleolimnol 14:253–268

Behling H (2002) South and southeastern grasslands during Late Quaternary times: a synthesis. Palaeogeogr Palaeoclimatol Palaeoecol 177:19–27

Behling H, Dupont L, Shafford HD, Wefer G (2007) Late quaternary vegetation and climate dynamics in the Serra da Bocaina, southeastern Brazil. Quatern Int 161:22–31

Behling H, Safford HD (2010) Late-glacial and Holocene vegetation, climate and fire dynamics in the Serra dos Órgãos, Rio de Janeiro state, southeastern Brazil. Glob Chang Biol 16(6):1661–1671

Bigarella JJ, Becker RD (1975) International symposium on the quaternary. Boletim Paranaense de Geociências 33

Bittencourt F, Follador M, Pereira V et al (2017) Climate vulnerability Index: a case study for the city of Belo Horizonte, Brazil. In: In: Climate change adaptation in Latin America: managing vulnerability, fostering resilience - climate change management series. Springer, Cham. In Press

Brandão AMPM (1997) As chuvas e a acão humana: Uma infeliz coincidência. In: Rosa LP, Lacerda WA (eds) Tormentas Cariocas. COPPE/UFRJ, Rio de Janeiro, pp 21–38

Camarinha PI, Debortoli NS, Hirota M (2015) Índice de vulnerabilidade aos desastres naturais relacionadas às secas: no contexto de mundaças climáticas, produto II: Relatório com os resul-

tados e discussões sobre o índice de vulnerabilidade e avaliação dos impactos relacionados às secas para as diferentes regiões do Brasil, Florianópolis
Chou SC, Lyra A, Mourão C et al (2014) Assessment of climate change over South America under RCP 4.5 and 8.5 downscaling scenarios. Am J Clim Chang 3(05):512
Coelho Netto AL (1999) Catastrophic landscape evolution in a humid region (SE Brazil): inheritances from tectonic, climatic and land use induced changes. In: Supplementi di Geografia Fisica e Dinamica Quaternaria III (T.3). Comitato Glaciologico Italiano, Torino, pp 21–48
Coelho Netto AL, Avelar AS, Fernandes MC, Lacerda WA (2007) Landslide susceptibility in a mountainous geoecosystem,Tijuca Massif, Rio de Janeiro: the role of morphometric subdivision of the terrain. Geomorphology 87:120–131
Coelho Netto AL, Sato AM, Avelar AS et al (2011) January 2011: the extreme landslide disaster in Brazil. In: Proceedings of the second world landslide forum, Rome, 3–7 Oct 2011
da Costa Nunes AJ (1969) Landslides in soils of decomposed rock due to intense rainstorms. In: Proceedings of the 7th international conference on soil mechanics and foundation engineering, Mexico City, vol 2, pp 547–554
Dantas ME, Coelho Netto AL (1995) Impacto do ciclo cafeeiro na evolução da paisagem geomorphológica no medio vale do Rio Paraiba do Sul. Cad Geosciências 15:22
de Oliveira PE, Behling H, Ledru M-P et al (2005) Paleovegetação e paleoclimas do Quaternário do Brasil. In: Gouveia Souza CR, Suguio K, Santos Oliveira AM, Oliveira PE (eds) . O Quaternário do Brasil, Ribeirão Preto, pp 52–74
de Ploey J, Cruz O (1979) Landslides in the Serra do Mar, Brazil. Catena 6:111–122
Duarte de Barros W (1956) A erosão no Brasil. Ministério da viação e obras públicas serviço de documentação
Dykes AP (2002) Weathering-limited rainfall-triggered shallow mass movements in undisturbed steepland tropical rainforest. Geomorphology 46:73–93
EM-DAT (n.d.) The emergency events database - Université catholique de Louvain (UCL) - CRED, D. Guha-Sapir - www.emdat.be, Brussels, Belgium. Accessed 30 Aug 2017
Escobar H (2015) Drought triggers alarms in Brazil's biggest metropolis. Science 347(6224):812
Fernandes MC, Laguens JVM, Coelho Netto AL (1999) O processo de ocupação por favelas e sua relação com os eventos de deslizamentos no Maciço da Tijuca. Anuário do Instituto de Geociências/UFRJ 22:22–36
Ferreira Filho JBS, Moraes GI (2015) Climate change, agriculture and economic effects on different regions of Brazil. Environ Dev Econ 20(1):37–56
Follador M, Ferreira Filho JB, Brito M (2015) Avaliação de impactos econômicos de riscos climáticos e proposta preliminar de medidas de adaptação a nível nacional – setor segurança alimentar. Comissão Econômica para a América Latina e o Caribe – Nações Unidas, Brasília
Follador M, Rocha A, Vaz C et al (2016) Impactos Biofísicos Potenciais da Mudança do Clima na Mata Atlântica. Ministério de Meio Ambiente MMA, Brasília
Fundação Coppetec and Laboratório de Hidrologia e Estudos de Meio Ambiente (2014) Elaboração do plano estadual de recursos hídricos do Estado do Rio de Janeiro: R3-A – Temas técnicos a eventos críticos, RT-03 – Vulnerabilidade a eventos críticos, vol 1
Furian S, Barbiéro L, Boulet R (1999) Organisation of the soil mantle in tropical southeastern Brazil (Serra do Mar) in relation to landslides processes. Catena 38:65–83
Getirana ACV (2015) Extreme water deficit in Brazil detected from space. J Hydrometeorol 17(2):591–599
Godoy JM, Oliveira AV, Almeida AC et al (2012) Guanabara bay sedimentation rates based on 210Pb dating: reviewing the existing data and adding new data. J Braz Chem Soc 23(7):1265–1273
Haug GH, Hughen KA, Sigman DM et al (2001) Southward migration of the Intertropical Convergence Zone through the Holocene. Science 293:1304–1308
Hiruma ST, Modenesi-Gauttieri MC, Riccomini C (2013) Late quaternary colluvial deposits in the Bocaina Plateau, southeastern Brazil highlands: age and palaeoenvironmental consequences. Boreas 42:306–316

Kirchner A (2014) Rekonstruktion der spätpleistozänen und holozänen Landschaftsgenese im Guapi-Macacu Einzugsgebiet, Rio de Janeiro, Südostbrasilien. PhD thesis, University of Leipzig. http://www.qucosa.de/recherche/frontdoor/?tx_slubopus4frontend[id]=14709

Kirchner A, Nehren U (2017) Comment on: "Linked variations in sediment accumulation rates and sea-level in Guanabara Bay, Brazil, over the last 6000 years" by Alberto G. Figueiredo Jr., Mauro B. de Toledo, Renato C. Cordeiro, José M.O. Godoy, Fabiano T. da Silva, Sérgio C. Vasconcelos, Ricardo A. dos Santos [Palaeo 3 415 (2014) 83–90]. Palaeogeogr Palaeoclimatol Palaeoecol 482:114–116

Kirchner A, Nehren U, Behling H, Heinrich J (2015) Mid- and late Holocene fluvial dynamics in the tropical Guapi-Macacu catchment, Southeast Brazil: the role of climate change and human impact. Palaeogeogr Palaeoclimatol Palaeoecol 426:308–318

Lange W, Sandholz S, Viezzer J et al (2018) Ecosystem-based approaches for disaster risk reduction and climate change adaptation in Rio de Janeiro state. In: Nehren U, Schlüter S, Raedig C, Sattler D, Hissa H (eds) Strategies and tools for a sustainable rural Rio de Janeiro. Springer International Publishing, Rio de Janeiro

Lima TA, Macario KD, Anjos RM et al (2002) The antiquity of the prehistoric settlement of the central-south Brazilian coast. Radiocarbon 44(3):733–738

Machado de Mello F, Machado R, Bilal E (2014) Predisposing factors for shallow landslides susceptibility in Southeastern Rio de Janeiro. Brazil Rom J Mineral Deposits 87(2):55–60

Marengo JA (2014) O futuro clima do Brasil. Revista USP 103:25–32

Marengo JA, Nobre CA, Seluchi ME et al (2015) A seca e a crise hídrica de 2014-2015 em São Paulo. Revista USP 106:31–44

Modenesi-Gauttieri MC (2000) Hillslope deposits and the quaternary evolution of the altos campos — Serra da Mantiqueira, from Campos do Jordão to the Itatiaia Massif. Rev Bras Geosci 30(3):504–510

Moss R, Babiker M, Brinkman S et al (2008) Towards new scenarios for analysis of emissions, climate change, impacts, and response strategies. IPCC Expert Meeting Report on New Scenarios. Intergovernmental Panel on Climate Change, Geneva, p 25

Motta D (2016) Estudo avalia o risco de desertificação no estado do Rio de Janeiro, FAPERJ. http://www.faperj.br/?id=3091.2.7. Accessed 21 Mar 2017

Nehren U (2008) Quartäre Landschaftsgenese und historische -degradation in der Serra dos Órgãos, Rio de Janeiro. PhD thesis, Univerity of Leipzig. http://www.cuvillier.de/flycms/de/html/30/-UickI3zKPS7xcUg=/Buchdetails.html

Nehren U, Kirchner A, Heinrich J (2016) What do yellowish-brown soils and stone layers tell us about Late Quaternary landscape evolution and soil development in the humid tropics? A field study in the Serra dos Órgãos, Southeast Brazil. Catena 137:173–190

Nehren U, Kirchner A, Sattler D et al (2013) Impact of natural climate change and historical land use on landscape development in the Atlantic Forest of Rio de Janeiro, Brazil. Anais Acad Brasil Ciênc 85(2):311–332

Nehren U, Lange W, Kupka S (2017) Ökosystembasierte Maßnahmen zur Risikominderung und Anpassung an den Klimawandel in Südostbrasilien. Geogr Rundsch 4:26–31

Nehren U, Sudmeier-Rieux K, Sandholz S et al (eds) (2014) The ecosystem-based disaster risk reduction case study and exercise source book. CNRD/PEDRR. http://postconflict.unep.ch/publications/DRR_CASE_STUDIES_&_EXERCISES.pdf

Nobre CA, Marengo JA, Seluchi ME et al (2016) Some characteristics and impacts of the drought and water crisis in Southeastern Brazil during 2014 and 2015. J Water Resource Prot 8:252–262

Pereira DR, Martinez MA, Pruski FF, da Silva DD (2016) Hydrological simulation in a basin of typical tropical climate and soil using the SWAT model part I: calibration and validation tests. J Hydrol: Regional Studies 7:14–37

Peterson LC, Haug GH (2006) Variability in the mean latitude of the Atlantic Intertropical Convergence Zone as recorded by riverine input of sediments to the Cariaco Basin (Venezuela). Palaeogeogr Palaeoclimatol Palaeoecol 234:97–113

Prado LF, Wainer I, Chiessi CM et al (2013) A mid-Holocene climate reconstruction for eastern South America. Clim Past 9:2117–2133

Ribeiro de Oliveira R, Stingel Fraga J, MacLeod Hickie M (2018) Environmental history, forests and landscape uses in Rio de Janeiro state. In: Nehren U, Schlüter S, Raedig C, Sattler D, Hissa H (eds) Strategies and tools for a sustainable rural Rio de Janeiro. Springer International Publishing, Rio de Janeiro

Rodrigues JM, Behling H, Giesecke T (2016) Holocene dynamics of vegetation change in southern and southeastern Brazil is consistent with climate forcing. Quat Sci Rev 146:54–66

Sandholz S, Lange W, Nehren U (2018) Governing green change: ecosystem-based measures for reducing landslide risk in Rio de Janeiro. Int J Disaster Risk Reduct. in press

Schuster RL, Salcedo DA, Valenzuela L (2002) Overview of catastrophic landslides of South America in the twentieth century. In: Evans SG, DeGraff JV (eds) Catastrophic landslides: effects, occurrence, and mechanisms. Reviews in engineering geology 15. The Geological Society of America, Boulder, pp 1–34

Seliger R, Sattler D, Soares da Silva A et al (2018) Rehabilitation of degraded sloped pastures - lessons learned in Itaocara. In: Nehren U, Schlüter S, Raedig C, Sattler D, Hissa H (eds) Strategies and tools for a sustainable rural Rio de Janeiro. Springer International Publishing, Rio de Janeiro

Semmel A, Rohdenburg H (1979) Untersuchungen zur Boden- und Reliefentwicklung in Süd-Brasilien. Catena 6:203–217

Silva Noelli F (2008) The Tupi expansion. In: Silverman H, Isbell WH (eds) Handbook of South American archaeology, Chap. 33. Springer, New York, pp 659–670

Soares da Silva A, Seliger R, Sattler D, Heinrich J (2018) Soil degradation in Southeast Brazil: a challenge for restoration and rehabilitation. In: Nehren U, Schlüter S, Raedig C, Sattler D, Hissa HR (eds) Strategies and tools for a sustainable rural Rio de Janeiro. in press

UFSC – Universidade Federal de Santa Catarina, Centro Universitário de Estudos e Pesquisas sobre Destastres (2013) Atlas Brasileiro de Desastres Naturais 1992–2012

Valentin C, Poesen J, Li Y (2005) Gully erosion: impacts, factors and control. Catena 63:132–153

Wang XF, Auler AS, Edwards RL et al (2007) Millennial-scale precipitation changes in southern Brazil over the past 90,000 years. Geophys Res Lett 34:L23701. https://doi.org/10.1029/2007GL031149

Wanner H, Beer J, Bütikofer J et al (2008) Mid- to Late Holocene climate change: an overview. Quat Sci Rev 27:1791–1828

Wilken RD, Moreira I, Rebello A (1986) 210Pb and 137Cs fluxes in a sediment core from Guanabara bay, Brazil. Sci Total Environ 58:195–198

Chapter 21
Disaster Risk Management in the State of Rio de Janeiro

Beate Frank, Rosa Maria Formiga-Johnsson, Viviane Japiassú Viana, and Silvia Marie Ikemoto

Abstract This chapter describes and analyzes the recent evolution of disaster risk management in the state of Rio de Janeiro as evidenced through empirical observation, a bibliographic review, and a desk review. Firstly, it describes the two most serious types of hazards – floods and landslides – and shows which territories are more susceptible to risks. Then it introduces the political and institutional context at the state and municipal levels (primarily the civil defense agencies) and details their evolution in reaction to several disasters that have occurred over time. In 2011, the mega disaster in the Serrana region marked an evolution in policies in the state of Rio de Janeiro and Brazil. It triggered a paradigm shift from solely disaster management practices, focused on response and recovery, to broader risk management initiatives focused on risk assessment, prevention, mitigation, preparedness, response, and recovery. Several institutions started to concentrate on risk management, especially in risk assessment and preparedness, for the two types of major hazards. The State Civil Defense, the Environmental State Institute, and the State Department of Mineral Resources were particularly relevant institutions in this context.

Keywords Disaster risk management · Flooding · Landslides · Institutional capacity · State of Rio de Janeiro

---

B. Frank (✉)
Regional University of Blumenau (FURB), Blumenau, Santa Catarina, Brazil

R. M. Formiga-Johnsson
State University of Rio de Janeiro (UERJ), Rio de Janeiro, Brazil

V. Japiassú Viana
Veiga de Almeida University (UVA) and Faculty of Technology Senac Rio, Rio de Janeiro, Brazil

S. M. Ikemoto
Environmental State Institute of Rio de Janeiro (INEA), Rio de Janeiro, Brazil

U. Nehren et al. (eds.), *Strategies and Tools for a Sustainable Rural Rio de Janeiro*, Springer Series on Environmental Management,
https://doi.org/10.1007/978-3-319-89544-1_21

21.1 Introduction

Disaster risk management (DRM) is a permanent process of analyzing, planning, decision-making, and implementing a diversity of measures to (a) identify, prevent, and reduce the possibility of disasters occurring, (b) respond appropriately should one occur, and (c) recover livelihoods, services, and systems following any such occurrence (EIRD 2004; Narváez et al. 2009; CIF-OIT 2012). This concept is an extension of the disaster management approach and emerged from recognizing that disaster risks originate from societies with unsustainable patterns of development (Maskrey 1993; PNUD 2004; UN 2011). Risks are the result of hazards in territories where communities and infrastructure are both exposed and vulnerable. As long as social, environmental, and infrastructure-related vulnerabilities persist, disaster risks will also be prevalent. Owing to the social production of risk, some are critical of the denomination "natural disasters" also used in Brazil, since only hazards are, in fact, natural.

The DRM approach incorporates this broader understanding of risk. It is a part of sustainable development management and has been promoted since 1999 by the International Strategy for Disaster Reduction (EIRD 2004) and more recently as part of the Hyogo Framework for Action (UNISDR 2011) for the period 2005 to 2015 and the Sendai Framework (UNISDR 2015) for the period 2015 to 2030.

DRM may be represented in six components: (1) risk assessment, (2) risk prevention, (3) risk mitigation, (4) preparedness for the occurrence of disasters, (5) response to disaster, and (6) recovery (Fig. 21.1).

The state of Rio de Janeiro (RJ) is in the list of the six states of Brazil with the greatest number of disasters, having suffered several extreme events associated with flooding and landslides in the last decade (CEPED 2013). However, only recently has DRM been recognized for its strategic relevance in sectoral policies for the environment and civil protection and defense (INEA 2014; SEDEC 2015). The main exception to this general configuration is the city of Rio de Janeiro, which, due to its high risk of landslides, initiated a risk management strategy in 1966 through the creation of the Geo-Rio Institute. The purpose of the institute was to perform inspections and determine measures to guarantee public safety. That public authority's pioneering work gained recognition as a geotechnical agency of excellence, since its activities drastically reduced the number of deaths associated with landslides[1] in the city of Rio de Janeiro.

This chapter hones in on the significant transformations in policy that took place as a result of a mega disaster in the Serrana region of RJ in 2011, which caused almost 900 deaths and left more than 300,000 people homeless. Instead of emphasizing purely on response and recovery ("disaster management"), "disaster risk management" started to be embraced. Here, we describe how DRM has evolved in RJ and analyze if the way that it has been managed is far from or close to the concept of "disaster risk management."

[1] http://www.rio.rj.gov.br/web/smo/geo-rio

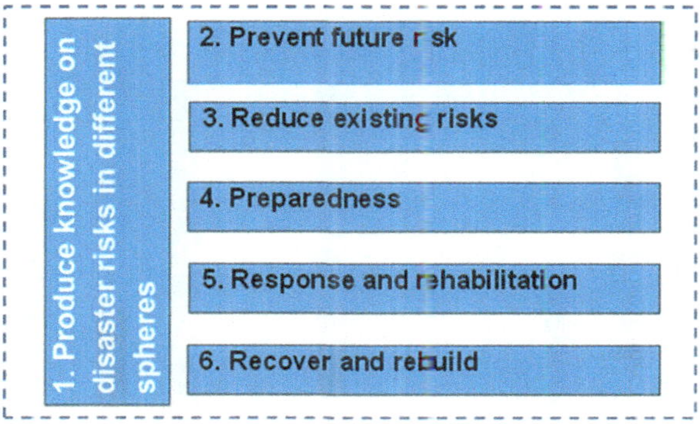

Fig. 21.1 Key processes in disaster risk management. (Source: Adapted from Narváez et al. 2009)

Exploratory research methods were used (Botelho and Cruz 2013), including bibliographic research, desk review, and empirical observation. The authors consulted reports, official documents, databases, and the websites of the main institutions involved in the research topic, giving priority to documents less than 10 years old. The results are presented in the style of a qualitative narrative: the main hazards in RJ, the institutional configuration in the sector, the main initiatives and changes in risk disaster management, and final considerations.

21.2 Natural Hazards and Disasters in the State of Rio de Janeiro

An analysis of the federal database of recognized emergencies or public calamities (MI 2016b) reveals that floods and landslides are the most frequently occurring natural disasters in the state, representing 86% of all registered occurrences between 2003 and 2016 (Fig. 21.2). Brazilian legislation defines a disaster as "the outcome of adverse, natural, or human-caused events from a situation of vulnerability, causing serious disruption to the functioning of a community or society, including extensive human, material, economic or environmental losses and damages, and exceeding any capacity to contain the events by individual means" (MI 2012).

Municipal civil defense agents have confirmed the prevalence of floods and landslides (SEDEC 2015) (Fig. 21.3). Of the state's 92 municipalities, 37 declared landslides as the main hazard, while 43 others declared that landslides were among the five main hazards. As for floods, only one municipality did not cite that disaster as one of the five main hazards (SEDEC 2015).

The frequent occurrence of floods and landslides in the state is due, to a large part, to its high pluviometric indices, rugged terrain, and hydrographic particularities (compare Nehren et al. 2018). The rivers of the state's sierra present elevated hydraulic speeds, which are marked by transient floods with heightened erosive

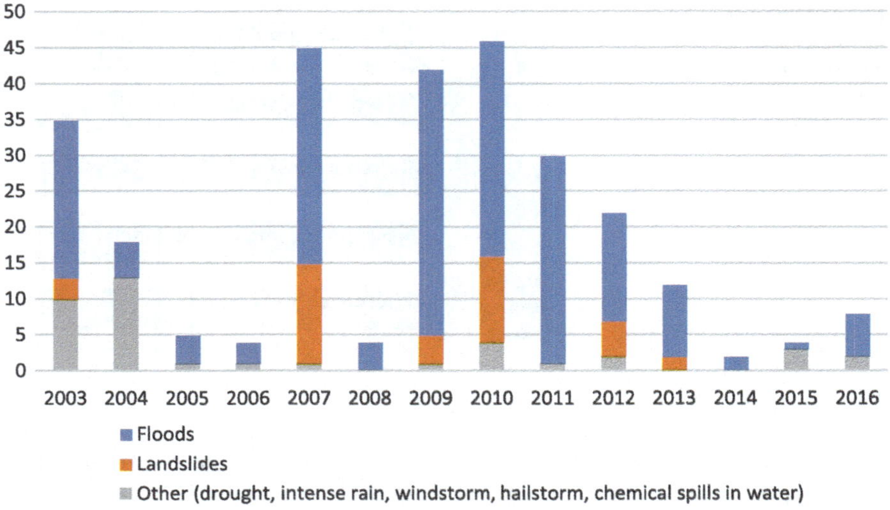

Fig. 21.2 Number of natural disaster occurrences in the state of Rio de Janeiro, January 2003 to 2016. (Source: Based on data from National Civil Defense Secretariat – SEDEC (MI 2016b))

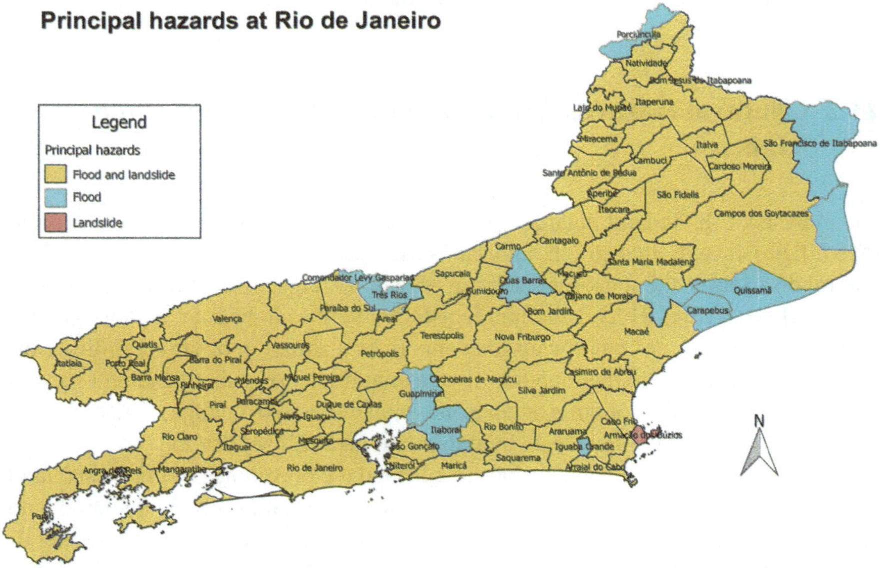

Fig. 21.3 Principal hazards in the state of Rio de Janeiro. (Source: Authors, based on data from Civil Defense College (SEDEC 2015))

capacity (Canedo et al. 2011). According to the geo-environmental map of RJ (*Mapa Geoambiental do Estado do Rio de Janeiro*; CPRM 2000), almost 40% of land in the state is highly susceptible to erosion, landslides, and rockfall, 37% is moderately susceptible to erosion, and only 8% presents low susceptibility. Sedimentation plains make up 12% of the state's lands.

The impacts of such events are compounded by several factors. Populations are increasingly exposed to hazards owing to their irregular occupation of hillsides, riverbanks, and floodplains. Their vulnerability is increased by the proliferation of self-constructed houses and sub-housing units, precarious sanitary conditions, and the inhabitants' overall low socioeconomic profile (INEA 2014). Due to these factors, the existing risks are of much greater consequence in urban areas.

In the period 1991 to 2012, almost 80% of all recorded flood and landslide disasters in Brazil occurred in RJ. In the decade 2000, the respective number of occurrences was 21.7 times greater than in the decade 1990. This shows a 3.6 times greater increase in floods and landslides than the rate of all recorded disasters of both periods in general.

In recent years, several disasters of notable magnitude have taken place: landslides in Angra dos Reis and Ilha Grande (01/01/2010), a flood in the city of Rio de Janeiro (05/04/2010), landslides in Niterói (Morro do Bumba, 07/04/10), and especially the mega disaster that affected seven municipalities in the Serrana region (11/01/2011). The average number of affected individuals in the period from 2007 to 2010 was 500,000 persons/year (MI 2016b), excluding the disaster that occurred on 05/04/2010, which affected 6 million people.

The flash floods and landslides in the Serrana region in January of 2011 caused the greatest number of deaths in Brazil since 2000. With an estimated 865 deaths, more than 16,000 left homeless and almost 305,000 people affected in the sierra's seven municipalities (SEDEC 2012), it was considered one of the world's ten worst landslide disasters in the last 111 years (CRED 2011). The losses and economic damages incurred by the municipalities were also extraordinary, being estimated at R$ 4.8 billion (IBRD 2012). Notably, this amount does not take into account the impacts sustained by the education and health sectors, for which there are no estimates.

This mega disaster occurred due to a combination of critical, natural factors and adverse meteorological conditions, associated with chronic, historical processes of environmental degradation and irregular human settlements on steep slopes, riverbanks, and hilltops (INEA 2014). According to the Ministry of Environment (MMA 2011), such areas, which are protected by law as Permanent Preservation Areas (Áreas de Preservação Permanente – APPs), were precisely those most affected by the disaster. According to that report, of the 657 landslides analyzed, 92% involved anthropic alteration, and only 8% occurred in preservation areas with native vegetation. Accordingly, the report highlighted that a great proportion of impacts and damages could be minimized if anthropic intervention did not occur in APPs. That conclusion reinforces the importance of preventive measures that work toward harmonizing legislation for land preservation and land management practices.

21.3 Policies for Disaster Risk Management and Institutional Framework

Following the disaster in 2011, the federal government promoted important changes in the legal, political, and institutional spheres in favor of disaster risk management at the national and state level.

One year after the disaster, the federal government approved a legal framework for civil protection and defense (Federal Law n° 12.608/2012): the creation of a new national policy and new institutions (Viana 2016). The National Civil Protection and Defense Policy (PNPDC in Portuguese) gave strong priority to risk reduction measures for natural disasters. Innovative measures were introduced that favored greater coordination between disaster risk management and other social and environmental policies. The abovementioned law delegates the authority to assess and monitor at-risk areas to the federal and state governments, as well as the responsibility of supporting municipalities in performing preventive urban planning. The law also establishes that preventive and mitigative measures must be adopted by all three levels of government.

The Federal Government program, *Programa 2040*, has been part of the last two "Pluriannual Plans" (for the periods 2012–2015 and 2016–2019) and is geared toward risk and disaster management. This program has had a significant evolution in policy inasmuch as it involves several federal governmental entities: the Ministries of National Integration and of Cities, the National Water Agency (ANA), the Geological Survey of Brazil (CPRM), the Brazilian Institute of Geography and Statistics (IBGE), and the National Center for Monitoring and Early Warning of Natural Disasters (CEMADEN). The program's measures emphasize adaptation and reduction of vulnerabilities rather than focusing only on tracking hazards (MI 2016a).

At the state level, in RJ the three central institutions for disaster risk management are the Sub-Secretariat of Civil Defense (SEDEC/RJ), the State Environmental Institute (INEA), and the State Department of Mineral Resources (DMR). While the SEDEC has extensive practical experience in administration related to disasters, INEA and DRM possess technical and scientific expertise concerning risks. The number of professionals that work in disaster risk management in each of the three entities is also rather different (see Table 21.1). Together, these are examples of obstacles in achieving greater integration between those institutions.

SEDEC/RJ, which has existed since 1996, has the Military Firefighter Corps (CBMERJ) as a subordinate entity. CBMERJ's main responsibility is search-and-rescue, and it also promotes preventive measures related to building security. Together, SEDEC/RJ and CBMERJ represent the state's capacity in preparing for and responding to disasters. In 2012, SEDEC/RJ began a change from the disaster management to the risk management approach, with emphasis on risk assessment and preparation. SEDEC/RJ's work plans can be outlined in the following groups: (a) preparedness, (b) response actions, (c) recovery, and (d) technical/operational support. Concerning the latter group, a notable measure was the creation of the

Table 21.1 Characteristics of institutions for disaster risk management

	SEDEC/RJ	INEA	DMR
Number of staff	16,042 (firefighters)	1130	90
Number of staff for disaster risk management	504	82	10

Data from 2013

Center for Monitoring and Early Warning of Natural Disasters (Cemaden/RJ), whose technical team puts out warnings of disasters (SEDEC 2017).

The INEA, created in 2007, has the purpose of administering state policies concerning the environment, water resources, and forestry. In the context of geo-hydrometeorological risks, INEA provides information and executes projects and works. Its DRM operations consist of (a) prevention, mitigation, response, and reconstruction and (b) monitoring and warning. An analysis of its operations reveals that the institution does not apply a systemic approach to risk management (GITEC/IP/CODEX 2014a).

The purpose of the DMR is to manage and promote the sustainable use of mineral resources, oil resources, and underground water resources. Its activities have included geological risk prevention since 2009. The DMR also plays the role of providing information aimed at supporting SEDEC/RJ and municipalities. Its two groups of operations are: (a) preventive and (b) operational and instructional.

After the disaster in the Serrana region, INEA and DMR significantly increased their production of studies assessing and analyzing flood- and landslide-related risks. INEA also augmented its monitoring network and hydrometeorological alert system and developed new projects to track and mitigate floods. Moreover, the disaster in the Serrana region revealed that the then-current model of institutional and operational management of disaster risk management was inefficient. The need to harmonize and integrate procedures between institutions becomes evident. Thus, INEA, SEDEC/RJ, and DMR coordinated a study aimed at elaborating the terms of an institutional and operational restructuration for geo-hydrometeorological disaster risk management in RJ and achieving greater coordination between the three institutions (GITEC/IP/CODEX 2014a, 2014b). Although the study was concluded in 2014, by 2017, none of its recommendations had been implemented.

The study included an assessment of municipal capacity for risk management (GITEC/IP/CODEX 2014b), which demonstrated that, in 2013, 82 out of the state's 92 municipalities generally:

- Invested in improving their technical and administrative capacity for civil defense
- Developed more measures for prevention and mitigation than for preparedness and recovery
- Developed little measures for coordination and participation

A weighted analysis of this data revealed that, in 2013, 22 municipalities presented satisfactory results regarding their approach to DRM. The municipality of Rio de Janeiro ranked first and far ahead of all others, validating its long experience in DRM (GITEC/IP/CODEX 2014b).

21.4 Main Initiatives and Changes in Disaster Risk Management

The most relevant measures implemented in RJ following the disaster in 2011, which stemmed from recent political and institutional changes, are detailed below.

21.4.1 Assessment and Analysis

A large proportion of the data, studies, and assessments has been produced within the past decade. The available data is recent (disasters after the year 2000), which hampers the ability to assess the effects of floods and landslides from previous years and their magnitude. Gaining a more realistic understanding of the related risks is, thus, also hindered.

As a part of the "Programa 2040," CPRM is designing maps that detail susceptibility to gravitational mass movements and floods (1:25,000) for the municipalities of RJ that are considered priorities. Maps have already been completed for 50 of the state's municipalities (CPRM 2017). CEMADEN and CPRM (2017) also design maps outlining landslide risks (vulnerability, susceptibility, and aptness for urbanization). Many of those maps possess scales that do not provide significant detail (>1:25,000), but they have nevertheless been used in support of other instruments for planning and urban and environmental management (Viana 2016).

Since 2010, DMR-RJ has analyzed landslide risks. The institution has created products at several scales, such as maps outlining susceptibility, geotechnical aspects, and risks. It has even created maps on imminent risks (for the majority of RJ's municipalities) and remaining landslide risks. Maps detailing imminent risk of landslide have been elaborated for 67 of RJ's municipalities, and maps detailing remaining risks have also been created for the municipalities affected by the disaster in the Serrana region (Teresópolis, Nova Friburgo and Sapucaia) (DRM 2017).

INEA has collaborated in identifying and assessing flood risks and producing maps that portray susceptibility, danger, and risk of floods in some river basins or particularly critical sections of certain rivers (INEA 2015). The information provided is capable of defining criteria to clear riverbanks and resettle populations residing in those areas. In addition, INEA created a methodology to assess natural susceptibility to flooding (Napoleão et al. 2016). Notably, flood risk assessments were developed for critical river stretches of four (of a total nine) of the hydrographic regions in the state of RJ (Piabanha, Rios Dois Rios, Médio Paraíba do Sul, and Baía de Guanabara) (INEA 2017).

The state water resources plan (PERHI-RJ) includes a report on vulnerability to extreme events containing important diagnostics of the most critical areas for geo-hydrometeorological disasters (INEA 2014). The report was elaborated with data obtained from assessments of critical occurrences of floods and inundations, landslides, droughts, and fires—despite the existing registers of such events being limited and incomplete.

While significant advances have been made recently, the knowledge needed to reduce the disaster risks of floods and landslides is still lacking. Susceptibility and vulnerability assessments must be performed to integrate and complement existing studies. Priority should be given to river basins with more dynamic land use and high-intensity rainfall indices. Notably, information systems and databases on risks and disasters are not available.

21.4.2 Prevention and Mitigation

Driven by the consequences of the mega disaster that occurred in 2011, RJ passed State Law 6.442 in 2013, which reinforces an essential requirement of the national policy that master plans and other regulatory instruments for land use and occupation must incorporate studies and assessments of risk factors and flood and landslide risks. This law is considered an advance for land management as it means that the knowledge produced on disaster risks should be collated for the purpose of the master plans, with municipalities being designated as the main prevention agents. Yet, in many municipalities risks have still not been studied. Only in a few isolated cases has the effective application of the related legislation been assessed during the revision process of such master plans.

With the aim of reducing flood risks, projects of flood control and environmental recovery have been developed in the areas affected by the disaster of 2011 in the Serrana region. Examples include the creation of environmental protection areas, river parks and riverbank recovery to improve drainage, recovery of riverbank areas, and the proposal of further measures to avoid the resettlement of at-risk areas (INEA 2015).

21.4.3 Preparedness

Preparedness includes hydrometeorological monitoring and warning, the establishment of early warning and response protocols, training of responsible authorities, and field exercises. Such precautions are taken to be able to respond to the probable occurrence of a hazard.

Since 2008, great advances have been made in the implementation and expansion of hydrometeorological monitoring networks. Protocols and systems have been established for early warning of floods and landslides, such as INEA's flood warning system (Bahiense et al. 2015) and the CEMADEN/RJ. In 2014, two meteorological radar systems were also installed to allow real-time monitoring of rainfall in the territory of the state.

Early warning systems against landslides operate based on the correlation between rainfall and landslides in diverse scenarios of risk (DRM 2016). They also involve monitoring by community agents and distress sirens to warn the population of at-risk areas.

Amid the state's economic crisis, from 2016 onward, there have been reports of hydrometeorological systems and early warning and alarm systems being partially or fully impaired (Fantti 2017; O Globo 2017). Such circumstances underscore the challenges in providing adequate operation and maintenance to these systems.

21.4.4 Recovery

The catastrophe in the Serrana region forced the relevant public authorities to protect the unaffected part of the population, to reestablish a situation of normalcy, and to restructure the main locations affected to diminish disaster risks. The measures undertaken include the following:

- Hydraulic sections from the damaged bridges were inspected and measured.
- Diverse areas were cleared.
- River channels were recomposed.
- Flood control and environmental recovery projects were undertaken in the most affected municipalities, including river dredging and canalization.
- New bridges were built.
- River parks were created.
- Families formerly living in areas of imminent flood risk were resettled (Júnior et al. 2013).

State Decree 43.415/2012 established guidelines for the relocation of the families affected by disasters and payment of a social allowance until definitive resettlement.

21.5 Final Considerations

RJ's history attests to the territory's high risk of loss of life and material damage from flood and landslide disasters. Vulnerability is particularly high in urban areas, where more than 96% of the state's population resides.

The mega disaster that occurred in 2011 evidently marked the evolution of policies in this sector—in RJ and in Brazil. Perceptions of the problem and of possible solutions were expanded, moving past a focus exclusively on disasters to encompass the management of disaster risks as well. Important legal and institutional changes occurred at the federal and state level, demonstrating an incorporation of this new vision. Notably, a profusion of initiatives took place to produce data, assess and survey risks, and standardize regulatory instruments for land use and occupation. Thus, the institutions made their operations increasingly reflect an optic of risk management, based mainly on the two most prevalent types of hazards (floods and landslides).

However, the effectiveness of these initiatives in reducing existing risks and preventing future ones in every municipality's territory must still be assessed. An evaluation of emerging initiatives in the municipality of Nova Friburgo (the one most affected by the disaster in 2011) revealed that while many advances had been made, overlapping authorities, duplicated efforts, communication problems between institutions, and a lack of methodological planning to harmonize the different initiatives were prevalent (Viana 2016). For example, several government entities undertook risk mapping with the aim of establishing restrictions and guidelines for land use and occupation. However, the scales, methodologies, and technical criteria adopted for each of those documents were sometimes not well-understood or accepted by the municipality's technical team. Thus, the materials ended up being suboptimally used for urban planning efforts that aimed at reducing risk disasters. In sum, many important measures were implemented, but adjustments must be made to integrate them in an effective and efficient local risk reduction strategy.

In general, although much has been accomplished in terms of risk reduction, the federal institutions concerned and the government of RJ still focus a great deal of their efforts on confronting disasters with more robust and integrated measures for preparedness, response, and reconstruction.

Finally, it is recognized that the challenges of disaster risk reduction are still great in technical terms and mainly in terms of governance. These challenges are greater still in the nation's current context of economic crisis, which engenders worries for the continuity of measures that have been already initiated and operationalized. The interruption of those measures could directly impede the capacity of the state and its municipalities to manage disaster risks.

References

Bahiense JM, Júnior JEFF, Costa LF, Chargel LT (2015) Monitoramento hidrológico quantitativo no estado do Rio de janeiro: importância, histórico e modernização. Paper presented at the XXI Simpósio Brasileiro de Recursos Hídricos. Distrito Federal, Brasília

Botelho J, Cruz V (2013) Metodologia Científica. Pearson Education do Brasil, São Paulo

Canedo P, Ehrlich M, Lacerda WA (2011) Chuvas na Região Serrana do Rio de Janeiro: Sugestões para Ações de Engenharia e Planejamento. Programa de Engenharia Civil, Instituto Alberto Luiz Coimbra de Pós-Graduação e Pesquisa em Engenharia – COPPE/UFRJ. Rio de Janeiro, Rio de Janeiro

CEPED – Centro Universitário de Estudos e Pesquisas sobre Desastres (2013) Atlas Brasileiro de Desastres Naturais: 1991 a 2012. 2. ed. rev. ampl. Florianópolis: CEPED UFSC, 2013. https // s2id.mi.gov.br/paginas/atlas/. Accessed 22 Ago 2017

CIF-OIT – Centro Internacional de Formação da Organização Internacional do Trabalho (2012) O risco de desastres: origem, avaliação, redução e prevenção no quadro de desenvolvimento local sustentável. Curso de Redução do Risco de Desastres e Desenvolvimento Local Sustentável, Unidade Didática 2. Programa Delnet de Apoio ao Desenvolvimento Local

CPRM – Serviço Geológico do Brasil (2000) Mapa Geoambiental do Estado do Rio de Janeiro. In: Projeto Rio de Janeiro. CPRM, Rio de Janeiro

CPRM – Serviço Geológico do Brasil (2017) Cartas de Suscetibilidade a Movimentos Gravitacionais de Massa e Inundações. http://www.cprm.gov.br/publique/Gestao-Territorial/Geologia-de-Engenharia-e-Riscos-Geologicos/Cartas-de-Suscetibilidade-a-Movimentos-Gravitacionais-de-Massa-e-Inundacoes-3507.html#riojaneiro. Accessed 10 Apr 2017

CRED – Centre for Research on the Epidemiology of Disasters (2011) Ranking of disasters. http://www.cred.be/publications. Accessed 20 May 2011

DRM – Departamento de Recursos Minerais (2016) Diagnóstico de Risco a Escorregamento no Estado do Rio de Janeiro e Plano de Contingência. Núcleo de Análise e Diagnóstico de Escorregamentos – NADE. Relatório Geotécnico. Rio de Janeiro, Niterói, pp 2016–2017

DRM – Departamento de Recursos Minerais (2017) Núcleo de análise e diagnóstico de escorregamentos. http://www.drm.rj.gov.br/index.php/risco-geologico. Accessed 10 Apr 2017

EIRD – Estratégia Internacional para Redução de Desastres (2004). Vivir con el Riesgo. Informe mundial sobre iniciativas para la reducción de desastres. Volumen I. Nova York y Ginebra: Publicación Naciones Unidas

Fantti B (2017) Sirenes de alerta de chuvas estão sem funcionar em 180 comunidades do estado. Jornal o Dia (versão eletrônica). http://odia.ig.com.br/rio-de-janeiro/2016-12-16/sirenes-de-alerta-de-chuvas-estao-sem-funcionar-em-180-comunidades-do-estado.html. Accessed 10 Apr 2017

GITEC/IP/CODEX – Consórcio GITEC Brasil/ IP/ CODEX Remote (2014a) Concepção do arranjo institucional e operacional para a gestão de risco de desastres no Estado do Rio de Janeiro. Produto 2: Caracterização das instituições técnico-científicas envolvidas na gestão de risco de desastres no Estado do Rio de Janeiro. Contrato SEPLAG-RJ 9/2013 para Serviços de Consultoria

GITEC/IP/CODEX – Consórcio GITEC Brasil/ IP/ CODEX Remote (2014b) Concepção do arranjo institucional e operacional para a gestão de risco de desastres no Estado do Rio de Janeiro. Produto 3: Avaliação da capacidade de atuação dos municípios na gestão de risco de desastre. Contrato SEPLAG-RJ 9/2013 para Serviços de Consultoria

IBRD – International Bank for Reconstruction and Development (2012) Avaliação de Perdas e Danos: Inundações e Deslizamentos na Região Serrana do Rio de Janeiro – Janeiro de 2011. Relatório elaborado com o apoio do Governo do Estado do Rio de Janeiro. Distrito Federal, Brasília: IBRD

INEA – Instituto Estadual do Ambiente (2014) Plano estadual de recursos hídricos do estado do Rio de Janeiro: Vulnerabilidade a eventos críticos fundação COPPETEC. Rio de Janeiro

INEA – Instituto Estadual do Ambiente (2015) Redução de risco de desastres associados a inundações. Boletim Águas & Território. Rio de Janeiro

INEA – Instituto Estadual do Ambiente (2017). Prevenção e controle de inundações: mapa de risco a inundações da Região Serrana. http://200.20.53.3:8081/Portal/MegaDropDown/ProgramaseProjetos/Projetos/PROJ_MAPAREASRISCOINUNDRGSERRA&lang=. Accessed 10 Aug 2017

Júnior Jeff, Ikemoto SM, Santos Morm, Johnsson RMF (2013) Gestão de riscos de inundações e a tragédia da Região Serrana: ações, resultados, desafios e perspectivas futuras. Paper presented at the XX Simpósio Brasileiro de Recursos Hídricos. Bento Gonçalves, Rio Grande do Sul

Maskrey A (1993) Los desastres no son naturales. LA RED. Red de Estudios Sociales en Prevención de Desastres en América Latina, Ciudad de Panamá

MI – Ministério da Integração Nacional, Brasil (2012) Instrução Normativa n° 01, de 24 de agosto de 2012. Estabelece procedimentos e critérios para a decretação de situação de emergência ou estado de calamidade pública pelos Municípios, Estados e pelo Distrito Federal, e para o reconhecimento federal das situações de anormalidade decretadas pelos entes federativos e dá outras providências

MI – Ministério da Integração Nacional, Brasil (2016a) Projeto de Cooperação Técnica BRA/IICA/13/001 – Interáguas – MI: Atualização do conteúdo dos Manuais Técnicos para atender ao Sistema Nacional de Proteção e Defesa Civil. Produto 1:2. (non published)

MI – Ministério da Integração Nacional, Brasil (2016b) Série Histórica de Reconhecimentos Realizados 2003-2016. Secretaria Nacional de Defesa Civil – SEDEC. http://www.mi.gov.br/web/guest/reconhecimentos-realizados. Accessed 16 Apr 2017

MMA – Ministério do Meio Ambiente, Brasil (2011) Áreas de Preservação Permanente e Unidades de Conservação & Áreas de Risco. O que uma coisa tem a ver com a outra? In: Relatório de Inspeção da Área atingida pela tragédia das chuvas na Região Serrana do Rio de Janeiro. MMA, Brasília

Napoleão PRM, Ikemoto SM, Ferreira CEG, Sapienza JA (2016) Indicações metodológicas para o mapeamento da suscetibilidade às inundações aplicadas à bacia dos rios Guapi-Macacu/RJ. Paper presented at the I Congresso Brasileiro de Redução de Riscos de Desastres. Curitiba, Paraná

Narváez L, Lavell A, Ortega GP (2009) La gestión del riesgo de desastres: un enfoque basado en procesos. Proyecto Apoyo a La Prevención de Desastres em La Comunidad. Andina – PREDECAN, Lima, Peru

Nehren U, Kirchner A, Lange W et al (2018) Natural hazards and climate change impacts in the state of Rio de Janeiro: a landscape historical analysis. In: Nehren U, Schlüter S, Raedig C, Sattler D, Hissa H (eds) Strategies and tools for a sustainable rural Rio de Janeiro. Springer International Publishing, Springer

O Globo (2017) Centro de prevenção de desastres naturais fecha devido à crise do RJ. http://g1.globo.com/rio-de-janeiro/noticia/2016/06/centro-de-prevencao-de-desastres-naturais-fecha-devido-crise-do-rj.html. Accessed 10 Apr 2017

PNUD– Programa de las Naciones Unidas para el Desarollo (2004) Un informe mundial para La Reducción de Riesgos de Desastres – Un desafío para o desarrollo. USA: John S. Swift Co.

SEDEC – Secretaria Estadual de Defesa Civil (2012) Notificação Preliminar de Desastre (NOPRED) e da Avaliação de danos (AVADAN). Departamento Geral de Defesa Civil. Rio de Janeiro

SEDEC –Secretaria Estadual de Defesa Civil (2015) Mapa de ameaças naturais do estado do Rio de Janeiro 2014/2015. Escola de Defesa Civil – ESDEC Rio de Janeiro https://wwwmindmeistercom/pt/376772296/mapa-de-amea-as-naturais-do-estado-do-rio-de-janeiro-2014. Acessed 10 Aug 2017

SEDEC – Secretaria Estadual de Defesa Civil (2017) Conheça o CEMADEN. Centro Estadual de Monitoramento e Alerta de Desastres Naturais – CEMADEN-RJ. http://cemadenrj.defesacivil.rj.gov.br. Accessed 10 Apr 2017

UN – United Nations (2011) Global assessment report on disaster risk reduction. Revealing risk, redefining development. Summary and main findings. Information Press, Oxford, p 2011

UNISDR – United Nations International Strategy for Disaster Reduction (2011) Hyogo Framework for Action 2005–2015: building the resilience of nations and communities to disasters. https://www.unisdr.org/2005/wcdr/intergover/official-doc/L-docs/Hyogo-framework-for-action-english.pdf. Accessed 10 Jun 2017

UNISDR – United Nations International Strategy for Disaster Reduction (2015) Sendai framework for disaster risk reduction 2015–2030. http://www.wcdrr.org/uploads/Sendai_Framework_for_Disaster_Risk_Reduction_2015-2030.pdf. Accessed 10 Jun 2017

Viana VJ (2016) Gestão de risco de desastres no Brasil: leitura das estratégias locais de redução de riscos de inundação em Nova Friburgo, RJ. Staty University of Rio de Janeiro State - UERJ, Tese

Chapter 22
Ecosystem-Based Approaches for Disaster Risk Reduction and Climate Change Adaptation in Rio de Janeiro State

Wolfram Lange, Simone Sandholz, Jennifer Viezzer, Martin Becher, and Udo Nehren

Abstract In the mountain ranges (*Região Serrana*) of the Rio de Janeiro state (RJ), natural hazards such as landslides, mudslides, and flooding are recurrent, causing considerable human and economic losses. These events are of natural origin, but landscape degradation contributes significantly to increased disaster risk. Extreme climatic events that triggered past disasters are predicted to increase in frequency and intensity in the future. The approaches of ecosystem-based adaptation (EbA) and ecosystem-based disaster risk reduction (Eco-DRR) aim at the sustainable management, conservation, and restoration of ecosystems to cope with the adverse effects of climate change and reduce the impacts of natural hazards. In this context, this chapter provides a comprehensive overview of climate change adaptation (CCA) and disaster risk reduction (DRR) policies at national and state level, before analyzing the linkages between CCA/DRR and biodiversity and ecosystem management. Finally, EbA/Eco-DRR measures that are currently planned or have

W. Lange (✉)
University of Cologne, Department of Geography, Köln, Germany

S. Sandholz
United Nations University – Institute for Environment and Human Security (UNU-EHS), Bonn, Germany
e-mail: sandholz@ehs.unu.edu

J. Viezzer
Federal Ministry of Environment (Ministério do Meio Ambiente – MMA), Brasília, Brazil
e-mail: jennifer.viezzer@mma.gov.br

M. Becher
Deutsche Gesellschaft für Technische Zusammenarbeit GmbH (GIZ), Eschborn, Germany
e-mail: martin.becher@gmx.net

U. Nehren
TH Köln – University of Applied Sciences, Institute for Technology and Resources Management in the Tropics and Subtropics (ITT), Köln, Germany
e-mail: udo.nehren@th-koeln.de

U. Nehren et al. (eds.), *Strategies and Tools for a Sustainable Rural Rio de Janeiro*, Springer Series on Environmental Management,
https://doi.org/10.1007/978-3-319-89644-1_22

already been implemented are presented, and the further potential and limitations of EbA/Eco-DRR in RJ are discussed.

Keywords Natural hazards · Disaster risk reduction · Climate change adaptation · Ecosystem-based approaches · Rio de Janeiro state

22.1 Introduction

In the state of Rio de Janeiro (RJ), natural hazards such as landslides, mudslides, and floods are recurrent, causing considerable human and economic losses (see Nehren et al. 2018). Natural and climatic factors, such as pronounced topography as well as rock formations and weathering mantels that are prone to slipping, promote the occurrence of such events. These natural risk factors are, however, considerably reinforced by human interventions. The Atlantic Forest (Mata Atlântica), which contributes to stabilize the slopes, has seen large-scale deforestation since European colonization and thus significantly increased the risk of flooding and landslides (Heinrich et al. 2010; Nehren et al. 2013). On top of this, extreme climatic events already known to trigger disasters are predicted to increase in number and intensity in the future (Marengo 2014).

To cope with such challenges, disaster risk reduction (DRR) in RJ state relies mainly on technical safeguarding and early warning systems, whereas the implementation of concrete measures for climate change adaptation (CCA) are still in an early stage. Ecosystem-based measures for DRR (Eco-DRR) or adaptation (EbA) have only rarely been planned and implemented although they have a high potential for reducing underlying risk factors and contribute to more sustainable development in general due to their multiple co-benefits. Therefore, according to Emerton et al. (2016:38), "ecosystems should be treated, counted and invested in as an integral part of climate-compatible development infrastructure — as a stock of facilities, services and equipment which are needed for the economy to function, grow, adapt and maintain its resilience in the face of climate change and other hazards". The growing importance of EbA and Eco-DRR as well as the recognition of their benefits are proven by their acknowledgment in two major global initiatives, the Sendai Framework for Disaster Risk Reduction and the Sustainable Development Goals (SDGs).

This chapter is a science-policy paper based on a review of the relevant scientific literature and policy documents as well as own observations. Section 22.2 describes the concepts of EbA and Eco-DRR as a basis for further understanding of the chapter. Section 22.3 provides an overview of general policies at the national and state levels (RJ) regarding CCA and DRR focusing on ecosystem-based policies. For a better understanding of the linkages between CCA/DRR, land, and ecosystem management in RJ, Sect. 22.4 shows how deforestation and overexploitation contributes to increased disaster risk. Section 22.5 describes existing ecosystem-based approaches in RJ and analyzes the potential of these measures before concluding the chapter with an overview of existing challenges on broader implementation of ecosystem-based policies and measures in Southeastern Brazil.

22.2 Ecosystem-Based Approaches for Reducing Risks and Adapting to Climate Change

EbA and Eco-DRR both aim at the sustainable management, conservation, and restoration of ecosystems to cope with the adverse effects of climate change or to reduce the impacts of natural hazards (Munang et al. 2013; Renaud et al. 2013).

"Ecosystem-based adaptation is the use of biodiversity and ecosystem services as part of an overall adaptation strategy to help people to adapt to the adverse effects of climate change" (CBD 2009:41). The goal of EbA is to reduce the vulnerability of people and ecosystems against climate change impacts to maintain and increase resilience (Lo 2016). For example, maintaining or restoring mangroves along coastlines could reduce the risk of coastal flooding, storms, coastal erosion, and saline intrusion and supports adaptation to sea level rise (Barbier 2016; Guannel et al. 2016).

Eco-DRR can be defined as "the sustainable management, conservation, and restoration of ecosystems to reduce disaster risk, with the aim of achieving sustainable and resilient development" (Estrella and Salismaa 2013:30). Eco-DRR approaches are often called no-regret strategies as they provide additional co-benefits that traditional, mostly engineering-based approaches, do not usually offer (Renaud et al. 2016; Daigneault et al. 2016). Common examples are the reforestation of slopes to reduce landslide risks or the use of wetlands as retention areas to reduce flood risks (cf. Lange et al. 2014; Brink et al. 2016). Both EbA and Eco-DRR overlap in making use of well-managed and functioning ecosystems, their goods and services, to help people and communities adapt to the adverse impacts of climate change and disasters, often with similar measures, despite the different focal areas (Fig. 22.1). Both allow for hybrid approaches that link natural with engineered approaches, e.g., by combining dykes with mangrove areas for coastal protection (van Bohemen 2012; UNEP et al. 2014). In addition, they also share a multi-sectoral and cross-scale nature, requiring cross-disciplinary approaches and thus a multi-actor approach.

A growing number of case studies prove the suitability and comparable advantages of Eco-DRR and EbA measures (Renaud et al. 2016, Doswald and Estrella 2015). Some of these studies point out that, in addition to their comparably lower costs and often longer durability, they can support local livelihoods and mitigate climate change impacts if planned and maintained in a proper way (Renaud et al. 2013; Renaud et al. 2016; Daigneault et al. 2016), e.g., in the case of protected areas (Murti and Buyck 2014; Dudley et al. 2015).

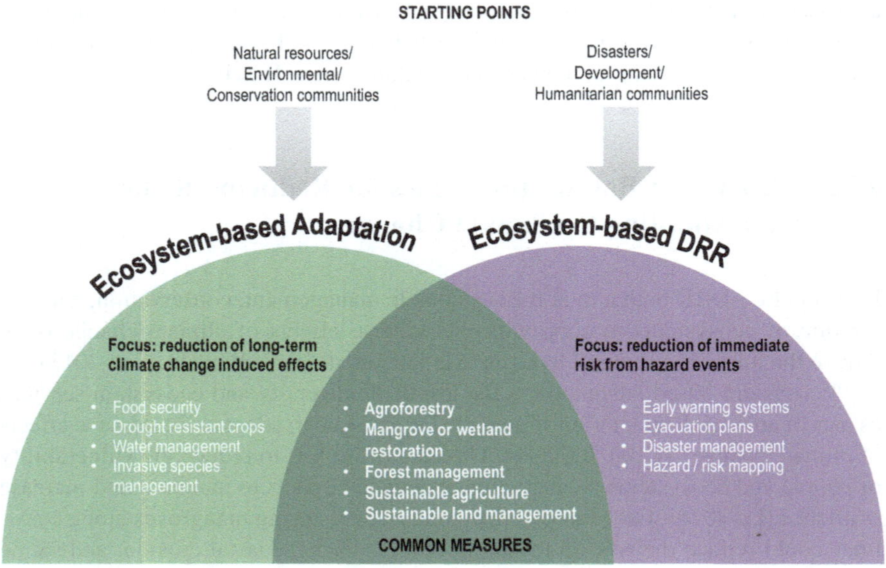

Fig. 22.1 EbA and Eco-DRR starting points, focal areas, and common measures. (Lange et al. 2014)

Among the challenges that have to be considered is the duration until an ecosystem-based solution is fully effective. Furthermore, such approaches can also fail, e.g., if non-suitable species are used for reforestation. Furthermore, responsible authorities have to be aware of their potential as well as the necessary skills to apply such methods. In addition, the multidisciplinary and multi-sectoral nature of Eco-DRR and EbA is challenging for planning which usually follows a sectoral approach (Renaud et al. 2013; Renaud et al. 2016). However, it may be possible to overcome such challenges by working through development planning and initiatives that serve as an "umbrella" framework for multi-sectoral engagement (Estrella & Salismaa 2013).

22.3 CCA and DRR Policies at National and State Level

CCA and DRR, and more so EbA and Eco-DRR, are relatively new topics in Brazil. On the one hand, a major boost in the formulation of policies targeting the reduction of natural and environmental hazards has come from international discussions and frameworks. At the same time, new policies were also formed in response to disasters that happened in the country, especially the landslides in the *Região Serrana* of RJ in 2011.

As a reaction to this latter event, in 2012 the National Congress legally approved implementation of the National Policy on Civil Protection and Defense (*Política Nacional de Proteção e Defesa Civil* – PNPDEC, Law n.12.608/2012), which establishes responsibilities at the national, state, and municipal level. A milestone of this legislation was the paradigm shift from disaster response to incorporating disaster risk prevention and mitigation in disaster risk management. However, although risk professionals confirm that this was the right step, risk prevention and mitigation strategies still lack implementation. According to national authorities, this has mainly to do with a lack of knowledge, as state and municipal disaster specialists were until now mainly trained in disaster response and are not familiar with most risk prevention and mitigation approaches, with the exception of technical safeguarding and early warning systems.

The implementation of the PNPDEC itself strongly depends on political will and financial possibilities. While national and state authorities received a strong financial and staff boost in the years after the release of the PNPDEC, the current financial crisis has also left its mark on the disaster management sector, cutting financial and human capacities. On the municipal level, the assignment of human and financial resources depends even more on the current political agenda. In those cities that were most affected by the 2011 events and where disaster events occur repeatedly, the memory of destruction is still fresh. Therefore, DRR is still high on the agenda. In other municipalities, such as Duque de Caxias, the Municipal Department of Civil Defense was drastically reduced in budget and staff after the change of government in early 2017.

Beyond DRR policies, CCA has been on the national political agenda since 2009, when the Nacional Climate Change Policy (*Política Nacional sobre Mudança do Clima* – PNMC, Law 12.187/2009) was established. Among others, it defines the elaboration of the National Adaptation Plan (*Plano Nacional de Adaptação* – PNA), which was instituted in May 2016, after a substantial participatory process. The PNA defines adaptation guidelines for 11 different sectors, including biodiversity, disaster management, and cities.

Most sectoral strategies in the PNA (10 out of 11) include EbA as a crosscutting approach. However, the document points out that the main bottleneck hindering EbA from becoming a mainstream topic in governmental and private sector plans is the lack of evidence of its efficiency. Therefore, many EbA-related activities of the PNA should first focus on generating cost-benefit information on EbA, identifying areas for its implementation to reduce climate risks and on monitoring and evaluation.

The most common EbA measure cited in the PNA is conservation and restoration of riparian and other types of forests, which are closely linked with related national policies. An example is the implementation of the Forest Code (*Lei de Proteção da Vegetação Nativa*, Law 12.651/2012), which establishes permanent preservation areas (*Áreas de Preservação Permanente* – APPs) and legal reserves (*Reservas Legais* – RLs) in private areas, besides the public protected areas system (Piratelli et al. 2018; Guzmán Wolfhard and Raedig 2018). APPs aim to protect water bodies and deep slopes. In addition, landowners have to reserve a minimum percentage of their land, between 20% and 80% depending on the biome, for native vegetation, constituting their RLs. When this vegetation is not existent, landowners are required to restore it in the short to medium term. In addition, the National Policy for National Vegetation Recovery (*Política Nacional de Recuperação da Vegetação Nativa*, Decree 8.972/2017) is fundamental. Another example are the Municipal Plans for the Conservation and Recuperation of the Mata Atlântica (*Plano Municipal de Conservação e Recuperação da Mata Atlântica*) for which the handbook for its elaboration recommends to research the vulnerability of the ecosystems and the services they offer for people for CCA and to propose measures for CCA (Dutra et al. 2013).

One of the most important instruments for implementing the Brazilian Forest Code is the Rural Environmental Register (*Cadastro Ambiental Rural* – CAR), which collects data, among others, of APPs and RLs of the whole country. Although these areas are not defined by a climate risk analysis and were not conceived with the intention to reduce climate vulnerability, they can show positive adaptation effects and could therefore be declared as EbA measures. Thus, the CAR is also an important tool for planning and implementing EbA in Brazil.

The same argument is also followed in the Guandu River basin, located in RJ. In 2009, the river basin committee, together with the state government, civil society, and research institutions started a large process to restore key areas with native vegetation. Big companies, such as Latin America's largest beverage producer Ambev, quickly joined the initiative. The goal is to guarantee adequate water quality and water supply of the river basin that is also key to ensuring water security in the metropolitan area of Rio de Janeiro. However, implementing organizations argue in this case as well that their measures are also reducing vulnerability to climate change and can therefore be labelled as EbA.

22.4 Linkages Between CCA/DRR, Land, and Ecosystem Management in Rio de Janeiro State

Alliance Development Works et al. 2012 pointed out that environmental degradation is an important factor that reduces the adaptive capacity of societies to deal with natural hazards. In the same year, the special report on "Managing the Risks of Extreme Events and Disasters to Advance Climate Change Adaptation" (IPCC 2012) demonstrated the close linkages between natural hazards and extreme events in the context of CCA. On page 69 of the report, it is explicitly mentioned that "the intensity or recurrence of hazard events can be partly determined by environmental degradation and human intervention in natural ecosystems." This statement can certainly be transferred to RJ.

Several research studies have shown that deforestation and overexploitation, in particular due to historical exploitation cycles, modern land use intensification, and urbanization processes (Fig. 22.2), have caused severe land degradation that in turn

Fig. 22.2 Deforestation for housing in Teresópolis. (Photo credit: Wolfram Lange)

has increased vulnerability to natural hazards and climatic extreme events (compare Nehren et al. 2018). Reforestation and rehabilitation are therefore key concepts in the context of EbA and Eco-DRR to maintain or restore protective ecosystem services and thereby enhance landscape resilience. However, so far most reforestation and land rehabilitation initiatives in RJ either focus on biodiversity conservation (see Guzmán Wolfhard and Raedig 2018; Piratelli et al. 2018), erosion prevention and soil restoration (Naegeli de Torres et al. 2018; Seliger et al. 2018), or agricultural management (see Sect. 22.2 in this book), while specific Eco-DRR/EbA measures are underdeveloped. Table 22.1 shows goals and ecosystem-based approaches in the five key sectors: (a) agricultural management, (b) water resources management, (c) ecosystem management and biodiversity, (d) climate change adaptation, and (e) disaster risk reduction. Forest ecosystems including agroforestry systems play a crucial role in each of these sectors. However, until now there has been little cross-sectoral cooperation on both the state and municipal levels to achieve the specific sectoral goals with joined ecosystem-based strategies and measures. For the municipality of Rio de Janeiro (metropolitan region), Lange et al. (2018) demonstrated that there is a particular gap between those authorities responsible for urban planning, conservation, and disaster risk management. Moreover, there is a lack of inter-municipal exchange and coordination, so that there are disproportionally high administrative expenses in each of the 91 municipalities of the state (see Frank et al. 2018).

After the 2011 event, a study by the Brazilian Ministry of Environment (MMA 2011) demonstrated the relation between APPs and environmental conservation units versus risk areas. Based on aerial and satellite imagery, areas affected by the heavy rainfall clearly showed the unequivocal linkage between damages due to the

Table 22.1 Key sectors related to the environment: overall goals and ecosystem-based approaches in rural Rio de Janeiro

	Agricultural management	Ecosystem management and conservation	Water resources management	DRR	CCA
Overall goals	Maintain soil fertility and ensure food security	Provide ecosystem goods and services and preserve biodiversity	Secure water supply and guarantee good water quality	Reduce disaster risk for people and their assets	Adapt to the medium and long-term impacts of climate change
Ecosystem-based approaches used in Rio de Janeiro state (at least as pilot measures)	Introduce agroforestry systems	Establish nature protection zones, conservation corridors, and areas of permanent protection (APPs)	Protect rivers and streams by preserving or restoring riverine forests (APPs)	Stabilize landslide-prone slopes by terracing and reforestation	Reforest larger areas to improve the regional climate and mitigate climate extremes
	Implement reforestation and rehabilitation measures on degraded lands	Foster the implementation of private natural heritage reserves (RPPNs)	Protect headwater regions by preserving or restoring the natural vegetation (APPs)	Protect flood-prone areas by maintaining or restoring riverine forests and creating retention room	
	Preserve key ecosystems and habitats that protect water resources and soils				

force of the water, landslides and mudslides, and the APPs. A major finding of the study is that more than 92% of the landslides within the researched area occurred where some type of anthropic alteration had taken place and only 8% in areas where natural vegetation predominated. However, as pointed out in Nehren et al. (2018), natural vegetation does not provide complete protection from landslides.

22.5 Status of EbA and Eco-DRR Measures in Rio de Janeiro State

So far, disaster risk management in RJ relies mostly on technical measures, i.e., protection dams, dykes, water reservoirs, rainwater retention basins, and river bank stabilization against flooding. The risk of landslides and sludge flows in settlements is mitigated with protective walls, the installation of drainage systems, and the sealing of sloped areas with shot concrete (Fig. 22.3). Along main traffic routes, large-scale hanger rides and geotextiles are used (Fig. 22.4). In some particularly vulnerable areas, early warning systems and evacuation plans have also been implemented.

In rare cases, ecosystem-based measures have been used, but are not yet considered large scale. Reforestation measures for flood protection are carried out sporadically, and measures for river restoration are initiated with the creation of retention areas. For protection against landslides, some of the endangered slopes are reforested and partly terraced. For example, the company for agricultural research in RJ (Pesagro-Rio) constructed a slope stabilizing measure for reducing landslide occurrence in the municipality of Nova Friburgo, which had been highly affected in 2011. After terracing the slope with sediment barriers and stabilizing it with geotextiles, grass and other *Leguminosae* were planted to generate a quick soil cover. After a closed vegetation cover had been formed, predominantly indigenous trees were planted but also some nonnative trees with the purpose of generating an economic benefit for the local communities. An agroforestry system was installed in the upper

Fig. 22.3 Protective structures with shotcrete, drainage pipes, and drain ditch in an informal settlement in Rio de Janeiro. (Photo credit Udo Nehren)

Fig. 22.4 Geotextiles for initial slope stabilization on a roadside in Teresópolis. (Photo credit: Wolfram Lange)

Fig. 22.5 During the extreme rainfall event in the Serrana Region in 2011, many slopes collapsed. Shortly after the event, Pesagro-Rio implemented slope stabilization and restoration measures, such as the one shown in the figure above in the municipality of Nova Friburgo. The slope was stabilized using palisades (**a**); later vetiver (*Chrysopogon zizanioides*) and other grasses and single trees were planted (**b**). (Photo credit: Aluísio Granato de Andrade and Tiago A. Chaves)

part of the area to improve the soil conditions and to reduce surface drainage. Maize, beans, lychee, and palm trees among others are being cultivated as they generate additional income (Nehren et al. 2017; Fig. 22.5).

Ecosystem-based measures for landslide risk reduction were also considered in the urban areas of RJ. In the municipality of Rio de Janeiro, for example, the main measure to date is the *Mutirão Reflorestamento*, the reforestation program of the municipal environmental agency, which has afforested more than 2000 ha of

Fig. 22.6 After a landslide, this slope bordering an informal settlement in Rio de Janeiro is being terraced and reforested. The measure is part of the program *Mutirão Reflorestamento* in 2015. (Photo credit: Simone Sandholz)

degraded land (Lange et al. 2018, Fig. 22.6). With the support of trained staff on the basis of voluntary community work, particularly steep slopes near informal settlements (favelas) were afforested for landslide risk reduction, among many other purposes. However, this program could be better oriented to disaster risk reduction, and many actors involved in disaster risk mitigation still focus mainly on gray measures such as protective walls, shotcrete, drainage systems, and classic precautionary activities like early warning systems (Sandholz et al. 2018).

The project Biodiversity and Climate Change in the Atlantic Forest (2013–2018) was proposed by a partnership between the Brazilian and German ministries of the environment (MMA and BMUB), through the International Climate Protection Initiative (IKI), with the technical cooperation of the German International Cooperation Agency (GIZ) and the financial cooperation of the German Development Bank (KfW). It seeks to mainstream EbA into planning at national, state, and local levels. The region of the Central Fluminense Protected Areas Mosaic, located in the *Região Serrana*, is one of its demonstration sites. In this context, the staff of the municipality of Duque de Caxias, directly bordering the city of Rio de Janeiro, were contacted in 2015 to implement a pilot project. The territorial planning department was fully aware of climate change and its impacts and sought ways for incorporating these considerations in the revision of its municipal master plan. From this starting point, the project and the municipality jointly planned and implemented a series of steps in order to mainstream EbA in urban planning.

The initial step was capacity building. First, the project facilitated the participation of two planning experts of the municipality in a Training of Trainers course on EbA. That was the basis for training additional staff within the department and to create an interdepartmental working group on adaptation to climate change. This resulted in a large success, as climate change became the first topic in years to convene different municipal departments on a regular basis, e.g., the departments of civil defense, planning, and health, therefore also facilitating interdepartmental dialogue. Secondly, in-depth diagnosis tools were applied using participatory processes. Using land use maps as a basis, local actors – experts, technicians, and local leaders – were interviewed on the relevance of land use and cover types and specific areas in the municipality as well as hotspots for ecosystem services, such as riparian vegetation, mangroves, and forests. Also in a participative approach, a vulnerability assessment was performed. Based on previously gathered information and the results of two multi-stakeholder workshops, experts were able to identify vulnerability hotspots and impact chains detailing drivers for vulnerability.

The results of these assessments will help the department of urban planning to review its municipal master plan, manage land use conflicts, and discuss different scenarios for resilient and sustainable urban development. Ultimately, the goal is to enhance productive activities while at the same time preserving key natural capital.

22.6 The Potential and Limitations of EbA and Eco-DRR in Rio de Janeiro State

Despite the promising cases of ecosystem-based approaches in RJ, these kinds of measures are only isolated cases or pilot studies in geographically limited areas, so their full potential has not been realized. However, at the national level, the government makes efforts for forest protection and restoration. In particular with the implementation of APPs on steep slopes (>45%), on hilltops, along rivers, and in headwater regions, the landscape can be stabilized. This means that APPs directly contribute to risk reduction and climate change adaptation. Further information and specific data on climate effects and disaster risk could help labelling these efforts as EbA and Eco-DRR measures, therefore supporting conservation and restoration initiatives to gain acceptance also in other sectors in addition to the environment.

Pilot projects such as the one in Duque de Caxias are also key to gaining experiences and evidence on EbA and Eco-DRR, in order to mainstream these topics on a national level. However, to increase the resilience of the landscape, packages of measures are needed which include, for example, large-scale reforestation for regulating the regional climate and the restoration of degraded pasture land (Seliger et al. 2018). The implementation of protection forests against landslides and rockfalls as well as retention areas for reducing flooding should be intensified. Synergies with biodiversity protection should also be used (see Scarano and Ceotto 2015) such as the design of biological corridors in a way that they simultaneously take over DRR functions. However, this requires a better coordination between disaster protection and environmental planning authorities.

A problem that ecosystem-based measures still face is less acceptance among decision-makers because their efficiency and cost-benefit ratio are difficult to determine and most of the measures are only effective in the long term. As ecosystem-based measures most often require the commitment of the local population, awareness raising campaigns and environmental education should incorporate these concepts to increase community perception for better engagement (Lange et al. 2014). There is also a lack of control mechanisms and appropriate environmental monitoring systems. From this point of view, the implementation and long-term monitoring and evaluation of pilot cases as the one in Duque de Caxias are key in order for EbA to be upscaled. Finally, change of land use is a major challenge as, for example, floodplains are intensively used for agriculture and settlements, and therefore relocation of settlements to create retention areas are costly and could generate major social and economic conflicts.

Overall, the potential for EbA and Eco-DRR is high in RJ. Existing projects and initiatives that already consider ecosystem services for CCA and DRR – whether purposefully or as a side benefit – can be strengthened and might serve as best practices to convince more actors and decision-makers in the region.

References

Alliance Development Works, United Nations University – Institute for Environment and Human Security, The Nature Conservancy (2012) World Risk Report 2012. Alliance Development Works, Berlin

Barbier EB (2016) The protective service of mangrove ecosystems: a review of valuation methods. Mar Pollut Bull 109(2):676–681

Brink E, Aalders T, Adam D et al (2016) Cascades of green: a review of ecosystem-based adaptation in urban areas. Glob Environ Chang 36:111–123

CBD – Secretariat of the Convention on Biological Diversity (2009) Connecting biodiversity and climate change mitigation and adaptation: report of the second ad hoc technical expert group on biodiversity and climate change. Montreal, Canada: Technical Series No. 41

Daigneault A, Brown P, Gawith D (2016) Dredging versus hedging: comparing hard infrastructure to ecosystem-based adaptation to flooding. Ecol Econ 122:25–35

Doswald N, Estrella M (2015) Promoting ecosystems for disaster risk reduction and climate change adaptation: opportunities for integration. United Nations Environment Programme, Geneva

Dudley N, Buyck C, Furuta N et al (2015) Protected areas as tool for disaster risk reduction, a handbook for practitioners. MOEJ and IUCN, Tokyo/Gland

Dutra CM, de Cordeiro SHT C, Cordeiro LA, Deitenbach A (2013) Roteiro para a elaboração dos Planos Municipais de Conservação e Recuperação da Mata Atlântica. Série Biodiversidade 48. Brasília/Brazil: MMA

Emerton L, Huxham M, Bournazel J, Priyantha Kumara M (2016) Valuing ecosystems as an economic part of climate-compatible development infrastructure in coastal zones of Kenya and Sri Lanka. In: Renaud FK, Sudmeier-Rieux K, Estrella M, Nehren U (eds) Ecosystem-based disaster risk reduction and adaptation in practice. Springer. Advances in Natural and Technological Hazards Research, Switzerland

Estrella M, Saalismaa N (2013) Ecosystem-based Disaster Risk Reduction (Eco-DRR): An Overview. In: Renaud F, Sudmeier-Rieux K, Estrella M (eds) The role of ecosystem management in disaster risk reduction. United Nations University Press, Tokyo

Frank B, Formiga-Johnsson RM, Viana VJ, Ikemoto SM (2018) Disaster risk management in the State of Rio de Janeiro. In: Nehren U, Schlüter S, Raedig C, Sattler D, Hissa H (eds) Strategies and tools for a sustainable rural Rio de Janeiro. Springer International Publishing, Rio de Janeiro

Guannel G, Arkema K, Ruggiero P, Verutes G (2016) The power of three: coral reefs, seagrasses and mangroves protect coastal regions and increase their resilience. PLoS One 11(7):1–22

Guzmán Wolfhard L, Raedig C (2018) Connectivity conservation management: connecting private protected areas. In: Nehren U, Schlüter S, Raedig C, Sattler D, Hissa H (eds) Strategies and tools for a sustainable rural Rio de Janeiro. Springer International Publishing, Rio de Janeiro

Heinrich J, Nehren U, Sattler D (2010) Waldfragmente des brasilianischen Küstengebirges – Entwicklungsdynamik und aktuelle räumliche Muster. Geogr Rundsch 9:34–41

IPCC (2012) Managing the risks of extreme events and disasters to advance climate change adaptation. A Special Report of Working Groups I and II of the Intergovernmental Panel on Climate Change. Cambridge University Press, Cambridge/New York

Lange W, Cavalcante L, Dünow L et al (2014) HumaNatureza^2 = Proteção Mútua. Percepção de riscos e adaptação à mudança climática baseada nos ecossistemas na Mata Atlântica, Brasil. SLE Publication Series Humboldt Universität zu Berlin, pp 255

Lange W, Sandholz S, Nehren U (2018) Strengthening urban resilience through nature: the potential of ecosystem-based measures for reduction of landslide risk in Rio de Janeiro. Working Paper. Cambridge, MA: Lincoln Institute of Land Policy (accepted – forthcoming)

Lo V (2016) Synthesis report on experiences with ecosystem-based approaches to climate change adaptation and disaster risk reduction. Technical Series No. 85. Secretariat of the Convention on Biological Diversity, Montreal. https://www.cbd.int/doc/publications/cbd-ts-85-en.pdf. Accessed 22 Mar 2017

Marengo JA (2014) O futuro clima do Brasil. Revista USP 103:25–32

MMA - Ministério do Meio Ambiente (2011) Áreas de preservação permanente e unidades de conservação & áreas de risco. O que uma coisa tem a ver com a outra? Relatório de Inspeção da área atingida pela tragédia das chuvas na Região Serrana do Rio de Janeiro. Série Biodiversidade. MMA, Brasília, Brazil, p 41

Munang R, Thiaw I, Alverson K, Liu J, Han Z (2013) The role of ecosystem services in climate change adaptation and disaster risk reduction. Curr Opin Environ Sustain 5(1):47–52

Murti R, Buyck C (eds) (2014) Safe havens: protected areas for disaster risk reduction and climate change adaptation. IUCN, Gland

Naegeli de Torres F, Richter R, Cardoso Fidalgo EC (2018) Multi-criteria site prioritization for land rehabilitation in the Guapi-Macacu Watershed, Rio de Janeiro. In: Nehren U, Schlüter S, Raedig C, Sattler D, Hissa H (eds) Strategies and tools for a sustainable rural Rio de Janeiro. Springer International Publishing, Rio de Janeiro

Nehren U, Kirchner A, Lange W et al (2018) Natural hazards and climate change impacts in the state of Rio de Janeiro: a landscape historical analysis. In: Nehren U, Schlüter S, Raedig C, Sattler D, Hissa H (eds) Strategies and tools for a sustainable rural Rio de Janeiro. Springer International Publishing, Rio de Janeiro

Nehren U, Kirchner A, Sattler D et al (2013) Impact of natural climate change and historical land use on landscape development in the Atlantic Forest of Rio de Janeiro, Brazil. Anais Acad Brasil Ciênc 85(2):311–332

Nehren U, Lange W, Kupka S (2017) Ökosystembasierte Maßnahmen zur Risikominderung und Anpassung an den Klimawandel in Südostbrasilien. Geogr Rundsch 4:26–31

Piratelli AJ, Marquez Piña-Rodrigues FC, Raedig C (2018) Integrating biodiversity conservation into agroecosystem management: Using birds to bring conservation and agricultural production together. In: Nehren U, Schlüter S, Raedig C, Sattler D, Hissa H (eds) Strategies and tools for a sustainable rural Rio de Janeiro. Springer International Publishing, Rio de Janeiro

Renaud F, Sudmeier-Rieux K, Estrella M (2013) The role of ecosystems for disaster risk reduction. United Nations University Press, Tokyo

Renaud FK, Sudmeier-Rieux K, Estrella M, Nehren U (eds) (2016) Ecosystem-based disaster risk reduction and adaptation in practice. Springer. Advances in Natural and Technological Hazards Research

Sandholz S, Lange W, Nehren U (2018) Governing green change: ecosystem-based measures for reducing landslide risk in Rio de Janeiro. Int J Disaster Risk Reduct. https://doi.org/10.1016/j-ijdrr.2018.01.020

Scarano FR, Ceotto P (2015) Brazilian Atlantic forest: impact, vulnerability, and adaptation to climate change. Biodivers Conserv 24:2319–2331

Seliger R, Sattler D, Soareas da Silva A et al (2018) Rehabilitation of degraded sloped pastures - lessons learned in Itaocara. In: Nehren U, Schlüter S, Raedig C, Sattler D, Hissa H (eds) Strategies and tools for a sustainable rural Rio de Janeiro. Springer International Publishing, Rio de Janeiro

UNEP, UNEP-DHI, IUCN et al (2014) Green infrastructure guide for water management: ecosystem-based management approaches for water-related infrastructure projects. UNEP, Geneva

van Bohemen H (2012) (Eco) System thinking: Ecological principles for buildings, roads and industrial and urban areas. In: van Bueren EM, van Bohemen H, Itard L, Visscher H (eds) Sustainable urban environments. Springer, Dordrecht, pp 15–70

Chapter 23
Integrated Participatory Methodologies for Disaster Risk Reduction: Tools to Analyze Complex Systems Through Participatory Processes in Brazil

Alicia Bustillos Ardaya, Mariele Evers, and Lars Ribbe

Abstract A crucial element for risk governance and disaster risk reduction in recognized international frameworks involves inclusiveness of stakeholders. Due to the dimensions of risk governance processes in Brazil, the complexity of the institutional arrangements in the state of Rio de Janeiro, and the dynamicity of the local context, a combination of different local and regional participatory methods is necessary to evaluate the participation and community-based disaster risk management that promotes it. This study describes five main participation processes applied in water-related risk areas of Nova Friburgo, Sao Fidelis, and Rio de Janeiro municipalities and implements it through combined participation methods. Exemplified focus groups and participatory workshops are described, both combining five methods: the required planning and organization, adaptation for rural and peri-urban areas, tools for an effective participatory mapping and a network analysis, digitalization of participatory maps, etc. The influences, advantages, disadvantages, and inputs of the different methodologies are analyzed and compared. Combining methods requires time, resources, and constant work; nevertheless, it helps all

A. Bustillos Ardaya (✉)
Department of Geography, Bonn University, Köln, Germany

Center for Development Research ZEF, Bonn University, Köln, Germany

TH Köln - University of Applied Sciences, Institute for Technology and Resources Management in the Tropics and Subtropics (ITT), Köln, Germany
e-mail: alicia@uni-bonn.de

M. Evers
Department of Geography, Bonn University, Köln, Germany
e-mail: mariele.evers@uni-bonn.de

L. Ribbe
TH Köln - University of Applied Sciences, Institute for Technology and Resources Management in the Tropics and Subtropics (ITT), Köln, Germany
e-mail: lars.ribbe@th-koeln.de

U. Nehren et al. (eds.), *Strategies and Tools for a Sustainable Rural Rio de Janeiro*, Springer Series on Environmental Management,
https://doi.org/10.1007/978-3-319-89644-1_23

stakeholders understand complex systems and actively participate in decision-making. For various levels of participation, a combination of quantitative and qualitative methods allows more interaction between stakeholders and different perspectives for deeper evaluation of participation and related aspects.

Keywords Risk governance · Participatory mapping · Exchange workshop · Cognitive mapping

23.1 Introduction

In January 2011, floods and landslides in the mountainous region of the state of Rio de Janeiro (RJ) caused around 900 deaths and 300,000 affected people, as confirmed by official data (World Bank 2012); other sources suggest that actual losses were 8–10 times greater (Custódio and Corrêa 2012; Canejo 2015), being classified as the worse disaster in the century (BBC 2011; da Mata Siqueira 2015). Floods, landslides, droughts, and other water-related disasters are increasing problems strongly linked to poor governance, marginalization, extreme poverty, lack of infrastructure, information, and education (Adikari and Yoshitani 2009; Guha-Sapir et al. 2015; Munich Re 2016). Improving risk governance (RG) for disaster risk reduction (DRR) is a challenge that must include natural, political, and social systems in its analysis (IRGC 2010) and requires several approaches that explore diverse processes and actors (Goldin 2014). For inclusion and collectiveness, participatory approaches have proven to be very helpful in understanding, accepting, and managing complex socio-environmental systems (UNISDR), contributing constant search of additional improvements, insights, and learning for RG and DRR (Dyer et al. 2014; Vacik et al. 2014:307; Folhes et al. 2015; Hassenforder et al. 2016:504).

According to the EU Water Framework Directive recommendations (WFD 2010), public participation can be defined as "a process allowing people to influence the outcome of plans and working processes." Besides the fact that participation also implies risks and can be resource intense (Evers 2012), it can result in positive behavioral changes for all stakeholders, improving decision-making; promoting active involvement, exchange, and empowerment; building trust and commitment among them; and creating connections in the knowledge-power and management constitution (Clark-Ginsberg 2017:431). In European and North American countries, participation methods are being applied (Evers 2012; Brown et al. 2017); however, information is needed about systematic processes and methods to be adapted for countries with local limitations and less resources. Several methods aim for better involvement of more stakeholders, but diversification of methods at implementation is recommended to better understand, model, and find solutions (Vacik et al. 2014:308). Evaluating some of the existing participatory methodologies would facilitate future planning for governmental and nongovernmental organizations in similar regions: eligibility, combination, and implementation of participatory processes.

In this chapter we present a combination of participatory methodologies to assess and promote participation in DRR processes and evaluate them with possibilities and limitations. An insight of citizens' selection, consultation, involvement, communication tools, decision-making, management, and feedback will be given, exemplified with five main participative methods used in the field: questionnaires, interviews, focus groups (FG) including participatory mapping, timeline and prioritization exercises, and participatory workshops (PW) including a multi-voting system and product schemes. The methodologies will be cross analyzed with criteria modified from participatory approaches, but the results of the methods will not be given considering the length limitations of this chapter.

The study case area and the five main participation methods used in the field and theories will be described in Sect. 23.2. Section 23.3 will present the framework and the table with the weighted methodologies and processes and will describe the methodologies in the order of the presented criteria, defining important steps and exemplifying them with the study case. Section 23.4 will give conclusions plus further steps for new processes.

23.2 Research Area and Methods

Between August 2014 and December 2016, data was collected mainly in four municipalities in RJ (Fig. 23.1) through five main methods (Table 23.1). Semi-structured interviews for experts and authorities working at the main institutions related to DRR were the starting point and a reference for the formulation of the questionnaires. Questionnaires for population in risk areas consisted of four main parts: (1) general demographic data; (2) previous experiences with natural disasters and adopted mitigation measures; (3) risk perception, e.g., severity and likelihood; and (4) coping appraisal. Participant observation was applied during other meetings, workshops, discussions, and courses for involved stakeholders.

FG in different neighborhoods in risk areas were developed for three aspects: (1) timelines of events to introduce and encourage participation, (2) participatory mapping for georeferencing and collective construction, and (3) problem identification, voting, and prioritization to agree on main concerns related to DRR. Finally, for a systematized construction of processes and roles through dialog and information exchange between civil society and public institutions, three PW were conducted in 2014, 2015, and 2016. Main bottlenecks, overlapping and missing processes along 3 years, could be found, building examples for bottom-up and top-down approach.

A total of 26 interviews, 391 questionnaires, 11 FG, and 3 PW were conducted. In addition, 25 events around DRR were observed under the participant observation method by the first author of this study. An average of nine participants took part on each FG; most of them previously had taken part on questionnaires organized by the same team. All participants were invited to the yearly workshop (Table 23.1).

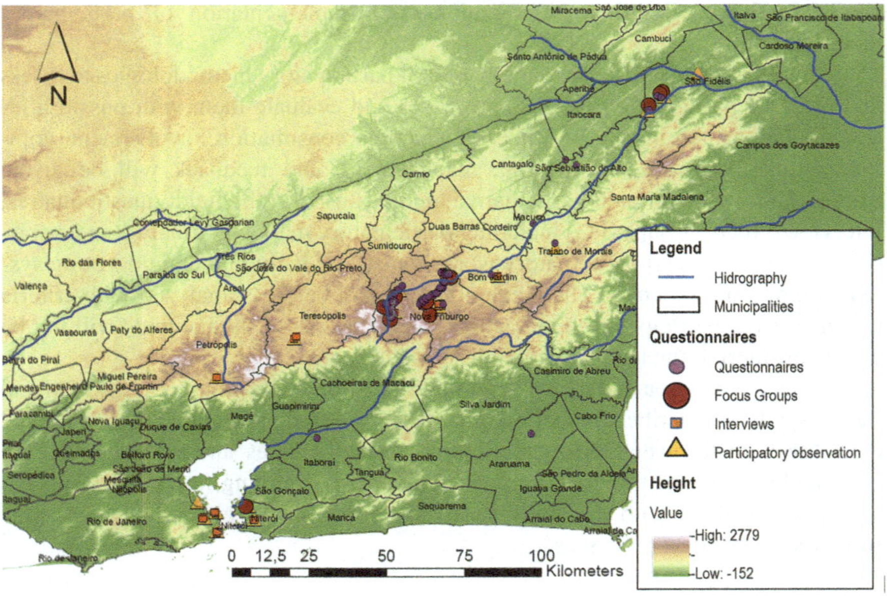

Fig. 23.1 Research area and methods used at the respective scale in the field

Table 23.1 Description of the methods implemented

Main method	Methods used	Method theory	Quantity / population included	Duration (each)	Planning and description
Participatory workshops PW	Four Rs framework, cognitive mapping, policy exercises	Brewer (1986); Eden (1992); Dubois (1998)	3 / 58	1 day	Contact with speakers, contact with participants, preparation of activities, etc. cognitive structure of knowledge built in a map and divided in rights, responsibilities, revenues, and relationship. Formulation of policies in the third workshop
Focus groups FG	Participatory mapping, timeline, voting, and prioritization	Kangas et al. (2006); Ares and Risler (2013); Schensul (2013)	11 / 62	3 h	Contact with the participants, elaboration and printing of maps and labels, and preparation of the material. Collective creation of social and risk maps, local history, and problem-solving through prioritization

(continued)

Table 23.1 (continued)

Main method	Methods used	Method theory	Quantity / population included	Duration (each)	Planning and description
Interviews	Influence matrix, stakeholder analysis	Fontana and Frey (1994); Ulrich and Probst (1995); Grimble (1998)	26 / 26	1–2 h	Preparation of the subjects to be discussed, constant contact for appointment, confirmations, and feedback. Mapping of actors and their influence, clear role of stakeholders, and their connections
Questionnaires	Questionnaire design, Q method	Taylor-Powell (1998); Steelman and Maguire (1999); Walonick (2003):6	391/ 391	0.5–1 h	Preparation and validation of the questionnaires, recognition of the houses under risk and repetition schedule. Main aspect for questionnaire development and data processing methods according to data taken
Participant observation	Stakeholder advisory committees	DeWalt and DeWalt (2002); McGurk et al. (2006); Neuman (2006)	25/ ~300	1 h to 5 days	Availability for conferences, workshops, and meetings, networking for information about events. Specific considerations for participant observation.

Above all recommendations given in the literature, the use of several methodologies requires organizational skills responding to a framework and a perspective for all interdisciplinary aspects building the criteria on which the methodologies will be evaluated.

23.3 Participation Processes

Taking communication and information, i.e., sharing and creation as a central role (IRGC 2010), four levels of participation (Evers 2012) are combined to the levels of authority and power (Fung 2006) in a loop (Fig. 23.2). In order to increase participation, benefiting individuals involved on the process and the process per se, a continuously looping system is implemented. In most of the methods presented in Table 23.1, communication is crucial and a central aspect for RG (Renn and Walker

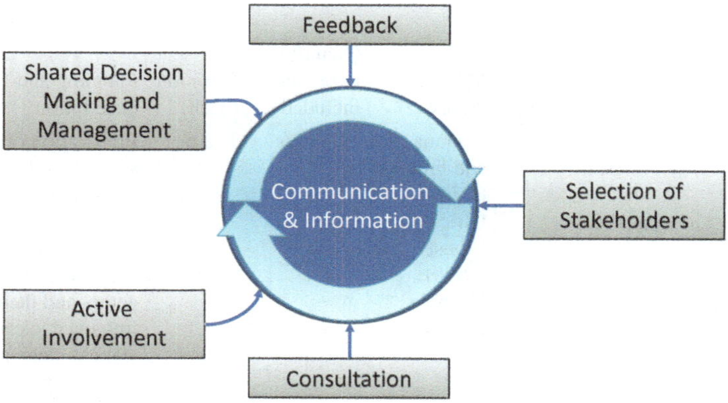

Fig. 23.2 Participation looping process for diversification of methods

2008). Communication tools vary from e-mails, phone calls, and even WhatsApp messages for arrangement, organizational aspects, and follow-ups to mostly direct communication for all described methods.

This continuous process diversifies methods to understand participation at various levels and to constantly select stakeholders according to availability and interests. Diverse methods had different influence on participation levels, but at the same time, participation levels influenced on the methods (Table 23.2). In this case, selection and consultation processes are mostly covered by interviews and questionnaires. Active involvement requires multilateral dialog; in FG and PW, ideas can be created, constructed, and exchanged. In order to improve shared decision-making and management, collective understanding of the problems and dialog between population living in risk areas and decision-makers needs to be built. Feedback is crucial for constant improvement. Results, i.e., conclusions and solutions in all processes, are input for other methods. It is expected that connections and networks built grow further. In the presented case, there are specific conditions that enabled or constrained connections that will be presented in order of participation levels presented in Fig. 23.2.

23.3.1 Selecting Stakeholders

The scope of participants in questionnaires, interviews, and participant observation is embracing (general public). Besides involving more population, questionnaires are individual processes, and planning is minimized once the structure is defined. Questionnaires may be taken in several not consecutive days, and data processing may be fast and systematic. PW and FG are more exclusive, e.g., experts/representatives, and depend more on space, time, and budget possibilities. In this case, more stakeholders/participants hinder interaction and individual decision.

Table 23.2 Weighted connections between methods and participation levels (3 = a lot; 2 = somehow; 1 = little; 0 = nothing) in Nova Friburgo, Brazil

How much do(es)… Contribute/influences on …	Participatory workshops PW	Focus groups FG	Interviews	Questionnaires	Participant observation	Selecting stakeholders	Consultation processes	Active involvement	Shared decision-making and &management	Feedback	Communication and information	Trust
Participatory workshops PW	0	3	2	1	2	3	2	3	3	1	2	3
Focus groups FG	0	0	2	3	2	3	2	3	2	1	2	2
Interviews	0	0	0	2	1	1	2	0	0	1	1	1
Questionnaires	0	0	2	0	1	0	1	0	0	0	0	0
Participant observation	2	2	1	1	0	0	0	0	0	0	0	0
Selecting stakeholders	0	0	2	2	1	0	0	0	0	0	0	0
Consultation processes	1	2	3	3	1	1	0	0	0	1	2	2
Active involvement	3	3	1	1	0	0	2	0	0	2	3	2
Shared decision-making and management	3	3	0	0	0	1	2	3	0	1	3	3
Feedback	3	2	1	0	0	0	1	2	1	0	3	3
Communication and information	3	3	2	1	1	1	3	3	2	3	0	3
Trust	3	3	2	1	0	0	1	2	3	3	3	0

In order to start with a local approach, we identified houses on risk areas near to rivers and conducted a door-to-door survey. The process required many visits at different hours to reach different population, e.g., day/night workers. At the end of the questionnaire, those surveyed were invited to a FG in their neighborhood. Response level in the study area was of 56% of the intended households to be surveyed. The most common limitation in the selection of stakeholders is a population not willing to participate. Main reasons in this case are time limitations; lack of incentives, e.g., monetary; social or perceived effectiveness on changes; mistrust on the institutions organizing and managing the mechanisms; or bad experiences with previous processes. Credibility to institutions is low; those surveyed affirmed that several non-concluded projects and visits from governmental institutions or NGOs had no visible impacts or results.

For interviews the approach was institutional, selecting main institutions working on DRR and interviewing main representatives following an official institutional arrangement (Consórcio GITEC / IP/ CODEX REMOTE 2013). After each interview, following the snowball sampling method, more contacts were collected. Every previously interviewed person enabled disposition for an interview with new contacts. In this way, state institutions will lead to local-level institutions and actors. More than 11 institutions working on DRR and representatives from the 6 local associations were interviewed after 2 cancellations and more than 15 postponements or delegations to other co-workers. All questionnaire, interview, and/or FG participants were invited to the PW held during a weekday. Although food and transport were covered for PW and FG, some representatives of civil society were not able to leave their positions for a day to attend this PW, which led to a limited representation.

23.3.2 Consultation Process

The focus of the consultation processes is to get stakeholders perception. This is characterized by medium level of interactivity and no responsibility from the stakeholder. Three documents provide an impression of consultation processes at three levels: state, municipal, and individual. The institutional arrangement organized all processes and responsibilities of main institutions working on DRR through a participation process (ConsórcioGICR 2013). In a smaller scale, before the renovation of the municipality master plan, the township of Nova Friburgo offered local consultation sessions in several neighborhoods (PMNF 2016), allowing the population to express important concerns to be considered for future development. Community meetings for consultation offered a possibility for follow-ups and participation in local projects. Nevertheless, participants complained about their suggestions not being considered, and the developers of the municipal plan complained about the suggestions being individualistic and for the sake of their own land or property. The perception of the population living in risk areas was measured through questionnaires, relating local knowledge with institutional trainings (Bustillos Ardaya et al. 2017). In individual consultations feedback and interaction are restricted, and answers are limited to or biased by predefined possibilities, yet they offer a picture of the perception, motivation, and simplified history of almost 400 dwellers.

The consultation process in semi-structured interviews developed by the first author follows a basic structure that can be managed by participants according to perceived importance and preferences. In order to get into DRR subjects, people recall previous experiences with floods and landslides (2011, 1997, 1985, etc.) and important dates in their community, e.g., when roads, electricity, or the Internet was installed, constructions, etc. This qualitative consultation process helps both the interviewed to get into the subject and the interviewer to understand the previous dynamic experienced by individuals and collective memory in case of FG and PW. Collective consultations allow discussion about priorities and collective wellbeing. In this case, consultation process in FG is better enabled through participatory mapping that allows problems to be located and possible solutions to be geographically referenced.

23.3.3 *Involvement*

Involvement requires contribution of stakeholders in planning, analyzing, and solving problems (Evers 2012). Participant observation in different events related to DRR and RG allows the understanding of involvement level of institutions and organizations: dialogs between stakeholders, information sharing, and relationships among several networks. Since 2011 the constant work of DRR-related institutions with the Japan International Cooperation Agency (JICA) has created an institutional network where communication, capacity building, and participation were continuous. According to a JICA representative, although focused only in public state institutions and very centered on Japanese methodologies, all workshops, trainings, and events organized by the JICA strengthened this network. Interviews and questionnaires can trigger analysis and questions ways of collaboration and possible solutions at an individual level. On the other hand, is collective involvement of the participants that may present a problem integrally and create discussion. Participant's selection and participation level will influence on the outcomes if stronger opinions are imposed. Participatory mapping provides advantages for involvement, and nowadays tools like GIS and Internet-based tools create high validity on perception of the population (Brown et al. 2017). Geographic information and instruments improve local work and improve decision capacity of the citizens (Usón et al. 2016). Nonetheless, involvement of local population in rural and peri-urban areas, where access to technologies and the Internet is limited, requires adaptation of existing methods. The adapted version of risk mapping can be replicated at low costs and is precise enough to be translated later to digital maps or shapefiles.

Maps for FG consisted of simplified satellite pictures of the neighborhoods printed on A0 paper sheets where only roads, houses, and river paths are visible as simple lines (Fig. 23.3). Different from suggested participatory mappings that start on blank sheets, these elements simplify location and digitization processes, making them geographically accurate. During the FG, existing elements, e.g., infrastructure, crops, springs, and problems in the area, e.g., landslides, droughts, water contamination, loss of springs, and manipulation of the river, are marked with illustrated labels and discussed. Labels simplify the process and encourage involvement: drawing or writing on the maps requires extra skills that not all participants have. Maximum water level experienced in floods and normal overflow are delineated together with evacuation routes and possible shelters. Discussion is encouraged as a promoter of involvement, information, and perception sharing in the group. Additionally, this process organizes personal ideas, shapes them collectively, and creates a product to be shared. Maps are later digitalized, made available online, printed onsite for validation, and compared to the flood risk maps created by INEA and the PMNF in nine evaluated areas using ArcGIS.

In order to encourage involvement, the observations, suggestions, and discussions during the participatory mapping should be considered for decision-making and management. Although the PW has the main goal of connecting institutions and dwellers for decision-making, by being an independent initiative, all results could only be taken as suggestions.

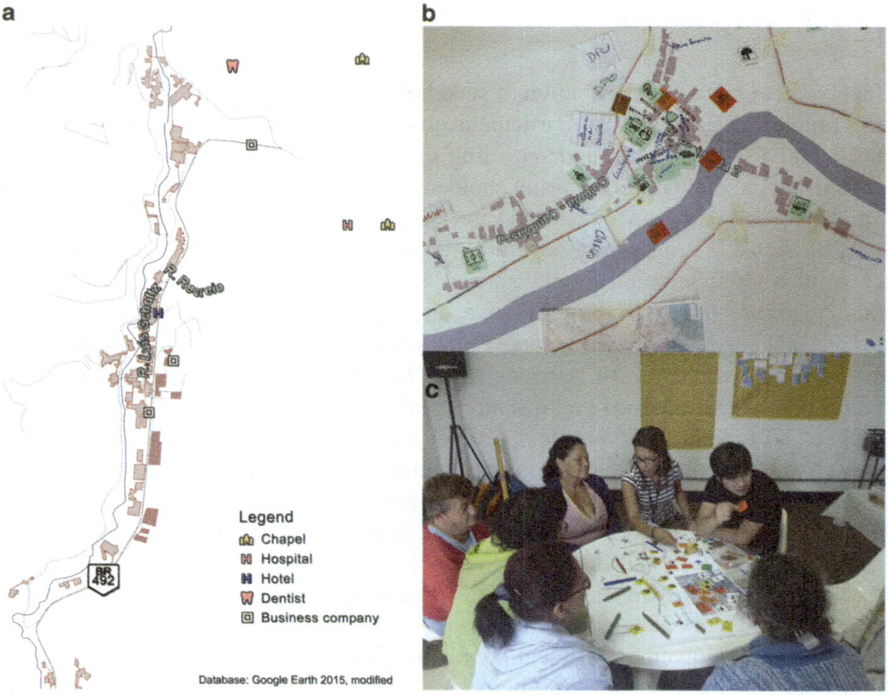

Fig. 23.3 A0 sheets with roads, houses, and rivers location (**a**) being filled by the participants (**b**, **c**). Photos: Conrado Werneck Pimentel

23.3.4 Shared Decision-Making and Management

The third phase of the FG discusses future perspectives and simulates management, dividing it in spatial and organizational. Elements needed in the community, e.g., hoarding locations, contention walls, or water plants, are pointed and located in the map in the best possible setting. Organizationally required aspects, e.g., better contact with institutions, strengthening in groups or skills, and potentialities, e.g., courses to offer, connections with organizations, or knowledge in bureaucratic processes, are discussed, visualized, and prioritized. A voting mechanism allows participants to debate, set priorities as a group, and has a clear outcome that can be easily transmitted in workshops or meetings as processed and validated information.

In a space for dialog between local population and institutional representatives, the PW exchanged perception at local and regional scale based on a systematic vision construction method. After the presentations of local organizations and governmental institutions, processes for DRR are mapped identifying rights, responsibilities, revenues, and relationships (four R's). A multi-voting dynamic locally created (Bustillos Ardaya 2015) is implemented to allow discussion and get the collective priority of all working groups. Mixed discussion tables are built where priorities to improve DDR are individually written and located in the table. The received priorities are analyzed, and a priority scale is added. The meeting and divergence points are discussed, and the table must agree in five priorities. Discussion allows debate and common points between civil society and public institutions that otherwise have few opportunities for dialog.

Finally, an additional decision-making exercise is implemented: after identifying overlapping and missing processes for DRR between governmental and local institutions (Fig. 23.4), alternatives are discussed for policy development. The exercise allows stakeholders to clarify activities and products of their institution related to DRR, including inputs and outputs necessary for each institution. They differentiate between processes before, during, and after an event (under or above the pointed line) and the outcomes for the civil society or public power (red and green cards).

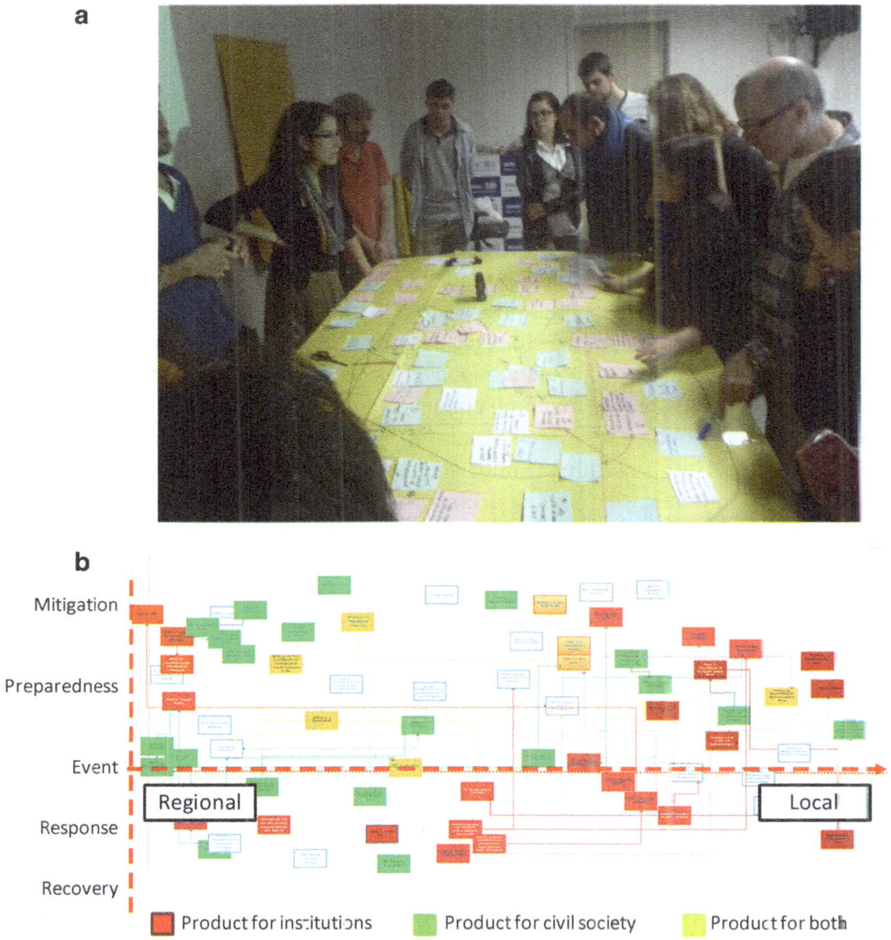

Fig. 23.4 Products of institutions and organizations in Nova Friburgo (a). Workshop result (b). (Photo: Conrado Werneck Pimentel)

Subsequently, they connect their activities and products with the agencies/institutions/stakeholders that use them (arrows) and suggest both new uses for the data and new sources for their inputs (pointed arrows). Having a general vision of the most connected products, e.g., risk maps and prevention sites, highlighted the main necessary elements for DRR attributed to local population and attributed to public power.

Despite being a theoretical and technical approach, the PW and the FG contribute significantly to the decision-making and management, communication, and trust. As affirmed by Clark-Ginsberg (2017), placing institutional experts and local leaders ends up blurring the lines between them and between researchers and subjects.

23.3.5 Communication, Feedback, and Trust

Changing environments and solving complex problems require dialog, exchange, and deliberation to assess knowledge and causal connections in a top-down and bottom-up approach (Fekete 2012; Mauelshagen et al. 2014). Communication is a cross-subject and is related to all previously listed elements. Official documents as the US National Research Council's "Red Book" (Stern and Fineberg 1996) and the Royal Society Study Group report (Royal Society Study Group on Risk Assessment 1992) recognize communication important for risk reduction. The Hyogo Framework for DRR (ISDR 2005) gives high priority to risk and hazard warning through communication. Furthermore, in the study case taken, it triggered social change, ameliorating multi-stakeholder understanding as Fekete describes (Fekete 2012). The dialogs during participatory mapping enhanced social identification, environmental values, in-group injunctive norms, and self-stereotypes, improving interagency coordination as Smith clarifies (Smith and Dowell 2000). In addition, during the workshops communication made decision-making more inclusive (Horita and Albuquerque 2013) and more effective (Kapucu and Garayev 2011). Innovation was also required for information diffusion of FG and PW; use of WhatsApp groups has proven to be more effective than e-mails in terms of response quantity and velocity.

Done appropriately, feedback may improve communication and build trust. Feedback should be asked at all steps and in both ways. On selection, stakeholders must be informed about the objectives, limitations, and main outcomes expected. After every consultation, involvement, and management session, it is important to ask for and give feedback and create opening spaces to update the group and closing sessions evaluating the method. Considering the several activities and concepts developed around DRR in different cities, information on the ongoing subjects, the results, and upcoming dates is important by opening every session. Feedback for FG and PW was above average, but on interviews it was clear that no big changes are expected after this kind of activities.

All methods used increase participation levels in a small scale; more work on a regular basis is required to improve these processes. Trust requires more time, constancy, resources, and involvement from all sides. Because of the duration of the field research, continuity had to be interrupted, and results could only be suggested to the government. Since limitations in time and reach were clarified to all participants at the beginning, the results are expected to vary from a governmental process. Assuring continuity and accountability of the results at a governmental level (and accomplishing it) may considerably increase trust.

23.4 Conclusions and Further Steps

Empowerment of the local population and collective decision-making is crucial in DRR, especially in complex natural and social systems like RJ. This chapter describes a series of participatory methods in five main methodologies with enablers and constrainers for participation and presents a study case where all are combined. Even when there are multiple possibilities to combine methodologies, exploring one of them illustrates the contributions of each method to the process. While participant observation, questionnaires, and interviews are better at selecting stakeholders and consulting them individually, group interactions are needed for further steps. Different activities in the focus groups and the participant workshops allow better communication and active involvement, improving the shared decision-making and therefore the management.

The consultation process in local households showed that the lack of trust on the institutions is evident. Even though the path to follow to reach visible change in DRR systems in a bureaucratic country like Brazil is long and tedious, participatory processes allow the construction of stronger networks that increase trust or reduce mistrust and improve knowledge sharing among local stakeholders. The PW allowed local stakeholders to understand the system, point on specific improvement possibilities, and through discussion understand the main concerns of institution representatives. Although being suggestion for governmental institutions and even being time and resource intense, there is great potential in the use of these methodologies combined to increase participation: the increase of communication among stakeholders and the use of local information both for public power and research create stronger networks that contribute to the development of DRR and RG among others.

A follow-up on the study case is needed to analyze how the dynamics at bottom-up and top-down level continue to work given all externalities in the actual context. More research in methods that may contribute to empowerment of local population will allow a spectrum of possible methods for specific topics. Future case studies or methodology applications may require different adaptations on the methods. Population access to technology will define communication and working tools, knowledge will determine the translation necessity, and the frequency of the loop will be set by resources, participation willingness, and time availability. In order to increase trust and improve decision-making capacity, a closer work with decision-makers at all levels is needed. After the consultation process, the most appropriate methods should be selected according to the needs of the main stakeholders.

Acknowledgments The authors would like to thank the Center for Natural Resources and Development, CNRD, for providing a PhD scholarship to conduct this research and the project INTECRAL for its logistical support.

References

Adikari Y, Yoshitani J (2009) Global trends in water-related disasters: an insight for policymakers. International Centre for Water Hazard and Risk Management ICHARM, United Nations Educational, Paris 07 SP, France

Ares P, Risler J (2013) Manual de mapeo colectivo. Recursos cartográficos críticos para procesos territoriales de creación colaborativa. Buenos Aires: Tinta Limón

BBC (2011) Brazil flood and mudslide deaths rise as rescue goes on – BBC News. http://www.bbc.com/news/world-latin-america-12180079. Accessed 16 Mar 2017

Brewer GD (1986) Methods for synthesis: policy exercises. In: Clark WC, Munn RE (eds) Sustainable development of the biosphere. Cambridge University Press, Cambridge

Brown G, Strickland-Munro J, Kobryn H, Moore SA (2017) Mixed methods participatory GIS: an evaluation of the validity of qualitative and quantitative mapping methods. Appl Geogr 79:153–166. https://doi.org/10.1016/j.apgeog.2016.12.015

Bustillos Ardaya A (2015) Metodología para a priorização construtiva de elementos para a redução de riscos. Interview with Thiago Mello. Nova Friburgo RJ. Voice recorder

Bustillos Ardaya A, Evers M, Ribbe L (2017) What influences disaster risk perception? Intervention measures, flood and landslide risk perception of the population living in flood risk areas in Rio de Janeiro state, Brazil. Int J Disaster Risk Reduct. https://doi.org/10.1016/j.ijdrr.2017.09.006

Canejo A (2015) N° de vítimas em tragédia no RJ pode ser 10 vezes maior, dizem entidades. Globo. http://g1.globo.com/rj/regiao-serrana/noticia/2015/08/entidades-apontam-subnotificacao-de-vitimas-da-tragedia-de-2011-na-serra.html. Accessed 22 Jul 2016

Clark-Ginsberg A (2017) Participatory risk network analysis. A tool for disaster reduction practitioners. Int J Disaster Risk Reduct 21:430–437. https://doi.org/10.1016/j.ijdrr.2017.01.006

Consórcio GITEC/IP/CODEX REMOTE (2013) Concepção do arranjo Institucional e operacional para a gestão de risco de desastres no estado do Rio de Janeiro

Custódio A, Corrêa T (2012) Tragédia na Região Serrana teve mais mortos que o divulgado. Extra. http://extra.globo.com/noticias/rio/tragedia-regiao-serrana-2011/tragedia-na-regiao-serrana-teve-mais-mortos-que-divulgado-5685873.html. Accessed 22 Jul 2016

DeWalt KM, DeWalt BR (2002) Participant observation: a guide for fieldworkers. AltaMira Press, Walnut Creek, CA

Dubois O (1998) Capacity to manage role changes in forestry. Introducing the 4R's framework. In iied. Forest Participation Series; 1. ISSN 10266887

Dyer J, Stringer LC, Dougill AJ et al (2014) Assessing participatory practices in community-based natural resource management: experiences in community engagement from southern Africa. J Environ Manag 2014:137–145

Eden C (1992) On the nature of cognitive maps. J Manag Stud 29(3):261–265. https://doi.org/10.1111/j.1467-6486.1992.tb00664.x

Evers M (2012) Participation in flood risk management. An introduction and recommendations for implementation. Karlstad: Centrum för klimat och säkerhet, Karlstads universitet (Rapport/Centrum för klimat och säkerhet, 2012:1)

Fekete A (2012) Safety and security target levels. Opportunities and challenges for risk management and risk communication. Int J Disaster Risk Reduct 2:67–76. https://doi.org/10.1016/j.ijdrr.2012.09.001

Folhes RT, Dutra de Aguiar AP, Stoll E et al (2015) Multi-scale participatory scenario methods and territorial planning in the Brazilian Amazon. Futures 73:86–99. https://doi.org/10.1016/j.futures.2015.08.005

Fontana A, Frey J (1994) Interviewing the art of science, Pp. 361-76 in The Handbook of Qualitative Research, edited by N. a. Y. L. Denzin. Thousand Oaks: Sage Publications.

Fung A (2006) Varieties of participation in complex governance. Public Adm Rev 66(s1):66–75. https://doi.org/10.1111/j.1540-6210.2006.00667.x

Goldin I (2014) The butterfly defect. How globalization creates systemic risks, and what to do about it. Edited by Mike Mariathasan. Princeton University Press, Princeton

Grimble R (1998) Stakeholder Methodologies in natural resource management. Socioeconomic Methodologies. Best Practice Guidelines. Chatham, UK:Natural Resources Institute

Guha-Sapir D, Below R, Hoyois Ph (2015) EM-DAT: The CRED/OFDA International Disaster Database. Université Catholique de Louvain. Brussels – Belgium. http://www.emdat.be/database. Accessed 28 Sept 2016

Hassenforder E, Ducrot R, Ferrand N et al (2016) Four challenges in selecting and implementing methods to monitor and evaluate participatory processes: example from the Rwenzori region, Uganda. J Environ Manag 180:504–516. https://doi.org/10.1016/j.jenvman.2016.05.019

Horita FE, Albuquerque JP (2013) An approach to support decision-making in disaster management based on Volunteer Geographic Information (VGI) and Spatial Decision Support Systems (SDSS). Supporting disaster management using VGI and SDSS

IRGC (2010) An introduction to the IRGC Risk Governance Framework, from the 1 IRGC White Paper No1 "Risk Governance – Towards an Integrative Approach', IRGC, Geneva, 2005.

Kangas A, Laukkanen S, Kangas J (2006) Social choice theory and its applications in sustainable forest management – a review. Forest Policy Econ 9(1):77–92. https://doi.org/10.1016/j.forpol.2005.02.004

Kapucu N, Garayev V (2011) Collaborative decision-making in emergency and disaster management. Int J Public Administration 34(6):366–375. https://doi.org/10.1080/01900692.2011.561477

da Mata Siqueira AM (2015) Conflitos em contextos de desastres relacionados com as águas. In Riscos de desastres relacionados à água

ISDR (2005) Hyogo Framework for Action 2005–2015: Building the Resilience of Nations and Communities to Disasters. Extract from the final report of the World Conference on Disaster Reduction. http://www.unisdr.org/files/1037_hyogoframeworkforactionenglish.pdf. Accessed 14 May 2016

Mauelshagen C, Smith M, Schiller F et al (2014) Effective risk governance for environmental policy making; a knowledge management perspective. Environ Sci Pol 41:23–32. https://doi.org/10.1016/j.envsci.2014.04.014

McGurk B, Sinclair J, Diduck A (2006) An assessment of stakeholder advisory committees in forest management: case studies from Manitoba, Canada, Society & Natural Resources An International Journal, Volume 19, 2006 - Issue 9 pp 809–826

Munich Re (2016) Worst natural disasters statistics | Munich Re. https://www.munichre.com/en/reinsurance/business/non-life/natcatservice/significant-natural-catastrophes/index.html. Accessed 28 Sept 2016

Neuman WL (2006) Social research methods: Qualitative and quantitative approaches

PMNF (2016) Plano diretor da prefeitura municipal de Nova Friburgo. Nova Friburgo RJ. https // www.meioambientedigital.pmnf.rj.gov.br. Accessed 17 Jan 2017

Renn O, Walker KD (2008) Global risk governance. Concept and practice using the IRGC framework. Dordrecht, the Netherlands: Springer (Book series on global risk governance, 1)

Royal Society Study Group on Risk Assessment (1992) Risk: analysis, perception and management / report of a Royal Society Study Group. The Royal Society, London

Schensul JJ (2013) Focused Group Interviews. Enhanced Ethnographic Methods

Smith W, Dowell J (2000) A case study of co-ordinative decision-making in disaster management. Ergonomics 43(8):1153–1166

Steelman TA, Maguire LA (1999) Understanding participant perspectives. Q-methodology in national forest management. J Pol Anal Manage 18(3):361–388. https://doi.org/10.1002/(SICI)1520-6688(199922)18:3<361::AID-PAM3>3.0.CO;2-K

Stern PC, Fineberg HV (1996) Understanding risk. Informing decisions in a democratic society. National Academy Press, Washington, DC

Taylor-Powell E (1998) Program development and evaluation: questionnaire design: asking questions with a purpose

Ulrich H, Probst G (1995) Anleitung zum ganzheitlichen Denken und Handeln. Ein Brevier für Führungskräfte. 3., erw. Aufl. Edited by Gilbert J. B. Probst. Bern: Haupt

Usón TJ, Klonner C, Höfle B (2016) Using participatory geographic approaches for urban flood risk in Santiago de Chile. Insights from a governance analysis. Environ Sci Policy 66:62–72. https://doi.org/10.1016/j.envsci.2016.08.002

Vacik H, Kurttila M, Hujala T et al (2014) Evaluating collaborative planning methods supporting programme-based planning in natural resource management. J Environ Manag 144:304–315. https://doi.org/10.1016/j.jenvman.2014.05.029

Walonick DS (2003) Survival statistics. StatPac, Minneapolis

WFD (2010) The EU Water Framework Directive information sheet. With assistance of European Commission

World Bank (2012) Avaliação de perdas e danos. Inundações e Deslizamentos na Região Serrana do Rio de Janeiro – Janeiro de 2011. http://mi.gov.br/pt/c/document_library/get_file?uuid=74dde46c-544a-4bc4-a6e1-852d4c09be06&groupId=10157. Accessed 09 Nov 2015

Chapter 24
Soil Degradation in Southeast Brazil: A Challenge for Restoration and Rehabilitation

Antonio Soares da Silva, Roman Seliger, Dietmar Sattler, and Jürgen Heinrich

Abstract The Southeast region of Brazil represents the main economic center of the country. Its use and occupation date back to the beginning of colonization by European countries in the sixteenth century. This long period of economic activities led to the current scenario of land degradation. Since the time of the first settlements, native forests have been removed mainly by the cultivation of coffee in the Paraíba do Sul River Valley and West of São Paulo state and also by mining in Minas Gerais. The deforestation exposed fragile soils, developed on sandstones (São Paulo and Triângulo Mineiro), and low fertility soils, developed on igneous and metamorphic rocks (Rio de Janeiro, Minas Gerais and Espírito Santo), resulting in the increase of erosion processes. In some areas there are many deep gullies, associated with agriculture and the expansion of cities. In other areas, sheet erosion, gullies, and ravines occur frequently due to coffee cultivation and the introduction of livestock. The northwest part of Rio de Janeiro state is a region which deserves to be highlighted due to the serious threat of desertification caused by lower annual rainfall averages and the intense compaction of soils under excessive grazing. Sustainable and strategically sound restoration and rehabilitation activities need to be implemented to stop the ongoing process of land degradation.

Keywords Tropical soils · Soil degradation · Paraíba do Sul River · Spring protection

A. Soares da Silva (✉)
State University of Rio de Janeiro – UERJ, Institute of Geography, Laboratory of Physical Geography, Rio de Janeiro, RJ, Brazil

R. Seliger · D. Sattler · J. Heinrich
Leipzig University, Institute of Geography, Physical Geography and Environmental Research Johannisallee 19, Leipzig, Germany
e-mail: roman.seliger@uni-leipzig.de; sattler@uni-leipzig.de; jhein@rz.uni-leipzig.de

U. Nehren et al. (eds.), *Strategies and Tools for a Sustainable Rural Rio de Janeiro*, Springer Series on Environmental Management,
https://doi.org/10.1007/978-3-319-89644-1_24

24.1 Introduction

Land degradation is a global phenomenon, defined as "(...) the reduction in the capacity of the land to provide ecosystem goods and services, over a period of time, for its beneficiaries" (FAO 2013). Several estimates of the global extent of land degradation are based on varying approaches, indicating that 20–25% (Clausing 2011) up to 35% (Bai et al. 2008; Nachtergaele et al. 2011) of all used land can be considered as degraded.

Bai et al. (2008) assessed land degradation using long-term (1981–2003), remotely sensed normalized difference vegetation index (NDVI) data at country level. Based on their study, 1882 million km^2 or 22.11% of Brazil's territory are subject to degradation. The Brazilian Institute of Environment and Renewable Natural Resources (IBAMA) first defined land degradation as occurring "when native vegetation and fauna has been destroyed, removed or displaced, the fertile soil is lost, removed or eroded and the hydric system has been altered." The revision now defines that land degradation "takes place when the capacity for natural physical, biological and chemical adaption is lost and makes socio-economic development unfeasible" (IBAMA 1990).

In Southeast Brazil, degradation can be basically addressed at two levels: firstly, the clearing of Atlantic Forest for farmland and urban areas, dating back to the sixteenth century and still occurring today, causing tremendous deforestation, a multitude of environmental risks, and erosion of ecosystem services (Nehren et al. 2018). Secondly, the unsustainable use of arable and pasture land caused by overexploitation and inadequate land use is the main driver of soil degradation, aggravated by a progressing climate change toward dryer conditions. Our main goal in this work is to analyze historical and recent developments of soil degradation in Southeast Brazil and to briefly discuss the challenges for its sustainable rehabilitation.

24.2 A Brief History of Southeast Brazilian Occupation

The level of land degradation in the southeastern region of Brazil has been reported by several authors (Moura 1994; Sato et al. 2007; Costa and Voltolini 2013; Silva and Botelho 2014). The lands that today comprise the states of Rio de Janeiro (RJ), São Paulo (SP), and Minas Gerais (MG) were historically occupied by economic cycles related to agriculture (coffee and sugarcane) and mining since the eighteenth century (Marquese 2008; Nehren et al. 2013).

Coffee cultivation in the Southeast region of Brazil began in Rio (Abreu 1987), whose profits financed part of the urbanization process of the then Brazilian capital. After the end of this cycle in the Tijuca Massif (current Tijuca National Park), coffee arrived in the Paraíba do Sul Valley using the same practices that caused the degradation of the land in the state capital. Large farms located in the Paraíba do Sul Valley were responsible for the removal of tropical forest that protected the soils from the heavy summer rains. Over the years, coffee plantations productivity decreased as a direct consequence of the lack of proper soil management (Silva and Botelho 2014). In the search of new lands to continue planting, coffee cultivation spread to SP, MG, Parana, and Espírito Santo (ES) states. After the abolition of slavery, farms began to use the labor force of European immigrants who arrived in Brazil in the early twentieth century. These workers had better knowledge of land management and soil depletion processes, and the problems which occurred in Paraíba do Sul Valley no longer took place. The twentieth century is also characterized by the expansion and diversification of the agricultural system, mainly in SP toward the Midwest region of Brazil.

After the coffee market collapse in the early twentieth century, landowners dedicated themselves to cattle ranching, and large former coffee plantations were converted into pasture. By that time, coffee farms were concentrated in SP, and cattle ranching was expanding in MG and RJ. However, the lack of pasture management promoted soil degradation as well. Moreover, there was no recovery of the forest cover, and the cattle started to graze freely on the hills, a situation that can be seen even today. Therefore it can be concluded that the sequence of land uses through a combination of coffee and livestock is the basis of land degradation found today. Further degradation drivers are associated with climate characteristics, soil, and relief.

24.3 Soils of "Oeste Paulista," "Triângulo Mineiro," and Other Areas of the Southeast Brazil

In geological terms there are three large lithologic domains in the Southeast that have great influence on soil characteristics: crystalline basement, Paraná Magnetic Province, and post-volcanic sediments (see Fig. 24.1) (Marques and Ernesto 2004). Covering part of the states of SP and MG, the rock units associated with the Parana Basin (basalts and sandstones) are the source material of Ferralsols, Nitisols, Acrisols, and Arenosols, whose texture varies from clayey in the Ferralsols, Nitisols, and Acrisols to very sandy in the Arenosols. The pedological covers that develop on these parent materials are deep to very deep, with potentially high fertility due to high levels of magnesium found in these soils.

In other areas of MG, RJ, and ES, lithologies associated with the Neoproterozoic Provinces and the São Francisco Craton (Cordani et al. 2009) are predominant. The rocks are metamorphic and igneous, with a high content of feldspar and quartz. Due to the historic and actual climatic characteristics of tropical Southeast Brazil with mean annual precipitations above 1200 mm and mean annual temperatures from 20 to 24 °C (Alvares et al. 2014), the rocks are deeply weathered, and the soils show textures that vary from sand to clay. Normally, Ferralsols and Acrisols predominate, but in mountain sectors Cambisols and Leptosols occur as well.

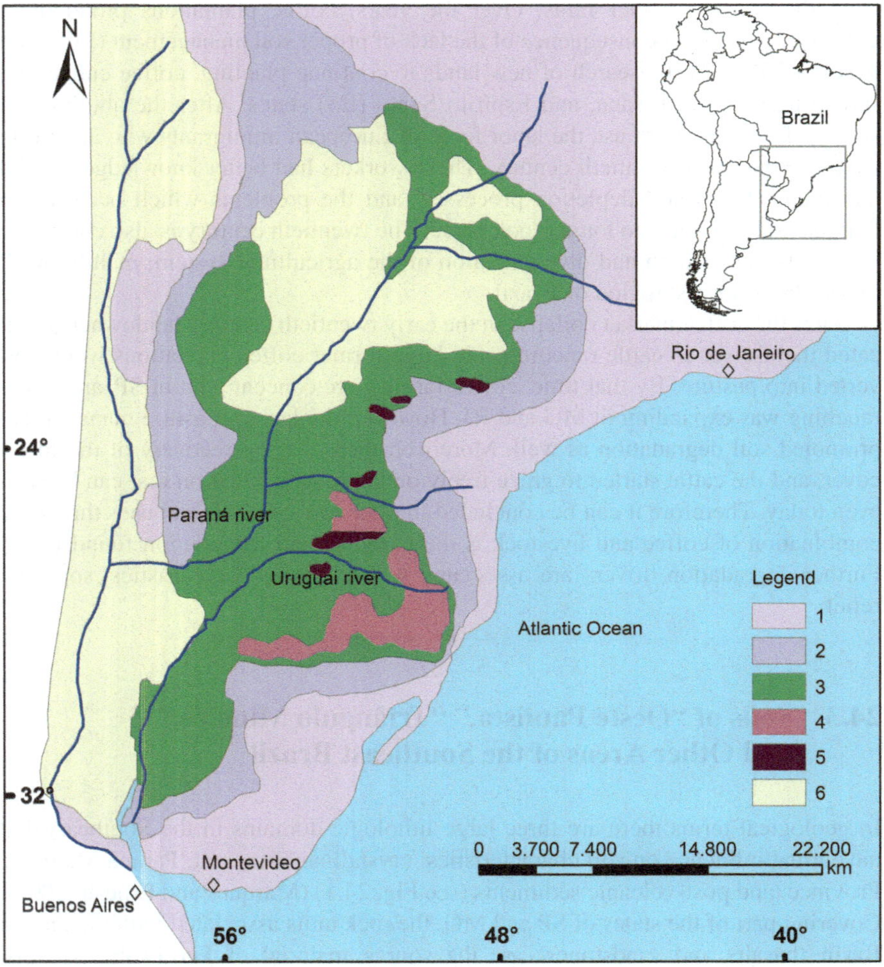

Fig. 24.1 Simplified map of the Paraná Basin modified from Marques and Ernesto (2004). (1) crystalline basement, (2) pre-volcanic sediment, (3) basic and intermediate volcanic rocks of the Paraná Magnetic Province, (4) acid rocks of Palmas type of the Paraná Magnetic Province, (5) acid rocks of Chapecó type of the Paraná Magnetic Province, and (6) post-volcanic sediments

24.4 Soil Degradation Processes

Soil degradation processes in the Paraná Basin are different from those found in other areas of Southeast Brazil. The combination of rainfall concentrated between November and March and deep soils results in large gullies, sometimes exceeding hundreds of meters in extent (Baccaro 2009).

Some pedological covers exhibit strong textural and structural discontinuities that result in variations in the hydraulic dynamics of soils. Vidal-Torrado et al. (1999), Furquim et al. (2013), and several other authors have studied pedological covers in the transformation from Ferralsols to Acrisol in SP and associated them with soil destabilization processes by gully erosion. With the development of the textural horizon B by illuviation, a zone of hydraulic discontinuity was created, and it often generated a lateral interflow that removed the entire sandy surface horizon.

In the soils which have developed on sandstones, this process is more pronounced. According to Baccaro (2009), areas in MG with moderately dissected relief are undergoing intense gully erosion due to soils composed of sandy materials, predominantly fine sand (60–70%), without coherence and with thickness ranging from 3 to 8 m. Vertical incision of the gullies cease when more compact and more resistant layers are reached due to changes in the rock or when the water table is found. From this point, ongoing erosion progresses in lateral direction (Baccaro 2009). The main drivers of these processes are the removal of vegetation and the inappropriate use of land (Sobral et al. 2015).

The rehabilitation of such strongly eroded areas requires excessive effort and often exceeds the financial capacity of the owners. There are only few actions that aim at the recovery of these areas, which mainly focus on the monitoring and control of the lateral and headward expansion of the gullies. In other regions of Southeast Brazil, gully erosion processes do not follow the same pattern as found in the soils of the Paraná Basin. Although the climate allows the formation of Acrisols and Ferralsols, there are no extremely deep gullies, except in the Paraíba do Sul Valley (Guerra and Botelho 2001).

Along the Serra do Mar mountain region of SP, RJ, and ES, mass movements are frequent. The weathering of the crystalline Precambrian basement has resulted in sandy loamy regoliths with a maximum thickness of 15 m on steep (40–60%) to very steep (70–90%) slopes. These soil mantles are exposed to annual precipitation rates from 2000 to 2500 mm, with a rainy season from September to March concentrating 70% of the annual rainfall (Cruz 1990; Furian et al. 1999). The high volume of rainfall creates geochemical conditions for the development of gypsic horizons of low permeability coincident with the saprolite (Furian et al. 1999; Silva et al. 2016). These horizons act as impermeable layers that after heavy rains can result in catastrophic mass movements.

According to Furian et al. (1999), this is the main mechanism of soil formation. When the weathered rock mantle is removed by mass movements, rock is again exposed to weathering, and a new soil begins to form. The excessive amount of rain almost completely leaches the silica present in the feldspars, leaving the saprolite rich in aluminum hydroxide (gibbsite).

The heavy rains that fall on the Atlantic slope of the Serra do Mar are due to the mountain range which acts as a physical barrier to the humid winds coming from the ocean. Rainfall decreases on the slope facing inland though remains concentrated in the summer months. In the Paraíba do Sul Valley and in the Northwest region of RJ, the annual average rainfall rarely exceeds 1200 mm (Lumbreras et al 2004; Brandão et al. 2016). As the precipitation decreases, the mass movements

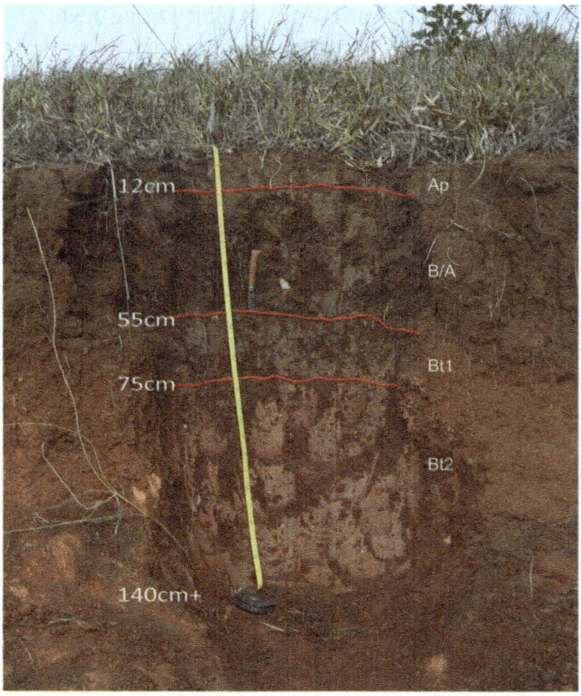

Fig. 24.2 Profile of Red Acrisols. Except for Ferralsols, the other soils in the state of Rio de Janeiro are deep averagely. The Red Acrisols have strong textural contrast which may be related to the development of gullies. (Photo: R. Seliger 2016)

become rarer, and superficial erosive processes predominate, influenced by the soils characteristics, relief, and land use.

In the middle Paraíba do Sul Valley, strongly inclined slopes create an environment conducive to the formation and occurrence of gullies (Coelho Netto 1994; Guerra and Botelho 2001; Moura and Silva 2001; Oliveira 2009). In addition to frequently occurring colluvial soils, intensive coffee cultivation in the nineteenth century was the factor that contributed most to accelerated erosion processes in that area.

In Northwest RJ the relief in the geological depression zone does not favor the development of gullies, though they are not completely absent. In this case, the combination of Acrisols and the predominant use in grazing results in strong laminar erosion processes, which remove most of the surface horizons. Only in Red Acrisols deeper incisions are found (Figs. 24.2 and 24.3). Linear erosion processes are almost always associated with the trails formed by the passage of livestock (Fig. 24.4).

Fig. 24.3 The gullies that appear on the slopes are moderately deep and are restricted to the Red Acrisols. There are no studies on the correlation between these soils and the gullies, although this relationship can be established with the strong textural contrast between horizons A and B. (Photo: A. Silva 2016)

Fig. 24.4 In areas of occurrence of the Red-Yellow Acrisol, the predominant erosive process is sheet erosion. However, due to incorrect management, ravines associated with the most intensely used places of passage of the animals, such as porches and place of access to water, are observed. (Photo: A. Silva 2007)

24.5 Soil Characteristics in the Serra do Mar, Paraíba do Sul Valley, and Northwest Fluminense Region

Soils represent the interface between the agents of weathering, source material, and time. RJ has soils that are the result of long-time weathering evolution, relatively high rainfall in almost the entire territory, and lithologies derived of acid magmas.

The acidic rocks had few nutrients to pass on to the soil, and the combination with high precipitation resulted in strongly leached soils with few exchangeable bases. The processes related to soil degradation are not only due to its physico-chemical characteristics. Relief of land may contribute, such as concave-convex morphology and declivity around 20 degrees may also contribute to explain some of the mechanisms that led to soil deterioration in the state.

There are three geomorphological domains in the state: coastal plains, mountain alignments (Serra do Mar and Serra da Mantiqueira), and the depression of the Paraíba do Sul River Valley. The coastal plains are not discussed in this chapter, since most have been occupied by city expansion (Dantas 2000).

The cliffs facing the ocean and the top of the plateau are always humid and have deep and chemically developed soils, usually Ferralsols. The steep to very steep slopes of the mountain cliffs created natural limitations for an intensive land use. Agriculture is practiced in the few plains not yet occupied by the cities, in the fluvial plains in the mountainous region (Serra do Mar) and in the areas where cultivation can be adapted to the hilly relief. All the rest of the territory is occupied by pastures. Loureiro and Guerra (2013) point out that in the municipality of Macaé, where declivity permits, soils are occupied by pastures that result in degradation processes in the form of gullies. This fact was verified by Saraiva et al. (2016) in the São João River Basin (RJ). Even on smoothly undulated lowland hills, ranching is the main cause for the development of severe laminar erosion and ravines on Red Acrisols and Red-Yellow Acrisols.

In areas of higher precipitation, Ferralsols are the main type of soil, though Cambisols may also occur. Chemically, the difference of these soils is in the quantity of primary minerals, mostly weatherproof and often occurring on the same slope. Cambisols occur near the hill tops and at steeper slopes and the Ferralsols from mid slope downward. Ferralsols are usually of colluvium origin due to frequent mass movements (Dourado et al. 2012; Silva et al. 2016). The difference between these soils is the presence of gypsic horizons that develop at the saprolite level (Furian et al. 2002; Silos et al. 2011). Geochemically, the passage of the feldspars to the gibbsite is due to the high rainfall in the higher portions of the Serra do Mar (Delvigne 1965). In the direction of the Paraíba do Sul Valley, rainfall decreases in quantity, concentrated in summer months (December to March), and the relief has a smoother slope (Ab'Saber 2001), and though the Ferralsols are present, there is a predominance of Acrisols.

These soils have high levels of aluminum and hydrogen, and low levels of exchangeable cations, but it is on these soils where a major part of Brazil's economy thrived in the eighteenth and nineteenth centuries. Coffee cultivation was developed succeeded by cattle farming (Lumbreras et al. 2004; Silva 2011) all founded on these low-fertility soils.

The geological depression of the Paraíba do Sul Valley extends from SP to the north of RJ. These are the lands most affected by degradation processes. According to Moura and Silva (2001), the relief evolution process is responsible for the emergence of gullies and ravines. These erosive forms would be reactivations of paleodrainages that were clogged during the quaternary. When the vegetation was removed for the development of the coffee and livestock cycle, it caused an imbalance in the landscape, and the result was the appearance of gullies.

The geological depression of the North and Northwest of RJ is not characterized by gully formation. The explanation for this phenomenon may be related to the thickness of the soils, rarely exceeding 2 meters, though laminar erosion process still occurs in almost all municipalities.

Lumbreras et al. (2004) divided the Northwest region of RJ (Northwest Fluminense) in three macro pedo-environments. The region presents high vulnerability to degradation, mainly due to its rugged relief, the high textural gradient of the soils, reduced vegetation cover, and the use of inadequate management practices and soil conservation. The main erosion forms are of the severe laminar type i.e., rills and gullies.

A division was proposed by the authors based on the soil-relief relationship. Roughly speaking, two types of Acrisols predominate: Red Acrisol (eutrophic) and Red-Yellow Acrisol (alic) (Fig. 24.5). The first macro pedo-environment covers

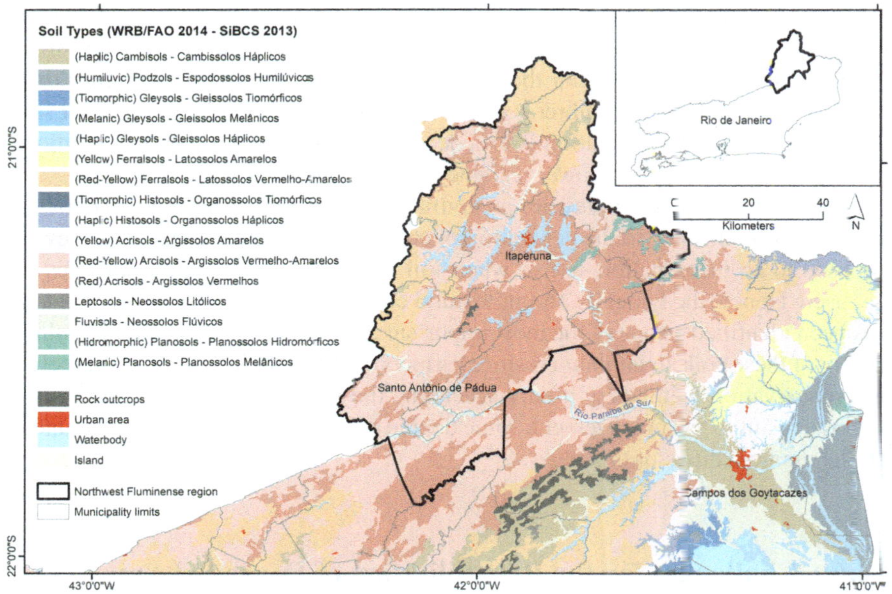

Fig. 24.5 A soil map of the Northwest Fluminense region shows predominance of Acrisols. Ferralsols and Cambisols occurring in the Serra do Mar mountain range at higher altitudes under more extensive precipitation. At the northern limits of Rio de Janeiro state, there are Ferralsols at the foothills of the Serra do Caparó mountain range. Acrisols developed in a depressed area between two elevated regions

areas with dissected relief and high natural fertility soils but with medium to low water availability. Greater soil fertility would make the region propitious for the development of agrarian activities, but water deficiency though relief makes it difficult to implement agricultural projects.

Vegetation reflects the climatic characteristics of the atmosphere and soil. In most parts of the Northwest Fluminense region, natural vegetation is seasonal semi-deciduous tropical forest, but in some drier places, deciduous forest predominates (Lumbreras et al. 2004). Land use is characterized by extensive cattle ranching, usually conducted on degraded pastures, requiring greater technological investment and resources.

The soils of this environment are mostly eutrophic Red Acrisols, eutrophic Red-Yellow Acrisols, and Planosols (eutrophic). In spite of being eutrophic, the base saturation of these soils is not high, being at most 72 $cmol_c kg^{-1}$ in the A horizon (Lumbreras et al. 2004). Normally, these soils are endoeutrophic, since the values of V are higher in the textural B horizon. In geological terms these soils developed on rocks of the Paraíba do Sul Complex whose lithologies contain biotite as one of the main minerals. This may give them their reddish color and higher fertility, possibly favored by the low precipitation and consequently reduced mineral leaching. Some of these soils have 386 g kg^{-1} of clay in the horizon A and 713 g kg^{-1} in the textural horizon B at 34 cm of depth. These features may be decisive for the occurrence of surface erosive processes due to an abrupt textural change.

The second and third macro pedo-environments defined by Lumbreras et al. (2004) are located at higher, more humid relief positions that favor the development of Red-Yellowish Ferralsols and Red-Yellowish Acrisols. These soils tend to be thicker and more leached than the Red Acrisols, and their base saturation is 45 $cmol_c kg^{-1}$ at maximum (Lumbreras et al. 2004). These soils have greater water storage capacity, and the vegetation is less deciduous. As a result of these conditions, the environment presents a lower tendency to superficial erosive processes.

Soils with low-activity clay are predominantly kaolinitic. The characteristics of the Acrisols contribute to the generation of erosive laminar processes when they are exposed to an incompatible use or when soil and water conservation practices are not used. Hence, the typically abrupt textural change between the horizons A and B_t has been identified as one of the most important determinants for the natural vulnerability of these soils to laminar erosion. Farmers who use these soils without including conservation methods will only accelerate the removal of the surface horizon, and as a result of mismanagement, some degree of soil deterioration can be found in all the municipalities of this region.

24.6 Challenges for the Rehabilitation of Degraded Areas

Signs of soil degradation can readily be found throughout RJ in all soils and types of relief. It is a challenge for researchers to reconcile land use and their soils for small farmers and restore areas sensitive or vulnerable to erosive processes. In the

Serra do Mar region, both movements of soil and rock caused by high precipitation and strong slope are common, and gullies in deep soils such as Ferralsols require mechanical interventions for their recovery (Loureiro and Guerra 2013). The authors used geotextiles to stabilize gully walls in the city of Macaé (RJ) and obtained a soil cover of over 80% after 79 days, though this is one of the rainiest regions of the state, favoring the development of vegetation as a soil cover.

There are many actions of restoration and recovery of degraded areas in the state (e.g., afforestation and engineered slope stabilization), but they are all isolated case studies since there is overall action plan to recover and restore large contiguous areas.

Overall there are many areas showing severe soil degradation in southeast Brazil, and the financial cost of their recovery and restoration is generally prohibitive to farmers. The Paraíba do Sul River Valley (East Valley) and the regions of North and Northwest of the state are those with the biggest challenges related to soil conservation and rehabilitation of degraded soils (Guerra and Botelho 2001; Sattler et al. 2017; Silva et al. 2017a, 2017b). Though the Brazilian government has classified and legally requires areas of permanent preservation (APPs) as defined by its forest code (Brazilian Federal Law 12.651, 2012) for hilltop, riparian, and water source vegetation cover, many municipalities have much to restore, and it is necessary to establish a restoration prioritization among these areas.

The authors propose that springs and the banks of the first order channels should be the first ones to be restored. These areas correspond to recharge areas, increase the supply of water to smallholders, and reduce the sediment load of the river channels.

Some governmental actions such as the Rio Rural Program (Hissa et al. 2018) and the implementation of the National Rural Registration System (CAR) and related stipulations (Laskos et al. 2016) have been applied in order to encourage landowners to adopt soil and water conservation practices, mainly through reforestation for springs and low-order watercourses protection; however basic environmental education for the rural population is needed to raise general awareness for land degradation and its overall negative social-economic and environmental effects.

References

Ab'Saber AN (2001) Megageomorfologia do território brasileiro. In: Cunha SB, Guerra AJT (Org.). Geomorfologia do Brasil. Ed. Bertrand Brasil. Rio de Janeiro, pp 71–105

Abreu M (1987) A evolução urbana do Rio de Janeiro. Ed. IPLANRIOZahar. Rio de Janeiro, p 147

Alvares CA, Stape JL, Sentelhas PC et al (2014) Koeppen's climate classification map for Brazil. Meteorol Z 22(6):711–728

Baccaro CAD (2009) Processos erosivos no domínio do cerrado. In: Guerra AJT, Silva AS, Botelho RGM (Org.). Erosão e conservação dos solos: conceitos, temas e aplicações. Ed. Bertrand Brasil. Rio de Janeiro, pp 195–227

Bai ZG, Dent DL, Olsson L, Schaepman ME (2008) Proxy global assessment of land degradation. Soil Use Manag 24:223–234

Brandão CB, Silva AS, Miranda RAC, Guerra AJT (2016) Determination of climatological profile of Santo Antônio de Pádua municipality, Rio de Janeiro state and its applicability to degraded areas recovery. Anu Inst Geocienc 39(1):5–12
Clausing P (2011) Peak Soil: soil destruction and the food crisis – The loss of fertile land and how to avoid it. Local land and soil news no.38/39 II, The Bulletin of the European Land and Soil Alliance (ELSA) e.V
Coelho Netto AL (1994) Hidrologia de encosta na interface com a Geomorfologia. In: Guerra AJT, Cunha SB (Org.) Geomorfologia: uma atualização de bases e conceitos. Ed. Bertrand Brasil. Rio de Janeiro, pp 93–148
Cordani UG, Teixeira W, D'Agrella-Filho MS, Trindade RI (2009) The position of the Amazonian Craton in supercontinents. Gondwana Res 15:396–407
Costa L, Voltolini JC (2013) Impacto do pisoteio de gado sobre a floresta ripária no vale do Paraíba, Tremembé, SP. Revista Biociências 19(1):93–103
Cruz O (1990) Contribuição geomorfológica ao estudo de escarpas da Serra do Mar. Rev IG São Paulo 11(1):9–20
Dantas EM (2000) Geomorfologia do estado do Rio de Janeiro. CPRM, pp 96–168, Brasília
Delvigne J (1965) Pédogenèse em Zone Tropicale: La formation des minéraux secondaires em milieu ferralitique. Dunod. Paris, pp 177
Dourado F, Arraes TC, Silva MF (2012) O megadesastre da região serrana do Rio de Janeiro - as causas do evento, os mecanismos dos movimentos de massa e a distribuição espacial dos investimentos de reconstrução no pós-desastre. Anu Inst Geocienc 35:43–54
FAO (2013) Land degradation assessment in drylands – methodology and results, Food and Agriculture Organization, Rome, p 56
Furian S, Barbiero L, Boulet R (1999) Organisation of the soil mantle in tropical southeastern Brazil (Serra do Mar) in relation to landslides processes. Catena 38:65–83
Furian S, Barbiero L, Boulet R et al (2002) Distribution and dynamics of gibbsite and kaolinite in an Oxisol of Serra do Mar, southeastern Brazil. Geoderma 106:83–100
Furquim SAC, Coltrinari L, Ferreira RPD et al (2013) Lamellae formation processes in tropical soils in southeastern Brazil. Catena 107:15–25
Guerra AJT, Botelho RGM (2001) Erosão dos solos. In: Guerra AJT, Cunha SB (Org.). Geomorfologia do Brasil. Ed. Bertrand Brasil. Rio de Janeiro, pp 181–227
Hissa H, Teixeira N, Costa M et al (2018) Sustainable rural development in Rio de Janeiro state: the Rio rural program. In: Nehren U, Schlüter S, Raedig C, Sattler D, Hissa H (eds) Strategies and tools for a sustainable rural. Springer, Rio de Janeiro
IBAMA (1990) Manual de recuperação de áreas degradadas pela mineração: técnicas de revegetação. MINTER, Brasília, p 96
Laskos AA, Cazella AA, Rebollar PBM (2016) O Sistema nacional de cadastro rural: História, limitações atuais e perspectivas para a conservação ambiental e segurança fundiária. Desenvolvimento e Meio Ambiente 36:189–199. https://doi.org/10.5380/dma.v36i0.39124
Loureiro HAS, Guerra AJT (2013) Recuperation of degraded areas by erosion: a case study on a pilot area, using banana tree geotextile fibers, Macaé Municipality, Rio de Janeiro state. In. VII Congresso Estadual de Agronomia. Available at: https://www.researchgate.net/publication/312213492
Lumbreras JF, Carvalho Filho A, Motta PEF et al (2004) Macropedoambientes da região Noroeste Fluminense – uma contribuição ao planejamento ambiental. Embrapa Solos Documentos 64:21
Marques LS, Ernesto M (2004) O magmatismo toleítico da Bacia do Paraná. In: Mantesso-Neto V et al (eds). Geologia do Continente Sul-Americano: evolução da obra de Fernando Flávio Marques de Almeida. Ed. Beca São Paulo, pp 245–263
Marquese RB (2008) Diáspora africana, escravidão e a passagem da cafeicultura do vale da Paraíba oitocentista. Amanack Braziliense 7:138–152
Moura JRS (1994) Geomorfologia do Quaternário. In: Guerra AJT, Cunha SB (Org.). Geomorfologia uma atualização de bases e conceitos. Ed. Bertrand Brasil. Rio de Janeiro, pp 335–364

Moura JRS, Silva TM (2001) Complexo de rampas de colúvio. In: Erosão dos solos. In: Cunha SB, Guerra AJT (Org.). Geomorfologia do Brasil. Ed. Bertrand Brasil. Rio de Janeiro, pp 143–180
Nachtergaele R, Biancalani R, Petri M (2011) Land degradation. SOLAW Background Thematic Report 3
Nehren U, Kirchner A, Sattler D, Turetta A, Heinrich J (2013) Impact of natural climate change and historical land use on landscape development in the Atlantic Forest of Rio de Janeiro, Brazil. Anais Academia Brasileira de Ciências 85(2):311–332
Nehren U, Kirchner A, Lange W et al (2018) Natural hazards and climate change impacts in the state of Rio de Janeiro: a landscape historical analysis. In: Nehren U, Schlüter S, Raedig C, Sattler D, Hissa H (eds) Strategies and tools for a sustainable rural. Springer International Publishing, Rio de Janeiro
Oliveira MAT (2009) Processos erosivos e preservação de áreas de risco de rosão por voçorocas. In: Guerra AJT, Silva AS, Botelho RGM (Org.). Erosão e conservação dos solos: conceitos, temas e aplicações. Ed. Bertrand Brasil. Rio de Janeiro, pp 57–99
Saraiva VIC, Silva AS, Santos JPC (2016) Uso do mapa de solos como subsídio para a definição de áreas de suscetibilidade à erosão na bacia hidrográfica São João, Lagos e Una. Geouerj 29(2):354–373
Sato AM, Avelar AS, Coelho Netto AL (2007) Hidrologia de encosta numa cabeceira de drenagem com cobertura de eucalipto na bacia do rio Semarias: médio vale do rio Paraíba do Sul. Anais do I Seminário de Recursos Hídricos da Bacia Hidrográfica do Paraíba do Sul: o Eucalipto e o Ciclo Hidrológico, Taubaté, IPABHi, pp 147–154
Sattler D, Seliger R, Nehren U et al (2017) Pasture degradation in South East Brazil: status, drivers and options for sustainable land use under climate change. In: Leal Filho W, Gallo E, Coelho Netto AL (eds), Climate change in Latin America: managing vulnerability, fostering resilience (in press)
Silos VR, Polivanov H, Santos LV et al (2011) Micromorfologia aplicada a estudos geológico-geotécnicos. Revista Brasileira de Geociências 41:256–262
Silva AS (2011) Mineração e Áreas de Preservação Permanente (APPs) em Santo Antônio de Pádua – RJ. Sociedade e Natureza 23(2):173–185
Silva AS, Botelho RGM (2014) Degradação dos solos no estado do Rio de Janeiro. In: Guerra AJT, Jorge MCO (Org.) Degradação dos Solos no Brasil. Ed. Bertrand Brasil, Rio de Janeiro, pp 261–292
Silva AS, Guerra AJT, Polivanov H, Fullen MA (2016) Soil structural indicators of hillslope destabilization in the Serra do Mar mountain range (Rio de Janeiro state, Brazil). Nat Hazards 81(2):1177–1191
Silva AS, Guerra AJT, Polivanov H, Fullen MA (2017a) Potential of rock dust and sewage sludge to improve soil fertility in degraded Brazilian pastures. In: Extended abstracts of the 1st world conference on soil and water conservation under global change, Lleida, 12–16 June 2017
Silva AS, Polivanov H, Guerra AJT et al (2017b) Model for degraded land restoration on regions under hydric stress, in Rio de Janeiro state, Brazil. In: Extended abstracts of the 1st world conference on soil and water conservation under global change, Lleida, 12–16 June 2017
Sobral AC, Peixoto ASP, Nascimento VF et al (2015) Natural and anthropogenic influence on soil erosion in a rural watershed in the Brazilian southeastern region. Reg Environ Chang 15(4):709–720
Vidal-Torrado P, Lepsch IF, Castro SS, Cooper M (1999) Pedogênese em uma sequência Latossolo-Podzólico na borda de um platô na Depressão Periférica Paulista. R Bras Ci Solo 23:909–921

Chapter 25
Rehabilitation of Degraded Sloped Pastures: Lessons Learned in Itaocara, Rio de Janeiro

Roman Seliger, Dietmar Sattler, Antonio Soares da Silva, Gabriel Campos Pereira da Costa, and Jürgen Heinrich

Abstract The municipality of Itaocara is characterized by a pasture-dominated landscape (mainly dairy farming) at moderately to strongly inclined slopes. The major drivers for pasture degradation are poor and erosion-susceptible soils, a changing regional climate towards less frequent but stronger precipitation events, unsustainable pasture management practices as well as low awareness for degradation phenomena among farmers. Today, almost all sloped pastures in the NW of Rio de Janeiro state show various degradation levels, most noticeable in distinctive erosion forms such as rills, cattle tracks and gullies. If no rehabilitation measures and no sustainable pasture management will be applied to these fragile, historically strongly modified ecosystems, pastures will soon degrade to a degree and extent, where land use will not be profitable anymore and rural population might lose their main source of income. The paper presents the implementation, monitoring and management of a low-cost rehabilitation measure applied on a medium degraded sloped pasture in rural Itaocara. Slope-parcelling hedgerow terraces of various native tree species combined with erosion diminishing bioengineered measures and soil amelioration practices aim both at strengthening pasture resilience against degradation and enabling an extensive rotational pasture management at the same time. The measure will be discussed in the context of other advantageous vegetative and structural pasture rehabilitation approaches under given regional environmental and socio-economic conditions.

R. Seliger (✉) · D. Sattler · J. Heinrich
Leipzig University, Institute of Geography, Physical Geography and Environmental Research, Leipzig, Germany
e-mail: roman.seliger@uni-leipzig.de; sattler@uni-leipzig.de; jhein@rz.uni-leipzig.de

A. Soares da Silva · G. C. P. da Costa
State University of Rio de Janeiro – UERJ, Institute of Geography, Laboratory of Physical Geography, Rio de Janeiro, RJ, Brazil

U. Nehren et al. (eds.), *Strategies and Tools for a Sustainable Rural Rio de Janeiro*, Springer Series on Environmental Management,
https://doi.org/10.1007/978-3-319-89644-1_25

Keywords Sloped pastures · Degradation · Erosion · Rehabilitation · Unsustainable land use

25.1 Introduction

Pastures cover over 50% of the area in the state of Rio de Janeiro (RJ), followed by forest (26%) and arable land (4%, mainly sugar cane) (SEA and INEA 2011). The rural area is mostly structured by a large number of micro- (<22 ha) and small-sized (22–88 ha) properties and a low number of big-sized (>330 ha) properties (INCRA 2006). An average stocking rate of 1.9 AU/ha in combination with a hilly undulating topography, extreme precipitation events (20% of the seasonal total precipitation in 1 day (Liebmann et al. 2001)), poor and erosion-prone soils formerly under Atlantic Forest and unsustainable, low-productive agriculture systems (Hissa 2014) makes pastures highly vulnerable and less resilient to degradation. Today, cattle farming (mainly diary) is besides arable farming the main income source of rural population in the northwest of RJ (Northwest-Fluminense region). Even if farmers are partly aware of occurring degradation and erosion processes on sloped pastures, they strongly identify with cattle farming and perceive it as a 'tradition' to be continued. It is of great importance to act against ongoing pasture degradation, aiming at stabilizing and improving threatened areas towards sustainable productivity and conservation (Phalan et al. 2016). Only those measures which are pragmatic, at low cost and low manpower and meet farmers' acceptance at the same time can succeed in obtaining more resilient pastures. Here, depending on the degradation degree, appropriate measures can range from sustainable pasture management, rehabilitation and recovery, as exemplarily demonstrated at one combined PRM at a medium to severely degraded pasture in the region of Itaocara.

25.2 Soils Under Degraded Pastures

As a result of the clearing vast parts of the natural Atlantic Forest in RJ, soils such as Acrisols, Ferralsols and Cambisols have suffered from advanced weathering due to exposed sloped position, humid environment and inappropriate land use (Silva et al. 2018). Strong precipitation events with subsequent surface runoff and interflow have been causing enormous loss of soil material and available surface organic material, destabilizing slopes even more. Acrisols under sloped pastures show generally a thin Ap horizon of granular or blocky angular structure and soft to hard consistency. A transition horizon passes over to the argic, Bt horizon with hard to very hard consistency, high clay contents (low activity clays), low base saturation, low porosity and high bulk densities (IUSS Working Group WRB 2015). Here, conditions under grass cover are usually better than under exposed soils (e.g. along cattle tracks), even if significance of change is often low (Costa and Silva 2015). Finer soil textures are easier and further transported downslope, even at low water speed (Hjulström 1935). These slow and long-term processes lead to a distinct separation of soil textures in surface soil from hill top to foot slope, indicating ongoing soil erosion processes. The more degraded a pasture, the more clay has usually been lost in upper and accumulated in bottom slope areas, while sandy parts remain in upper slope area causing more acid conditions there, at often toxic level. Excessive cattle trampling causes continuous destruction of soil protecting vegetation cover, high compaction of the soil surface and decrease of soil infiltration, causing stronger erosive forces of surface runoff, reinforced by medium to strongly inclined slopes. Pasture areas with higher cattle density and frequency such as around shadow trees, water places or along fences often represent initial forms of further sheet erosion, rills, gullies or deeper cattle tracks.

The red-yellow Acrisol under PRM's sloped pasture has developed on quartz-rich bedrock and shows severe degradation signs. Along slope transect, soil conditions slightly improve from upper to foothill slope with higher soil organic material content and base saturation, better nutrient availability, higher porosity and lower bulk density (Table 25.1). Clay content increases only slightly in accumulation areas (S5). From Ap towards Bt horizons, the most significant changes are an increase of clay content and a decrease of organic material, whereas bulk density and soil porosity change only slightly. The base saturation in Bt horizon is higher than in Ap horizon. This could be explained by a low managed pasture in the past that has caused a strongly nutrient-depleted Ap horizon.

25.3 Bioindicators for Pasture Degradation

As pastures are man-made landscapes, initial vegetation cover is mostly man-made too. The majority of pastures in Southeast Brazil are not directly converted (cleared) Atlantic Forest areas but abandoned former sugarcane or coffee plantations. African fodder grasses such as a wide variety of *Brachiaria* species and hybrids, *Panicum* spp., *Pennisetum purpureum* Schumach. and *Melinis minutiflora* P. Beauv., are the main pasture grasses, occasionally planted together with native *Paspalum maritimum* Trin. Along with the sown pasture grasses, a more or less typical set of accompanying weeds such as *Desmodium incanum* (Sw.) DC, *Sida spinosa* L., *Mimosa pudica* L., *Sidastrum micranthum* (A. St.-Hil.) Fryxell and *Crotalaria spp.* can be found especially in the dryer, sloped pastures of Southeast Brazil. Some drought-tolerant and heliophilous pioneer tree and shrub species like *Baccharis dracunculifolia* DC., *Psidium guineense* Sw., *Moquiniastrum polymorphum* (Less.) G. Sancho and *Lantana camara* L., the latter one an invasive shrub from Central America, are also frequently to be found as part of spontaneous pasture vegetation (Lima et al. 2006). A study on vegetation structure and composition at sloped pastures of varying degradation stages carried out by Reith (2016) using 15 10×10 m sample plots (six plots within/near PRM, see Fig. 25.1) confirmed the typical weed and woody

Table 25.1 Macromorphological, physical and chemical soil parameters along PRM pasture slope transect from hill top (S1) to bottom (S6) – sample position in Fig. 25.1; undisturbed soil samples taken from soil profiles *under dry conditions

Prof./pos.	Horizon/depth	Structure	Consistency*	Sand [g/kg]	Silt [g/kg]	Clay [g/kg]	BD [g/cm^3]	PO [%]	BS [%]	OM [g/kg]
S1 (upper slope)	Ap (0–12)	Granular	Slightly hard	660	210	130	1.44	45.30	55	22.24
	Bt1 (26–72)	Blocky angular	Very hard	330	250	420	1.47	44.60	68	7.07
S2 (T3)	Ap (0–17)	Granular	Soft	740	130	130	1.29	48.40	48	20.69
	Bt1 (40–81)	Blocky angular	Extremely hard	220	240	540	1.37	49.90	50	7.07
S3 (T2)	Ap(B) (6–30)	Blocky angular	Soft	650	120	230	1.52	41.40	38	18.27
	Bt (47–100+)	Blocky angular	Hard	320	370	310	1.35	48.70	53	7.07
S4 (T1)	Ap (0–20)	Granular (lumpy)	Slightly hard	690	170	140	1.58	36.62	45	15.17
	Bt (65–83)	Blocky subangular	Hard	550	220	230	1.79	31.81	56	7.07
S5 (toe slope)	Ap (5–25)	Granular	Slightly hard	620	210	170	1.41	42.00	61	19.83
	B (25–45)	Blocky subangular	Slightly hard	520	270	210	1.53	41.40	73	11.72
S6 (bottom)	A (0–32)	–	–	420	450	130	–	–	100	26.55
	Cg (50–120+)	–	–	610	90	300	–	–	100	8.97

T terrace, *BD* bulk density, *PO* porosity, *BS* base saturation, *OM* organic material

species composition but revealed significant differences regarding the abundance and cover patterns of species. Pastures which have been abandoned for only 4 years already show important initial succession with increasing shrub layer cover. The recently abandoned pastures consequently showed only half of the grass species cover when compared to active pastures. In degraded active pastures with high content of bare soils, abundant spontaneous regeneration of woody species in the herbaceous layer can be found. Species especially like *Psidium guineense* Sw. and *Lantana camara* L. seem able to easily outcompete pasture grasses in reconquering bare soils. Usually, spontaneous woody plant colonization is perceived by farmers as a sign of degradation. This perspective has also been confirmed in questionnaire-based surveys on dairy cattle farming and pasture degradation within the INTECRAL

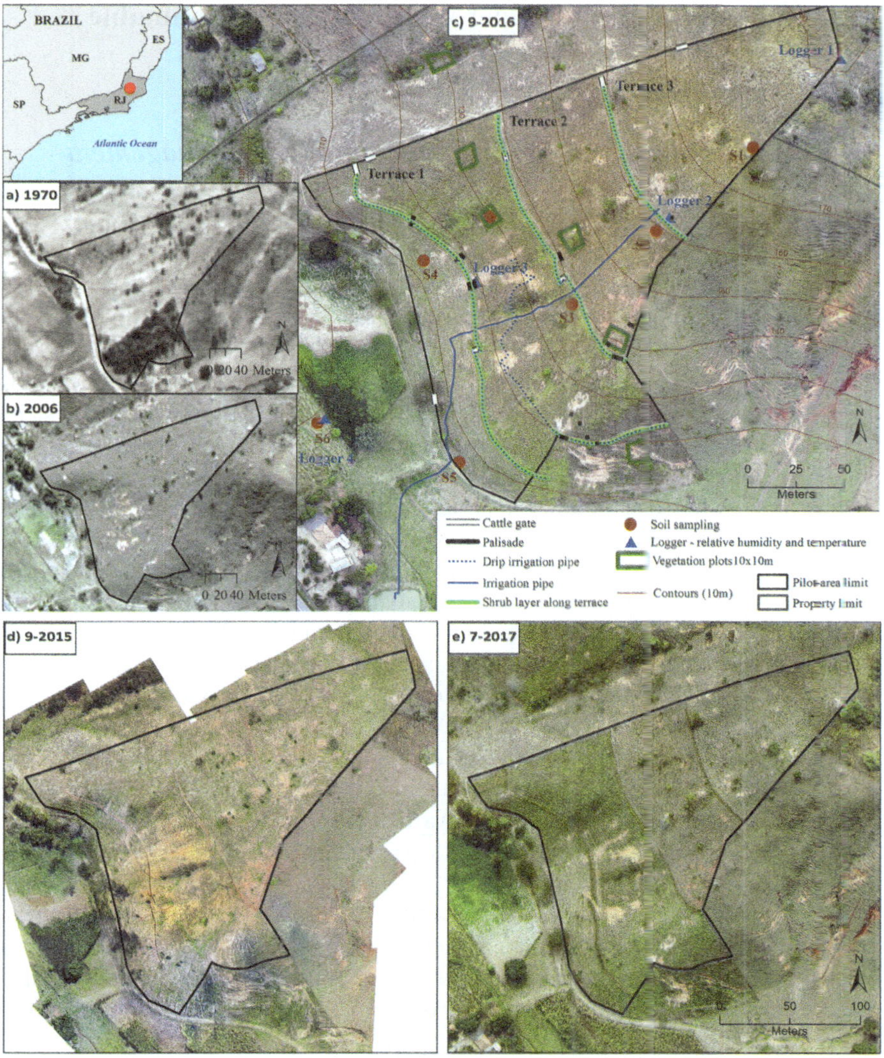

Fig. 25.1 Implemented PRM shown on drone orthophoto mosaic of (**c**) September 2016. The last forest fragment in the south corner of the pilot area visible in the (**a**) 1970 orthophoto (DEME 1970) was cleared until (**b**) 2006 (IBGE 2006); drone orthophoto mosaics of (**d**) September 2015 and (**e**) July 2017, revealing exposed soils and dominant erosion forms such as cattle tracks, rills and gullies, most frequent at steep middle slope position in 2015 and a significant increase of grass cover after 2 years of applied PRM (used drone: DJI Phantom-4, image resolution ± 10 cm, orthophoto mosaic processed with Pix4Dmapper Pro by Pix4D 2017)

project region. Hence, insufficient pasture management and soil compaction caused by overstocking favour invasion of spontaneous woody vegetation of native and non-native origin and indicate progressive pasture degradation.

25.4 Slope Parcelling Hedgerow Terraces for Sustainable Extensive Cattle Farming: Lessons Learnt from the Itaocara Case Study

25.4.1 Site Selection, Implementation and Post-management

The implementation of a modular PRM in the municipality of Itaocara is aimed both at stabilizing a degraded sloped pasture and a continuation of dairy cattle as main income source of farmers in this region. The PRM incorporates structural (terraces) and vegetative (shrub lines) measures and a subsequent extensive pasture management (rotation pasture). The chosen representative PRM area had to match several selection criteria (reasons in brackets): (i) micro- to small-sized property, which is the dominant property class in rural northwest RJ; (ii) pasture on moderately steep or steep slopes, as more than 50% of pastures in the region are on slopes with slope gradients from 8.5° to 24°; (iii) pasture with medium degradation degree which enables pragmatic rehabilitation approaches at reasonable costs, manpower and time for implementation; (iv) good visibility from main road for raising PRM awareness and ease environmental education/trainings with farmers and school classes; (v) access to pond water and electricity for plant irrigation; (vi) access/work permit for farmer's property and farmer's commitment to PRM post-management and (vii) well-organized local authorities such as Emater-Rio and Pesagro-Rio for logistic and social support. Five kilometres south-east of Itaocara city, the selected pilot area of 3.3 ha out of 22.5 ha total property size was fenced off on a moderately to steeply, south-west facing sloped pasture that has been used for dairy cattle over more than 40 years. The PRM was implemented within the dry season from June to August 2015 to diminish the risk of rain wash of the terrace constructions in the wet season. A subsequent 1-year pause of cattle grazing facilitated natural grass seeding and expansion over the slope.

Three slope parallel terraces (T1, foothill; T2, mid slope; T3, upper slope) at approximately every 20 m altitude difference (Fig. 25.1) were created using a micro-tractor with rotating tiller. Terraces are interrupted by cattle gates to enable access to all pasture lots for rotational pasturing. The terraces of a total length of 515 m were built 1 m wide and slightly tilted towards the slope (Fig. 25.2). For erosion control and terrace protection, 16 bamboo palisades were built in existing rills and gullies erosive relief forms. A total of 2433 tree saplings were planted comprising 5 native species provided as donation by a nursery, operated by CEDAE RJ (state water treatment company): *Chloroleucon tortum* (Mart.) Pitt. (*CT*), *Machaerium nyctitans* (Vell.) Benth (*MN*), *Schinus terebinthifolius* Raddi (*ST*), *Anadenanthera colubrina* (Vell.) Brenan var. *cebil* (Griseb.) Altschul (*AC*) and *Enterolobium contortisiliquum* (Vell.) Morong (*EC*). Planting pattern followed a scheme where five saplings per metre were planted in *CT-EC-CT-AC-ST* order in 20 cm distance to each other. After each 5 m, the first *CT* was substituted by one *MN*. On a 20 m section on the northwest side of the terraces, the five plants per metre scheme were changed to three plants per metre scheme with *EC-CT-ST* order

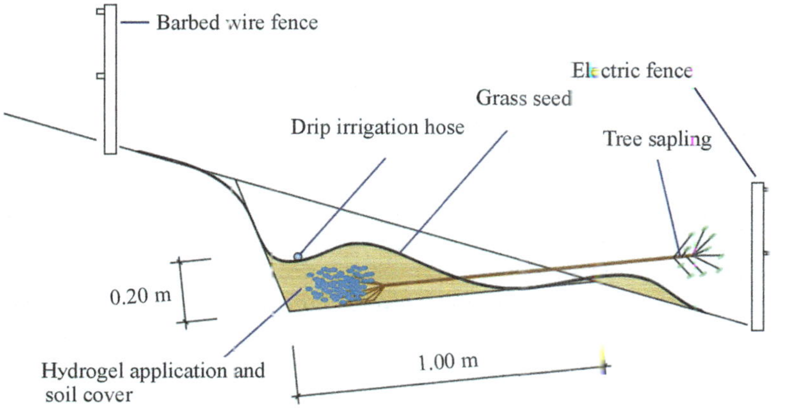

Fig. 25.2 Horizontal planting method of tree saplings along slope terraces. (Credit: Sven Dulleck, edited)

(33 cm distance). Saplings were laid almost horizontally on the terraces to stimulate sprouting of branches but also to facilitate planting procedure (Fig. 25.2). The dense and multispecies planting scheme aimed at identifying the best-performing species under given regional conditions and ensuring diverse (spiny/non-spiny, partly Leguminosae) and dense shrub fences even in case of higher mortality rates. To better cope with drought periods, 1 l of 'hydrogel' (water-retaining, biodegradable polymer, Forth©, Gel para plantio) was applied on each root ball for improving soil water retention before covering roots and trunk with soil. Installed drip irrigation along the terraces guarantees continuous water supply during initial and dry growth phase. Seeding *Brachiaria* grass along planted terraces serves as additional terrace stabilization but should be rarely applied in order to avoid too much competition for nutrient, water and light between tree saplings and *Brachiaria*.

Areas of exposed bare soil between terraces (pasture lots) were ox-ploughed and seeded with *Brachiaria* grass afterwards. In addition, grass sods were transplanted from dense grass PRM donation areas to bare soil areas. Soil pH was corrected with lime and classic NPK fertilization (composition, 12-6-12) complemented by a rock dust and sewage sludge mixture. After 3 years growth, a maintenance cut of tree saplings is recommended once a year at the beginning of the rainy season to keep shrub lines dense. Pasture lots should be cleared from spontaneous shrub growth. A 2-month interval of pasturing and pausing is recommended for area recovery. Cattle density on slope should be less than 2 AU/ha.

25.4.2 Monitoring

Tree sapling performance (plant growth, development, mortality) along the terraces and slope gradient has been monitored over an 18-month period (Sattler et al. 2018a). Generally speaking, mortality rate increases with higher slope position. While *S. terebinthifolius* shows highest mortality, *Ch. tortum* and *A. columbrina* cope best with PRM conditions. Dense hedge growth characteristics and low mortality favour *Ch. tortum* for PRM purposes even if growth speed is relatively slow. Multi-temporal analysis on density and distribution of vegetation cover enables to evaluate whether the PRM has fostered an increase of vegetation cover over time. Proportions and change of exposed soil and vegetation coverage (mainly grass and few shrubs, trees cover on pasture amounts to 3.58% on pasture) at pasture lots were monitored after PRM implementation from September 2015 to July 2017 (dry season) using geo-corrected high-resolution drone orthophoto mosaics and the supervised image classification tool ILMS (Selsam and Schwartze 2016). Areas of no data (image artefacts) and man-made changes such as constructed terraces and dug soil profiles were masked out from all input imagery before analysis. Two years after PRM implementation, vegetation cover on grazing lots has remarkably increased from 67.75% to 95.44%. *Brachiaria* is the dominantly occurring grass genus, undermined with patterns of Pernambuco grass (*Paspalum maritimum* Trin.). Lowest vegetation cover is observed on sun-exposed, concave slope areas with high surface runoff and areas of high stocking density. Temperature, relative humidity (RH), plant available nutrients (NPK), Al-toxicity, organic matter content, soil structure and texture strongly influence plant growth performance and differ from slope position and micro-geomorphology. Measurements on RH and temperature at four different positions (plane, middle, upper slope and crest; see Figs. 25.1 and 25.3) from August 2015 to August 2016 have identified June as the month of highest RH and lowest temperatures in average, while January is characterized by low RH and high temperatures. From midnight (00:00 h) to noon (12:00 h), RH diminishes constantly in average around 30% and temperature around 10°, with highest amplitudes during the second half of the day. Plane areas show higher humidity and lower temperatures over day and year, particularly during the dry season and in evening hours. Upper slope areas are facing the driest and hottest conditions, followed by middle slope positions. With regard to plant growth, best RH and temperature conditions can be found in the bottom, followed by foothill (near T1) and middle slope positions. Upper slope areas (near T3) provide the poorest conditions. Only during midday hours do hilltop areas show preferable conditions with higher RH and lower temperatures compared to other slope areas. At other day times, hilltops represent unfavourable conditions similar to upper slope positions.

Following a slope gradient, best condition for grass growth can be found at bottom areas, followed by foothill areas. Middle and upper slope areas represent worst soil conditions, underpinned by weak vegetation cover and severe erosion. Hilltop areas are anyway prohibited for cattle farming according to the Brazilian Forest Code (Brazilian Federal Law 12.727 2012). In terms of soil conditions but also slope stability, only gley soils in the humid and nutrient-rich valley and, partly,

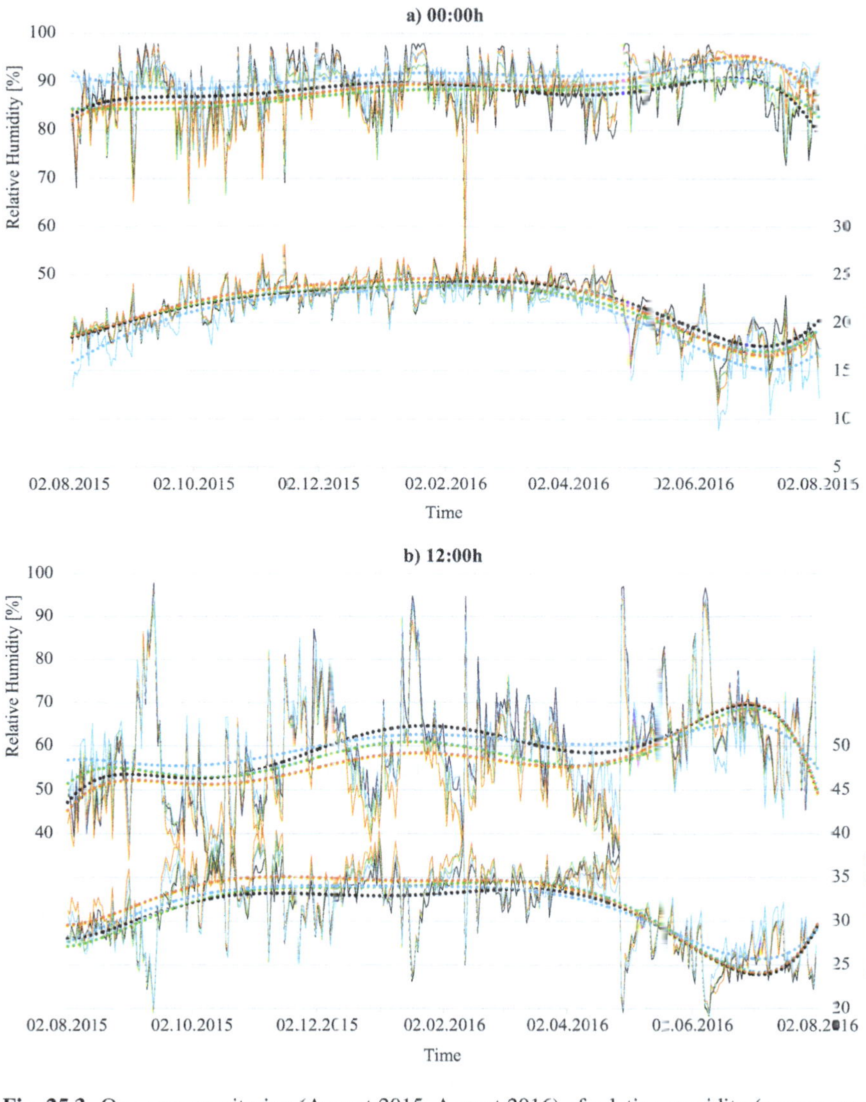

Fig. 25.3 One-year monitoring (August 2015–August 2016) of relative humidity (upper curves) and temperature (lower curves) at (**a**) midnight, 00:00 h and (**b**) noon, 12:00 h as absolute values (solid lines) and trend lines (dotted lines) along the pilot area transect from hilltop to bottom (black line, logger 1, crest; green line, logger 2, upper slope; orange line, logger 3, middle/foothill slope; blue line, logger 4, bottom) – logger position in Fig. 25.1, used device: HOBO ProV2-U23, Onset Computer Corporation (2016); the corresponding annual precipitation rate of 1131 mm was slightly above the average total annual value of 1073 mm; Aw-climate according to Köppen classification

accumulation toe-slope areas (red-yellowish Acrisol with colluvium) can be stated as favourable location for (intensive) cattle ranging. Measurements of accumulated soil material behind palisades (slope-wards) have proven its erosion-controlling effect by diminishing soil deposition and breaking surface runoff forces. Analysis on changes in rill and cattle track geometry over an 18-month period and investigations on distribution and density of termite mounds and its relevance to pasture degradation is still under investigation.

25.4.3 Lessons Learned

Pasturing on slopes, in particular with intensively roaming dairy cows, is problematic, both from an ecologic and economic perspective. Pasture properties such as soil conditions, carrying capacity, humidity, temperature as well as grass and water availability follow a gradient, usually deteriorating with increasing slope angle and length. Factors like steep and long pasture slopes combined with high grazing pressure and unsuitable pasture management make pastures more vulnerable against degradation, and the ecological pressure on landscape and ecosystem services becomes higher. The PRM presented demonstrates one approach towards a more sustainable and stable management on sloped pastures. Grass cover closure and hence pasture amelioration can be achieved by the following measures: (i) hedgerow terraces for interrupting the erosive power of surface runoff, (ii) horizontal ploughing and grass seeding on bare soil areas, (iii) paused pasture lots within a pasture rotation concept and a sustainable management with appropriate stocking densities and (iv) soil amelioration and pest control to enhance pasture productivity and stability. Hedgerow terraces stabilize the slope and improve soil fertility and biodiversity. *Brachiaria* is acceptable as pasture grass since it is well adapted to regional growth conditions, has a high productivity and is widely used by farmers. Furthermore, it does not invade afforested areas, even though it is impeding initial tree growth. In the long term, the presented PRM is seen as an interim solution, transferring cattle ranching from extensively managed slopes towards intensively managed plane areas with higher profitability and ecologic stability (Latawiec et al. 2014). Nevertheless, intensifying cattle ranching on the few alluvial plains available for pastures in RJ (SEA and INEA 2011: only 3% of RJ) remains a challenge. Moreover, a general lack of manpower and financial resources available in rural RJ slows down the introduction of new land use approaches.

25.5 Recommended Measures and Conclusions

Depending on the degree of degradation and specific site conditions, appropriate strategies against degraded sloped pastures range from suitable and sustainable management, pasture rehabilitation or recovery or a combination of these (Fig. 25.4).

Sustainable management	Pasture melioration and conservation	Slope-parallel ox-ploughing
		Organic fertilizer
		Seed of native grass species
		Preservation of surf. org. matter
		Shrub cutting
	Pasture management systems	Extensive rotational pasture
		Agrosilvicultural
		Silvopastoral
		Agrosilvopastoral
Rehabilitation (modular) and sustainable management (as above)	Simple bio-engineered measures	Palisades & fascines
		Grass sod transplantation
		Geo-textile or bio-fibre-mats
	Advanced bio-engineered measures	Slope-parallel (hedge) terraces
		Drainage channels and accumulation trenches
Recovery and conservation (fenced area)	Fallow & passive revegetation	Primary succession
		Secondary succession
	Active revegetation	Reforestation & afforestation
		Bio-corridors
		Compensatory measures
Increasing Degradation	Conservation (of forest fragments)	APP, RPPN, CAR
		Governmental incentive, PES
		Compensatory measures

Fig. 25.4 Strategies and measures to combat various degradation levels (green, no/low degradation; yellow/orange, medium; red, severe) on sloped pastures. Environmentally sensitive areas must be considered for recovery and conservation only. APP, permanent protection areas; RPPN, natural heritage private reserves; CAR, Rural Environmental Cadaster; PES, payment for ecosystem services

The more degraded the pasture, the more efforts and investments are usually needed to combat degradation. Strategies against pasture degradation contribute to (i) decelerating the degradation processes, (ii) regaining ecological stability and land productivity as a base for ecosystem services and farmer's income and (iii) preventing tipping points of landscape devastation which would lead to abandoned areas and subsequent migration of rural population to urban and industrial areas. Up until now, relatively few efforts to combat degraded sloped pastures have been practically realized in rural RJ, and even fewer have been published.

Slightly degraded pastures can be improved with some sound and easy to apply measures in combination with gentle management systems towards a more ecologically stable and profitable pasture. On sloped pastures, it is crucial to preserve organic matter on the surface of Acrisols and to prevent further erosion (IUSS Working Group WRB 2015). Liming, conventional fertilization supplemented by organic rock dust and sewage sludge mixture and use of Al-tolerant grass species additionally improve pasture quality. Overstocking with >2 AU/ha (Strassburg et al. 2014) and repetitive use of the same cattle trails need to be prevented in order to avert soil compaction and initial erosion forms. Agrisilvicultural, agrosilvopastoral and silvopastoral systems as discussed in Nair (2014) could contribute to more pasture diversity and income diversification through fruit and timber production. Projects like 'Balde Cheio' ('full bucket'), launched in 2011 by the Brazilian Agricultural Research Corporation (EMBRAPA), aim at introducing sustainable land management and agricultural best practices to small farmers by technology transfer and technical assistance (Silva et al. 2017). Furthermore, environmentally sensitive spring, riparian, recharge, hilltop and extremely steep (>45°) areas need to be excluded from pasture management and considered for recovery and conservation as permanent protection areas (APPs) or natural heritage private reserves (RPPNs) according to the Brazilian Forest Code requirements. Here, farmers could be financially rewarded within the frame of payment for ecosystem services (PES) (Hissa 2014) if areas are approved by Rural Environmental Cadaster (CAR).

Pastures at medium degradation level are recommended for rehabilitation measures accompanied by a sustainable post-management. Simple and advanced bioengineered measures with modular character are appropriate because implementation is flexible and adaptable to specific site conditions, costs, time and manpower settings. Detailed information on rehabilitation measure modules implemented at the pilot area and other degraded areas in RJ is provided among others in Tavares (2008), Chaves et al. (2012), Guerra and Jorge (2014) and Hebner et al. (2018). A good example is the PRM implemented near Raposo, RJ where a degraded area was terraced and reforested afterwards. Most projects in RJ focus on a stabilization of roadside flanks or landslide (prone) areas by applying terraces, drainage channels, palisades and biofibre mats for revegetation.

Severely degraded pastures should be 'rested' (rotation pasture, fallow) or completely excluded from further use to allow spontaneous or managed revegetation (recovery). These areas could then function as recharge or opportunity areas for carbon sequestration (Sattler et al. 2018b). Several studies in the Atlantic Forest biome have demonstrated fast growth rates of afforested trees even under less favourable site conditions (e.g. Rodrigues et al. 2009; Sattler et al. 2014). Reforestation could be realized and financed in the frame of compensatory measures, mandatory for companies and institutions with significant environmental impact.

Acknowledgments We thank the German Federal Ministry of Education and Research (BMBF) for funding the research and development project INTECRAL (033L162F) and the Rio de Janeiro sustainable rural development project (SEAPPA-RIO RURAL) for facilitating the survey. We thank the staff of Emater-Rio and Pesagro-Rio in Itaocara and Juliana M. Santos (University of Jena, Germany) and Giuseppe Latorraca for excellent logistic support during fieldwork. Many thanks to Hugh Eva from the Joint Research Centre of the European Commission for proofreading and critical advice.

References

Brazilian Federal Law 12.727 (2012) Lei n° 12.727, de 17 de outubro de 2012. http://www.planalto.gov.br/CCIVIL_03/_Ato2011-2014/2012/Lei/L12727.htm. Accessed 30 Mar 2017

Chaves TDA, Andrade AGD, Lima JAD, Portocarrero H (2012) Recuperação de áreas degradadas por erosão no meio rural. Manual técnico, 34. ISSN 1983-5671, p 21. Programa Rio Rural. Niteroi

Costa GCPD, Silva ASD (2015) Aumento da erosão pela compactação do horizonte superficial causada pelo pisoteio bovino em Santo Antônio de Padua – RJ. – 15° Congresso Brasileiro de Geologia de Engenharia e Ambiental, Rio Grande do Sul

DEME – Departamento Estadual de Minas e Energia (1970) Projeto Areas do Estado do Rio de Janeiro, executed by Aerofoto Natividade S.A., scale orthophoto: 1:20.000, analog orthophoto available at DRM-RJ -Serviço Geológico do Estado do Rio de Janeiro, Niteroi, scanned on 4 April 2016

Guerra AJT, Jorge MDCO (2014) Processos erosivos e recuperação de áreas degradadas. Oficina de Textos, São Paulo, pp 192

Hebner A, Kopielski K, Dulleck S et al (2018) Bioengineered measures for prevention of soil degradation as result of climate change in South East Brazil. In: Leal W, Esteves Freitas L (eds) Climate change adaptation in Latin America – managing vulnerability, fostering resilience. Springer International Publishing, Cham, pp 75–88

Hissa HR (2014) Financial sustainability for environmental services: rural development in micro-watersheds, Rio Rural, Brazil, In: FAO Case studies on Remuneration of Positive Externalities (RPE)/Payments for Environmental Services (PES)

Hjulström F (1935) Studies of the morphological activity of rivers as illustrated by the River Fryis. Bull Geol Inst Uppsala 25:221–527

IBGE – Instituto Brasileiro de Geografía e Estatística (2006) Projeto RJ-25, executed by Base Aerofotogrametria e Projetos S.A., scale orthophotomosaic: 1:25.000. http://downloads.ibge.gov.br/downloads_geociencias.htm#. Accessed 20 May 2016

INCRA – Instituto Nacional de Colonização e Reforma Agrária (2006) Relação de certificados de cadastro de imóveis rurais no município de Itaocara, Unidade Federal Rio de Janeiro, pp 113

IUSS Working Group WRB (2015) World reference base for soil resources 2014, update 2015. International soil classification system for naming soils and creating legends for soil maps. World soil resources reports no. 106. FAO, Rome

Latawiec AE, Strassburg BBN, Valentim JF et al (2014) Intensification of cattle ranching production systems: socioeconomic and environmental synergies and risks in Brazil. Animal 8(8):1255–1263

Liebmann B, Jones C, Carvalho LMVD (2001) Interannual variability of daily extreme precipitation events in the state of São Paulo, Brazil. J Clim 14:208–218

Lima HC, Pessoa SDVA, Guedes-Bruni RR et al (2006) Caracterização fisionômico-florística e mapeamento da vegetação da Reserva Biológica de Poço das Antas, Silva Jardim, Rio de Janeiro, Brasil. Rodriguésia 57(3):369–389

Nair PKR (2014) Agroforestry: practices and systems. In: Alfen, NKV (ed) Encyclopedia of agriculture and food systems - 2nd rev ed. Elsevier, London, pp 270–282

Onset Computer Corporation (2016) HOBO Pro v2 Loggers and HOBOware Version 3.7.8

Phalan B, Green RE, Dicks LV et al (2016) How can higher-yield farming help to spare nature. Science 351:450–451

Pix4D (2017) Pix4Dmapper Pro software, Educational Version 3.1.23, Lausanne

Reith J (2016) Einfluss bestimmter abiotischer und biotischer Faktoren auf die Vegetation degradierter Weiden im Munizip Itaocara (RJ), Brasilien – Deskriptive und statistische Analyse von vegetations-ökologischen Geländeaufnahmen. BSc Thesis, University of Leipzig, pp 105

Rodrigues RR, Lima RAF, Gandolfi S, Nave AG (2009) On the restoration of high diversity forests: 30 years of experience in the Brazilian Atlantic Forest. Biol Conserv 142(6):1242–1251

Sattler D, Murray LT, Kirchner A, Lindner A (2014) Influence of soil and topography on aboveground biomass accumulation and carbon stocks of afforested pastures in South East Brazil. Ecol Eng 73:126–131

Sattler D, Seliger R, Nehren U et al (2018b) Pasture degradation in South East Brazil: status, drivers and options for sustainable land use under climate change. In: Leal W, Esteves Freitas L (eds) Climate change adaptation in Latin America – managing vulnerability, fostering resilience. Springer International Publishing, Cham, pp 3–18

Sattler D, Raedig C, Hebner A, Wesenberg J (2018a) Use of native plant species for ecological restoration and rehabilitation measures in Southeast Brazil. In: Nehren U, Schlüter S, Raedig C, Sattler D, Hissa HR (eds) Strategies and tools for a sustainable rural Rio de Janeiro. Springer International Publishing, Cham

SEA and INEA – Secretaria de Estado do Ambiente and Instituto Estadual do Ambiente (2011) O estado do ambiente. Indicadores ambientais do Rio de Janeiro 2010, pp 160, Rio de Janeiro

Selsam P, Schwartze C (2016) Remote Sensing image analysis without expert knowledge – a web-based classification tool on top of Taverna workflow management system. World Multidisciplinary Earth Sciences Symposium, WMESS 2016, Prague

Silva RFBD, Rodrigues MDA, Vieira SA et al (2017) Perspectives for environmental conservation and ecosystem services on coupled rural – urban systems. Perspect Ecol Conserv 15(2):74–81

Silva ASD, Seliger R, Sattler D, Heinrich J (2018) Soil degradation in Southeast Brazil: a challenge for restoration and rehabilitation. In: Nehren U, Schlüter S, Raedig C, Sattler D, Hissa HR (eds) Strategies and tools for a sustainable rural Rio de Janeiro. Springer International Publishing, Cham

Strassburg BBN, Latawiec AE, Barioni LG et al (2014) When enough should be enough: improving the use of current agricultural lands could meet production demands and spare natural habitats in Brazil. Glob Environ Change Human Policy Dimens 28:84–97

Tavares SRDL (2008) Curso de recuperação de áreas degradadas: a visão da ciência do solo no contexto do diagnóstico, manejo, indicadores de monitoramento e estratégias de recuperação. Documentação 103. ISSN: 1517-2627, Embrapa Solos, Rio de Janeiro, pp 228

Chapter 26
Multicriteria Site Prioritization for Land Rehabilitation in the Guapi-Macacu Watershed, Rio de Janeiro

Friederike Naegeli de Torres, Ronny Richter, and Elaine Cristina Cardoso Fidalgo

Abstract Progressing land degradation is of increasing global concern because it deteriorates essential ecosystem features, functions, and services as observed in our study area in the highly fragmented Mata Atlântica biome, RJ, Brazil. The prioritization of sites for land rehabilitation is a frequent task for environmental managers and essential to develop urgently needed public policies. Hence, we present a GIS-based tool based on ecological principles for the selection of reforestation sites by integrating data on pasture degradation and landscape fragmentation. Thereby, the prioritized areas do not only represent opportunities for increasing forest cover and enhancing forest connectivity but also for reducing the area of degraded pastures. From our perspective, the inclusion of spatial information on degraded pastures is essential, since many of these areas have lost important ecological and economic function and should thus be prioritized for intervention. Data on degraded pastures and forest cover used in this study are derived from high- and medium-resolution satellite imagery. Other spatial data are derived from a digital elevation model

F. Naegeli de Torres (✉)
Institute for Geography, Department of Geoinformatics and Remote Sensing, Leipzig University, Leipzig, Germany
e-mail: friederike.naegeli@uni-leipzig.de

R. Richter
Institute for Geography, Department of Geoinformatics and Remote Sensing, Leipzig University, Leipzig, Germany

Institute for Biology, Systematic Botany and Functional Biodiversity, Leipzig University, Leipzig, Germany

German Centre for Integrative Biodiversity Research (iDiv), Halle-Jena-Leipzig, Leipzig, Germany
e-mail: ronny.richter@idiv.de

E. C. Cardoso Fidalgo
Empresa Brasileira de Pesquisa Agropecuária, EMBRAPA Solos, Rio de Janeiro, Brazil
e-mail: elaine.fidalgo@embrapa.br

U. Nehren et al. (eds.), *Strategies and Tools for a Sustainable Rural Rio de Janeiro*, Springer Series on Environmental Management,
https://doi.org/10.1007/978-3-319-89644-1_26

(DEM) with 20 m resolution. The multicriteria analysis for the data integration was conducted using ESRI ArcGIS and the open-source software ILWIS.

Keywords Mata Atlântica · Spatial multicriteria evaluation · Site prioritization · ILWIS · Rehabilitation

26.1 Introduction

Land degradation is of major global concern, especially in the context of climate change and food security. The ecological consequences of land degradation are diverse and lead among others to biodiversity loss, reduction of carbon storage capacities, and fertile soils as well as alteration of the water cycle. Rehabilitation of the land aiming at restoring important ecosystem services is therefore needed to counteract the negative consequences of degradation (Gurr et al. 2014). Hence, site selection for rehabilitation is a common task in decision-making in landscape planning for many purposes, e.g., environmental conservation plans and payments for ecosystem services programs (Malczewski 2006; Rodrigues et al. 2009; Orsi et al. 2011; Gjorup et al. 2016). The prioritization always depends on the selected criteria and on the preferences of the decision-maker, namely, the weights they decide to give to the single-selected criteria (Malczewski 2006).

In pasture-dominated landscapes, the as-is status of the pastures is to our understanding the key for proper selection of reforestation sites, in order to optimize the impacts of reforestation projects. Reforestation may thus not only aim at gains in forest area but also at improving connectivity and core areas as well as reducing the area of degraded pastures and soil loss and increasing the carbon storage. Hence, we consider the degradation status of the land for the selection of priority areas as an indispensable criterion for site selection. Based on this, the objectives of our study are (i) to offer a GIS-based methodology to identify rehabilitation sites of high priority that fulfill the defined criteria, aiming at an ecological optimization of the landscape through proper reforestation planning and (ii) to provide a map indicating the prioritized areas under ecological and legal principles.

26.2 Methods

26.2.1 Study Area

Our study area is located within the Guapi-Macacu watershed in the northwest of the metropolis of Rio de Janeiro. Formerly a forest-dominated ecosystem, the Guapi-Macacu watershed suffered from wood extraction and successive cycles of agricultural occupation since the sixteenth century (Dean 1997; Cabral 2004; Carneiro et al. 2012; Nehren et al. 2013). Today, the area is characterized by a high

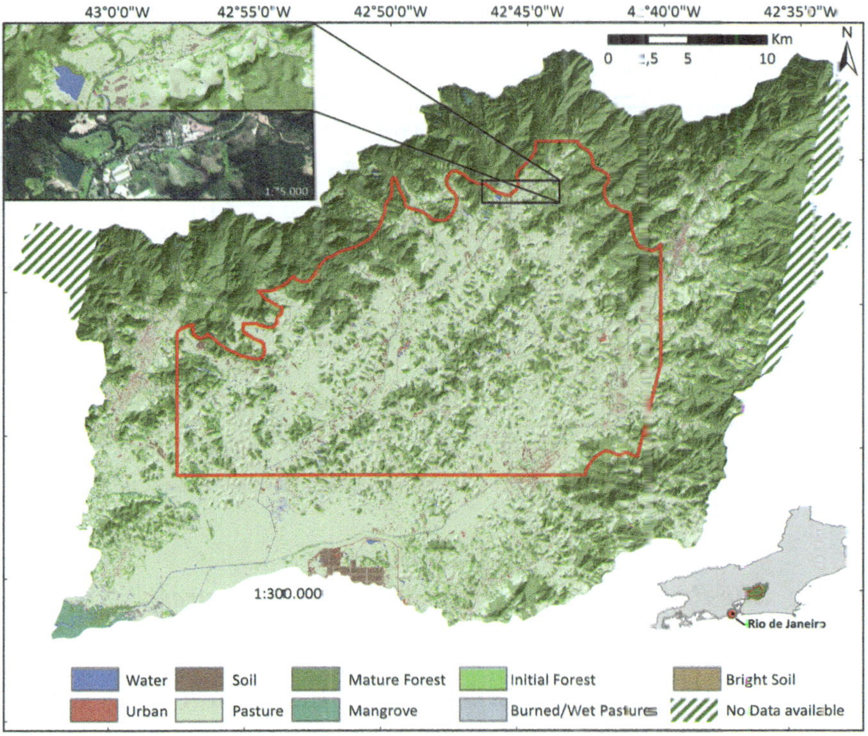

Fig. 26.1 Overview of the Guapi-Macacu watershed with the study area marked by a red polygon. Strong forest fragmentation is clearly visible in the lowlands of the watershed. The underlying land cover map (2011) with 5 m resolution was created based on RapidEye data with SWIR bands of SPOT 5 and Landsat 5 TM using RandomForests classifier in R

degree of forest fragmentation and pastures being the main land use type. However, due to inadequate management practices, most of them show light to strong signs of degradation (Fidalgo et al. 2014). Besides, many of the forest fragments are extremely small or have such an irregular shape that if assuming an edge effect of 100 m (Ribeiro et al. 2009), they do not even contain a core area. The vicinity, however, to one of the major and highly diverse remaining forest remnants of the biodiversity hotspot, Mata Atlântica protected by the Três Picos State Park, emphasizes the importance for conservation of the many small forest fragments located within this watershed as stepping stones (Nehren 2008; Nehren et al. 2013).

The irregular shape of the study area is due to the data availability. The area's size is of approx. 460 km^2 and encompasses the main part of the lowland of the Guapi-Macacu watershed, where most of the small-sized forest fragments are located (Fig. 26.1).

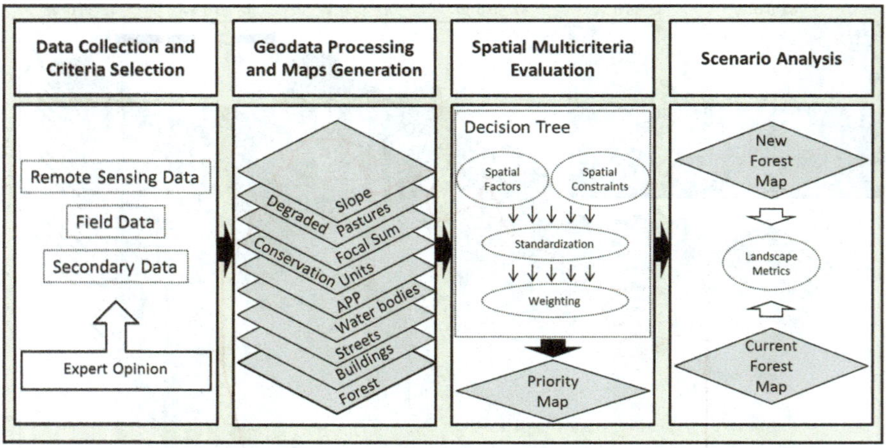

Fig. 26.2 Flow chart of applied methodology for the prioritization of sites for rehabilitation

26.2.2 Site Prioritization for Rehabilitation Measures

In this study we use spatial multicriteria evaluation (SMCE) to prioritize areas for rehabilitation as already applied in other studies (Roth et al. 2006; Orsi et al. 2011; Vettorazzi and Valente 2016). Therefore, we first selected criteria that are relevant to the selection process from an ecological and legal point of view and according related geodata (for details, see Sect. 26.2.2.1). In a next step, we processed the geodata and created raster maps. In a third step, we weighted the criteria and conducted the SMCE using the open-source software ILWIS, and last we developed different scenarios and made assumptions on the effectiveness of reforesting differently prioritized areas for forest patch quality and connectivity (Fig. 26.2).

26.2.2.1 Criteria Selection and Data Collection

For the selection of the criteria, we considered criteria selected in comparable studies (Roth et al. 2006; Orsi et al. 2011; Pedreira et al. 2011; Vettorazzi and Valente 2016), expert knowledge and experience from the field, but also the availability of data, which in our case was a limiting factor due to spatial inaccuracy or lack of appropriate spatial scale for the purpose of this study, e.g., digital soil maps at an appropriate scale or accurate cadastral data of rural properties and legal reserves. Since the aim of this study is to provide spatially explicit information on the priority of areas for reforestation according to the defined objectives, geodata with a spatial scale of 1:50,000 or lower was not considered. Based on the assumption that degraded pastures are not productive and face the increased soil loss and erosion

(Zhang et al. 2011; Dlamini et al. 2014) and due to the fact that meanwhile forest fragments in the Mata Atlântica biome are highly threatened but important for biological conservation (Joly et al. 2014), we identified the following criteria as highly important: (1) location of the areas within the landscape matrix with focus on its capacity to improve forest quality given by the focal sum of the forest cover (Roth et al. 2006), (2) degree of degradation of the pastures given by a degradation map, (3) the slope class, as degradation strongly correlates with the slope and is thus an indicator for areas prone to degradation. Additionally, five criteria were identified as spatial constraints, where reforestation is impossible: (1) current forest cover, (2) waterbodies, (3) roads, (4) buildings/settlements, and (5) areas in and outside highly protected areas (Conservation Units of Integral Protection, UCPIs, and Areas of Permanent Protection, APPs), respectively.

26.2.2.2 Geodata Processing

Land cover map In order to obtain the land cover, the RandomForest classifier (Breimann 2001) was applied using the open-source software R (R Core Team 2015) to an image product resulting from a combination of RapidEye images with the resampled SWIR bands of a SPOT 5 and a Landsat 5 TM image. For model construction, we randomly selected 60% of all pixels from each class for training and used the remaining 40% as test data. We repeated the classification task 25 times per band combination and applied a majority vote based on the 25 classification runs to further improve the final classification result. The resulting map (Fig. 26.1) has an overall accuracy of >98% with a spatial resolution of 5 m. A majority filter using the eight surrounding pixels was applied eight times repeatedly to the land cover map to smooth the classification output.

Forest cover From the land cover map described above, all pixels classified as mature or initial forest were merged into one forest class using ArcGIS. The raster was then converted into a vector data type, and the areas of the single fragments were calculated. All fragments sized smaller than 0.1 ha were subsequently removed, and the data was converted back to raster format.

Buildings and settlements Buildings and settlements were extracted from the land cover map, and the data was amplified by intense manual digitalization in Google Earth. The vector data (polygons) was converted into raster format. All areas identified as settlements or buildings were included as a spatial constraint.

Streets Streets were mapped in the field, and the data was amplified by manual digitization using aerial images from 2005/2006 from the Brazilian Institute of Geography and Statistics (Instituto Brasileiro de Geografia e Estatística – IBGE). For the vector data (lines), we created a buffer of 3 m and converted it to raster format. All areas identified as streets were included as spatial constraints.

Waterbodies From the land cover map, we extracted the waterbodies and waterbodies identified by INEA (Rio de Janeiro State Environmental Institute) plus waterbodies identified using Google Earth that were not included in one of the other two data sources.

Areas of Permanent Protection (Áreas de Preservação Permanente – APPs) APPs are defined by Forest Code and are composed of buffers of 30 m along the rivers of up to 10 m and slopes above 45°. This raster was created based on a shapefile provided by INEA indicating the river courses and the springs, as well as the slope map derived from the IBGE DEM with 20 m resolution. Hilltops (topos de morros) as another type of APP were not identified within the research area according to a GIS-based methodology that aims at calculating hilltops according to the National Forest Code (Castro Oliveira and Fernandes Filho 2013).

Conservation Units of Integral Protection (Unidades de Conservação de Proteção Integral – UCPI) One of the main categories of protected areas in Brazil is the conservation unit of integral protection which is restrictive to land uses. A conservation unit of integral protection map was created using the data of federal and state conservation units provided by INEA (GEPRO 2017). Três Picos State Park and Estação Ecológica Paraíso (INEA 2013) are the protected areas of this category in the study area.

Degraded pastures During a field campaign in 2011, 210 sites on pastures were mapped in the field attributing the stage of degradation as follows: no degradation visible; light degradation visible, e.g., small cattle tracks and little laminar erosion; moderate degradation visible, e.g., cattle tracks, laminar erosion, rill erosion (bras. *sulcos*), and initial and small gullies (bras. *ravinas*); and strong degradation visible, e.g., strong and broken cattle tracks, gully erosion (bras. *ravinas* and *voçorocas*). The degraded pastures were mapped by applying MESMA (multiple endmember spectral mixture analysis) using the open-source software Viper Tools 1.5 (available under https://sites.google.com/site/ucsbviperlab/viper-tools, last access 11/16/2017) to the image product used for the land use classification (Roberts et al. 1998, 2007; Dennison and Roberts 2003). Thereby the green vegetation fraction (GVF) was calculated on a per pixel basis. This information was then related to the degradation information of all pixels of the 210 polygons. GVF thresholds for each class were derived at the intersections of the GVF histograms calculated for each degradation classes. The thresholds were then used to classify the GVF image. In order to remove highly dynamic pastures from the classification, we masked all pixel that were identified as being dynamic using the mean absolute spectral dynamics (MASD) index (Franke et al. 2012) calculated on a time series of 5 RapidEye images from December 2010 to December 2011. The resulting degradation map with a resolution of 5 m indicates the following four stages of degradation (Fig. 26.3) – no degradation visible, light degradation visible, moderate degradation visible, and strong degradation visible – and shows an overall accuracy of near 80%.

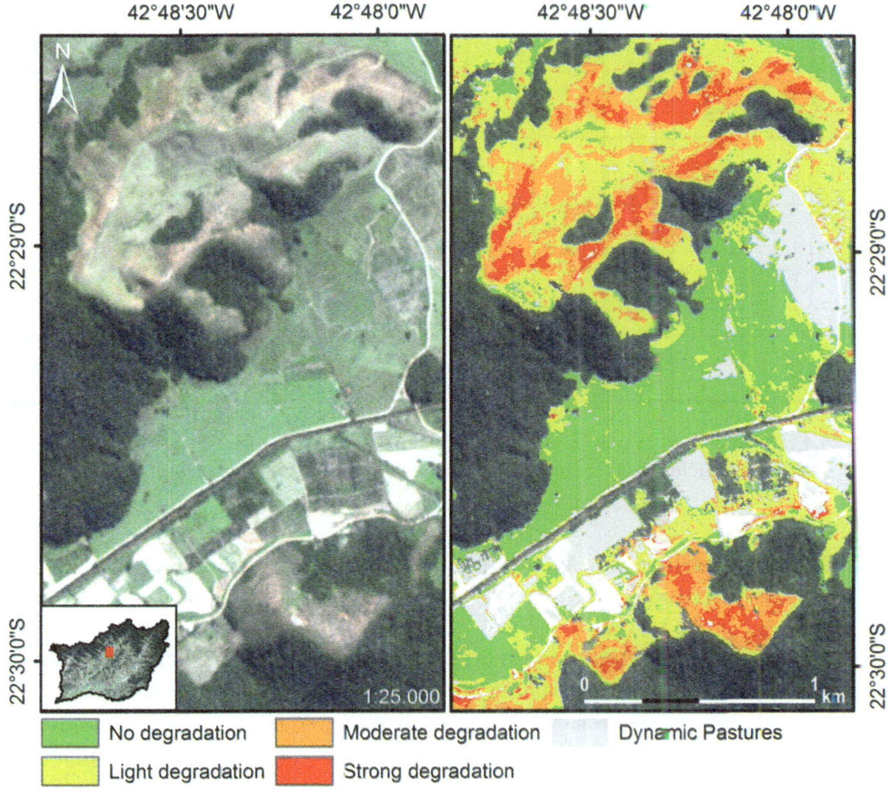

Fig. 26.3 Comparison of a subset of the degradation map (right) with the RGB-RapidEye image from 13/08/2011 (left)

Slope classes The slope was derived in degree from a DEM with 20 m spatial resolution calculated based on isolines with 10 m equidistance provided by the IBGE. As the slope angle has a major impact on pasture degradation processes and soil loss (Liu et al. 1994), we related the degraded pastures to eight slope classes. On slopes with more than 10°, the share of degraded pastures accounted for more than 75%, more than 28% of which are moderately to strongly degraded. Slope classes were thus defined based on these findings and legal aspects stated in the Brazilian Forest Code (Brazilian Federal Law 12651, May 25, 2012) as follows: 0–10°, 10–25°, 25–45°, and > 45°. These classes are also in accordance with the findings by Lepsch et al. 1991 who state that slopes above 9° inclination are easily eroded (except those being highly permeable including sandy soils).

Potential forest corridor (focal sum raster of the forest cover) Based on the forest cover raster, we applied the focal sum functionality of ArcGIS (Fig. 26.4), as suggested by Singleton and McRae (2013). A kernel of 99 was applied to the forest raster which equals 495 m. Thereby, a raster indicating how many forest pixels are located within the kernel of each pixel was created.

Fig. 26.4 Schematic graphic of the focal sum functionality. Areas filling gaps of the patches or those ones in their vicinity receive higher values and will thus be prioritized

26.2.2.3 Standardization, Weighting, and Spatial Multicriteria Evaluation

In our study we use spatial multicriteria evaluation (SMCE) to identify areas that should be prioritized for reforestation under ecological and legal principles. For the analysis, we use the open-access software ILWIS (52°North ILWIS Community 2015). All criteria were standardized to 0–1 in order to make them comparable (Table 26.1).

Criteria weighting was done by pairwise comparison according to Saaty (1980) using a nine-step scale (Table 26.2). The focal sum parameter and the degradation status of the pastures were set to be equally important, whereas both were set to be moderately more important than the slope classes. Thereby we obtained the following weights: focal sum = 0.43, status of pastures = 0.43, and slope classes = 0.14. Based on these weights, we derived the priority maps once for the area inside and once outside the protected areas.

26.2.3 Scenario Development

In order to test the impact of the identified priority areas on the existing forest cover, we created different scenarios based on the identified priorities. Therefore, we merged the results of the priority maps for in and outside the protected areas (UCPIs and APPs). We then extracted the areas with priority values of 0.8–1.0 (Scenario A) and merged these areas with the actual forest cover. We repeated this step for the areas 0.7–1.0 (Scenario B), 0.6–1.0 (Scenario C), and 0.5–1.0 (Scenario D). Based on these hypothetical forest rasters, we calculated different patch parameters for the

Table 26.1 Overview of the criteria and the standardization used (Spatial constraints: 1 = yes, 0 = no. Spatial factors: 1 = very important, 0 = not important)

Criterion	Criteria type	Standardization
Forest	Spatial constraint	Forest = 1/no forest = 0
Buildings	Spatial constraint	Buildings = 1/no buildings = 0
Streets	Spatial constraint	Streets = 1/no streets = 0
Waterbodies	Spatial constraint	Waterbodies = 1/no waterbodies = 0
UCPI_APPs	Spatial constraint	UCPI_APPs = 1/no UCPI_APPs = 0 (used to map the area outside the protected areas)
		UCPI_APPs = 0/no UCPI_APPs = 1 (used to map the area inside the protected areas)
Focal sum	Spatial factor	4900.5–9801 = 1/0–4900.5 = 0–1 (linear)
Degraded pastures	Spatial factor	1: No degradation = 0 2: Light degradation = 0.5 3: Moderate degradation = 0.75 4: Strong degradation = 1
Slope classes	Spatial factor	1: 0–10° = 0 2: 10–25° = 0.5 3: 25–45° = 0.75 4: >45° = 1

Table 26.2 Nine-step matrix developed by Saaty (1980) to identify the criteria weights by applying pairwise comparison

Index	Definition	Calculation
−4	Extremely less important	1/5
−3	Very strongly less important	1/4
−2	Strongly less important	1/3
−1	Moderately less important	1/2
0	Equally important	1/1
1	Moderately more important	2
2	Strongly more important	3
3	Very strongly less important	4
4	Extremely more important	5

scenario-based forest fragments using the open-source software FRAGSTATS 4.2 (McGarigal et al. 2012). We compared the mean values for each parameter of each scenario to those calculated for the actual forest fragments in 2011. For the edge-dependent parameter, we assumed an edge effect of 100 m according to Ribeiro et al. 2009 and findings by Thier and Wesenberg (2016). The following eight parameters for patch size, shape, and connectivity/degree of isolation were of special interest (for detailed information, please refer to the FRAGSTATS documentation (McGarigal et al. 2012)):

I. Class area (CA): a measure of total forest cover.
II. Mean parameter-area-ratio (PARA): a shape metric which relates the patch perimeter to its area.

III. Mean shape index (SHAPE): another shape metric that relates the patch perimeter to its area with an adjustment for a standardized measure.
IV. Mean contiguity index (CONTIG): a metric of patch shape which measures the spatial connectedness of cells within a patch.
V. Total core area (TCA): a measure of the sum of the core areas of each forest patch.
VI. Mean core area index (CAI): where CAI is the percentage of each patch, that is core area, and mean CAI is the average of all CAI of the landscape.
VII. Mean proximity index (PROX): an aggregation metric that measures for each patch, the sum of patch area divided by the distance squared between the focus patch and other patch, within a specified distance.
VIII. Mean Euclidean nearest neighbor (ENN): another aggregation metric. ENN measures the distance between two patches.

26.3 Results

26.3.1 Prioritization of Areas for Rehabilitation

In our analysis we focused on the prioritization of areas in and outside protected areas based on three main criteria: (i) the as-is state of the pastures, (ii) the importance of the area for connectivity or improving patch shape, and (iii) the inclination of the slope. Of the whole study area (approx. 460 km^2), i.e., around 250 km^2 were taken into account for the SMCE. Of these 250 km^2, 20% (approx. 50 km^2) are protected by law as UCPI and APPs and must be preserved. That is the reason why we decided to analyze them separately.

26.3.2 Characteristics of the Prioritized Areas

Compared to the areas within the protected areas, there are relatively more areas of moderate to high priorities outside the protected area according to our criteria (Fig. 26.5).

Figures 26.6 and 26.7 show maps of the identified priority areas outside (Fig. 26.6) and inside protected areas, i.e., areas within UCPIs and APPs: buffers of river and spring plus areas above 45° slope (Fig. 26.7). In both cases we identified several high-priority areas in close vicinity to forest patches. Meanwhile high-priority sites outside protected areas can be found more or less equally distributed within the watershed, especially in close vicinity to the larger fragment in the central part of the watershed as well as at the edges of the PETP. Most of the high-priority areas inside the protected areas are located within the UCPI. On the assumption that most of the high-priority sites within the protected areas seem to be mainly located within the UCPI is from our understanding due to the fact that most

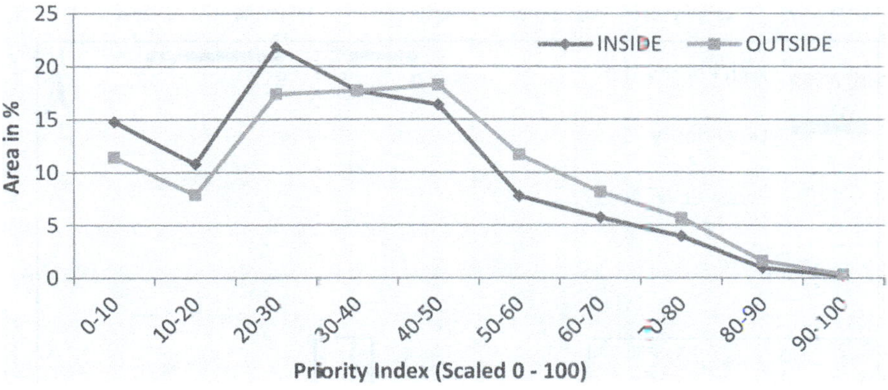

Fig. 26.5 The graph illustrates the distribution of priority values in % for the areas in and outside the protected areas (UCPI/APPs)

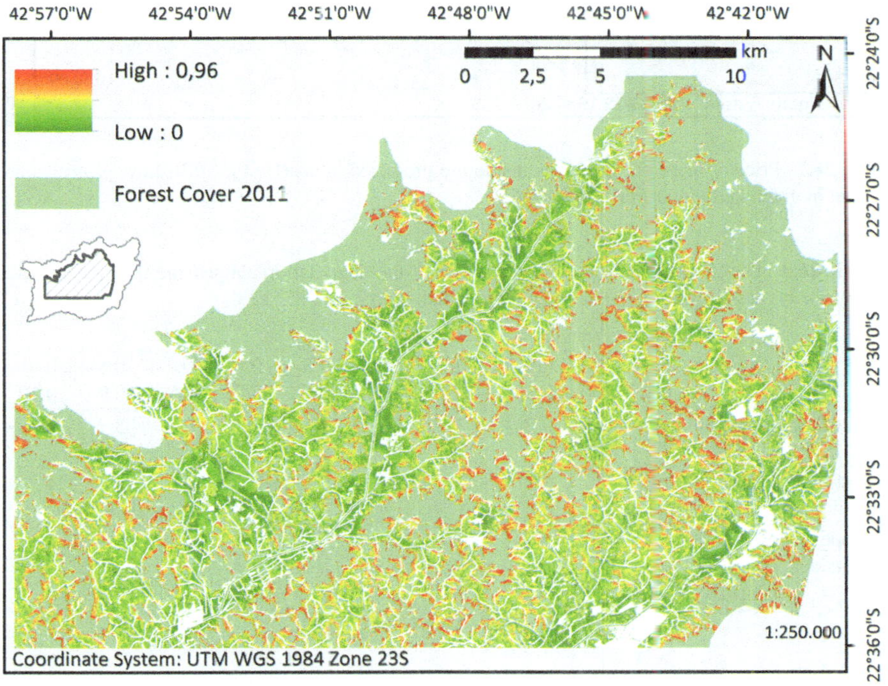

Fig. 26.6 Priority maps for the areas outside the protected areas (UCPI.APP). Low priorities are given in green and highest priority in red

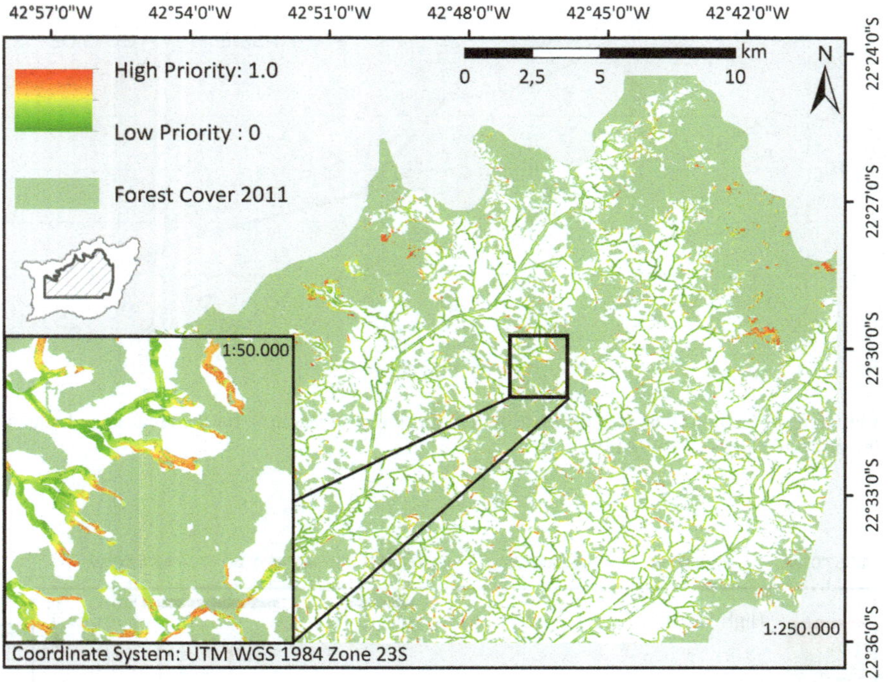

Fig. 26.7 Priority maps for the areas inside the protected areas (UCPI/APP). Low priorities are given in green and highest priority in red

Table 26.3 The table shows the class majorities of the degradation classes and the slope classes per priority class for areas in and outside protected areas

	Priority classes									
	0.0–0.1	0.1–0.2	0.2–0.3	0.3–0.4	0.4–0.5	0.5–0.6	0.6–0.7	0.7–0.8	0.8–0.9	0.9–0.10
Pasture degradation (outside)	1	1	2	2	2	2	2	2	3	4
Pasture degradation (inside)	1	1	2	2	2	2	2	2	3	4
Slope class (outside)	1	1	1	2	2	2	2	2	3	4
Slope class (inside)	1	1	1	1	1	1	1	1	2	2

of the APP area is located along rivers and that within the APP area there seem to be less moderately to strongly degraded pastures than outside the protected areas.

We calculated the class majority of the degraded pastures and the slope classes per priority class for both areas to get a further insight of the identified priorities. Table 26.3 indicates the different majority values for in and outside protected areas per priority class. The areas of the highest priorities (0.8–1.0) are mainly located on

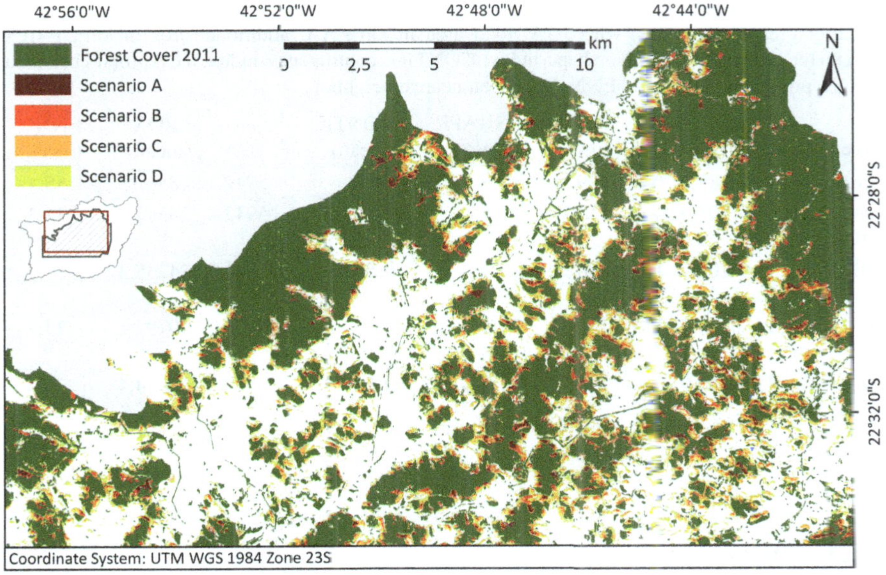

Fig. 26.8 Overview of the different scenarios developed for the study area

moderately to strongly degraded pastures on slopes >25° for areas outside protected areas and on slopes 10–25° for areas inside protected areas. This fact may support the conclusion that most of the high-priority areas inside protected areas are located within UCPIs, as the UCPI area is located at the foothills of the Serra dos Órgãos with steeper slopes than the APP area.

26.3.3 Hypothetical Improvement of Forest Quality

For a more detailed comprehension on the impact of possible reforestation in prioritized areas according to our priority maps, we compared patch parameter of the original forest cover 2011 with what-if scenarios of the forest cover when being enlarged by afforesting certain prioritized areas (Fig. 26.8).

The results in Table 26.4 indicate clearly that afforesting the prioritized areas with a priority index of 0.8–1.0, with only 4.3 km^2 would have the relatively largest impact on area, patch shape, and proximity parameters compared to the other scenarios. This would benefit the forest patches enormously, even though no absolute gain in total core area (assuming an edge of 100 m) will be achieved. A gain in total core area is only recognizable from scenario (B) to (D).

Table 26.4 Comparison of the forest parameter calculated for the original forest cover 2011 and the four created scenarios, where TA, total area in km²; AA, additional area in km²; PARA, parameter-area-ratio; SHAPE, shape index; CONTIG, contingency index; TCA, total core area; PROX, proximity index; and ENN, Euclidean nearest neighbor

Scenario	TA	AA	PARA (mean)	SHAPE (mean)	CONTIG (mean)	TCA	PROX (mean)	ENN (mean)
Forest 2011	188.6	0.0	1097.0	1.8	0.8	5942.2	4836.1	55.4
(A) Forest 11 (0.8–1.0)	192.9	4.3	3170.0	1.5	0.6	5942.4	23577.4	34.3
(B) Forest 11 (0.7–1.0)	205.7	17.1	4002.9	1.4	0.5	5966.0	34235.3	27.9
(C) Forest 11 (0.6–1.0)	224.4	35.8	4530.1	1.4	0.4	5978.2	40792.1	23.4
(D) Forest 11 (0.5–1.0)	250.9	62.3	4928.3	1.3	0.4	6045.5	61634.4	20.6

26.4 Discussion

The presented methodology was especially developed for pasture-dominated environments with strong forest fragmentation. From our understanding, highly degraded areas, being located on steep slopes prone to further degradation and in close vicinity to forest fragments, where they would improve the patch shape and/or the connectivity between fragments when being reforested, should be considered priority for rehabilitation. We believe that our approach properly identified those areas that match the mentioned criteria.

In this study we used only high-resolution data on the degradation status of the pastures, the focal sum raster of the forest cover, and the slope classes with appropriate resolutions to select small priority sites. For the focal sum raster of the forest cover, we used a Kernel of 99 (495 m). This kernel has been proper for our study area, as it identified successfully important sites for improving forest shape and connectivity. Further studies might however be required to systematize best kernel sizes for different landscapes. The slope map used in our study was derived from a 20 m DEM. Improved resolution of the source DEM could also improve the identification of steep slopes. In general, it was difficult to obtain high-resolution data for the study area, one reason why most of the data was produced by the author to fit the requirements. Missing data, such as high-resolution soil map or data on the soil susceptibility to erosion, could have been very interesting supportive data. Soil map was only available with a scale of 1:250,000, but we recommend not mixing the scales. We also would have liked to consider information on the landownership, as landowners must conserve 20% of their land as forests. Unfortunately, this data is not available. Recently obtained spatial information from the Rural Environmental Survey (Cadastro Ambiental Rural – CAR) is also not ready to use at the moment due to high spatial inaccuracies and missing data. From an ecological perspective, however, it is important to identify areas of high ecological value in the first place.

The integration of socioeconomic information may be added subsequently. However, this does not mean that socioeconomic criteria are not necessary. Its integration with the criteria presented in this work is highly encouraged.

Prioritizing areas always depends on the selected criteria, the data quality, availability and scale, and the preferences of the decision-maker who gives the weights to the criteria. Thus, the final priority maps may always differ and should mainly be regarded as a supportive information to environmental and landscape planners. According to Malczewski (1996), the final decision should be taken based on different alternatives prepared by several groups or individuals of different interests. Also thresholding and classifying the priority values are again a quite subjective task. We decided to use equal classes of steps of 0.1. However, there are many possibilities to classify the final results depending on the purpose, e.g., using jerks to create equally area size classes.

We are convinced that the presented approach could further be used to adapt schemes for payments for ecosystem services by giving different incentives to areas of different priorities in order to increase the willingness of the landowners to rehabilitate areas of high ecological importance, although some more research might be needed to properly transfer our methodology into a legal framework.

26.5 Conclusion

The above-described method identifies high-priority areas within and outside conservation units (UCPIs and APPs) for increasing forest cover and enhancing forest shape and connectivity. Using patch parameter to make further statements on the quality of the prioritized areas showed to be very effective. The information obtained from the scenario comparison proves the effectiveness of our methodology, the selected criteria, and the weights given for the specific aim of our study.

Related to the issue of data availability and quality we mentioned in the methods and the discussion, we would like to emphasize the importance of proper geoprocessing and the development of properly georeferenced geodata, as they are indispensable for any kind of spatial analysis. Thus, the development of a high-resolution geodatabase based on surveying technology should be of highest priority to support environmental management and public policies.

Acknowledgments This work was supported by the German Ministry of Education and Research (BMBF) through the DINARIO [FKZ 01LB0801B] and INTECRAL [FKZ 033L162F] projects and the Heinrich Böll Foundation. We would like to thank the Project Planet Action (an Astrium Geo Initiative) for providing SPOT 5 image data and RESAweb for providing RapidEye 3A imagery. Landsat 5 TM imagery was provided by the US Geological Survey (USGS). We also like to thank the Instituto Estadual do Ambiente (INEA) and the Instituto Brasileiro de Geografía e Estadística (IBGE) for the provision of supportive geodata and EMBRAPA Solos, Rio de Janeiro and the Programa RioRural (Superintendência de Desenvolvimento Sustentável, Secretaria de Agricultura e Pecuária, Governo do Rio de Janeiro) for all the support provided during field work. Moreover, we gratefully acknowledge the scientific support to the land use and pasture degradation classification provided by Prof. Dr. Michael Vohland (Leipzig University).

References

52°North ILWIS Community (2015) Integrated Land and Water Information System - ILWIS. Version 3.8.5, Released September 3rd, 2015. Under GPL

Breimann L (2001) Random forests. Machine Learn 45(1). Kluwer Academic Publishers):5–32. https://doi.org/10.1023/A:1010933404324

Cabral DC (2004) Produtores rurais e indústria madeireira no Rio de Janeiro do final do século XVIII - Evidências Empíricas Para a Região do Vale do Macacu. Ambiente & Sociedade 7(2):125–145

Carneiro MJT, Latini JL, Coelho TD et al (2012) Histórico do processo de ocupação das bacias hidrográficas dos rios Guapi-Macacu e Caceribu. Embrapa Solos. Série Documentos/ Embrapa Solos, Rio de Janeiro, p 152

Dean W (1997) A ferro e fogo: a história da devastação da Mata Atlântica brasileira. Companhia das Letras, Rio de Janeiro

Dennison PE, Roberts DA (2003) Endmember selection for multiple endmember spectral mixture analysis using endmember average RMSE. Remote Sens Environ 87(2–3):123–135. https://doi.org/10.1016/S0034-4257(03)00135-4

Dlamini P, Chivenge P, Manson A, Chaplot V (2014) Land degradation impact on soil organic carbon and nitrogen stocks of sub-tropical humid grasslands in South Africa. Geoderma 235-236:372–381. https://doi.org/10.1016/j.geoderma.2014.07.016

Fidalgo ECC, Thompson D, Naegeli de Torres F, Mendonça RN (2014) Comparative analysis between estimated soil loss and occurrence of erosion: a case study in Guapi-Macacu watershed, RJ. In: Moura MF, Pellegrino GQ, Rodrigues LN (eds) Impactos da agricultura e das mudanças climáticas nos recursos hídricos: anais. Campinas, SP, 25 a 27 de março 2014. Embrapa, Brasília, DF

Franke J, Keuck V, Siegert F (2012) Assessment of grassland use intensity by remote sensing to support conservation schemes. J Nat Conserv 20(3):125–134. https://doi.org/10.1016/j.jnc.2012.02.001

GEPRO (2017) Unidades de Conservação da natureza no estado do Rio de Janeiro. Map and digital data.. http://geproinearj.blogspot.com.br/p/nucleos.html. Accessed 23 Oct 2017

Gjorup AF, Fidalgo ECC, Prado RB, Schuler AE (2016) Análise de procedimentos para seleção de áreas prioritárias em programas de pagamento por serviços ambientis hídricos. Rev Ambient Água 11(1):225–238

Gurr GM, Johnson AC, Liu J (2014) Land use: restoration and rehabilitation. Reference Module in Food Science – Encyclopedia of Agriculture and Food Systems 139–142

INEA (2013) Parque estadual dos três picos: Plano de manejo. Resumo executivo. Instituto Estadual do Ambiente, Rio de Janeiro. http://r1.ufrrj.br/lmbh/pdf/ensino/MAS/2017-I/MAS%20Dinamica%2003%20(texto)%20(2017%201).pdf. Accessed 20 Apr 2017

Joly CA, Metzger JP, Tabarelli M (2014) Experiences from the Brazilian Atlantic Forest: ecological findings and conservation initiatives. New Phytol 204:459–473

Lepsch IF, Bellinazzi JRR, Bertolini D, Espíndola CR (1991) Manual para levantamento utilitário do meio físico e classificação de terras no sistema de capacidade de uso, 4th edn. Sociedade Brasileira de Ciência do Solo, Campinas, SP

Liu BY, Nearing MA, Risse LM (1994) Slope gradient effects on soil loss for steep slopes. Trans ASAE 37(6):1835–1840

Malczewski J (1996) A GIS-based approach to multiple criteria group decision-making. Int J Geogr Inf Syst 10:955–971

Malczewski J (2006) GIS-based multicriteria decision analysis: a survey of the literature. Int J Geogr Inf Sci 20(7):703–726

McGarigal K, Cushman SA, Ene E (2012) FRAGSTATS v4: spatial pattern analysis program for categorical and continuous maps. Computer software program produced by the authors at the University of Massachusetts, Amherst.. http://www.umass.edu/landeco/research/fragstats/fragstats.html. Accessed 23 Oct 2017

Nehren U (2008) Quartäre Landschaftsgenese und histroische -dedradation in der Serra dos Órgaos, Rio de Janeiro. Dissertation, Universität Leipzig, Leipzig
Nehren U, Kirchner A, Sattler D et al (2013) Impact of natural climate change and historical land use on landscape development in the Atlantic Forest of Rio de Janeiro, Brazil. An Acad Bras Cienc 85(2). https://doi.org/10.1590/S0001-37652013000200004
Oliveira GC, Fernandes Filho EI (2013) Metodologia para delimitação de APPs em topos de morros segundo o novo Código Florestal brasileiro utilizando sistemas de informação geográfica. Anais XVI Simpósio Brasileiro de Sensoriamento Remoto – SBSR, Foz do Iguaçu, PR, Brasil, 13.-18. of april 2013, INPE
Orsi F, Geneletti D, Newton AC (2011) Towards a common set of criteria and indicators to identify forest restoration priorities: an expert panel-based approach. Ecol Indic 11:337–347
Pedreira BCC, Fidalgo ECC, Abreu MB (2009) Mapeamento do uso e cobertura da terra da bacia hidrográfica do Rio Guapi-Macacu, RJ. Anais XIV Simposio Brasileiro de Sensoriamento Remoto - SBSR, Natal, RN, Brasil, 25.-30. of april 2009, INPE
R Core Team (2015) R: a language and environment for statistical computing. R Foundation for Statistical Computing, Vienna
Ribeiro MC, Metzger JP, Martensen AC et al (2009) The Brazilian Atlantic Forest: how much is left, and how is the remaining forest distributed? Implications for conservation. Biol Conserv:1141–1153. https://doi.org/10.1016/j.biocon.2009.02.021
Roberts DA, Gardner M, Church R et al (1998) Mapping chaparral in the Santa Monica Mountains using multiple endmember spectral mixture models. Remote Sens Environ:267–279. https://doi.org/10.1016/S0034-4257(98)00037-6
Roberts DA, Halligan KQ, Dennison, PE (2007) VIPER tools user manual (version 1.5). University of California at Santa Barbara.. https://sites.google.com/site/ucsbviperlab/viper-tools. Accessed 16 Nov 2017
Rodrigues RR, Brancalion PHS, Isemhagen I (eds) (2009) Pacto pela restauração da Mata Atlântica: referencial dos conceitos e ações de restauração florestal. LERF/ESALQ: Instituto BioAtlântica, São Paulo
Roth RE, Zhu A-X, Holbus E et al (2006) An automated approach to site selection for ecological restoration in fragmented landscapes. Geogr Inform Sci 12(2):98–105
Saaty TL (1980) The analytic hierarchy process: planning setting priorities, resource allocation. McGraw-Hill International, New York
Singleton PH, McRae BH (2013) Chapter 10: Assessing habitat connectivity. In: Craighead FL, Convis CL (eds) Conservation planning: shaping the future. ESRI Press, Redlands
Thier O, Wesenberg J (2016) Floristic composition and edge-induced homogenization in tree communities in the fragmented Atlantic rainforest of Rio de Janeiro, Brazil. Trop Conserv Sci 9(2):852–876
Vettorazzi CA, Valente RA (2016) Priority areas for forest restoration aiming at the conservation of water resources. Ecol Eng 94:255–267. https://doi.org/10.1016/j.ecoleng.2016.05.069
Zhang G, Kang Y, Han G et al (2011) Grassland degradation reduces the carbon sequestration capacity of the vegetation and enhances the soil carbon and nitrogen loss. Acta Agric Scand Sect B Soil Plant Sci 61(4):356–364. https://doi.org/10.1080/09064710.2010.495079

Part V
Environmental Governance and Economic Instruments

Chapter 27
Evolution of Public Policies and Local Innovation in Landscape Conservation in Rio de Janeiro

Peter H. May, Leonardo S. Fernandes, and Vanesa Rodríguez Osuna

Abstract This chapter reviews the evolution of state policies and institutions responsible for land use policy and conservation incentives to landowners in Rio de Janeiro, with emphasis on the areas involved in implementing agro-environmental programs. We refer to some of the efforts underway to create private patrimony reserves, biological corridors, and PES instruments based on water resource protection and sustainable land use practices. We discuss the impacts of those policy interventions on regional efforts to motivate landscape protection and restoration along with possible alternatives. The discussion is couched within the framework of national and global debates regarding participatory landscape management and poverty amelioration in developing countries.

Keywords Conservation incentives · Environmental governance · PES · Ecological ICMS · Forest Code · Rio de Janeiro

P. H. May (✉)
CPDA/UFRRJ, Seropédica, Brazil

Earth Institute, Columbia University, New York, NY, USA

L. S. Fernandes
PPED/IE/UFRJ, Rio de Janeiro, RJ, Brazil

V. Rodríguez Osuna
CUNY Advanced Science Research Center, New York, NY, USA
e-mail: vanesa.rodriguez.osuna@asrc.cuny.edu

U. Nehren et al. (eds.), *Strategies and Tools for a Sustainable Rural Rio de Janeiro*, Springer Series on Environmental Management,
https://doi.org/10.1007/978-3-319-89644-1_27

27.1 Review of State Policies and Institutions

27.1.1 History of Rural Territorial Expansion and Forest Conversion in Rio de Janeiro

Rio de Janeiro (RJ), like Brazil overall, is no longer an agricultural state. In fact, Rio's economy and population have long been urbanized, since the state assumed the status of capital of the imperial government in the early nineteenth century, after the proclamation of independence in 1822 and of the republic from 1891 through the creation of the Federal District in Brasília in the late 1950s. Despite its cosmopolitan role in national governance, the state was historically an important producer of sugarcane, coffee, and citrus. Intensive coffee cultivation based on slave labor occupied a substantial share of the hinterland throughout the nineteenth and early twentieth centuries. Nevertheless, the impoverishment of soils in the Paraíba River Valley, allied with the high cost of slave labor and the increased coffee productivity in São Paulo (SP) – benefitting from improved techniques brought by European immigrants – led to the decadence of coffee production in RJ. To substitute for coffee, RJ was left only with pastures on heavily eroded soils bereft of nutrients, barely able to support the few head of cattle that now occupy most of the formerly fertile hillsides.

Agroecological concerns are by no means a recent phenomenon in RJ. The intense degradation of the original Atlantic Forest that covered the state, as it was put to the torch to respond to the coffee boom (Dean 1995), preoccupied scientist and imperial advisor José Bonifacio whose writings lamented the impoverishment of the state's natural capital in his writings (Pádua 2002). In the 1850s, repeated water shortages in the capital caused by denuded forestlands cut for coffee plantations in the Tijuca highlands prompted emperor Dom Pedro II to request the aid of Major Gomes Archer, who with the aid of a group of slaves produced hundreds of thousands of tropical tree seedlings with which they reforested risky hillsides in what is now the Tijuca Forest National Park (Parna Tijuca) (Drummond 1996).

Such magnanimous gestures were rare, however, and during most of RJ's history, the continued decimation of the Atlantic Forest is notable. In 2011, SOS Mata Atlântica and INPE registered a total loss of original forest area of 82.3% in the state since colonization, declining from 4.4 million to 820,000 ha (SOSMA and INPE 2015). Although historical, there had been substantial forest clearing since the 1980s, when measurements began. The rate of deforestation has however been reduced dramatically since the early 1990s (to some extent a reflection of efforts to "green" the state after Rio 92) (Fig. 27.1). Of late, indeed, the process has reversed and become one of forest regenerations. Although in RJ this process is not as pronounced as in other states in the biome, 4092 ha of the area originally under forest had regenerated after having been deforested during the period analyzed. Natural regeneration has been shown to be one of the most effective and least costly means to restore degraded lands in Brazil (Rodrigues

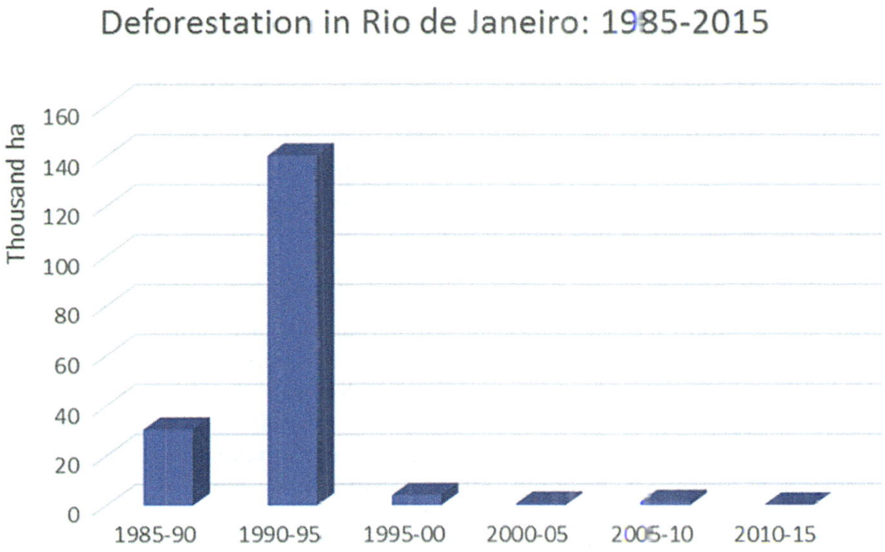

Fig. 27.1 Deforestation in Rio de Janeiro: 1985–2015. Instituto Estadual do Ambiente-INEA elaboration based on data available in SOSMA and INPE (2015)

et al. 2009; Rezende et al. 2015) now adopted on a grand scale to meet the national INDC of 12 million ha of forest restoration by 2030 presented officially as part of national commitments under the Paris Climate Accord (Federative Republic of Brazil 2016).

Needless to say, this regenerative process and the maintenance of forest remnants in the state can only be accomplished through interventions on private lands in rural landscapes. Those areas that were sufficiently well preserved to become part of the national system of protected areas have already been so designated, since the 1930s, when the Tijuca Forest National Park was created.[1] Currently, the state hosts 462,542 hectares in 55 public protected areas (comprising nearly 58% of all forests in the state mapped by SOSMA and INPE 2015). These areas form a complex mosaic representative of the various ecosystems that make up the state, from dunes, mangroves, and *restingas* along the coastline to cloud forests in the mountainous areas of central and southern RJ that make up part of the Serra do Mar that stretches from neighboring Espírito Santo to Santa Catarina in Southern Brazil (Rizzini 1997; Scarano 2002). The National System of Protected Areas (SNUC), enacted in 2000, provides for a range of public and private land use designations for protective purposes. The state has been an important participant in the creation of 150 private natural patrimony reserves (RPPN) that although small in size (total

[1] http://www.inea.rj.gov.br/cs/idcplg?IdcService=DOC_INFO_BY_NAME&dDocName=INEA0123058. This link provides information on the statewide conservation units and does not include area in private natural patrimony reserves (RPPN).

area = 12,700.13 ha, averaging <100 ha) have an important role in protecting critical habitat (CNRPPN 2017). Furthermore, a number of private property areas in buffer zones of public parks and reserves were designated as environmental protection areas (APAs) which, although they typically provide much less protection due to reduced public staffing or support, establish permissible land uses and stimulate participatory decision-making processes. Around three-fourths of remaining forest in RJ lies on private lands; thus, conservation outside integrally protected areas is crucial to prevent further biodiversity loss.

At a biome level, all states in the Atlantic Forest form part of the UNESCO-designated Biosphere Reserve of the Atlantic Forest (RBMA) (see Fig. 27.2 for areas of RJ within the RBMA), a loose federation of governments and NGOs that seeks to promote protection of remaining biodiversity and forest restoration throughout the biome. More recently, a confederation of academic and NGO groups throughout the biome has formed a pact for the restoration of the Atlantic Forest whose first product was a map of areas in need of restoration (Pacto 2009; Pinto et al. 2014). A recently formed Coalition on Climate, Forests and Agriculture (Coalízão Clima, Floresta e Agricultura 2017) made up of business, government, and NGO segments monitors efforts to comply with the Paris accord.

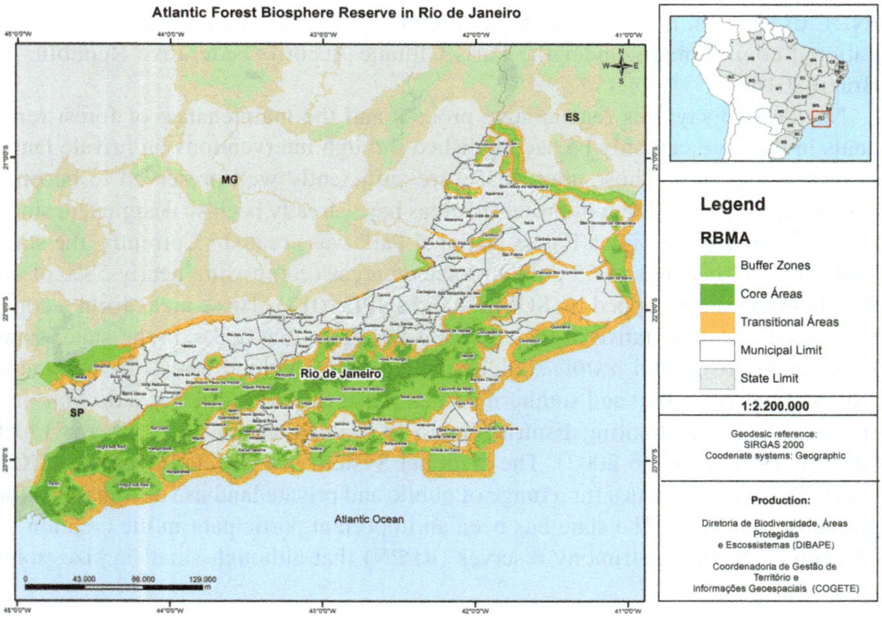

Fig. 27.2 Atlantic Forest Biosphere Reserve in Rio de Janeiro (INEA 2015)

27.1.2 Evolution of Agroecological Institutions and Policies

Environmental policy in RJ has been historically administered through command and control instruments such as licensing, pollution standards, fines, and plant closures. This approach has evolved during the past decade to incorporate a number of economic instruments that have begun to be implemented primarily on a pilot basis (payment for environmental services or PES) but also at a statewide level ("Ecological" ICMS), described in greater detail below. Initially, environmental policy was managed by the state's Foundation for Environmental Engineering (FEEMA), concerned solely with the so-called "brown" agenda, for which it became a reference in environmental pollution control. However, in later years, the organ lost some of its prestige since many of its highest qualified personnel had retired, while its activities became dedicated primarily to command and control, including public sanitation. Depollution of Guanabara Bay in metropolitan Rio de Janeiro became one of the state's most visibly incomplete agendas after billions in treatment expenditures and a cleanup effort showcased in preparations for the 2016 Olympics turned into an utter failure (Romero and Clarey 2014).

Control over industrial residuals – also a responsibility of FEEMA – was also sources of corruption and negligence, with the major revenue generator of the state, Petrobras and its affiliates, getting away with nearly all oil spills, persistent organic pollutants, and CO_2 emissions from deep water drilling and methane flaring for decades. Water resources were managed in a haphazard way for water supply and sanitation services (principally provided by the parastatal company CEDAE) and for hydropower and electrical distribution (respectively, furnished by the parastatal company Furnas and by Light, now controlled by Électricité de France-EDF), as well as the flood prevention and stream channeling actions under the responsibility of the state superintendency of rivers and lakes (SERLA). A weak and understaffed State Institute of Forests (IEF) was responsible for policing forest degradation and managing the state's protected areas (Silveira-Filho 2012).

Several factors, not least of which a series of environmental disasters moved these institutions closer together, leading the State Secretarial of the Environment (SEA) to create its subsidiary Institute of Environment (INEA) in 2009. The latter "superagency" in the mold of IBAMA at the Brazilian federal level integrated the functions of the several state environmental agencies previously responsible for environment and natural resource management: FEEMA, SERLA, and IEF. Fused by law does not mean they are always coordinated in practice, and many of the previously mandated functions of the three executive agencies continue to operate in separation.

The "green" agenda, including agro-environmental and forest governance in RJ has evolved from a fragmented and uncoordinated stage in the 1980s in which only IEF had a coherent role to play in forest conservation and protected area management, to the current institutional arrangements, in which land use policy is somewhat more coherently addressed and linked with water resource management.

Agricultural development, on the other hand, is the domain of the state's Secretary of Agriculture, Livestock, Fisheries, and Supply (SEAPPA), whose actions are generally not well coordinated with those of SEA.

In 1997, legislation at the federal level (Law 9.433) established a national system for water resource management derived from experience in France, with a management structure at the river basin level. This legislation began a revolution in natural resources management in Brazil, which is nevertheless still far from complete. River basin commissions and agencies established under the law were charged with planning and instituting a system of water charges (*cobrança*), grounded in a recasting of water as an economic resource rather than being a common pool. The legislation was timely in that many of the nation's water resources were already facing scarcities by the 1990s, resulting in the need to ration electricity, which in Brazil is primarily generated by hydroelectric facilities. The water management law sought to avoid growing conflicts over competing uses of water resources for power, pollution dilution and transport, drinking water, and irrigation, as well as industrial requirements.

In fact, however, only a few interstate rivers whose management was charged to the national water agency, ANA, have been able to establish the institutional and technical capacities necessary for planning permissible use, allocating resource flows and withdrawals, and instituting registry and charge schemes that would reflect the ensuing scarcity. On one such river, the Paraíba do Sul, which flows through the states of SP, Minas Gerais (MG), and RJ, the charge system was instituted in the year 2003 and generated R$ 142.7 million in revenues up to 2016 (ANA 2017). Water revenues are allocated to water resource management, restoration of water quality, protection of catchments, and administrative costs, in efforts to relieve the intense pressures faced by the basin for dwindling clean water supplies, crucial to major urban centers such as SP and RJ itself. In RJ, a statewide charge system was created in 2004 (Law 4.274/2004). In the same year, the state created a Water Resources Fund (FUNDRIH) that has already collected more than R$ 236.3 million through water user charges.

Steps toward an integrated landscape approach to management of mosaics of private and public lands and water resources in the Atlantic Forest was severely undermined in 2012 through legislation (Law 12.727) named the Native Vegetation Protection Law, to downgrade the degree of protection of natural vegetation on private lands instituted by the original national Forest Code. Initially passed in 1964, the Forest Code established that a proportionate area of natural vegetation on each private property, differentiated by biome, be protected in a legal reserve (LR) and that lands in several classes important to watershed management (streambanks, steep slopes, and hilltops) be preserved as permanent preservation areas (APPs). Areas that require protection and/or restoration to meet the Code's requirements as LR in the Atlantic Forest amount to 20% of each property by area. APPs were additive to LRs. While the original law provided important protections (to native vegetation, biodiversity, and ecosystem services), it was only haphazardly enforced (Brancalion et al. 2016).

After an acrimonious battle that pitched environmentalists against ruralists in the national congress, a solution was negotiated that – while it removed many areas previously protected – has provided tools to strengthen monitoring of forest remnants on private lands. The law now requires that each landowner registers their property and identify its respective LRs and APPs (totaling 20% combined) on a satellite image. Each landowner then commits to a restoration plan that will meet this objective. The law also provides for some flexibility in protection, enabling property owners to mix native and exotic species in restoring APPs and LRs and to trade in environmental reserve certificates (CRA) as a tradeable development rights approach to protection, similar in some respects to the Transfer of Development Rights schemes in wide use in the USA (May et al. 2015). PES, carbon credits, and rural credit differentials were also identified in the law as potentially compatible means to compensate landowners for their compliance.

27.2 Conservation Incentives and Compensation Instruments

Although environment-development conflicts are common in Brazil, as described above, federal legislation while suffering recent drawbacks is still considered to provide for considerable flexibility in achieving society's objectives. The current section describes several of the most innovative experiences in applying national legal frameworks to state and local instruments that conciliate environmental and productive processes.

27.2.1 Ecological ICMS in Rio de Janeiro

Besides the Native Vegetation Protection Law, tools that have begun to be implemented throughout the country. Ecological Fiscal Transfers (EFT), adopted through allocation criteria for intergovernmental fiscal transfers, redistribute tax revenue from higher to lower levels of government to compensate revenue lost in areas subject to protection (Ring and Schröter-Schlaack 2011). Under the 1988 Brazilian Constitution, each state is permitted to allocate 25% of the value-added tax (ICMS) revenue according to its own internal priorities. EFTs in Brazil, where this innovative fiscal instrument was born, redistribute a share of these public revenues according to nature conservation or other environmental indicators established through state "Ecological ICMS" regulations. This policy tool is intended to increase the economic attractiveness of creating new protected areas as well as to improve the management of existing reserves (Veiga 2008).

RJ was a latecomer to this instrument, initially established in the state of Paraná in the early 1990s, and largely credited with the creation of many new state and local protected areas, particularly environmental protection areas (APAs) and RPPNs. This then triggered larger proportional value-added tax revenue outlays from the state to local levels in those municipalities which harbored them (May et al. 2002). In RJ, perceiving the importance of this instrument as an incentive to local conservation, municipalities that created their own protected areas were double-weighted in the allocation formula created by the 2007 "Green ICMS" enabling statute (Law No. 5010). Municipal governments were also rewarded by their investments in solid waste management and watershed protection. The Final Index of Environmental Conservation (IFCA), which indicates the share of Green ICMS that is due to each municipality, is composed of five thematic sub-indices with different weights: garbage disposal (20%), sewage disposal (20%), watershed protection (10%), remediation of landfills (5%), and protected areas (45%).

The environmental component was incorporated gradually into the distribution formula for the Green ICMS, having increased from 1% in 2009 to 2.5% by 2011. In 2017, the total predicted value of the Green ICMS had attained R$220.5 million, with the municipalities of Cachoeiras de Macacu (R$10.1 million) and Silva Jardim (R$8.6 million) receiving the largest shares.[2] These municipalities, whose territory is largely dedicated to protected areas, received considerable additional revenues, contributing to their ability to deliver public services and strengthening local resolve to protect remaining forests.

27.2.2 The Native Vegetation Protection Law in Rio de Janeiro

Under the 2012 revisions to the Forest Code, although the law permitted a reduction in area protected and allowed existing "consolidated" uses, landowners were obligated to register their properties and identify LRs and APPs on satellite images, so they could be monitored over time for compliance with the Code. In RJ, extension agents were mobilized with funding in part obtained from water use charges to assist small farmers to register their properties in the rural environmental cadaster (CAR) that will serve as the basis for environmental management. Without their assistance, many properties would otherwise be outside the system. It is expected that the CAR will serve as the underpinning for compensatory instruments to landowners who protect remaining Atlantic Forest fragments on their properties or in other parts of the state (Brancalion et al. 2016).

One of the instruments that was specifically leveraged by the new Forest Code – in fact already contemplated in its prior version in 1997 but rarely employed in practice – is that of LR trading. Similar in concept to the "cap and trade" schemes associated with climate mitigation, the so-called *Cota de Reserva Ambiental* (environmental reserve credit or CRA) allows landowners whose properties have been cleared beyond the 20% minimum area required by the law to remain in vegetative cover in private properties in the Atlantic Forest, to satisfy the law's requirements by purchasing CRA created by reserves on properties that exceed those limits. This instrument thus provides an incentive to comply with the Forest Code (Soares-Filho

[2] Estimate of Green ICMS distribution for 2017 provided by the General Superintendency of INEA/SEA. Communication by email, 08 August 2017.

et al. 2014), since in economic terms, it is advantageous for reserve-deficient landowners to pay surplus reserve owners up to the value of their opportunity costs had they been required to remove LRs from production. In order for CRA to enter into a state trading scheme, however, the law requires the preparation of restoration plans at the state and property level, as well as legal procedures which will add transaction costs to trading and may end up making it impractical. To ignite the market for CRA, several organizations have begun to provide trading services to landowners to facilitate transactions, helping to locate partners and facilitating arbitrage (May et al. 2015).

27.2.3 PES in Rio de Janeiro

Landowners have long complained that the Forest Code's demands placed them at a disadvantage with countries that did not require forests to be set aside from production for conservation purposes. They also argued, correctly, that a uniform proportional set-aside on a property by property basis would not necessarily assure conservation of critical habitats or protect watersheds fundamental to regional hydrologic equilibrium. Furthermore, there had been no mechanism built into the Forest Code's command and control strategy that provided incentives for landowners to contribute positively toward conservation goals. Instead fines were meted out, often to the small and powerless rather than to those who could manipulate the legal system to their advantage. RJ, with a strong historical environmental degradation footprint and high development pressures, is estimated to have a vegetation restoration potential of 939,800 hectares to comply with the Forest Code as previously formulated (Pacto 2009). Recent mapping in all municipalities of the entire Atlantic Forest biome at 1:20,000 found that the deficit in hydric APPs is 526,318 ha (FBDS 2017), showing both that the previous estimates may have been excessive (though they combined both RLs and APPs) but also that the current legislation removed many such forests from protection (Soares-Filho et al. 2014).

To correct the rigidities of the prior legislation, the revisions in the Forest Code fostered the adoption of economic instruments that could motivate landowners to adopt environmentally appropriate practices, such as the CRA. At the same time, numerous initiatives to promote PES schemes began in watersheds stemming from local government and producer association partnerships with nongovernmental organizations, river basin commissions, and both state and federal government agencies. The focus of these schemes varies according to their purpose; some aim at supporting and improving ecosystem management (through carbon sequestration and storage, watershed protection, landscape aesthetics, and biodiversity protection), while others also endeavor to achieve broader goals (e.g., poverty reduction, regional development).

As of 2012, 41 Payments for Watershed Services projects had either been implemented or were in preparation across an area of approximately 40,000 ha in the Atlantic Forest biome (Guedes and Seehusen 2011). Several such projects developed in RJ are described here.

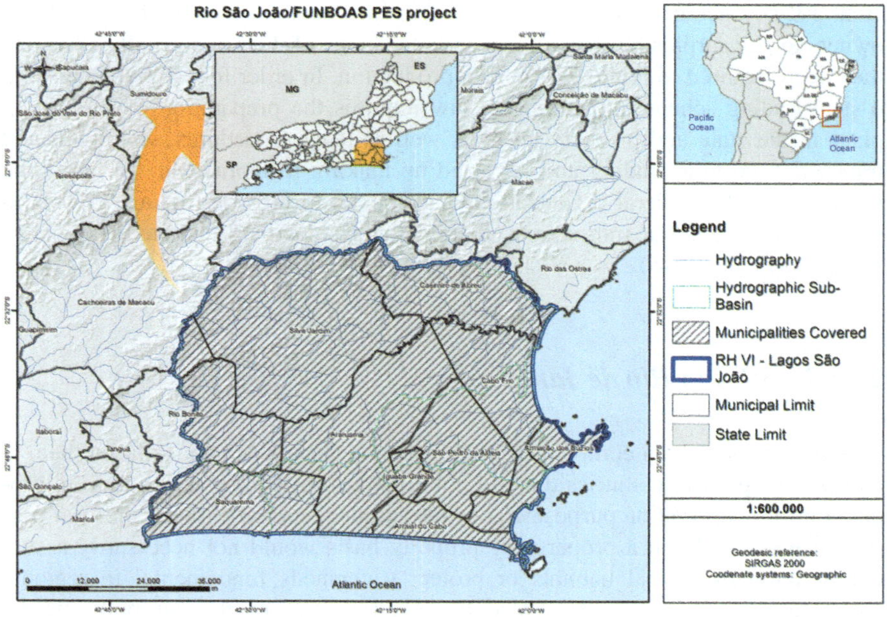

Fig. 27.3 Location of the Rio São João PES project. Authors' elaboration based on conception in Fulgêncio (2012)

In 2007, a pilot initiative was started in the Guandu watershed, which supplies 80% of potable water and 25% of electricity to metropolitan Rio de Janeiro and benefits at least seven million people (ANA 2007). This initiative was launched in the Rio das Pedras sub-basin, in Rio Claro municipality (Fig. 27.3), whose total area of 5227 ha includes the principal headwaters of the Piraí River (Pereira et al. 2012; Castello Branco 2015). These streams are estimated to supply 15% of water received by the Guandu River. Lands in the area are also high in biodiversity – acting as the nucleus of the Biosphere Reserve surrounding the Cunhambebe State Park and lying within the territory of the Alto Piraí Environmental Protection Area (Ibid.).

This scheme was made possible by the efforts of the municipal government of Rio Claro to contract with local landowners (environmental service providers). They in turn receive support and technical assistance by the Instituto Terra de Preservação Ambiental (ITPA), a regional NGO, and The Nature Conservancy, with financing derived primarily from the Paraíba do Sul River basin authority, which allocated a portion of its water charge revenues to this purpose. Water user fees are aligned with RJ state government policy (Resolution No.42/2009). The high costs associated with restoration (average cost of R$ 33,000 per hectare for 1-year planting and 3 years of maintenance) were covered using compensation measures from private enterprises. Landowners within the municipality were provided with the option to develop land use plans to restore degraded streambank vegetation and wetlands, whose costs were footed by the program. They also received a payment, varying with conservation priority, from R$ 20 to 60/ha/year to participate in the program. Given that the principal rural economic activity in the region – dairy cattle – only had net opportunity costs of about R$ 22/ha/year, these payments were superior to net earnings from the alternative uses of land. In addition to a direct payment, landowners receive benefits to help restoration efforts (seedlings, fertilizer, wire, and clamps). In the same municipality, landowners willing to dedicate part of their land to permanent protection as RPPNs were rewarded with payments redirected from additional Green ICMS revenues coming to the municipality due to the increment in protected areas (Castello Branco 2015).

By 2013, a total of 4165 ha had been contracted for conservation practices, and an additional 495 ha subjected to restoration plantings and protection, both figures having exceeded their respective targets, while a tremendous institutional learning achievement is also notable.

The São João River basin in the Lake District to the north of the city of Rio de Janeiro was the focus of the second such local venture. This experience, which began in 2008 as an undertaking of a river basin consortium among local governments, served as a model for other initiatives in the state and throughout Brazil (Fig. 27.4). Under the guidance of the São João River basin and lakes committee, the partners created a water resource fund, called FUNBOAS (Fundo de Boas Práticas Socioambientais) that was instituted to provide compensatory payments to landowners who adopted best agricultural and forest conservation practices con-

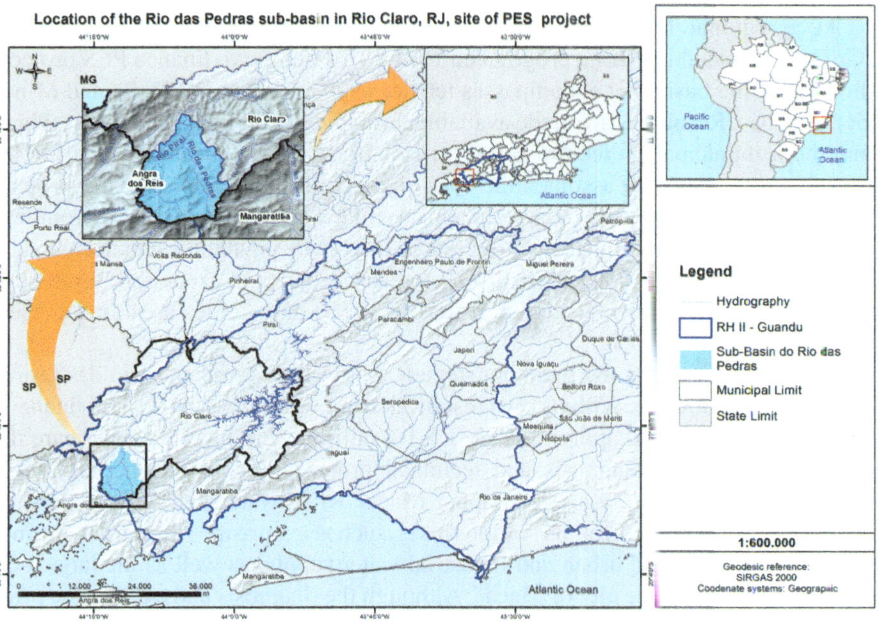

Fig. 27.4 Location of the Rio das Pedras sub-basin in Rio Claro, RJ, site of PES project. Authors' elaboration based on conception in Castello Branco (2015)

ceived to strengthen water provision to the Jaturnaiba reservoir, responsible for 75% of all water supplies in a basin heavily impacted by domestic and international tourism (CBLSJ 2009). A pilot project was begun in 2009 among occupants of lots in an agrarian reform settlement, Fazenda Cambucaes, which had settled 125 families. In this settlement, a total of R$ 30,000 was devoted to environmental sanitation and R$ 5000 to each family that showed progress in restoring degraded lands with tree plantations and agroforestry systems (Fulgêncio 2012). Further partnerships were initiated with producers in two other small watersheds in the river basin, Roncador and Lontra, but due to cessation of funding by the state to the river basin consortium, the program was suspended (Gabriel Kruschewsky, General Coordinator, Consórcio Intermunicipal Lagos São João – CILSJ – personal communication).

Based in part on this experience, in 2011 the state provided regulation for PROHIDRO as an instrument complementary with its water resource management policy. Through this program – placed under the responsibility of the State Water Resources Council – with the objective to carry out efforts to revitalize and conserve water resources through basin level management, a statewide PSA scheme was instituted as a public policy measure. A subprogram entitled PRO-PSA (State Program for PES) was established to enable retribution for environmental services generated in rural areas and public water catchments. The creation of PRO-PSA has permitted new initiatives to flourish. Although the statewide scheme is not yet formally implemented, several PES initiatives have been supported in the state, in seven of the state's nine hydrographic regions, under the auspices of local river basin commissions (Table 27.1).

At a regional level, the Paraíba do Sul River basin integration committee (CEIVAP) created in 2014 a program entitled PSA-Hídrico, to finance PES projects throughout the basin that encompasses territories of São Paulo (SP), RJ and Minas Gerais (MG). Resources are made available through bidding by local entities, which may be a municipal prefecture acting in isolation or associated with a public or private entity that then assumes the role of executing institution for the project. Municipalities, which need to have an approved PES law, can contract environmental service providers by means of notices and to execute payments, which can reach R$ 200/ ha/year. Altogether there are 20 forestry projects in execution, forecast to be of 3 years duration with a goal of restoring 420 ha of degraded areas and conservation of 350 ha of forest.

A future initiative in the planning stage is that of the Macaé and Rio das Ostras River basin committee. With its own resources from water use charges and financial support from the ANA, approximately R$ 1.4 million were invested to prepare the socio-environmental diagnosis and technical proposal for actions necessary for environmental conservation of the upper Macaé River watershed. This watershed has special importance for its various uses such as tourism, recreation, rafting, energy generation, and urban and industrial water supply, as well as the largest oil industry in Brazil in the city of Macaé. Although the diagnosis and a reference price for payments (R$ 240 ha/year) have been prepared, the PES mechanism has not yet been implemented.

Table 27.1 Main characteristics of PES initiatives in Rio de Janeiro

PES initiatives	Environmental services targeted	Actors involved	Implementation costs (R$)	Payments costs (R$)	Results to date
PES Guandu	Water	NGOs, CBH, government	1,265,074	379,423	134 ha of conservation and 520.08 ha of restoration implemented
PES Hídrico (CEIVAP)	Water	NGOs, CBH, government and municipality	5,303,776	113,646	0 from a 151 ha conservation target and 43 ha from a 134 ha restoration target
PES Macaé	Water	CBH, ANA, government	1,400,000 (diagnosis)	–	–
PES GEF Mata Atlântica	Biodiversity, climate, water	IDB, GEF, MCTI, FINATEC, government	35,317,900[a]	6,134,553[a,b]	No results yet but a target of 750 ha restoration, 1500 ha conservation and 1500 ha of production systems conversion

Source: Authors' elaboration

All initiatives are ongoing except the PES in Macaé, which is in the implementation phase. All promoted actions by all schemes are focused on forest conservation and restoration efforts

[a]Estimated values based on general project budget and considering an exchange rate of 3.04

[b]Value is based on payment estimation considering that this project is still in a formulation phase

As can be seen, PES projects underway to date have all been related to water resources. However, a new more encompassing project is underway, involving biodiversity and climate services. This is the project for the Recovery of Climate and Biodiversity Services for the Paraíba do Sul River basin in the Brazilian Atlantic Forest (PES GEF Mata Atlântica), involving the Federal Government (Ministry of Science, Technology and Innovation – MCTI); governments of the states of SP, RJ, and MG; the Global Environmental Facility (GEF); the Inter-American Development Bank (IDB); and the Foundation for Scientific and Technological Enterprises (FINATEC). Through this project, in RJ alone, more than US$ 14.4 million will be invested in the establishment of a PES scheme that will aim at the ecological restoration of 750 ha, conservation of 1500 ha of forest, and productive conversion of 1500 ha, through implementation of agroforestry and silvopastoral systems, forest consortia, and sustainable soil and water management practices. The actions will focus on the mid and low Paraíba do Sul regions. Additional projects in the planning stages include areas within the Guapiaçu-Macacu watershed (Strobel et al. 2007; Guedes and Seehusen 2011; Rodríguez Osuna et al. 2014), which supplies potable water to almost 2.5 million inhabitants of five municipalities, including the city of Niteroi (Pedreira et al. 2009).

Although compensation for environmental service provision has attracted considerable interest on the part of local and state governments, it has been difficult for them to sustain ongoing payments for such services, contributing to landowners becoming more reticent to engage in these schemes when they imply permanent reduction in area available for productive activities. Other aspects of PES projects that have led to decisions by farmers to abstain from participation include incomplete information on the benefits of such participation, as well as lack of concern for environmental issues. Project scale and the number of landowners enlisted at a watershed level often are insufficient to assure that environmental conditions are improved (Zanella et al. 2014).

27.3 Summary and Conclusions

Governmental institutions from agricultural and environmental sectors have traditionally not worked in an integrative and coordinated way toward the mission of aligning agricultural production and environmental conservation in RJ. This is likely to change where multiple stakeholders of different sectors participate in project design and implementation as is foreseen in the ongoing PES GEF project, which also goes beyond a focus solely on watershed services with the inclusion of biodiversity and climate-related environmental services.

Environmental policy in RJ has evolved from focusing merely on command and control instruments to more flexible conservation incentives including PES, "Ecological" ICMS, and potential trading in CRA, which suggest several opportunities to compensate landowners and municipal governments for their dedication of land to conservation, so contributing to sustainable development in this region.

For example, pilot PES initiatives have shown that maintaining such "green infrastructure" is strategic for fostering human water security and protecting biodiversity in RJ (da Silva and Wheeler 2017). Measures promoted by the discussed policies foster maintenance or increment in forest cover, which help regulate water provision essential for the state's millions of urban inhabitants. At the same time, such measures also reward rural landowners for applying sustainable land use practices mainly toward watershed protection.

Conservation incentives in place in RJ focus on the protection or restoration of native vegetation, especially high-priority areas in terms of water production (headwaters and riparian forests) and biodiversity richness. Ecosystem conservation and restoration measures in water critical areas can also reduce costs associated with treating potable water (i.e., forest cover reduces erosion transport and maintains lower turbidity levels that in turn reduce chemical treatment requirements) (Medeiros et al. 2011; Mansur et al. 2013). Services promoted by such incentives are especially important in times of need to secure a reliable and sustainable water provision, such as during water scarcity crisis and droughts as the one experienced in

2014 in RJ and SP, which created a dispute between these two states for the access to water from the Paraíba do Sul River basin.

Still, the deficit of forest in RJ is high, and restoration measures are expensive though considered cost-effective (Calmon et al. 2011); therefore, it is important to highlight creative approaches that bring funds from other resources with a mix of policy interventions. One example is found in the PES scheme in Rio Claro that, besides payments from water users, also relies on environmental compensation from private enterprises to pay for the project's restoration costs. This strategy is also being unlocked by the PES GEF project, considering that the state is now able to offer a "restoration grant" including contributions from environmental liabilities funds set up by private companies, which have the potential to be used to bolster project financial sustainability in time. This is an innovative way of overcoming ongoing concerns regarding the fiscal feasibility of sustaining payments for environmental services over time. In the near future, conservation initiatives could seek more involvement from the private sector to provide financial support for landscape conservation in RJ (e.g., through corporate sustainability efforts).

Several challenges associated with implementation of conservation incentive projects have been overcome with strategic partnerships that brought differentiated expertise at strategic moments, facilitated bureaucratic procedures, and brought critical financial resources to enable project implementation when facing bureaucratic hurdles or budgetary shortfalls. There are other favorable enabling conditions that contributed to the success of conservation agro-environmental policies in RJ. One is related to forums and network movements (Forum to create PSA-RJ and the Network for Water and Forests) that supported and promoted the elaboration of statewide legislation. However, evaluation of the impact of these policies is often missing or not well-grounded scientifically, or their period of implementation foreshortened by budgetary gaps, affecting our ability to show that investments in conservation incentives are yielding significant and more cost-effective ecological and social benefits.

References

ANA (2007) Plano estratégico de recursos hídricos das bacias hidrográficas dos Rios Guandu, da Guarda e Guandu Mirim. Relatório Gerencial. Agência Nacional de Águas. Superintendência de Planejamento de Recursos Hídricos, Brasilia-DF, p 63

ANA (2017) Arrecadação por usuários – Bacia do Paraíba do Sul. Agência Nacional de Águas.. http://www2.ana.gov.br/Paginas/servicos/cobrancaearrecadacao/BaciaPBS_ArrecadacaoeRepasse.aspx. Accessed 01 Jan 2017

Brancalion PHS, Garcia LC, Loyola R et al (2016) A critical analysis of the native vegetation protection law of Brazil (2012): updates and ongoing initiatives. Natureza & Conservação 14(1):1–15

Calmon M, Brancalion PHS, Paese A et al (2011) Emerging threats and opportunities for large-scale ecological restoration in the Atlantic Forest of Brazil. Restor Ecol 19(2):154–158

Castello Branco MR (2015) Pagamento por serviços ambientais: da teoria à prática. A Experiência do Projeto Produtores de Água e Floresta na região hidrográfica do Guandu. Instituto Terra de Preservação Ambiental (ITPA), Rio Claro-Rio de Janeiro, p 188

CBLSJ (2009) Resolução No. 023/2009. Fundo de Boas Práticas Socioambientais em Microbacias (FUNBOAS). Araruama – Rio de Janeiro: Comitê de Bacia Lagos São João. Câmara Técnica Permanente de Microbacias, Programa de Gestão Ambiental Participativa em Microbacias

CNRPPN – Consórcio de Confederação Nacional de Reservas Privadas de Patrimônio Natural (2017) Painel de Indicadores do CNRPPN. Versão Beta. https://datastudio.google.com/u/0/#/org//reporting/0B_Gpf05aV2RrNHRvR3kwX2ppSUE/page/J7k. Accessed 03 Oct 2017

Coalízão Clima, Floresta e Agricultura (2017). http://coalizaobr.com.br/2016/. Accessed 23 May 2017

da Silva JMC, Wheeler E (2017) Ecosystems as infrastructure. Perspect Ecol Conserv 15(1):32–35. https://doi.org/10.1016/j.pecon.2016.11.005

Dean W (1995) With broadax and firebrand: the destruction of the Brazilian Atlantic Forest. University of California Press, Berkeley

Drummond J (1996) The garden in the machine: an environmental history of Brazil's Tijuca forest. Environ Hist 1(1):1–83

FBDS – Fundação Brasileira para o Desenvolvimento Sustentável (2017) Mapeamento das áreas de preservação permanente dos biomas Cerrado e Mata Atlântica na escala 1:20.000. Projeto de Apoio à Implantação do CAR. Rio de Janeiro, Brasil. Available at www.fbds.org.br

Federative Republic of Brazil (2016) Intended nationally determined contribution towards achieving the objective of the United Nations framework convention on climate change. UN CSD

Fulgêncio LG (2012) Programas de pagamento por serviços ambientais na gestão dos recursos hídricos: a experiência do FUNBOAS na bacia Lagos São João – RJ. Dissertation, Universidade do Estado do Rio de Janeiro

Guedes FB, Seehusen SE (eds) (2011) Pagamentos por serviços ambientais na Mata Atlântica: lições aprendidas e desafios, Serie Biodiversidade 42, vol 2. Ministério do Meio Ambiente (MMA), Brasilia-DF

INEA (2015) Atlas das unidades de conservação da natureza do estado do Rio de Janeiro: Atlas of protected areas in Rio de Janeiro state. Metalivros. Instituto Estadual do Ambiente, São Paulo, p 171

Mansur H, Gumarães J, Klemz C (2013) Estudo de viabilidade de implantação do programa PRO-PSA na região hidrográfica do Guandu- Rio de Janeiro. Rio de Janeiro: The Nature Conservancy – TNC, Instituto Terra de Preservação Ambiental-ITPA, Conservação Internacional – CI, p 66

May PH, Veiga Neto F, Denardin V, Loureiro W (2002) Using fiscal instruments to encourage conservation: municipal responses to the 'ecological'value-added tax in Paraná and Minas Gerais, Brazil. In: Pagiola S, Bishop J, Landel-Mills N (eds) Selling forest environmental services: market-based mechanisms for conservation and development. Earthscan, London, pp 173–200

May PH, Bernasconi P, Wunder S, Lubowski R (2015) Environmental reserve quotas in Brazil's new forest legislation: an ex ante appraisal, Occasional paper 131. CIFOR, Bogor

Medeiros R, Young CEF, Pavese HB, Araújo FFS (2011) Contribuição das unidades de conservação brasileiras para a economia nacional. Brasilia-DF, UNEP-WCMC, p 44

Pacto (2009) Mapa de Áreas Potenciais para a Restauração Florestal. Pacto pela restauração da Mata Atlântica, São Paulo.. Available at: http://www.pactomataatlantica.org.br.. Accessed 23 May 2017

Pádua JA (2002) In: Zahar J (ed) Um sopro de destruição: pensamento político e crítica ambiental no Brasil escravista, 1786-1888, 2nd edn. Jorge Zahar Editor Ltda, Rio de Janeiro

Pedreira B, Fidalgo ECC, Abreu MB (2009) Mapeamento do uso e cobertura da terra da bacia hidrografica do rio Guapi-Macacu, RJ. In: Anais XIV Simposio Brasileiro de Sensoriamento Remoto. INPE, Natal, pp 2111–2118

Pereira GS, Vilar MB, Bustamante J, Ruiz M. (2012) Produtores de água e floresta – Guandu. In: Pagiola S, Von Glehn HC and Tafarello D (eds) Experiências de pagamentos por serviços ambientais no Brasil. Secretaria do Meio Ambiente/Coordenadoria de Biodiversidade e Recursos Naturais, São Paulo, pp 67–83

Pinto S, Melo F, Tabarelli M et al (2014) Governing and delivering a biome-wide restoration initiative: the case of Atlantic Forest restoration pact in Brazil. Forests. Multidisciplinary Digital Publishing Institute 5(9):2212–2229

Rezende CL, Uezu A, Scarano FR, Araujo DSD (2015) Atlantic forest spontaneous regeneration at landscape scale. Biodivers Conserv 24(9):2255–2272. https://doi.org/10.1007/s10531-015-0980-y

Ring I, Schröter-Schlaack C (eds) (2011) Instrument mixes for biodiversity policies. POLICYMIX report. Helmholtz Centre for Environmental Research – UFZ, Leipzig, p 209. Available at: http://policymix.nina.no

Rizzini CT (1997) Tratado de Fitogeografia do Brasil: aspectos ecológicos, sociológicos e florísticos. Âmbito Cultural Edições Ltda, Rio de Janeiro

Rodrigues RR, Lima RAF, Gandolfi S, Nave AG (2009) On the restoration of high diversity forests: 30 years of experience in the Brazilian Atlantic Forest. Biol Conserv 142(6):1242–1251

Rodríguez Osuna V, Börner J, Nehren U et al (2014) Priority areas for watershed service conservation in the Guapi-Macacu region of Rio de Janeiro, Atlantic Forest, Brazil. Ecol Process 3(16):1–21

Romero S, Clarey C (2014) Note to Olympic sailors: don't fall in Rio's water. New York Times May 18

Scarano FR (2002) Structure, function and floristic relationships of plant communities in stressful habitats marginal to the Brazilian Atlantic rainforest. Ann Bot 90(4):517–524

Silveira-Filho TB (2012) A política florestal estadual do Rio de Janeiro: ação e inação do estado entre 1975-2011. Master's Thesis, Federal Rural University of Rio de Janeiro, Graduate Program in Environmental and Forest Sciences

Soares-Filho B, Rajão R, Macedo M et al (2014) Cracking Brazil's Forest code. Science 344(6182):363–364

SOSMA, INPE (2015) Atlas dos remanescentes florestais da Mata Atlântica: período 2014–2015. Fundação SOS Mata Atlântica. Instituto Nacional de Pesquisas Espaciais, São Paulo, p 60

Strobel JS, De Souza C, Da Motta RS et al (2007) Critérios Econômicos para a Aplicação do Princípio do Protetor-Recebedor: Estudo de Caso do Parque Estadual dos Três Picos, Serie Técnica 11. Conservation Strategy Fund, Rio de Janeiro, p 66

Veiga FC (2008) A Construção dos Mercados de Serviços Ambientais e suas Implicações para o Desenvolvimento Sustentável no Brasil. Universidade Federal Rural do Rio de Janeiro

Zanella MA, Schleyer C, Speelman S (2014) Why do farmers join payments for ecosystem services (PES) schemes? An assessment of PES water scheme participation in Brazil. Ecol Econ 105:166–176

Chapter 28
Promoting Sustainable Agriculture, Boosting Productivity, and Enhancing Climate Mitigation and Adaptation Through the RIO RURAL Program, Brazil

Vanesa Rodríguez Osuna, Peter H. May, Joyce M. G. Monteiro, Roland Wollenweber, Helga Hissa, and Marcelo Costa

Abstract The Brazilian agricultural sector is a major source of greenhouse gas emissions, but climate-smart practices combined with degraded land restoration can result in a more resilient landscape contributing to integrated climate change mitigation and adaptation efforts. The Sustainable Rural Development Program of Rio de Janeiro (RIO RURAL) has been supporting the transition of degraded rural areas to sustainable productive systems by providing technical assistance and incentives to small-scale family farmers. RIO RURAL promotes reforestation and sustainable agriculture practices, which can boost productivity as well as carbon stocks in the agricultural landscape. Using estimates of carbon mitigation potential for such practices, we identified methodologies eligible for certification in the voluntary markets. We estimated transaction, implementation, and certification

V. Rodríguez Osuna (✉)
CUNY Advanced Science Research Center, New York, NY, USA
e-mail: vanesa.rodriguez.osuna@asrc.cuny.edu

P. H. May
CPDA/UFRRJ, Seropédica, Brazil

Earth Institute, Columbia University, New York, NY, USA

J. M. G. Monteiro
Brazilian Enterprise for Agricultural Research -Embrapa Solos, Rio de Janeiro, Brazil
e-mail: joyce.monteiro@embrapa.br

R. Wollenweber
Business Development and Innovation, TÜV Rheinland, Cologne, Germany
e-mail: roland.wollenweber@de.tuv.com

H. Hissa · M. Costa
The Sustainable Rural Development Program in Micro-Watersheds of the State of Rio de Janeiro – Rio Rural, Horto, Fonseca, Niterói/Rio de Janeiro, Brazil

U. Nehren et al. (eds.), *Strategies and Tools for a Sustainable Rural Rio de Janeiro*, Springer Series on Environmental Management,
https://doi.org/10.1007/978-3-319-89644-1_28

costs and calculated potential revenues associated with RIO RURAL's activities. This did not only allow us to discuss the constraints and identify opportunities and co-benefits from RIO RURAL's contribution to climate mitigation, adaptation, and environmental integrity but also to food security as it targets family farms. We propose a bundling approach to carbon, where multiple benefits are measured and certified including water, food systems, as well as social and cultural benefits. This would allow accessing resources from both mitigation and adaptation programs in addition to markets that value ecosystem integrity as well as water and food security.

Keywords Family farming · Cost-effectiveness · Sustainable agriculture · Low-emission agriculture · Rio de Janeiro · Co-benefits

28.1 Introduction

Land use change and agriculture are the major sources of greenhouse gas (GHG) emissions in Brazil (Escobar 2015). In the state of Rio de Janeiro (RJ), agriculture plays a key role in supplying products to nearly 16 million inhabitants (IBGE 2010), mostly concentrated in the metropolitan region. Furthermore, 80% of the food production in RJ originates from small-scale agriculture (RIO RURAL 2014, Fig. 28.1), which accounts for ~90% of all rural properties (INEA 2013).

RIO RURAL[1] supports the transition of degraded rural areas to sustainable productive systems by providing technical assistance and economic incentives to small-scale family farmers. Practices promoted by RIO RURAL include the protection of springs and streams, establishment of agroforestry systems, and sustainable agriculture practices (see Hissa et al. 2018 in this volume).

Low emissions agriculture combined with the restoration of degraded land can result in a more resilient landscape that contributes to climate change mitigation and adaptation efforts. The objective of this chapter is to assess RIO RURAL's potential to generate climatic solutions as well as other co-benefits. We first present the main linkages between agriculture, climate change, and the concept of low-emission agricultural systems (Sect. 28.2). Section 28.3 gives an overview of RIO RURAL's program and their promoted activities; Sect. 28.4 summarizes legal frameworks and mitigation policies relevant to the agriculture, land use change, and forestry sectors. Section 28.5 introduces emission trading in the compliance and voluntary markets, and Sect. 28.6 presents the processes required to assess the mitigation potential of agricultural projects eligible for certification. Based on the identification of RIO RURAL's activities with mitigation potential, we have assessed the feasibility of carbon finance to provide incentives for the program's promoted practices (Sect. 28.7). We discuss constraints and identify opportunities associated with RIO RURAL's contribution to climate mitigation, adaptation, and environ-

[1] The Sustainable Rural Development Program in Micro-Watersheds of the State of Rio de Janeiro.

Fig. 28.1 Small-scale agricultural systems in Rio de Janeiro

mental integrity (e.g., recovery of degraded pastures by enriching soil carbon stocks) as well as to food security as it targets family-oriented rural properties in agricultural landscapes across the state of RJ (Sects. 28.7 and 28.8).

We found that carbon finance can be an important complementary source of funds but that it cannot support changes in agricultural practice alone. To this end, a change in the consumption patterns to value local food production, equitable distribution of returns related to such production, and ecocertification with differential pricing for reduced external costs to the society is required (Komives and Jackson 2014). Carbon financing may be significant for large-scale integrated crop/livestock production systems. Small-scale diversified systems often do not generate sufficient additional carbon mitigation to justify the effort. However, the assessment and inclusion of other co-benefits (e.g., the promotion of water conservation and quality, productivity improvements, moderation of extreme events, and biodiversity conservation) can make RIO RURAL's contribution more feasible, financially. Additional benefits of the program include enhancing ecosystem integrity and fostering water and food security.

28.2 Agriculture, Climate Change, and Low-Emission Agricultural Systems

Three major links exist between climate change (CC) and agriculture. First, the agricultural sector is highly affected by the impacts of CC, especially in relation to increased mean temperature, higher variability in temperature and rainfall patterns, changes in water availability, as well as more intense and frequent extreme events. These changes affect crop production, food distribution, and supply (FAO 2013) but also generate significant social and economic costs (Young et al. 2015).

Second, agricultural GHG emissions occur directly (e.g., production and use of synthetic fertilizers and livestock emissions) or indirectly due to deforestation and land use/cover change. The sources of agricultural GHG emissions include carbon dioxide (CO_2), methane (CH_4), and nitrous oxide (N_2O) resulting from biological decomposition processes in the soils and during digestion by livestock (enteric fermentation). Other processes that generate emissions are rice production, burning of biomass, manure management, feed production, processing and transport, and energy consumption (Gerber et al. 2013; Tubiello et al. 2014).

Third, agricultural practices can influence carbon sequestration processes since carbon can be absorbed from the atmosphere and restored in plant biomass and soils. The soils represent the most important long-term organic carbon reservoir in terrestrial ecosystems (Scharlemann et al. 2014). Sustainable land management practices can thus improve productivity and soil health but also store carbon. Some farming practices foster soil carbon losses, which degrade soil quality and reduce productivity (Lal et al. 2004). On the other side, soil organic carbon losses can be mitigated through management approaches specifically targeting the soil carbon stocks in agricultural systems (Smith and Olensen 2010).

Considering these links, there is a clear need for highly productive but low-emission agricultural systems that use inputs more efficiently (captured in the concept of *climate-smart agriculture*, see WB/CIAT/CATIE 2014). Such systems can lead to more resilient rural landscapes with increased carbon sinks and reduced GHG emissions per agricultural product or/and unit of land, contributing to CC mitigation efforts but also to food security and poverty alleviation (FAO 2013; Wheeler and von Braun 2013). For example, in the livestock sector, practices that reduce emissions depend on the improvements of production efficiency in the animal and herd levels, such as using better feeding practices to reduce enteric and manure emissions and improving husbandry and health management to decrease the unproductive part of the herd and their related emissions (Rojas-Downing et al. 2017). Additional measures include manure management practices that enable recovery and recycling of nutrients as well as a more efficient water and energy use along the supply chains (Brandt et al. 2017).

28.3 The RIO RURAL Program

RIO RURAL supports the transition of degraded rural areas to sustainable productive systems by providing technical assistance and incentives to small-scale farmers based on a micro-watershed level approach in RJ. This program aims at increasing improved agricultural land with more sustainable production systems, reaching 814,000 ha in the period 2010–2018 (RIO RURAL 2014).

The program area embraces 72 municipalities and 366 micro-watersheds across RJ, giving priority to areas with highest poverty incidence. The program area has a

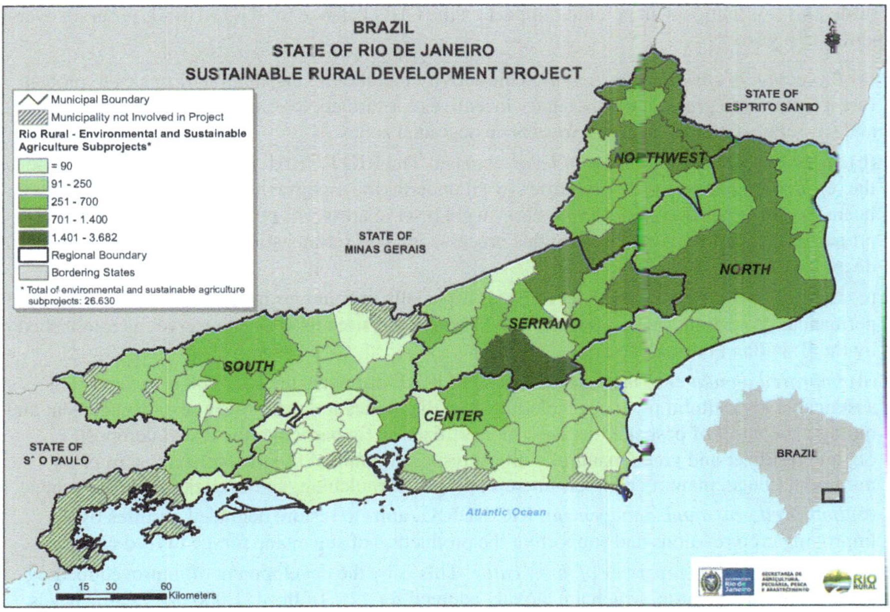

Fig. 28.2 Map of environmental and sustainable subprojects supported by RIO RURAL

high concentration of family-oriented and small-scale agricultural production. RIO RURAL expects to benefit 48,000 rural families, mostly family-oriented and small-scale farmers (RIO RURAL 2014). Figure 28.2 and Table 28.1 illustrate the type, number, and geographic concentration of RIO RIURAL's supported subprojects, including environmental (incentives to conserve forested areas around springs and streams[2]) and sustainable agriculture practices.

28.4 Legal Frameworks and Policies Related to Climate Change Mitigation and the Agriculture and Forestry Sectors

In terms of legislation, both the "*Forest Code*" (Law 12.651/2012) and the Atlantic Forest Law impose certain limitations on agricultural production. The latter prohibits the conversion of secondary forest to other land uses and the Forest Code establishes that 20% of rural properties' area need to be maintained under permanent forest cover as a "*legal reserve* (LR)," in the Atlantic Forest biome (INEA 2013).

Additionally, this legislation also established the *Areas of Permanent Preservation* (APPs), which determine a minimum vegetation area to protect riverbanks and springs to conserve water resources and prevent soil erosion. APPs embrace both

[2] For further details on economic incentives, see May et al. (2018).

Table 28.1 Activities with potential impact on the GHG balance for RIO RURAL program over a period of 6 years

(a) *Protection of springs and streams*. This activity includes (i) installation of fences to protect forest from cattle grazing and monetary incentives for farmers to cease exploiting these zones and (ii) reforestation with native species on degraded zones
(b) *Support to the establishment of legal reserves*. The RIO RURAL program intends to support the legal compliance of rural properties by (i) undertaking topographic surveys, environmental licensing, and notarization of "in-process" legal reserve areas, (ii) providing incentives to properties that have not entered into this process, and (iii) reforestation of native vegetation on degraded zones
(c) *Expansion of agroforestry*. RIO RURAL especially encourages its development in areas of permanent protection, such as those around springs and streams or legal reserves as established by the Brazilian Forest Code
(d) *Improved annual crop management*. RIO RURAL promotes the adoption of several sustainable agricultural practices including crop diversification; integrated pest management and biological control of pest and diseases; bio-fertilization (in particular the use of compost, organic fertilizer and green manure); soil analysis, and rational use of fertilizers; zero or minimum tillage, planting contour, intercropping, and mulching; and irrigation management
(e) *Improved grassland management*. RIO RURAL aims to restore degraded pastures by improving crop rotations and supporting the production of sugarcane forage to feed cattle
(f) *Improved feeding practices of dairy cattle*. This is by the development of improved feeding practices for dairy cattle, which are already adopted for 12% of the 421,000 dairy cattle heads counts in the Rio de Janeiro state
(g) *Use of lime to fight soil acidification and sustainable use of agrochemicals*. The project supports the use of lime to combat soil acidification and a more sustainable use of agrochemicals in cropland and grassland management

Modified from Branca and Medeiros (2010)

riparian preservation areas (RPAs) and *hilltop preservation areas* (HPAs) on hilltops, steep slopes, and elevations >1,800 m (Soares-Filho et al. 2014).

Small-scale properties, ranging from 20 to 140 ha in RJ, are not obliged to restore 20% of LR in their properties if the deforestation occurred before 2008, and RPAs can also be included in their calculation of the LR area. In this way, the restoration requirement is reduced. The restoration of APPs in rural properties is mandatory and differs according to the property size, where smaller properties have lower requirements[3] (INEA 2013).

The Brazilian INDCs[4] at Paris included efforts toward zero illegal deforestation by 2030, compensation for the GHG emission from legal suppression of vegetation, restoration, and reforestation of 12 M ha of forests and the enhancement of forest management systems (Federative Republic of Brazil 2016). Brazil also committed

[3] 10% of property area for rural properties <2 fiscal modules and 20% for those with an area > 2 and 4 fiscal module (INEA 2013).

[4] Intended Nationally Determined Contributions (INDC) are public disclosures of countries' intended post-2020 climate action, which were adopted at the UN Framework Convention on Climate Change (UNFCCC) Conference of the Parties in Paris.

to strengthen the Low Carbon Emission Agriculture program (known as the ABC) as the core strategy for sustainable agriculture development including the restoration of additional 15 M ha of degraded pasturelands by 2030 and the establishment of 5 M ha of integrated cropland-and-livestock-forestry systems, by 2030 (Federative Republic of Brazil 2016). The ABC program seeks to reduce emissions by promoting the best practices in agriculture, targeting medium to big farmers. For small-scale farmers and/or family farmers, Brazil's Federal Government has established the national agroecology policy and its corresponding "PRONAF agroecology" credit program (MDA 2016a).

28.5 Emission Trading: Compliance and Voluntary Markets

There are two types of markets for the GHG emission reductions transactions. On one hand, the *compliance markets* are ruled by the UNFCCC framework, where private and public institutions of industrialized countries are encouraged to promote reduction emission projects in developing countries, through the Clean Development Mechanism (CDM).

The compliance market works under the cap and trade mechanism, where a limit or "cap" is set for the total amount of GHG emissions that can be emitted by companies, power plants, and other installations. Emission trading enables the transaction of allowances and credits[5] in markets and can be considered in national mitigation targets.

On the other hand, *voluntary carbon markets (VCM)* operate outside the compliance market and have more flexible requirements (i.e., they allow for experimentation, new procedures, methodologies, and technological measures). This allows companies, governments, NGOs, and individuals to compensate their emissions through the purchase of offsets generated by this market.[6] VCM participants are enterprises and individuals interested in reducing their carbon footprint, gaining competitive advantage, or anticipating the implementation of compliance markets. In 2015, 84.1 $MtCO_2e$ in VCM emissions credits were transacted from voluntary buyers at the lowest ever recorded average price of US\$3.3/tCO_2e (Hamrick and Goldstein 2015).

Given the higher applicability of VCM to promote activities for smallholders, the lower transaction costs, and the higher flexibility in requirements when compared to those in the compliance market, we consider VCM to be relevant for RIO RURAL's activities (Phan et al. 2017).

[5] Compliance market credits are known as *certified emission reductions* – CERs.

[6] Voluntary carbon market credits are known as *verified (or voluntary) emission reductions* –VERs.

28.6 Processes and Methods to Assess the Mitigation Potential of Agricultural Projects Eligible for Certification

Methodologies are required to quantify the GHG benefits of a project and guide project developers to determine the project's boundaries, set baselines, assess additionality,[7] and quantify the GHG emission reductions. Baselines express the business-as-usual emission scenario and project emissions methodologies that are used to calculate a project's GHG benefits beyond established baselines.

Thus, as in other carbon markets, a set of project designs, monitoring, and reporting criteria are adopted in the VCM, according to which carbon offsetting activities and/or projects' environmental, social, and other co-benefits are verified. Various standards have evolved, each with its own certification processes and some emission registry services. The project cycle of one of the most commonly adopted certification and registry services, *Voluntary Carbon Standard* (VCS), is shown in Fig. 28.3.

To determine the potential to certify emission reductions from RIO RURAL (Table 28.1), methodologies from both compliance and voluntary markets are identified in Table 28.2. Each methodology has specific eligibility criteria, which needs to be considered when estimating a project's GHG balance.

A methodology, not part of CDM or VCS, considered helpful to assess the mitigation potential of all activities promoted by RIO RURAL, was EX-ACT. This tool, developed by FAO, estimates the net C balance of new investment programs and was developed using recommendations to establish the National GHG Inventories.[8] The main result is an estimate of the CO_2e balance associated with the adoption of land management improvement options when compared to a "business-as-usual" scenario (Bernoux et al. 2010; Cerri et al. 2010). We used the estimations of the mitigation potential of RIO RURAL's activities from Branca and Medeiros (2010), which compared emissions "without project's interventions" to those "with project activities," over 20 years (for more details, see Branca and Medeiros 2010).

We used these estimations, the area size dedicated to RIO RURAL's agricultural and forest restoration activities, and the most conservative price paid by voluntary buyers in 2015 (US$ 3.3 tCO_2e/ha[9]), to estimate the revenues from potential emission reductions. We then estimated the certification, implementation, and maintenance costs based on expert knowledge. With this revenue and cost information, we were able to assess the financial feasibility of carbon finance for RIO RURAL's promoted practices. Finally, we reviewed and analyzed the adaptation and other co-benefits generated from this program's practices.

[7] Additionality refers to demonstrating that emission reductions are real, permanent, and attributable only to the project and that emission reductions due to the project are additional to the reductions that would have occurred without a project.

[8] Intergovernmental Panel on Climate Change – IPCC (2006) – "2006 IPCC Guidelines for National Greenhouse Gas Inventories" Hayama: Task Force on National Greenhouse Gas Inventories. Intergovernmental Panel on Climate Change.

[9] This payment value refers to the reference average price under the voluntary market in 2015, which is the lowest recorded value and thus considered to follow a conservative valuation approach.

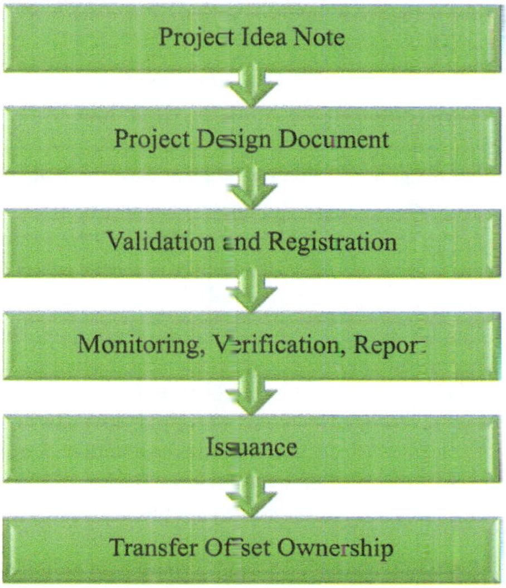

Fig. 28.3 Voluntary Carbon Standard Project Cycle

28.7 Results and Discussion

28.7.1 RIO RURAL's Activities with Mitigation Potential and Methodology Toolkit for Methodologies that Can Be Utilized for Certification

Branca and Medeiros (2010) identified that the protection of springs and streams and the support to establish LRs (Sect. 28.4) are the RIO RURAL's activities with the highest mitigation potential (Table 28.3). Agricultural activities with mitigation potential included improved annual crop management practices (i.e., crop diversification, integrated pest management, biological control of pest and diseases, agroecological systems), nutrient management (i.e., green manure, organic fertilizer, composting), tillage and residues management (i.e., contour cropping, minimum tillage), and irrigation management.

Independent of the methodology used to estimate the GHG balance of project's activities, a proof of additionality is required (i.e., that the smallholders' restoration efforts are beyond legal requirements, see Sect. 28.4). If the additionality criteria are met, increasing the magnitude of forest restoration activities within the scope of RIO RURAL would result in a higher mitigation potential. Other similar development projects in Santa Catarina, Brazil, showed a higher mitigation potential due to a wider area of forest protection activities, such as expanding agroforestry systems and rehabilitating degraded land (Branca et al. 2013).

Table 28.2 Methodology toolkit for RIO RURAL's program activities with climate change mitigation potential

Activity group	Potential activities to generate emission reductions	Methodology
Protection of springs and streams	Installation of fences to protect forest from cattle grazing and monetary incentives for farmers to cease exploiting these zones Plantation of native forest on degraded zones	VCS VM0015: Methodology for avoided unplanned deforestation
		AR-AMS0007: Afforestation and reforestation project activities implemented on lands other than wetlands
Support to the establishment of legal reserves	Undertaking topographic survey, environmental licensing and notarization of legal reserve Providing incentives to rural properties to establish legal reserves Reforestation of native vegetation on degraded zones	AR-AM0014: Afforestation and reforestation of degraded mangrove habitats
		AR-ACM0003: Afforestation and reforestation of lands except wetlands
Improved annual crop management	Foster agricultural practices: Crop diversification; Integrated pest management and biological control of pest and diseases; Bio-fertilization (compost, organic fertilizer and green manure) Soil analysis and rational use of fertilizers Zero and minimum tillage Planting contour, inter/relay cropping and mulching; Irrigation management	AMS-III.BF.: Reduction of N_2O emissions from use of nitrogen use efficient (NUE) seeds that require less fertilizer application
		AMS-III.A.: Offsetting of synthetic nitrogen fertilizers by inoculant application in legumes-grass rotations on acidic soils on existing cropland
		AMS-III.BE.: Avoidance of methane and nitrous oxide emissions from sugarcane pre-harvest open burning through mulching
		VCS VM0017: Adoption of sustainable agricultural land management
Improved grassland management	Restoration of degraded pastures by improving rotations (rotational pastures)	VCS VM0032 methodology for the adoption of sustainable grasslands through adjustment of fire and grazing
		VCS VM0026 sustainable grassland management
Improved feeding practices of dairy cattle	Supporting the production of sugarcane forage for cattle	Partially included in VCS VM001 methodology

Table 28.3 Estimated annual revenues from emission reduction credits of different RIO RURAL program activities

RIO RURAL program activity	Area size (ha)	Specific measures	Type of mitigation action	Estimated mitigation via EX-ACT (tCO_2e over 20 years)[a]	Annual revenues from emission reduction credits (US$/ tCO_2e)[b]
(a) Protection of springs and streams	900	Forest regeneration and plantation of native forests	GHG removal by sinks	517,166.0	85,332.4
(b) Support to the establishment of legal reserves	110				
(c) Expansion of agroforestry	1,100	Expansion of agroforestry on degraded grassland, resulting in changes in both biomass and soil C stock	GHG removal by sinks	267,257.0	44,097.4
(d) Improved annual crop management	4,110	Improved agricultural practices (higher C biomass and soil C stocks)	GHG emission avoidance	18,334.0	3,025.1
(e) Improved grassland management[c]	691	Improved pasture rotations, supporting the production of sugarcane forage (increase in C stock)	Displacement of a more-GHG-intensive output	19,437.0	3,207.1
(f) Improved feeding practices of dairy cattle: 421,000 herd size, increasing 7% of adoption of practices		Increased productivity (meat and milk), resulting in an overall reduction of CH_4 emissions per unit of product	GHG emission avoidance	39,769.0	6,561.8
					142,223.9

Based on Branca and Medeiros (2010)

[a]Branca and Medeiros (2010)

[b]Based on an estimated payment of US$ 3.3/ tCO_2e. RIO RURAL's activities (a) and (b) were calculated using the "forest regeneration EX-ACT module," (c) "other land use change and perennials modules," (d) "annuals module," (e) "grassland module," and (f) "livestock module"

[c]The use of lime against soil acidification and the sustainable use of agrochemicals are included in the module "inputs" for improved annual crop management practices

28.7.2 Estimated Costs and Revenues Associated to Obtaining Carbon Credits for RIO RURAL's Agricultural and Forest Restoration Activities

Considering the mitigation potential presented in the previous subsection and the average value paid by voluntary buyers in 2015, the estimated annual revenue that can be obtained by RIO RURAL's activities is US$ 142,223.9 (Table 28.3). The average costs for obtaining *Verified Emission Reductions* are estimated in Table 28.4, considering that these vary according to the methodology used (the required human resources matter more than the project size). The annual certification costs (Table 28.4) are estimated at US$ 29,644 for project development and validation, in addition to US$ 12,704 for annual verification, monitoring, and reporting (over 10 years). Certain activities (such as agroforestry) yield minimal or zero reductions and then achieve an increase in sequestration at a certain vegetation age, after which they stabilize. This means that most projects will not provide any payment until a certain duration after its implementation. In addition, implementation and maintenance costs are taken into account. In the current carbon projects in the Atlantic Forest, such costs (mean value per hectare) represent US$ 1,222 and US$ 348, respectively (Table 28.5).

28.7.3 Financial Analysis of the Mitigation Potential from RIO RURAL's Promoted Practices and Adaptation Benefits

Certification costs in the voluntary market are estimated at US$ 156,691 (Table 28.4). Variations will depend on the project's selected activities and corresponding methodologies. Furthermore, the annual implementation and maintenance costs also need to be added.

Table 28.4 Estimated costs related to certification schemes in the voluntary market

Estimated costs of certification schemes	US $a
Internal costs (including project development, feasibility study, data inquiry)	5,293.6
Carbon consultant for developing the project as a carbon project (i.e., writing a project design document (PDD), financial analysis, baseline determination)	10,587.2
Validation body	12,704.6
Administration costs and fees	1,058.7
Costs for development and validation of the project activity	29,644.1
Verification body for periodic verification (each year)	8,469.8
Carbon consultant for preparing the monitoring report (annually)	4,234.9
Total costs for annual verification of the GHG emission reductions over a verification period (10 years)	127,047.0

[a]Costs converted to USD using the exchange rate of 1EUR = 1.05872 US$

Table 28.5 Implementation and maintenance costs in carbon-related projects of the Atlantic Forest region

Project	Implementation costs (R$/ha)	Annual maintenance costs (R$/ha)
Banco de Carbono (Brasil)	7,000	1,000 (3 years)
Mapa dos Sonhos do Pontal do Paranapanema (SP)	5,000	1,800 (3 years)
Plantando Água (SP)	5,000	1,800 (3 years)
Neutralização de Emissões de Carbono (SP, MG)	7,000–9,000	1,800–2,500 (3 years)
Com Café (CE)-SAF	4,100 (coffee plantation + forest shading)	300
EcoCitrus (RS)-SAF	2,175 (enrichment of citrus trees with native forest)	n.d
Carbono, Biodiversidade e Comunidade (BA)	15,000	6,000 (3 years) + 1,500 (monitoring –30 years)
Projeto Floresta Viva (BA)	12,000	n.d.
Brazil Mata Viva (GO)	1,043	n.d.
Carbono, Biodiversidad e Renda (SP)	5,000	1,300 (3 years)
Carbono Seguro (SP)	n.d.	256 (30 years)
Projeto Ação contra o Aquecimento Global em Guaraqueçaba (PR)	US$ 230	US$ 45
Projeto de Restauração da Floresta Atlântica (PR)	US$ 350	US$ 65
Projeto Piloto de Reflorestamento em Antonina (PR)	US$ 300	US$ 60
Reflorestamento das Bordas dos Reservatórios da AES Tietê (SP)	11,000	Included in implementation value
Programa Desmatamento Evitado (PR/SC)	n.a.	500
Recomposição da Paisagem e SAFs (Café com Floresta, SP)	920	n.d.
Mean values	4,750 R$[a] (1,222 US$)	1,354 R$ (348 US$)

Modified from May (2011)

[a]Considering an exchange rate of 1 US$ = 3.88 R$

RIO RURAL's activities were found to have a positive but low annual average mitigation potential of 0.2 tCO_2e/ha (Branca et al. 2013), which valued at US$ 3.3/tCO_2e implied a return of only US$ 0.6/ha/yr. Even with the cost advantage for the registry and the certification of RIO RURAL's projects given their large area for operation, these returns are not sufficient to counteract other land use options.

Factoring these estimates, carbon payments alone are not likely to be a financially feasible option to incentivize emission reductions from RIO RURAL's promoted activities. This analysis assumes limited returns from low carbon payments (average price paid in 2015) but also the constrained area of activities that have high mitigation potential by RIO RURAL (reforestation activities).

Carbon payments per hectare are often rarely higher than what a landowner could gain from an alternative land use (opportunity costs), especially considering cash crops or profitable agricultural activities. This means that carbon finance is not likely to provide incentives to stop land use conversion, if they are based only on economic motivations (Goldstein and Gonzalez 2014). In practice, carbon finance seems to work as an incentive for those already inclined toward forest conservation (May 2011).

The possibility of certifying groups rather than individual farmers can reduce transactions costs and the inclusion of value added to products, such as certified coffee, also helps to make them more viable as financial alternatives. Project preparation expenses assumed by the state through extension or university personnel is another way to make these schemes work (Shames et al. 2012).

Even though we found a low financial mitigation potential for carbon finance alone, sustainable agriculture and productivity enhancement promoted by RIO RURAL has resulted in the gain of ecosystem services and led to effective mitigation actions, increase in landscape resilience, and the adaptive capacity of production systems and people against climate-related hazards.

Significant adaptation benefits result from RIO RURAL's promoted activities that include avoiding or minimizing land degradation, the risk of landslides, and other CC-inducing hazards associated with poor land use management. There have been environmental disasters with considerable economic and social costs stemming from these problems in recent years in Brazil. Economic losses, as a result of the extreme weather events in Brazil (i.e., flooding, flash floods, and landslides) between 2002 and 2012, have led to significant damage valued at between US$ 57.21 and 113.1 billion[10] (Young et al. 2015). The state of Rio de Janeiro reported that 45% of all national number of deaths are associated to such hazards. In times of extreme events, the poor people concentrated mostly in rural areas are more vulnerable to natural disasters and the impacts of CC, which led to rising inequalities (Scarano and Ceotto 2015; Young et al. 2015).

RIO RURAL's promoted activities therefore play an essential role in supporting the CC adaptation by increasing rural landscape resilience. Such efforts are attractive for financial support from international financial mechanisms. For example, the Green Climate Fund (GCF) as the operating entity of the financial mechanism of the UNFCCC has committed among its investment priorities to support activities that encourage low emission and climate resilient agriculture and that scale-up finance for forest and CC.[11]

[10] Using the exchange rate of 1 BR$ = 0.31795 US$ on the original reported data.

[11] Green Climate Fund (GCF) (2014) Funding. Available at: http://www.greenclimate.fund/ventures/funding/#how-it-works

28.7.4 Co-benefits (Water and Food Security, Biodiversity, and Socio-Cultural Contributions)

Water security has become a sensitive challenge that creates conflicts in high deforested areas of the Paraíba do Sul watershed between São Paulo and RJ (Scarano and Ceotto 2015). Therefore, forest restoration activities (particularly on the hilltop and riparian preservation areas) yield hydrological benefits, in terms of water conservation and quality, thus fostering human water security. Moreover, there is a growing market for watershed services in the Atlantic Forest, driven by many initiatives (see May et al. 2018). Some have received support from the private sector (e.g., Oásis Payment for Watershed Services project in São Paulo with funds from the Mitsubishi and Grupo Boticário foundations), which can provide new opportunities in RJ (Veiga and Gavaldão 2011). Additional hydrological regulation services provided by the riparian and hilltop vegetation areas include the moderation of extreme events (e.g., floods and landslides), which have severely hit RJ in the past (Joly et al. 2014).

Despite its outstanding biodiversity and high endemism, the Atlantic Forest is highly fragmented with just 12–16% of the original forest cover left (Ribeiro et al. 2009). The remnant forest is also threatened and has a very low protection level of 2.6%, which is way below the recommended minimum of 17% (Ribeiro et al. 2009; Soares-Filho et al. 2014). Therefore, restoration and reforestation activities are essential to protect native vegetation and biodiversity. Furthermore, landowners that restore forests beyond the requirements set by the Forest Code may be able to generate environmental reserve certificates (CRAs) (See May, Fernandes & Rodríguez Osuna 2018 in this volume), which could be a source of revenue for smallholders. These efforts provide a clear contribution to the costly and high restoration needs in RJ (PACTO 2009).

Family farming is essential for supplying food in RJ, which provides livelihoods, income, and jobs, and is responsible for around 68% of beans, 75% of manioc, 67% of maize, 55% of rice, and 52% of coffee (MDA 2016b). Furthermore, food production from family farming is key to the sustainable development of communities and healthy diets in RJ. To this end, RIO RURAL supports farmers to adopt organic farming practices (>200 families), which makes them eligible to an increased price for their products. This is to incentivize the supply of organic food products aligned with the rules of national programs.

The social empowerment approach to small-scale rural farmers by the RIO RURAL type of projects can be captured by certifiable social standards, such as the Social Carbon Standard. Others include the Gold Standard, Plan Vivo, which seek to quantify co-benefits from health and gender to the protection of biodiversity. Besides the emission reductions in 2015, global voluntary projects provided several co-benefits such as 8000 jobs, protected habitats for 376 endangered species, support to vulnerable groups, contributions to water security, built resilience to CC impacts, and regularization of land tenure. These beyond-carbon benefits have earned more attractive prices per ton. Nearly half of all forest carbon projects in 2015 engaged in these markets because of such co-benefits (Goldstein and Ruef 2016).

Other potential co-benefits (amenity values) include ecotourism activities associated to an increasing demand of urban settlers that value scenic beauty offered by the rural landscapes in RJ. Ecotourism in private reserves is clearly underdeveloped (de Vasconcellos and Castley 2014) but offers a myriad of opportunities that include income generation for biodiversity protection and benefits such as hiking, bird watching, and outdoor sports, in addition to inspiration for art, music, science, and religion.

Most co-benefits from carbon-related projects in the Atlantic Forest are generated as increased income and often relate to the conservation and restoration of biodiversity and ecosystems, including springs and stream protection. However, grouping diverse ecosystem services requires accounting, monitoring, and verification (May 2011). Consequently, we propose a bundling approach, where multiple benefits are measured and certified (Table 28.6). This would allow accessing resources from both mitigation and adaptation funds, in addition to the markets that value ecosystem integrity as well as water and food security.

28.8 Conclusions

RIO RURAL's forest restoration activities provide small-scale farmers with means to comply with the Forest Code and have the highest mitigation potential. We found that given the low price for conventional emission reduction credits (at the time of this analysis) and the high costs associated with implementation and certification, the estimated carbon revenues potentially generated from RIO RURAL are not financially viable if they are based solely on carbon finance. However, the adaptation benefits from RIO RURAL's actions are remarkable, taking into account how sustainable agriculture and forest restoration practices increase landscape resilience and farmers' adaptive capacity against the impacts of CC.

RIO RURAL's sustainable agriculture practices improve productivity and resource efficiency and contribute positively to food security and rural farming livelihoods. Forest restoration activities promote biodiversity conservation and healthy ecosystem functions and services (especially considering RJ's highly degraded landscape). Some of the most remarkable co-benefits of such promoted activities are watershed protection that foster water security.

The increased incidence of extreme events in RJ has caused significant social and economic costs in recent years. As a result, actions to avoid or mitigate future negative impacts of CC (adaptation benefits), especially on the rural poor, need to be taken seriously. Conflicts related to water insecurity (e.g., access to water between São Paulo and RJ), in addition to the pressures from a growing demand for resources, show the urgency to act now to move from a *shrinking hotspot to a climate hope spot* (Scarano and Ceotto 2015).

Table 28.6 Possible co-benefits associated to activities promoted by RIO RURAL

Activities	Food security (productivity)	Water quality	Water conservation	Soil quality	Air quality	Biodiversity and wildlife habitat	Scenic beauty/ amenity value	Moderation of extreme events
Protection of springs and streams		+	+	+	+	+	+	+
Support to the establishment of legal reserves		+	+	+	+	+	+	+
Expansion of agroforestry	+/−	+/−	+/−	+	+	+	+	+/−
Improved annual crop management	+	+/−	+	+		+		
Improved grassland management	+	+/−		+		+	+	
Improved feeding practices of dairy cattle	+			+				

All these activities are carried out with a social empowerment approach based on participatory management of micro-watersheds (social co-benefits)

We found that a bundling approach to carbon is most appropriate to address RIO RURAL's contribution to CC, which accounts for the various benefits this program provides in both mitigation and adaptation efforts, in addition to the co-benefits associated to increased water and food security. Furthermore, the smallholder social and participative approach of programs such as RIO RURAL add relevant social benefits that improve livelihoods and smallholders' adaptive capacity to cope with the impacts of CC.

Sustainable agriculture activities can also be considered as a contribution to Brazil's national forest restoration and low carbon emission agriculture goals set to be accomplished by 2030. These activities can also contribute to regional land restoration activities launched recently that are supported by public and private funds.

Carbon mitigation is only a small part of the potential social and environmental benefits to be gained from an integrated landscape approach focused on smallholder land use practices allied with the corridors between protected areas and that which take advantage of the Forest Code structures regarding riparian forests and spring protection. These benefits include the protection against the risks associated with CC, and permit better adaptation to these impacts on smallholders, and are not currently supported by any programs other than schemes such as RIO RURAL.

Acknowledgments We thank Denitsa Gaydarova-Itrib from TÜV Rheinland who kindly provided estimates of costs associated to certification and helped identify voluntary market methodologies associated to RIO RURAL's promoted practices. We appreciate the support from Carlos Marconi and all technical officers from the Rural Extension and Technical Assistance Agency (EMATER) of the Italva municipality and Marcos André Dias from RIO RURAL, who provided helpful insights to understand this program's operational scope. We also thank Zachary Tessler and Andrea Rodríguez Osuna for their support with proofreading abstracts of this chapter and Nathália Duarte for her contribution with the RIO RURAL area map.

References

Bernoux M, Branca G, Carro A et al (2010) Ex-ante greenhouse gas balance of agriculture and forestry development programs. Sci Agric 67(1):31–40

Branca G, Medeiros K (2010) Estimating mitigation potential of agricultural projects: an application of the EX-Ante carbon-balance tool (EX-ACT) in Brazil, Working paper LAC/03/10. Latin America, the Caribbean, East Asia and the Pacific Service Investment Centre. Division Food and Agriculture Organization of the United Nations, Rome

Branca G, Hissa H, Benez MC et al (2013) Capturing synergies between rural development and agricultural mitigation in Brazil. Land Use Policy 30:507–518

Brandt P, Kvakić M, Butterbach-Bahl K, Rufino MC (2017) How to target climate-smart agriculture? Concept and application of the consensus-driven decision support framework "targetCSA". Agric Syst 151:234–245

Cerri CC, Bernoux M, Maia SMF et al (2010) Greenhouse gas mitigation options in Brazil for land-use change, livestock and agriculture. Sci Agric 67(1):102–116

de Vasconcellos FP, Castley JG (2014) Ecotourism as a conservation tool and its adoption by private protected areas in Brazil. J Sustain Tour 22(4):604–625

Escobar H (2015) Brazil's climate targets fail to impress experts. Science. https://doi.org/10.1126/science.aad4637

FAO (2013) Climate-smart agriculture sourcebook. Food and Agriculture Organization of the United Nations, Rome, p 557

Federative Republic of Brazil (2016) Intended nationally determined contribution towards achieving the objective of the United Nations Framework Convention on Climate Change.

Gerber PJ, Steinfeld H, Henderson B et al (2013) Tackling climate change through livestock: a global assessment of emissions and mitigation opportunities. FAO, Rome

Goldstein A, Gonzalez G (2014) Turning over a new leaf: state of the forest carbon markets 2014. Washington, DC, p 87

Goldstein A, Ruef F (2016) View from the understory. State of forest carbon finance 2016. Forest Trends' Ecosystem Marketplace, Washington, DC, p 51

Hamrick K, Goldstein A (2015) Ahead of the curve: state of the voluntary carbon markets 2015. Forest Trends, Washington, DC, p 50

Hissa H, Teixeira N, Costa M et al (2018) Sustainable rural development in Rio de Janeiro state: the Rio rural program. In: Nehren U, Schlüter S, Raedig C, Sattler D, Hissa H (eds) Strategies and tools for a sustainable rural Rio de Janeiro. Springer International Publishing, Cham

IBGE (2010) Censo Demográfico 2010: População residente, por tipo de deficiência permanente. Instituto Brasileiro de Geografia e Estatística, Brazil

INEA (2013) Adequação Ambiental de Imóveis Rurais. Orientações Gerais. Instituto Estadual do Ambiente, Rio de Janeiro, p 46

Joly CA, Metzger JP, Tabarelli M (2014) Experiences from the Brazilian Atlantic Forest: ecological findings and conservation initiatives. New Phytol 204(3):459–473

Komives K, Jackson A (2014) Introduction to voluntary sustainability standard systems. Springer, Berlin/Heidelberg, pp 3–19

Lal R, Griffin M, Apt J et al (2004) Managing soil carbon. Science 304(5669):393

May PH (2011) Iniciativas de PSA de carbono florestal na Mata Atlântica. In: Guedes FB, Seehusen SE (eds) Pagamentos por Servicos Ambientais na Mata Atlanica: licoes aprendidas e desafios, Serie Biod. Ministerio do Meio Ambiente, Brasilia

May PH, Fernandes LS, Rodríguez Osuna V (2018) Evolution of public policies and local innovation in landscape conservation in Rio de Janeiro. Chapter 29. In: Nehren U, Schlüter S, Raedig C, Sattler D, Hissa HR (eds) Strategies and tools for a sustainable rural Rio de Janeiro. Springer International Publishing

MDA (2016a) Brasil agroecológico: Plano Nacional de Agroecologia e Produção Orgânica – Planapo: 216-2019. Ministério do Desenvolvimento Agrário, Brasilia DF, p 89

MDA (2016b) Agricultura familiar movimenta a economia local no Rio de Janeiro, Secretaria Especial de Agricultura Familiar e do Desenvolvimento Agrário. Ministry of Agricultural Development.. http://www.mda.gov.br/sitemda/noticias/agricultura-familiar-movimenta-economia-local-no-rio-de-janeiro%C0D. Accessed 04 July 2017

PACTO (2009) Mapa de Áreas Potenciais para a Restauração Florestal. Pacto pela restauração da Mata Atlântica, São Paulo. http://media.wix.com/ugd/5da841_c4c4a0c-5c32f44298a2808ef2df520f1.pdf. Accessed 28 Mar 2017

Phan THD, Brouwer R, Davidson MD (2017) A global survey and review of the determinants of transaction costs of forestry carbon projects. Ecol Econ 133:1–10

Ribeiro MC, Metzger JP, Camargo Martensen A et al (2009) The Brazilian Atlantic Forest: how much is left, and how is the remaining forest distributed? Implications for conservation. Biol Conserv 142:1141–1153

RIO RURAL (2014) Manual Operacional. Projeto desenvolviemento rural sustentável em microbacias hidrográficas do estado do Rio de Janeiro, Brasil. Niteroi-Rio de Janeiro, p 198

Rojas-Downing MM, Nejadhashemi AP, Harrigan T, Woznicki SA (2017) Climate change and livestock: impacts, adaptation, and mitigation. Climate Risk Manag 16:145–163

Scarano FR, Ceotto P (2015) Brazilian Atlantic forest: impact, vulnerability, and adaptation to climate change. Biodivers Conserv 24:2319–2331

Scharlemann JP, Tanner EV, Hiederer R, Kapos V (2014) Global soil carbon: understanding and managing the largest terrestrial carbon pool. Carbon Manag 5(1):81–91

Shames S, Buck LE, Scherk SJ (2012) Reducing costs and improving benefits in smallholder agriculture carbon projects: implications for going to scale. In: Wollenberg E, Tapio-Biström M-L, Grieg-Gran M, Nihart A (eds) Climate change mitigation and agriculture. Earthscan from Routledge, New York

Smith P, Olensen JE (2010) Synergies between the mitigation of, and adaptation to, climate change in agriculture. J Agric Sci. Cambridge University Press 148(5):543–552

Soares-Filho B, Rajão R, Macedo M et al (2014) Cracking Brazil's Forest Code. Science 344(6182):363–364

Tubiello FN, Salvatore M, Cóndor Golec RD et al (2014) Agriculture, forestry and other land use emissions by sources and removals by sinks, ESS working paper series no. 2. FAO Statistics Division, Rome, p 75

Veiga F, Gavaldão M (2011) Iniciativas de PSA de conservação dos recursos hídricos na Mata Atlântica. In: Becker Guedes F, Seehusen SE (eds) Pagamentos por Serviços Ambientais na Mata Atlânica: lições aprendidas e desafios, Serie Biodiversidade 42, vol 2, 2nd edn. Ministério do Meio Ambiente (MMA), Brasilia-DF, Brazil, pp 123–181.

WB/CIAT/CATIE (2014) Climate-Smart agriculture in Costa Rica, CSA country profiles for Latin America series. The World Bank Group, Washington, DC, p 12

Wheeler T, von Braun J (2013) Climate change impacts on global food security. Science 341(6145):508–513

Young CEF, Aguiar C, Neto de Souza E (2015) Valorando tempestades: Custo econômico dos eventos climáticos extremos no Brasil nos anos de 2002 – 2012. Observatório do Clima, São Paulo, p 21

Chapter 29
Promotion of Alternative Income Diversification in Rural Rio de Janeiro: Making Use of Innovation Resources

Omar Augusto Torres Zárate and Utz Dornberger

Abstract In the northwest region of Rio de Janeiro state, where many rural households base their income on unproductive farm activities, federal subsidies, and social help, young family members migrate, threatening the sustainable economic development of the region. As answer to the problem, the government of Rio de Janeiro has promoted the adoption of methodological and technological innovations in several micro-watersheds for the conservation and sustainable use of natural resources through the RIO RURAL program. A big challenge the program faces is to improve the quality of life in the countryside by conciliating these measures with increasing and stable revenues for the households. Since income diversification practices reduce the inherent risk of price volatility of commodities, this chapter offers an insight on the sources of income diversification of 200 rural households that benefited from the RIO RURAL program. The analysis shows that most households diversify or plan to diversify their income based on conventional strategies on-farm, or alternative off-farm employment, following the path that causes the original problems. In this context, the chapter describes an applied methodology for promoting alternative income diversification strategies in the region, taking advantage of the recently enriched innovation resources and involving young labor force. The results endorse the further application of the methodology in other regions.

Keywords Alternative income diversification · Innovation resources · New rurality · On-farm income diversification · Off-farm income diversification

O. A. Torres Zárate (✉) · U. Dornberger
International SEPT Program, Leipzig University, Leipzig, Germany
e-mail: dornberg@uni-leipzig.de

U. Nehren et al. (eds.), *Strategies and Tools for a Sustainable Rural Rio de Janeiro*, Springer Series on Environmental Management,
https://doi.org/10.1007/978-3-319-89644-1_29

29.1 Background

In contrast to the dominating assumption over the last decades of the twentieth century that rural producers in developing countries depend exclusively on their agricultural production for their livelihood and food security (Crole-rees and Consultants 2000), the importance of owning multiple businesses outside agriculture is recognized today as a crucial strategy for minimizing low-income risk in rural regions, where economies of scale can be reached very early (Carter 1999). Recent studies indicate that rurality nowadays is partially characterized by the temporary movement of high-income urbanites into rural areas for weekend or summer stays, as well as for the permanent movement of lower-income urban households into rural areas to build affordable homes. As a consequence, there is a growing demand in rural regions for the provision of personal services, leisure activities, and other services linked to other economic activities, e.g., energy production, tourism, environmental preservation services, as well as production aimed at niche markets (Graziano da Silva and Eduardo del Grossi 2001). Thus, diversification instead of specialization is the norm in the "new rurality," where it is increasingly difficult to establish the limit between rural and urban and where the consciousness of preservation of the biodiversity and the protection of the identity and cultural heritage of rural communities has been rising (Romero Avendaño 2010).

Within the Federal State of Rio de Janeiro (RJ), the northwest region depends mostly on unproductive farm activities and is characterized by a steady decline in the rural population, as well as by the aging of the workforce. Though the region has the highest agricultural production and the highest concentration of family small farms, it is one of the territories with the most pronounced rural poverty. Within this context, it could be helpful to promote the generation of income diversification strategies as an alternative to agriculture in the northwest region of RJ in order to diversify the risk of low income by taking advantage of the recently adopted innovations and the opportunities offered by the new rurality.

29.2 Methodology

In order to promote the generation and execution of alternative income diversification strategies using available innovation resources, it is necessary to understand how rural households generate and diversify their income and to distinguish what innovation resources they count with. Information was obtained from a field research conducted through a survey to 200 rural producers in the northwest region of RJ, who adopted methodological and technological innovations for environmental sustainability between 2008 and 2015 under the RIO RURAL program. Through purposive sampling, the survey collected quantitative and qualitative information regarding the availability of innovation resources, income generation, and income diversification activities and strategies.

Two municipalities were selected for carrying out the survey, Varre-Sai and Italva, as they are good representatives of the northwest region, having numerous small farms producing almost exclusively two commodities with volatile prices, namely, coffee and milk.

29.3 State of the Art

Most documents regarding rural innovation are approached from an agricultural point of view under the framework of agricultural innovation systems (AIS); however, in a more generalized rural context, i.e., outside agriculture, innovation literature concerning rural areas is very poor (Singh and Bhowmick 2015). Innovation is not easy to distinguish within rural development activities because it depends on the geographic, economic, social, and cultural situation of the territory.

The concept of diversification in the farm business literature differs from strategic management definitions of business diversification. While the latter commonly involves adding product, services, location, customers and markets to a business portfolio, there is no agreed standard definition for diversification in rural business literature. Some definitions suggest that a farm can be considered as diversified, if it uses its resources for activities different to the conventional production of crops and livestock to generate income (Hansson et al. 2013). The analysis of rural income diversification has been also difficult due to the lack of harmonized data (OECD 2003). Alternative income diversification activities have been differently conceptualized in the literature, e.g., as part-time farming, diversification, multiactivity, pluriactivity, rural nonfarm employment (RNFE), nonfarm income diversification, off-farm activities, off-farm employment, off-farm income, multiple business ownership, portfolio entrepreneurship in nonfarm sectors, and new rural income-generating activities, among others (Loyns and Kraut 1992; Reardon et al. 2001; Knickel et al. 2003; McNamara and Weiss 2005; Démurger et al. 2010). This overlay of approaches to rural income diversification has generated an overlapping of concepts that drive to significant differences in the interpretations and to permanent contradictions in the literature. The findings of this study are based on the concept framework developed by Torres Zárate and Dornberger (2017) which integrates the locational and sectoral approaches for the classification and analysis of rural income generation/diversification activities.

The relationship between innovation and diversification has been widely researched. The attempts to prove if one generates the other have driven to the general acceptance that there is an endogenous relationship between innovation and diversification (Rodríguez-Duarte et al. 2007). Though this issue has been widely regarded by the innovation and entrepreneurship literature, empirical studies on rural innovation are still rare (Singh and Bhowmick 2015). However, it is recognized that innovation can provide new ways to solve the problems of rural people and create new opportunities for entrepreneurship development (Singh and Bhowmick 2015).

The innovation resources comprise knowledge, competencies, and capital (Sundbo 2001). Knowledge, e.g., professional and technological, of the market and customers, etc. is a crucial element of creativity and affects areas of the creative process such as idea generation, determination of future vision, and project management. Considered equivalent with competitive advantage, knowledge is recognized as one of the most valuable resources within every domain of human life (Nonaka and Takeuchi 1995). Competencies refer to the ability to act in the different innovation roles, i.e., creativity, capacity to develop ideas, and decision to struggle for implementing the new idea (Sundbo 2001). Capital is nothing different to the total sum of the intermediate resources that participate in the different phases of the production of a good (Böhm-Bawerk 1890). Capital resources denote the financial and physical resources necessary for producing goods or services.

This work contributes by describing a procedure that approaches innovation resources and income generation/diversification in rural areas: first through the incorporation of all possible locational and sectoral approaches by the classification and analysis of income diversification activities in rural regions and second through the description of a methodology that facilitates the interaction between available innovation resources and the generation of alternative income diversification strategies.

29.4 Policy Context

In 2004, the northwest region was selected by the Ministry of Agricultural Development for the implementation of a participatory regional policy which includes the rural communities in hydrographic micro-watersheds into the diagnosis, planning, execution, and control phases of the sustainable rural development program RIO RURAL. Executed by the Superintendency of Sustainable Development of the Secretariat of Agriculture and Livestock of RJ (SEAPPA), RIO RURAL has World Bank funding and support from the Food and Agriculture Organization of the United Nations (FAO) for promoting the conservation and sustainable use of the natural resources in production systems. The aim is to improve the quality of life in the countryside, reconciling the income of rural producers with the conservation and sustainable use of natural resources. Between 2006 and 2017, RIO RURAL assigned resources to rural producers for the adoption of innovations that support the conservation and sustainable use of the natural resources in production systems of 407 micro-watersheds of RJ. In some territories of the northwest region, these measures have equipped micro-watersheds with new technologies that directly and indirectly contribute to the generation of new resources able to be used for developing income diversification strategies within the new rurality perspective.

Thanks to the cooperation of the German Federal Ministry of Education and Research (BMBF), through the INTECRAL project, the RIO RURAL program developed a strategy of action with communities from several regions in order to apply technologies aimed at the recovery of degraded areas, at the adoption of silvopastoral systems, at the monitoring of watersheds, and at the generation of income from alternative markets. The analysis and methodology described in this chapter were part of the INTECRAL Project.

29.5 Analysis of Findings or Evidence

The analysis of primary and secondary data obtained from households with recently adopted methodological and technological innovations in the northwest region revealed the following:

- The new rurality perspective amplifies the spectrum of alternative income diversification activities.
- The recent adoption of sustainability practices and land use systems nurtures the innovation resources of the region.
- Most of rural households base their income diversification on conventional on-farm activities or alternative off-farm employment.
- Most of alternative income diversification activities on-farm are not part of the new rurality perspective.
- Most of the rural households have no plans to diversify their income, and those with income diversification plans follow conventional on-farm strategies.
- The use of innovation resources to generate alternative income generation strategies on-farm can be promoted within the framework of the new rurality perspective.
- The new rurality perspective amplifies the spectrum of alternative income diversification activities.

29.5.1 The New Rurality Perspective Amplifies the Spectrum of Alternative Income Diversification Activities

The Latin American and Caribbean Network for Rural Extension Services, established in November 2010, sustains, among others, the existence of the "new rurality," a perspective that emphasizes the increasing difficulty to establish the limit between rural and urban, as well as on the rising consciousness of the multiple functions of agriculture. This agricultural multifunctionality refers to the preservation of the biodiversity and the protection of the identity and cultural heritage of rural communities (Romero Avendaño 2010). Agriculture is considered multifunctional when it has other additional functions to food and fiber production (UNCED 1992). Contrary to the productivity-driven approach of the "Green Revolution," the "new rurality" perspective highlights the importance of adopting sustainability practices over a merely focus on maximizing agricultural productivity. The adoption of sustainability practices generates externalities that support the multifunctionality of agriculture, which increases the opportunity to develop alternative income diversification strategies for farm households (Torres Zárate and Dornberger 2017) (Fig. 29.1).

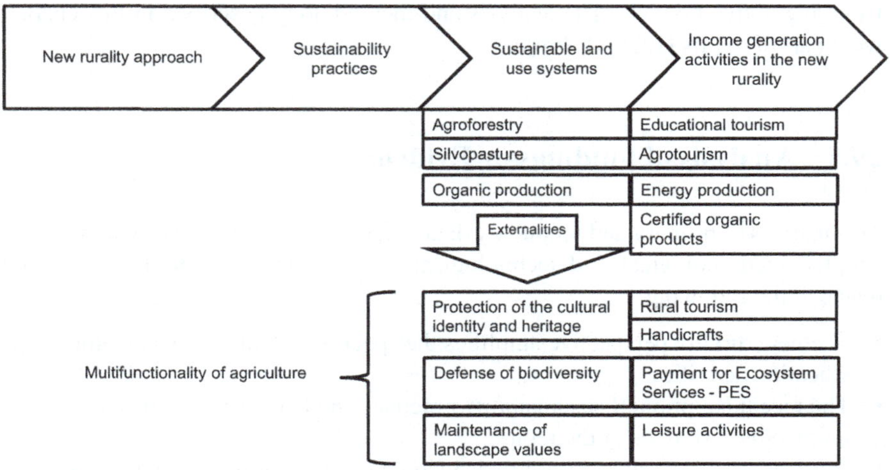

Fig. 29.1 The new rurality perspective

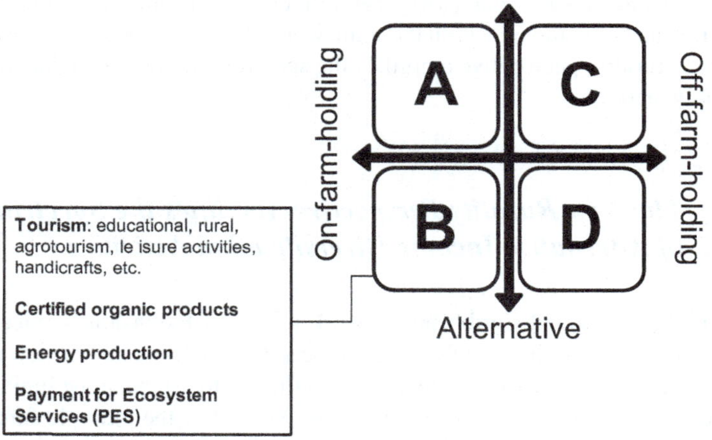

Fig. 29.2 Locational and sectoral classification of income generation activities in the new rurality

According to the locational and sectoral classification of rural income generation/diversification activities by Torres Zárate and Dornberger (2017), alternative income generation activities in the new rurality perspective correspond to quadrant B, being alternative to conventional farm activities and using the on-farm holding resources (Fig. 29.2).

Table 29.1 Top innovation resources in Varre-Sai and Italva

Innovation resources	Factors	Top resources
Knowledge	Professional	Children with studies (most of them emigrated)
	Technological	Recently adopted sustainability practices and land use systems
	Market and clients	91% of the households belong to some association
Competencies	Idea generation	Usual interaction between managers, relatives, and institutions
	Idea development	
	Decision to implement	
Capital	Financial	Easy access to subsidies and loans
	Physical resources	Resources for tourism

29.5.2 *The Recent Adoption of Sustainability Practices and Land Use Systems Nurtures the Innovation Resources of the Region*

The innovation resources comprise knowledge, competencies, and capital (Sundbo 2001). Based on the results of the survey to the 200 rural producers, the most important resources were classified according to the following structure (Table 29.1).

29.5.2.1 Professional and Technological Knowledge

The results of the survey show that only 24% of rural producers' children with a technical degree or superior work on-farm while 66% work off-farm.

Sixty percent of rural producers' children that emigrated, i.e., moved to other places and do not share partially or totally their income with the household, have a technical degree or superior (Fig. 29.3).

Between 2008 and 2015, the RIO RURAL program has promoted the adoption of environmental, productivity, and adding value sustainability practices among rural producers benefited by the RIO RURAL program in the northwest region. In the specific case of the 200 interviewed rural producers, 86% have adopted environmental and productivity sustainability practices, while 14% have adopted sustainability practices intended for adding value to their products (Fig. 29.4).

29.5.2.2 Knowledge of the Market

Fifty percent of the interviewed households know the final client of their products within the distribution chain, while the other 46% sell to intermediaries. However, 91% of the households clearly expressed their belonging to at least one association that represents their interests.

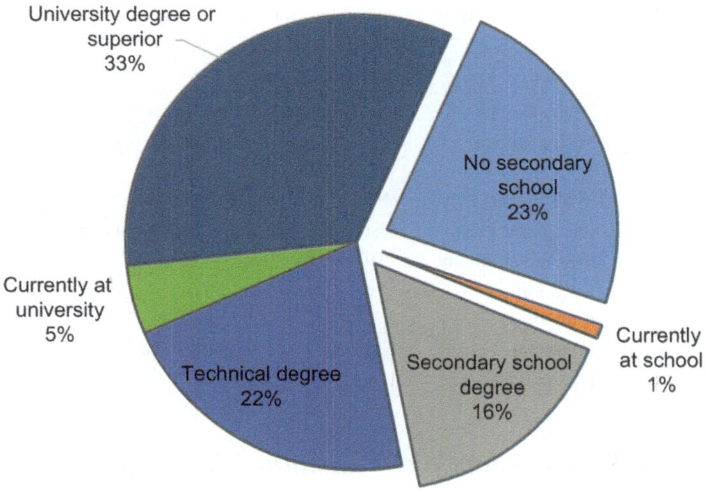

Fig. 29.3 School degree of emigrants

Fig. 29.4 Kind of sustainability practices adopted by the 200 interviewed households, through the RIO RURAL program

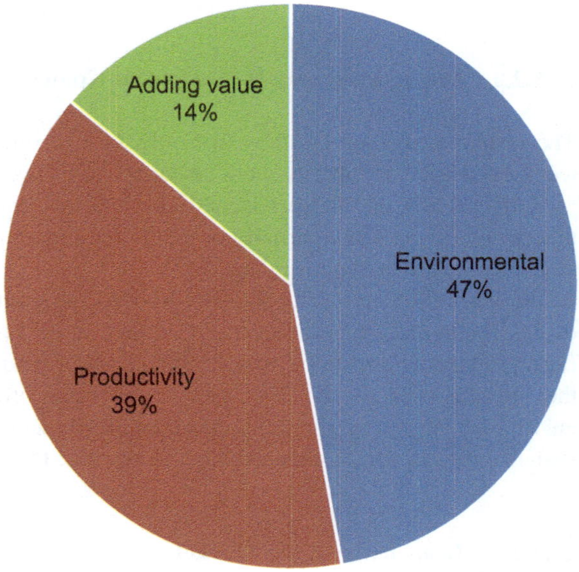

29.5.2.3 Innovation Competencies

RIO RURAL is the main source of innovation ideas and economic resources in the region, while the development and implementation of the ideas is a common teamwork between the households and staff of the extension offices. This process occurs during the application of the methodology that RIO RURAL established in the region for assigning its resources to promote the adoption of new sustainability-driven technologies.

29.5.2.4 Financial Capital

In addition to the subsidies granted by RIO RURAL since 2008, 77% of the interviewed households had access to loans or other subsidies, like the different credit lines of PRONAF, between 2008 and 2015. In general, there is good access to financial capital.

29.5.2.5 Physical Resources

Historical farms are considered by 22.5% of the rural producers as the most common touristic attractions of the region, while 9% mentioned natural landscape, and 8.5% highlighted traditional festivals of the region as touristic attractions.

29.5.3 Most of Rural Households Base Their Income Diversification on Conventional On-Farm Activities or Alternative Off-Farm Employment

Table 29.2 classifies the income generation activities of the interviewed households according to their locational and sectoral classification, as in Fig. 29.5.

According to the results, 59% of the rural households depend on one only conventional on-farm activity (Quadrant A). The second more common income source corresponds to alternative income generation activities off-farm (Quadrant D) where 32% of the interviewed households diversify their income through wage employment or as entrepreneurs.

Wage employment, i.e., working as freelancers or employees, is the main source of income in Quadrant D with 77% of all cases (Fig. 29.5).

Table 29.2 Locational and sectoral classification of rural income generation activities in Varre-Sai and Italva

Number of households	Quadrants of income generation activities	Percentage of 200 interviewed households (%)
118	A	59
50	A,D	25
9	A,B,D	5
8	A,B	4
5	A,C	3
4	B	2
3	C	2
1	A,B,C	1
1	A,C,D	1
1	B,D	1

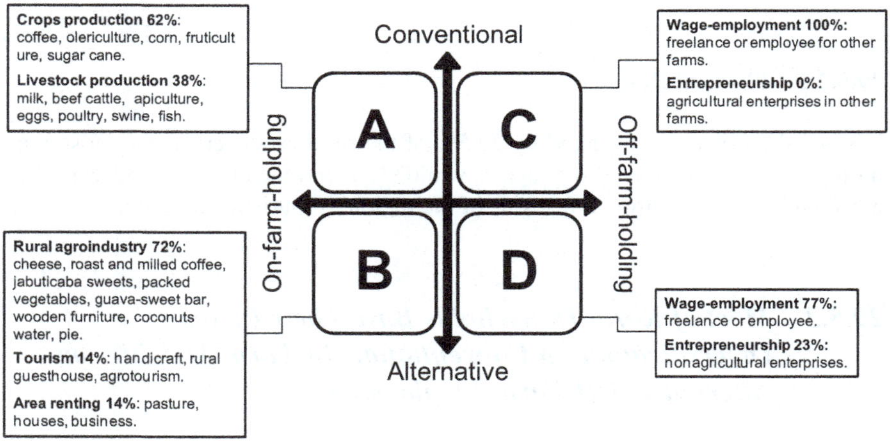

Fig. 29.5 Locational and sectoral classification of rural income generation activities in Varre-Sai and Italva (percentages per quadrant)

29.5.4 *Most of Alternative Income Diversification Activities On-Farm Are Not Part of the New Rurality Perspective*

As Table 29.2 shows, 13% of the interviewed households generate alternative income on-farm (Quadrant B). They diversify their income mostly through rural agroindustry and area renting for business, pasture, or living. Only 14% of households with activities in Quadrant B have economic activities in the tourism sector (Fig. 29.5).

Table 29.3 Locational and sectoral classification of planned rural income diversification strategies in the northwest region of Rio de Janeiro

Number of households	Quadrants of planned income diversification activities	Percentage of 200 interviewed households (%)
156	No plan	78
32	A	16
7	D	3.5
5	B	2.5
0	C	0

29.5.5 Most of the Rural Households Have No Plans to Diversify Their Income, and Those with Income Diversification Plans Follow Conventional On-Farm Strategies

As represented in Table 29.3, 78% of the rural producers did not have plans to diversify their income in the following years. Only 22% of the interviewed rural producers expressed their will to diversify their income, and 16% plan to diversify their income based on conventional on-farm activities (Quadrant A).

29.5.6 The Use of Innovation Resources to Generate Alternative Income Generation Strategies On-Farm Can Be Promoted Within the Framework of the New Rurality Perspective

Based on the results of the survey, an event was organized to motivate farmers to generate innovative entrepreneurship ideas for alternative income diversification while taking advantage of available innovation resources in their farms and the region at the same time.

A first workshop was conducted in the headquarters of RIO RURAL in order to identify and contact potential partners for engaging in the organization of a business idea contest. Delegates from RIO RURAL and PESAGRO-RIO took part in the workshop and helped to coordinate individual meetings with the identified potential partners. The meetings permitted to evaluate potential places for the event, to conform a team of experts for evaluating the competing ideas, and to find sponsors for the organization and awards for the winners.

Once the institutions cooperating in the organization of the event were defined, at least one representative of each institution (PESAGRO-RIO,[1] RIO RURAL,[2] SEBRAE,[3] IFF,[4] EMATER-RIO,[5] Embrapa Solos,[6] EMATER VARRE-SAI,[7] EMATER ITALVA,[8] Prefeitura de Varre-Sai[9]) took part on a second workshop to establish the essential assessing criteria for the competing ideas. Based on the joint analysis of the research results, economic subsectors where prioritized for developing strategies that could take advantage of the adopted sustainability practices promoted by the RIO RURAL program, as well as of their generated externalities. The active inclusion of young educated labor force in the business ideas contest was considered as crucial criteria.

A third workshop was accomplished in order to constitute the teams to compete at the business idea contest. The participants were rural producers from the region as well as students and graduates of the Federal Institute of Rio de Janeiro (IFF) in Bom Jesus do Itabapoana, the closest educational institution of technical studies in the region, which congregates much of the rural regional young labor force. The teams were constituted after a short presentation of the rural producers, where they explained their business ideas and/or described the kind of resources available in their farms. After interacting with the producers, the graduates and students in the public constituted and registered teams with at least one rural producer as member. Brazilian and German experts assessed the competing ideas according to the previously established criteria.

At the awards event, all teams presented their business ideas briefly to the public. Three winning teams were awarded according to the assessment of a team of experts, and one team was awarded according to the assessment of the public at the event. The winning business ideas correspond to the sectors of ecotourism, sustainable energy generation, consulting services on rural technologies, and rural tourism.

---

[1] Agricultural research company of the state of Rio de Janeiro: institution subsidizing public policies to make feasible technological solutions for the rural development of the state.

[2] Sustainable rural development program in hydrographic micro-watersheds: program financed by the World Bank under execution of the Ministry of Agriculture and Livestock of Rio de Janeiro – SEAPPA. The ultimate goal of the program is the improvement of the social, environmental, and economic conditions of the rural households of the region.

[3] Service of support to micro and small businesses in the state of Rio de Janeiro: It is a private nonprofit entity that aims to foster entrepreneurship and formalization of companies, using specific programs for developing the entrepreneurial culture and and stimulating the generation of employment and income.

[4] Federal Institute of Rio de Janeiro: Educational institution of technical and professional studies which congregates much of the young labor force in rural regions.

[5] Organization responsible for the technical assistance and rural extension in the state of Rio de Janeiro.

[6] It is the institution providing solutions for the sustainable development of Brazilian agribusiness, based on the study and rational use of soil heritage and its interactions with the environment.

[7] Organization responsible for the technical assistance and rural extension in Varre-Sai.

[8] Organization responsible for the technical assistance and rural extension in Italva.

[9] Municipal Government of Varre-Sai.

All winning teams were awarded with money prizes donated by the International SEPT Program of Leipzig University (Germany) for starting their projects, as well as with participation on entrepreneurship training programs and excursions to successful cases, sponsored by RIO RURAL.

29.6 Conclusions

The northwest region of RJ is rich in innovation resources, though most of the households depend on one only agricultural activity or diversify their income through conventional on-farm economic activities. Just a third of the researched households have alternative income generation activities, mostly through wage employment off-farm, i.e., not making use of their on-farm innovation resources. This path seems to be maintained in the future since only a quarter of the interviewed rural producers plan to diversify, mostly through conventional on-farm activities.

The study has shown that it is possible to promote the use of local innovation resources, including young educated labor force, within rural communities for generating alternative income diversification strategies on-farm. The positive results of the study endorse its replication in other rural municipalities and regions, where available innovation resources can be used within the framework of the new rurality perspective. Ignoring this opportunity would complicate the further identification of alternative income diversification opportunities that conciliate the sustainable use of natural resources with the improvement of the quality of life of rural communities.

From a theoretical point of view, this study might contribute to the further application of business study concepts in the context of agricultural smallholders.

Acknowledgments We would like to thank the German Ministry of Education and Research (BMBF) for supporting this work through the INTECRAL Project [FKZ 033L162F]. We would also like to thank the staff of the RIO RURAL program, PESAGRO-RIO, SEBRAE, IFF, EMATER-RIO, Embrapa Solos, EMATER VARRE-SAI, EMATER ITALVA, and the Prefeitura de Varre-Sai for all the support during the development of this study. Particularly, we would like to express our gratitude to Rosane Bendia Grazioli (EMATER VARRE-SAI), Marconi Resende (EMATER ITALVA), Elaine Cardoso Fidalgo (Embrapa Solos), and Daniel Coelho Ferreira (IFF – Bom Jesus do Itabapoana), for their unconditional and permanent support during our research and stay in Brazil.

References

Böhm-Bawerk EV (1890) Capital and interest. Macmillan and Co., London

Carter S (1999) Multiple business ownership in the farm sector: assessing the enterprise and employment contributions of farmers in Cambridgeshire. J Rural Stud 15(4):417–429

Crole-rees A, Consultants C (2000) Rural income sources and diversification : getting the assumptions right. Festschrift in honor of Professor Peter Rieder, pp 41–46

Démurger S, Fournier M, Yang W (2010) Rural households' decisions towards income diversification: evidence from a township in northern China. China Econ Rev 21:S32–S44. http://linkinghub.elsevier.com/retrieve/pii/S1043951X10000544. Accessed 18 Dec 2014

Graziano da Silva J, Eduardo del Grossi M (2001) Rural nonfarm employment and incomes in Brazil: patterns and evolution. World Dev 29(3):443–453. http://linkinghub.elsevier.com/retrieve/pii/S0305750X00001030

Hansson H, Ferguson R, Olofsson C et al (2013) Farmers' motives for diversifying their farm business – the influence of family. J Rural Stud 32:240–250. https://doi.org/10.1016/j.jrurstud.2013.07.002

Knickel K, van der Ploeg JD , Renting H (2003) Multifunktionalität der Landwirtschaft und des ländlichen Raumes: Welche Funktionen sind eigentlich gemeint und wie sind deren Einkommens- und Beschäftigungspotenziale einzuschätzen? In: Schriften der Gesellschaft für Wirtschafts- und Sozialwissenschaften des Landbaues e.V. pp 1–20

Loyns RM, Kraut M (1992) The family farm in the next decade: the positive role of part-time farming. Can J Agric Econ/Revue canadienne d'agroeconomie 40(4):591–604. http://doi.wiley.com/10.1111/j.1744-7976.1992.tb03721.x. Accessed Feb 2017

McNamara KT, Weiss C (2005) Farm household income and on- and off-farm diversification. J Agric Appl Econ 1(37):37–48

Nonaka I, Takeuchi H (1995) The knowledge-creating company: how Japanese companies create the dynamics of innovation. Oxford University Press, New York

OECD (2003) Note: Other gainful activities – pluriactivity and farm diversification in EU-27. In: Farm household income: issues and policy responses. OECD Publications Service, Brussels, pp 1–43. of the Note

Reardon T, Berdegué J, Escobar G (2001) Rural nonfarm employment and incomes in Latin America: overview and policy implications. World Dev 29(3):395–409

Rodríguez-Duarte A et al (2007) The endogenous relationship between innovation and diversification, and the impact of technological resources on the form of diversification. Res Policy 36(5):652–664

Romero Avendaño L (2010) Diseño de una agenda de extensión rural Latinoamericana para un desarrollo rural inclusivo: La extensión como parte de un sistema de innovación, providencia

Singh S, Bhowmick B (2015) An exploratory study for conceptualization of rural innovation in Indian context. Procedia Soc Behav Sci 207:807–815. http://www.sciencedirect.com/science/article/pii/S1877042815053045

Sundbo J (2001) The strategic management of innovation "a sociological and economic theory". Edward Elgar Publishing Limited, Cheltenham

Torres Zárate OA, Dornberger U (2017) Income diversification in rural regions: an integrative locational and sectoral approach, Leipzig

UNCED (1992) United Nations Conference on Environment & Development. Agenda 21, Rio de Janeiro, RJ, Brazil, 3.-14. of June 1992: 351

Chapter 30
Regional Marketing Strategies for Sustainable Production in Rio de Janeiro State

Laura Maria Noriega Gamarra, Claudia Raedig, and Sabine Schlüter

Abstract Environmental protection that enhances the resilience of rural landscapes depends ultimately on farmers' decisions. Enforcing environmental law is essential to maintain the ecological functionality within watersheds but represents high costs particularly for family farmers in rural Rio de Janeiro state, who are embedded in a highly competitive rural-urban context. The RIO RURAL program simultaneously addresses environmental sustainability at the micro-watershed level and competitive employment in the family farming sector. The program seeks incentives for more sustainable land use for family farmers. The INTECRAL project has identified pathways for incentives giving a marketable value to the environmental achievements of more sustainable landscapes provided by small-scale farmers. Several international trends for marketing eco-friendly farm products were identified and reviewed in terms of suitability for the local conditions of production, markets, and institutional environments. These trends refer to payment for ecosystem services (PES) for eco-friendly traditional farming as well as regional sustainability labels, including organic farming and carbon mitigation initiatives to appease consumers' conscience. The willingness to participate was identified for family farmers. RIO RURAL program plays a central role to introduce PES in potential marketing strategies and to facilitate label introduction based on already implemented control mechanisms at low costs for producers.

Keywords Carbon market · Organic farming · Payments for ecosystem services · Regional branding · Regional labeling

L. M. Noriega Gamarra (✉) · C. Raedig · S. Schlüter
TH Köln – University of Applied Sciences, Institute for Technology and Resources Management in the Tropics and Subtropics (ITT), Köln, Germany
e-mail: launoga@gmail.com; claudia.raedig@th-koeln.de; sabine.schlueter@th-koeln.de

U. Nehren et al. (eds.), *Strategies and Tools for a Sustainable Rural Rio de Janeiro*, Springer Series on Environmental Management,
https://doi.org/10.1007/978-3-319-89644-1_30

30.1 Introduction

National agricultural development policies around the world are currently focused on sustainable production balancing ecological, economic, and social requirements of agricultural production rather than only increasing productivity (Welch and Graham 1999; Sorrensen 2009; Newton et al. 2016). A further global trend is the development toward medium- and large-scale farming systems, which makes it more and more difficult for small farmers to produce at a competitive price (Hazell et al. 2006; Maessen et al. 2008).

In Brazil, the typical farming system corresponds to family agriculture in which households produce food for their own subsistence needs and to support local and regional markets (FAO 2002). Although the Brazilian government has attempted to develop the rural agriculture through the establishment of boards such as the National Family Farm Strengthening Program (PRONAF) and Rural Advisory Services (RAS), family farm systems are often struggling today to compete with medium- and large-scale farmers. Awareness of product quality is only increasing in large cities, and the demand for organic and regional products in supermarkets is growing (Campanhola and Valarini 2001; Dalcin et al. 2014). These products are sold with a higher price than nonorganic products in the cities, while in rural areas, there is no price difference visible.

The RIO RURAL program fosters the empowering of family farmers promoting their social and productive inclusion in sustainable rural development (see Hissa et al. 2018). RIO RURAL program's main goal is to raise agricultural productivity while promoting biodiversity conservation and reversing land degradation in critical areas of the Atlantic Forest biome in Rio de Janeiro state (RJ) (FAO 2013). Within the RIO RURAL program approach, small farmers are increasingly adopting sustainable natural resources management and conservation practices aiming the integration of agricultural and nonagricultural systems (FAO 2013). The major challenge is to put eco-friendly agricultural practices on an economically independent basis as a prerequisite for sustainability so that the success of the practices adopted can be established at the state level.

In many areas of RJ, the advantages of land use change have been widely accepted. Farmers' reports in Nova Friburgo in the *Serrana* region have demonstrated understanding with respect to protecting springs and water bodies, as well as forest regrowth on recharge areas and its contribution to higher sustainability of land use. The advantages of forest conservation are not as obvious in other areas of the state as in that specific region, in which environmental conservation in the vicinity of protected areas has been a more obvious target and has also contributed to touristic attractiveness. In order to comply with the most recent version of the Forest Law (see Piratelli et al. 2018), especially family farmers have to be supported in the adaptation of eco-friendly practices. Particularly in such cases when direct costs related to the implementation of the environmental law exist, economic incentives can facilitate the adaptation of eco-friendly agricultural practice.

Giving a marketable value to the environmental achievements in the form of incentives represents a strategy to promote sustainable land use. Currently, more regional products and rural tourism programs are accessing the market, and the region is seen as an example of integration and identity to rural development (Leader+ 2008). These marketable alternatives must seek a cost-benefit balance and a diversification of the income within the family agriculture model including conservation, greenhouse gas (GHG) emission reduction, and incentives for good agricultural practices. The objective of this research is to identify sustainable marketing concepts for eco-friendly family farm products, taking into account that a marketing strategy can constitute a set of collective actions led by local actors pursuing sustainable development and considering local production, markets, and institutional environments.

30.2 Pathways Analyzed

The concepts presented are based on literature review and the results of the INTECRAL research cooperation between Brazil and Germany (see Schlüter et al. 2018). The project catalyzed knowledge exchange among experts from both countries.

30.2.1 Certified Organic Farming

According to Badgley et al. (2006), the term "organic farming" refers to practices considered agroecological, sustainable, or ecological. These practices are based on natural processes (e.g., cover crops, manures, compost, crop rotation, intercropping, biological pest control) and intend to exclude the use of synthetic pesticides. Organic methods can contribute substantially to a more sustainable system of food production by feeding the current and future human populations while maintaining soil fertility (Badgley et al. 2006). Organic agriculture is generally considered to reduce external input costs due to the abatement in the use of pesticides and fertilizers; however, it requires the creation of cooperatives in order to ensure technical, marketing, and certification support to gain entry to the organic market (Campanhola and Valarini 2001; Oelofse et al. 2010).

Organic products can be certified with a guarantee label to provide the consumers with credibility, quality, and transparency regarding the practices and the principles used in the production (Campanhola and Valarini 2001). The certification is based on different standards and can be granted by different institutions; however, regulatory standards and certification models have made organic foods less accessible to small-scale producers and lower-income consumers (Sacchi et al. 2015). The costs associated with third-party certification can be a problem for small-scale farmers. For many family producers, the expected premium price for certified organic products does not compensate for the higher transaction costs for market participation and the losses faced during the transition phase from conventional to organic agriculture, usually characterized by productivity losses and excessive labor supply during the learning phase. Therefore, small producers have begun to rely on alternative quality and credibility assurance systems with strategies based on direct marketing that links producers and consumers demands without third-party intervention to differentiate their organic products (Sacchi et al. 2015). To date, there are two main systems: internal control systems (ICS) and participatory guarantee systems (PGS). While ICS include an external certification body, PGS are based on active participation of the stakeholders, need minimal paper work, and do not require a third-party certifier (IFOAM 2008). Both systems have significantly increased the potential of market access for small-scale producers but still represent a significant risk for small farmers.

Brazilian *Rede Ecovida de Agroecologia* represents one of the oldest labeling programs implemented in Brazil and led to the official recognition of PGS within national legislation (Sacchi et al. 2015; Rover et al. 2017). Brazilian legislation recognized two formal (1,2) and one informal system (3) for organic quality: (1) third-party certification, subject to the conformity assessment bodies (*Organismos de Avaliaçã o da Conformidade* or OAC); (2) participatory guarantee systems, subject to the participatory bodies for conformity assessment (*Organismos Participativos de Avaliaç ã o da Conformidade* or OPAC); and (3) organizations for social control (*Organizaço de Controle Social*) (Sacchi et al. 2015; Rover, et al. 2017).

The line of development for the organic market has been supported by many sides in RJ. For instance, the Brazilian Micro and Small Business Support Service (SEBRAE), in partnership with the RIO RURAL program, support family farmers in the conversion from traditional to organic agriculture and in all other aspects of farm administration (Oelofse et al. 2010). Still, the market share of organic products is a niche, characterized by high transaction costs due to a highly fragmented supply and demand structure. Costs for certification, even when reduced due to a group certification under a PGS system, pose a further challenge to family farmers. In addition, access to the market requires securing sufficient supply with products of an expected quality, which family farmers currently cannot guarantee (Blanc and Kledal 2012). Therefore, the RIO RURAL program finds itself at an impasse to open market channels for their beneficiary farmers.

In terms of environmental conservation, organic products, especially horticulture products, have very high positive impacts in relation to the conventional production systems. For beef and dairy cattle production systems, which are in terms of land cover the most area-demanding systems of the region, this is different (see Fischer et al. 2018). Since many aspects of good agricultural practices are lacking in the implemented systems (see Espinel et al. 2018), it is comparatively difficult to convert these into organic systems. Further, the value chain of organic meat or milk production would have to be completely transferred to an organic chain of custody. The bottleneck in the chain of custody for milk production is the dairy company, which has to take over coordinative and control functions and which has to implement organic standards itself. The required degree of integration cannot yet be achieved in the highly fragmented farm structures of the target group of the RIO RURAL program.

30.2.2 Payment for Ecosystem Services

The market-based concept of PES aims to change behaviors that are harming the environment by creating financial incentives for conservation purposes (Gómez-Baggethun et al. 2010; Garbach et al. 2014). In practice, PES incentivizes landowners to maintain and safeguard these services by adopting sustainable resource use and land management practices to minimize the loss of ecosystem services or to even improve the ecosystem service provision to eventually sustain human well-being (Wunder 2005; Pagiola and Platais 2007; FAO 2011; Ferraro 2011). The potential defenders of ecosystem services are referred to as sellers, whereas the prospective clients for the ecosystem services are seen as buyers (Wunder 2005).

Traditionally, the Northwestern region is the center of RJ's coffee production. One research focus of the INTECRAL project was to explore the willingness of coffee producers to change their agricultural practices to more eco-friendly ones. Based on 26 coffee producers interviewed in Varre-Sai and Porciúncula, 88.5% indicated that they were willing to change their usual agricultural practices. For the traditional center of horticultural production, the *Serrana* region, also the willingness to participate in a PES scheme was identified as 88%. This was based on 24 persons who were interviewed of which 42% indicated they were willing to change to more biodiversity-friendly agricultural practice even without financial incentives. Both studies revealed a lack of information concerning the environmental impacts of the different land-use practices as well as uncertainty on the magnitude of incentives necessary for the ecosystem sellers to change their practices.

Since the practices used by both groups of producers (coffee and horticulture) are different, a PES scheme that is introduced on a larger scale should have a modular structure, therefore, not only protecting different ecosystem services (with specific seller-buyer relationships for each ecosystem service) but also motivating a larger number of landowners to change to more eco-friendly behaviors.

In a further study carried out in the municipality of Varre-Sai, the willingness of landowners to protect parts of their lands covered with forest was identified in 53% of 18 persons interviewed. Landowners often feel a strong responsibility to safeguard forest land in their possession for the next generations. As a possible solution, in addition to the 20% protection requirement of the entire forest area of each private property in the Atlantic Forest biome (Legal Reserves, see Piratelli et al. 2018), the establishment of private protected areas for such forest stands exist. After a complex procedure, a forest area can be transformed to a protected area with an official permanent legal status known as "*Reservas Particulares do Patrimônio Natural*" – RPPNs (see Guzmán Wolfhard and Raedig 2018). This procedure involves considerable effort from the landowners' side and can take up to 5 years. In RJ the municipality of Varre-Sai has taken over the role model, dedicating a part of the municipality's tax income from ICMS (a tax on transactions on the circulation of goods and service delivery of interstate and inter-municipal transportation and communication) to support the landowners who want to establish such RPPNs. Several municipalities in the state are considering adopting the Varre-Sai model, and the RIO RURAL program could act as intermediary in an advisory capacity among them.

30.2.3 Carbon Credit Market

In order to support the global efforts to cope with climate change, the United Nations Framework Convention on Climate Change made carbon a tradable commodity within the emission trading market, known as the "carbon market," in which carbon credit units are measured in CO_2e (e = equivalent). In order to achieve national emissions targets, those who reduce, remove, or sequester carbon receive payment, and those who produce emissions can buy carbon credits to compensate them.

The carbon market can be divided into two categories: the regulatory and voluntary market. Agriculture has been basically excluded from both carbon markets, mainly because of the difficulty in the assessment of carbon emissions and reductions related to agricultural activities as well as the associated transaction costs for small farmers (de Pinto et al. 2010). Transaction costs in afforestation or reforestation projects also represent an important implementation barrier. The inclusion of such projects in the carbon market entails significant costs associated with project formulation, expert trainings in geoprocessing activities, carbon pool measurements, and monitoring (Nehren et al. 2012).

The GHG mitigation potential in agriculture activities comes from soil carbon sequestration, particularly through cropland management, grazing land management, and restoration of cultivated organic soils and of degraded lands (Smith et al. 2008). In rural RJ, the higher mitigation potential is associated with the protection of springs and streams and other activities supporting legal reserves (see Rodríguez Osuna et al. 2018). These practices are estimated to sequester 521,468 tCO_2e over 20 years, of which 4302 tCO_2e correspond to avoid deforestation of tropical rainforest and 517,166 tCO_2e to forest regeneration (Branca et al. 2013).

For the identification of the mitigation impact and quantification of the market potential, the project owner has to identify a potential buyer for the credits, carry out a feasibility study and a scientifically valid assessment of carbon reduction potentials based on an accurate baseline scenario in which the GHG emissions are defined in absence of the project. The whole process implies fulfilling strict requirements and applying methodologies of the regulatory or voluntary market standards, which implies higher transaction costs for the RIO RURAL program. According to Branca et al. (2013), considering an average carbon price of US\$3/tCO_2e, the activities promoted by the RIO RURAL program have an average mitigation potential of 0.19 tCO_2e/ha/yr., i.e., US\$ 0.57 per tCO_2e/ha/yr., which is insufficient to support transaction costs associated with carbon crediting. In addition, a 10-year carbon project implemented for activities related to protection of springs and streams, and the establishment of legal reserves would require an area of more than 150 ha in order to be economically feasible. For an estimation of the feasibility of the different activities by the RIO RURAL program in terms of net present value, please use the toolkit provided on the INTECRAL project website (see publications – handbooks, and tools; http://intecral-project.web.th-koeln.de).

30.2.4 Regional Branding and Labeling

In the context of world economy, "territorial competition" is gaining importance and becoming a trend (Hospers 2004). Many sellers offer different brands of the same products and compete for investment, tourism, labor, and political power, making local branding an opportunity to capture market shares (van Ham 2008). This region-focused strategy helps small producers to add value to their portfolio and to increase competitiveness by creating a more distinctive image for the region and by providing a place of connection for the consumer, thus creating brand recognition (van Ittersum 2001; Maessen et al. 2008; van Ham 2008). Regional branding also leads to an increase in the demand of local production, creates jobs, enhances the relationship between farmers, as well as develops rural economies. Cooperation and networks among the different stakeholders are very important for the success of a regional branding approach, which requires meeting the needs of the region according to their specific physical, social, cultural, institutional, and historical context (Messely et al. 2010).

Regional branding with low certification fees and also simpler certification procedures compared to, e.g., organic products, also implies environmental advantages: regional products are more efficient, transparent, have less logistical costs, and low CO_2 emissions (Kullmann and Wolfgang 2007; Kraus et al. 2014).

The first step for the creation of a regional brand requires increasing visibility by creating awareness and identifying endogenous development potential as cultural markets that include food, languages, crafts, folklore, historical sites, landscapes, and associated biodiversity among others (Ray 2001; Domínguez et al. 2013). As a second step, the development of products and services controlled by local people requires identification of resources besides institutional, organizational, or market innovations (Domínguez et al. 2013). Finally, regional branding requires an organizational structure that involves communities in a multi-stakeholder process supported by facilitating policy frames and strategies (Domínguez et al. 2013).

A renowned historical example from France is the top-down establishment of a classification system for the best Bordeaux wines of France in 1855 by Napoleon III that ranked these wines according to the respective vineyard's reputation and the usual trading price, which at that time was directly related to quality (Maessen et al. 2008).

In Germany, regional brands are often developed by associations formed by participants with economic, ecological, and social interest (Kullmann and Wolfgang 2007). The common organizational structure of regional branding organizations in Germany is a dual system consisting of a nonprofit organization and an economic entity. The first organization (in form of an association) is responsible for developing the framework, setting up goals, as well as raising awareness for the brand. The latter is in charge of production chains including logistic and everyday business, which can be in a form of cooperative or company. This company will be responsible for hiring an independent audit to control and monitor the products' quality. Such independent audits done by a third-party are an integral part for the trust building of prospective buyers of the branded products, since they ensure an independent way of measuring the quality of the labeled products.

One example of a successful regional label in Germany is the "Regionalmarke Eifel." The Eifel is a hilly area in western Germany, which has been developed as a destination for tourists interested in landscape beauty and for outdoors activities. "Regionalmarke Eifel" was among the first labels in Germany to include a national park (Eifel National Park, in 2015 designated as UNESCO Global Geo-park) in its regional branding scheme. This regional branding scheme consists of a dual system: (i) a local association for the development of the brand with representatives from various sectors, industry, agriculture, school, tourism, and natural parks, and (ii) Regionalmarke Eifel Ltd. (GmbH), which is a company responsible for assembling the brand producers and services, certifying products, and controlling the quality. To use this label, farmers must comply with the production criteria set by Regionalmarke Eifel. Nowadays, Regionalmarke Eifel products are distributed all around the region and sold in several large food stores. In addition, consumers identify themselves with the label principles, have a positive image of the Eifel, and visit the area to buy the regional products, promoting tourism at the same time (ICLEI 2008).

The landscape of RJ is a unique matrix of agricultural and natural areas with good prospects for the introduction of regional branding. Italva, for example, is working on the improvement of milk quality by making efforts to shorten the chain of custody. The cooperative of family agriculture and solidarity economy in Italva – Copafi – entered into the local market by selling their own milk and cheese directly in bakeries and small shops. The path between producers and final consumers has been shortened, leading to higher profits, and the Copafi brand is picking up in the local market due to the awareness and support of the community. Moreover, the RIO RURAL program interest is to support such small-scale region-focused initiatives, as well as to bring together different small producers in the state to establish a regional brand for rural RJ.

30.3 Discussion

Whereas organic farming as well as carbon mitigation yield environmental achievements but are economically not attractive for family farmers to date, there is a good reason to introduce regional branding in rural RJ.

First of all, this region can draw on impressive landscape beauty and has several biological reserves and protected areas particularly in its southern part. These assets can be used to develop a regional identity with which local producers as well as consumers feel a connectedness as in the Regionalmarke Eifel case. The RIO RURAL program represents an ideal starting point to establish an association between rural farmers and their cooperatives, local businesses, NGOs, and municipality networks for a regional label conception.

Although the introduction of a regional label for sustainable products in RJ does not require a legal framework, both environmental benefits and integrity of production need to be guaranteed to prospective consumers. In order to establish credibility of the regional brand, transparency and control of the production processes need to be included. Based on the German experience with Regionalmarke Eifel, the low-cost mutual control mechanism applied for the brand proved successful. For RJ's family farming systems such a PGS would be suitable, preferably with a modular structure to allow the distinction of different quality levels of products, including both conventional farmers with high-quality regional products as well as farmers with organic products.

Furthermore, the observed trend regarding the willingness to adopt a PES scheme by changing toward more biodiversity-friendly practices and forest protection indicated that a regional labeling could be complemented by the introduction of a modular PES scheme addressing different main land uses across RJ: in the northwestern region, PES for coffee producers and cattle farmers and the facilitation for the establishment of RPPNs could be coupled, whereas in the southern region, PES for horticulturists would contribute to a more sustainable family farming-oriented agricultural landscape of the entire state.

Due to the high transaction costs for small farmers to access the carbon market and the high area requirements for a carbon project feasibility associated with the activities promoted by the RIO RURAL program, carbon crediting holds little potential for implementation; however, the identification and communication of carbon emission (e.g., product carbon footprint) could add value to products when included in the marketing of the regional brand targeting climate change awareness. The same applies for organic production where including organic farming practices into the umbrella of a regional labeling concept will bring products to the market at higher prices in relation to the business as usual prices, minimizing transaction costs at the same time.

Factors that should be included in the development of a regional brand in general are: the size of the region, production criteria, the involvement of local business, political support at least on a local scale, marketing of regional quality, and the raising of awareness of consumers for regional products (Kullmann and Wolfgang 2007). The creation of a regional brand for RJ would further have to address the following challenges:

- Awareness promotion of the environmental benefits achieved by the regional branding: More eco-friendly agricultural practices will secure ecosystem functioning and thus ecosystem services in the long run and allow for a diverse, family farm-dominated agricultural landscape, as opposed to a uniform landscape dominated by medium and large-scale farm companies.
- Identification and development of potential products and services: Innovative products and services, e.g., ecological and agriculture tourism, constitute an alternative for rural development and contribute to the diversification of the local income. Whereas Rio de Janeiro city is already internationally known and benefits from tourism, the rural state with its scenic beauty holds high touristic potential.

- Integration of local actors and bringing together producers and consumers: The success of the regional brand and the product quality assurance requires the development of a community spirit and all parties' identification with the brand concept.
- Definition of a dual organizational structure: In order to build trust on the regional brand and to promote transparency to consumers, one entity should be responsible for creating the framework, branding objectives, and establishing production criteria. The other entity should control and monitor the product and service quality.

30.4 Conclusion

A marketing strategy for sustainable production in RJ has to assure a balance in the development of regional products and services (endogeneity) and the integration of stakeholders by promoting sustainable practices under a joint umbrella vision. Based on the different pathways analyzed, regional branding in combination with payment for ecosystem services and promotion of organic farming constitutes a potential marketing strategy to support sustainable development of rural RJ.

The integration of the different stakeholders and communities' interests around a regional brand would guarantee the development of new markets, thus adding value to local products and the landscape in order to enhance the conservation of vulnerable natural resources. The RIO RURAL program serves as the ideal primer toward this goal, since it has already established contacts with all necessary parties and can bring all stakeholders together. In addition, the program offers a unique opportunity to implement a recognizable label derived from their actual project label, which is already widely recognized in the rural environment. Since the RIO RURAL program has established participative project management procedures required for setting reasonable standards, the program can provide the necessary assistance to farmers in deciding on the quality and management standards. One important component of the marketing strategy would be to consolidate the organizational structure of the regional brand as long as the RIO RURAL program is ongoing as well as to promote the awareness of local branding benefits in the different stakeholder groups. The introduction of a regional sustainability label can act as a stepping stone toward a more complex certification scheme, which under current conditions is not feasible for family farmers.

In parallel to establishing the regional brand in rural RJ, the RIO RURAL program can incorporate the experiences gained into policy frameworks for promoting regional labeling strategies as a tool for sustainable development.

Acknowledgments The authors would like to thank the master student Maliya Phommasone for the information compiled in her semester project report and included in the analysis of local branding. Special thanks to Maria Luisa Espinel, Lorena Guzmán, Daniel Perozo, Pedro Rios, Edson Rodriguez, and Julia Suβmann for their contribution to the development of this chapter through their master thesis research performed within the BMBF-funded INTECRAL project.

References

Badgley C, Moghtader J, Quintero E et al (2006) Organic agriculture and the global food supply. Renewable Agric Food Syst 22(2):86–108. https://doi.org/10.1017/S1742170507001640

Blanc J, Kledal P (2012) The Brazilian organic food sector: prospects and constraints of facilitating the inclusion of smallholders. J Rural Stud 28(1):142–154. https://doi.org/10.1016/j.jrurstud.2011.10.005

Branca G, Hissa H, Benez MC et al (2013) Capturing synergies between rural development and agricultural mitigation in Brazil. Land Use Policy 30(1):507–518. https://doi.org/10.1016/j.landusepol.2012.04.021

Campanhola C, Valarini P (2001) A agricultura orgânica e seu potencial para o pequeno agricultor. Cadernos de Ciencia e Tecnologia 18(3):69–101

Dalcin D, Leal de Souza AR, de Freitas JB et al (2014) Organic products in Brazil. Br Food J 116(12):1998–2015. https://doi.org/10.1108/BFJ-01-2013-0008

Domínguez MD, Horlings L, Swagemakers S, Fernandez XS (2013) Place branding and endogenous rural development. Departure points for developing an inner brand of the River Minho estuary. Place Branding Public Diplomacy 9(2):124–140. https://doi.org/10.1057/pb.2013.10

Espinel ML, Schlüter S, Resende CM (2018) Towards good agricultural practices (GAP) in smallholders' dairy production systems from an animal welfare perspective. In: Nehren U, Schlüter S, Raedig C, Sattler D, Hissa H (eds) Strategies and tools for a sustainable rural Rio de Janeiro. Springer International Publishing, Cham

FAO (2002) Non-certified organic agriculture in developing countries. In: Scialabba N, Hattam N (eds) Organic agriculture, environment and food security, 4th edn. Sustainable Development Department, Rome

FAO (2011) Payments for ecosystem services and food security. FAO, Rome, p 281

FAO (2013) Financial sustainability for environmental services: rural development in micro watersheds Rio rural, Brazil. Case studies on remuneration of positive externalities, p 8

Ferraro PJ (2011) The future of payments for environmental services. Conserv Biol 25(6):1134–1138. https://doi.org/10.1111/j.1523-1739.2011.01791.x

Fischer SB, Pedraza Luengas A, Schlüter S, Oliveira Antunes LA (2018) From design to implementation: a participatory appraisal for silvopastoral systems. In: Schlüter S, Raedig C, Sattler D, Hissa H (eds) Nehren U, Strategies and tools for a sustainable rural Rio de Janeiro. Springer International Publishing, Cham

Garbach K, Milder JC, Montenegro M et al (2014) Biodiversity and ecosystem services in agroecosystems. In: Biodiversity and Ecosystem Services in Agroecosystems Encyclopedia of Agriculture and Food Systems, vol 2, pp 21–40. https://doi.org/10.1016/B978-0-444-52512-3.00013-9

Gómez-Baggethun E, de Groot R, Lomas PL, Montes C (2010) The history of ecosystem services in economic theory and practice: from early notions to markets and payment schemes. Ecol Econ 69(6):1209–1218. https://doi.org/10.1016/j.ecolecon.2009.11.007

Guzmán Wolfhard L, Raedig C (2018) Connectivity conservation management: connecting private protected areas. In: Schlüter S, Raedig C, Sattler D, Hissa H (eds) Nehren U, Strategies and tools for a sustainable rural Rio de Janeiro. Springer International Publishing, Cham

van Ham P (2008) Place branding: the state of the art. Ann Am Acad Pol Soc Sci 116(1):126–149. https://doi.org/10.1177/0002716207312274

Hazell P, Poulton C, Wiggins S, Dorward A (2006) The future of small farms: synthesis paper. Word Bank, November 2006

Hissa H, Teixeira N, Costa M et al (2018) Sustainable rural development in Rio de Janeiro state: the Rio rural program. In: Nehren U, Schlüter S, Raedig C, Sattler D, Hissa H (eds) Strategies and tools for a sustainable rural Rio de Janeiro. Springer International Publishing, Cham

Hospers G (2004) Place Marketing in Europe: the branding of the Oresund region. Intereconomics 39(5):271–279. https://doi.org/10.1007/bf03031785
ICLEI (2008) Eifel initiative for the future, Germany: urban-rural linkages enhancing European territorial competitiveness – mini case study on business clusters. Regional Policy - Northumbria University, p 1
IFOAM (2008) PGS guidelines: how participatory guarantee systems can develop and function, Germany, p 36
van Ittersum K (2001) The role of region of origin in consumer decision making and choice. Wageningen University, Netherlands, 10 Dec 2001
Kraus F, Cornelius M, Job H (2014) Biosphere reserves and their contribution to sustainable development: a value-chain analysis in the Rhön biosphere reserve, Germany. Zeitschrift für Wirtschaftsgeographie 58(1):164–180
Kullmann A, Wolfgang J (2007) Sustainable regional food marketing in German biosphere reserves and other model regions – from grassroots projects to professional organisations, p 9
Leader+ (2008) Leader+ magazine: leader achievements; a diversity of territorial experience. European Communities
Maessen R, Wilms G, Jones-Walters L (2008) Branding our landscapes: some practical experiences from the LIFESCAPE project. Paper presented at the 8th European ISFA Symposium, Clermont-Ferrand, 6–10 July 2008
Messely L, Dessei J, Lauwers L (2010) Regional identity in rural development: three case studies of regional branding. Appl Stud Agribusiness Comm 4(3–4):19–24
Nehren U, Loaiza T, Alfonso S et al (2012) Potentials for reforestation and forest conservation under CDM and REDD+ in the Atlantic Forest of Rio de Janeiro, Brazil. Geographies Inequal Lat Am 123(1):133–158
Newton P, Gomez A, Jung S et al (2016) Overcoming barriers to low carbon agriculture and forest restoration in Brazil: the rural Sustentável project. World Dev Perspect 4(1):5–7
Oelofse M, Høgh-Jensen H, Abreu LS et al (2010) Certified organic agriculture in China and Brazil: market accessibility and outcomes following adoption. Ecol Econ J 69(9):1785–1793. https://doi.org/10.1016/j.ecolecon.2010.04.016
Pagiola S, Platais G (2007) Payments for environmental services: from theory to practice. World Bank, Washington, 2007
de Pinto A, Mangalhaes M, Ringler C (2010) Potential of carbon markets for small farmers. Food Policy Research Institute, 2010
Piratelli AJ, Marquez Piña-Rodrigues FC, Raedig C (2018) Integrating biodiversity conservation into agroecosystem management: using birds to bring conservation and agricultural production together. In: Nehren U, Schlüter S, Raedig C, Sattler D, Hissa H (eds) Strategies and tools for a sustainable rural Rio de Janeiro, Springer International Publishing, Cham
Ray C (2001) Culture economies: a perspective on local rural development in Europe, Newcastle
Rodríguez Osuna V, May PH, Monteiro JMG et al (2018) Promoting sustainable agriculture, boosting productivity and enhancing climate mitigation and adaptation through the RIO RURAL program, Brazil. In: Nehren U, Schlüter S, Raedig C, Sattler D, Hissa H (eds) Strategies and tools for a sustainable rural Rio de Janeiro, Springer International Publishing, Cham
Rover O, Corrado de Gennaro B, Roselli L (2017) Social innovation and sustainable rural development: the case of a Brazilian agroecology network. Sustainability 9(1):1–14. https://doi.org/10.3390/su9010003
Sacchi G, Caputo V, Nayga RM (2015) Alternative labeling programs and purchasing behavior toward organic foods: the case of the participatory guarantee systems in Brazil. Sustainability 7(6):7397–7416. https://doi.org/10.3390/su7067397
Schlüter S, Nehren U, Sattler D, Raedig C (2018) The INTECRAL project. In: Nehren U, Schlüter S, Raedig C, Sattler D, Hissa H (eds) Strategies and tools for a sustainable rural Rio de Janeiro. Springer International Publishing, Cham
Smith P, Martino Z, Cai D et al (2008) Greenhouse gas mitigation in agriculture. Philos Trans R Soc Lond Ser B Biol Sci 363(1492):789–813. https://doi.org/10.1098/rstb.2007.2184

Sorrensen C (2009) Potential hazards of land policy: conservation, rural development and fire use in the Brazilian Amazon. Land Use Policy 26(3):782–791. https://doi.org/10.1016/j.landusepol.2008.10.007

Welch R, Graham R (1999) A new paradigm for world agriculture: productive, sustainable and nutritious food systems to meet human needs. Field Crop Res 60(1):1–10

Wunder S (2005) Payments for environmental services: some nuts and bolts. Occasional Paper No 42, Center for International Forestry Research, Jakarta 2005

Part VI
Conclusions

Chapter 31
Rural Rio de Janeiro: Over the Hills and Far Away?

Claudia Raedig, Helga Hissa, Sabine Schlüter, Dietmar Sattler, and Udo Nehren

31.1 An Old Story and a New Challenge

The urban-rural landscape between the three cities of Rio de Janeiro, São Paulo, and Belo Horizonte is the most densely populated part of the entire Brazil and its economic backbone. For the state of Rio de Janeiro (RJ), many rural areas are characterized by declining and aging population, while the urban areas still increase in population (George and McGranahan 2010; Gragnolati et al. 2011). Greater economic opportunities and other pull factors of urban areas draw rural migration toward the cities where they cause higher demands for food, water, and energy resources, which are met by rural landscapes' decreasing agricultural production. This development puts additional pressure on the natural resources, which already suffered from historical overexploitation (Chap. 2).

Overexploitation of natural resources in RJ is mirrored by developments, such as forest fragmentation, deforestation, soil erosion, and water pollution. But in the wake of these visible indicators, there are more consequences, such as biodiversity loss or forms of soil and water degradation that are less obvious but lead to long-term

C. Raedig (✉) · S. Schlüter · U. Nehren
TH Köln – University of Applied Sciences, Institute for Technology and Resources Management in the Tropics and Subtropics (ITT), Köln, Germany
e-mail: claudia.raedig@th-koeln.de; sabine.schlueter@th-koeln.de; udo.nehren@th-koeln.de

H. Hissa
Secretaria de Agricultura e Pecuária do Estado do Rio de Janeiro/Superintendência de Desenvolvimento Sustentável/Programa Rio Rural, Rio de Janeiro, Brazil
e-mail: helga.hissa@globo.com

D. Sattler
Leipzig University, Institute of Geography, Physical Geography and Environmental Research Johannisallee, Leipzig, Germany
e-mail: sattler@uni-leipzig.de

U. Nehren et al. (eds.), *Strategies and Tools for a Sustainable Rural Rio de Janeiro*, Springer Series on Environmental Management,
https://doi.org/10.1007/978-3-319-89644-1_31

deterioration of the ecosystem. Such hardly visible changes can affect the ecosystem dramatically, if they are connected to ecosystem functioning. Once disrupted, the ecosystem becomes more vulnerable to external pressures, such as climate extremes. Additionally, the provision of important ecosystem services is further reduced. The colonial exploitation cycles that had started with the extraction of pau brasil (*Paubrasilia echinata* Lam.) in the early sixteenth century and were followed by sugarcane cultivation, gold routes, and coffee cultivation had tremendous impacts on the landscape system, as they changed the former forest landscape into a highly fragmented agricultural landscape with scattered forest fragments (Chap. 20). Only in areas with land use limitations, which is particularly the case for the steep central parts of Serra do Mar mountain range, the development of agriculture and human settlements did not proceed. Consequently, larger forest areas are still found here today (Chap. 2). In contrast, particularly in the northwest and north Fluminense mesoregions, forest fragmentation is severe. In some areas, deforestation is the more appropriate term with erosion-battered hills converted into pastures clearly dominating the landscape. These hills frequently show large fractions of bare soil and numerous cattle tracks. The hills have already suffered from overstocking even though they barely feed just a few animals.

In the light of this development, unprecedented extreme climatic events put an additional challenge to this rural landscape. In recent years, extreme climatic events, such as the 2014–2017 Brazilian drought which strongly affected the Southeast of the country, showed the high vulnerability of the social-ecological system, which might even increase under the projected climate scenarios (Chaps. 20, 21, and 22).

31.2 The Setting for INTECRAL Project and RIO RURAL Program

For the reduction of resources exploitation, environmental legislation is imperative. To date, Brazil hosts one of the most complex environmental legal frameworks worldwide with the Forest Law being its best known pillar (Chap. 11). This law regulates forest protection on private lands and specifies that 20% of the forested property in the Atlantic Forest has to be protected. In addition, source areas, areas directly adjacent to water bodies, and hilltops on private properties are required as protected areas (Chap. 10). Whereas the Forest Law underwent considerable changes in the last version in 2012 and has been widely criticized for the weakening of the more strict preceding version, this development has to be seen in parallel to the establishment of the CAR (Rural Environmental Register), which is by now finished for RJ, and allows for the first time to control whether legal environmental requirements are met systematically (Chap. 10). Furthermore, by registering private properties, CAR lays the basis for future compensation of landowners as well as municipal governments if they change to more ecosystem-friendly land uses. However, successful application of the environmental legislation is dependent on the effectiveness of monitoring the implementation of environmental restoration plans. The legal framework will only take effect if the environmental authorities carry out legal examinations and sanction perpetrators. Environmental authorities require suitable routines to assess environmental status as part of monitoring of law compliance. For example, mitigation of the impact of hydromorphological alterations and the restoration to natural or seminatural conditions are legally stipulated by the Brazilian National Water Law ("Lei de Águas," Law 9.433/97) and by the Rio de Janeiro State Water Resource Policy (State Law 3.239/99, Chap. 12). To explore hydromorphological potentials and deficits, the status of streams has to be assessed. In order to develop a protocol for such an assessment, an approach used in Germany (LAWA-OS method) was transferred to conditions prevailing in RJ. Such a transfer of protocols which worked well in other countries could be an option for further environmental assessments, if suitable protocols are lacking.

While CAR is a tool for monitoring compliance to environmental legislation on the one hand, it allows for identification of beneficiaries of financial incentives for more ecosystem-friendly practices on the other hand. As a crucial step for environmental authorities, CAR inheres the potential to be an instrument to further evolve from focusing on command and control instruments to more flexible conservation incentives (including PES (payment for ecosystem services), "ecological" ICMS (state tax on circulation of goods and services), to be used for ecological purposes, and potential trading in CRA (environmental reserve quotas)), which provide several opportunities to compensate landowners and municipal governments for their dedication of land to conservation (Chaps. 11 and 27).

In a scenario of global climate change, environmental legislation often implies undesirable consequences. This is the case of the set of laws under way to stop the burning of sugarcane fields in preparation for manual harvesting. Since in Brazil greenhouse gas emissions are mainly caused by agricultural practices and deforestation, these laws are aimed at CO_2 mitigation, bringing a major cause for CO_2 emissions to a halt. However, this development leads to financial losses of farmers because without the burning of sugarcane, manual harvesting becomes overly expensive and might lead to a critical loss of competitiveness of the farmers (Chap. 9). Whereas large-scale sugarcane cultivation can make use of mechanized harvesting equipment, the new legislation might force family farmers to shut down. Therefore, no matter how desirable the stop of the burning practice is from the environmental point of view, technology has to be available to allow future sugarcane cultivation by smallholders, thus sustain family farming structures. If more sophisticated technologies are provided to farmers and jointly used, as in this case a sugarcane harvester tailored to the special demands of family farmers, this technology will allow for compliance with the law and preservation of the ecosystem, while at the same time family farming will remain economically viable.

Also important, Brazil has recently established a law to improve access to information (Law 12,527). Although apart from the environmental framework, this law significantly contributes to greater transparency and cooperation between environmental governmental agencies and scientists by regulating the constitutional right of access to public information (Chap. 16). Therefore, the availability of past and current environmental datasets which so far have often been difficult to access by both decision-makers and scientists will have a positive effect on upcoming management decisions and research.

A crucial future task will be to use legislation in a way that builds a framework in which environmental protection can be successfully linked with sustaining agricultural production in rural RJ. Since laws often lack easy comprehensibility, legislation to effectively reduce resources overexploitation also requires means to inform the persons concerned. Awareness building for environmental issues and knowledge about environmental processes can motivate a higher willingness to follow environmental legislation. Farmers who are informed about the benefits of resilient landscapes will more readily use ecosystem-friendly methods, particularly if incentives exist which motivate the introduction of these methods. This concept was the main strategy of the RIO RURAL program, as highlighted in Chap. 3. RIO RURAL was set up as program promoting the public policy to start a process of farmers' compliance with environmental legislation by combining capacity building, financial incentives, technical assistance, and adaptive technologies. The aim of the program's strategy was to support the transition to sustainable agriculture and to more resilient rural landscapes, raising awareness and engaging almost 50,000 people in RJ – farmers, rural women and youth, technicians, local managers, and policymakers. The micro-watershed participatory approach developed under RIO RURAL started a territorial planning process among the participants and improved their skills for the prioritization areas for environmental restoration. This approach was combined with the introduction of environmentally friendly, agroecological, and Good Agricultural Practices. On the one hand, the combined approach supported ecosystem services provision including agricultural productivity and quality, and on the other hand, it allowed for a better access of participants to markets, ensuring food security and environmental conservation at the same time.

This book is written in times of financial and political insecurity for the entire country of Brazil. Therefore, it will be a challenge to implement the recommendations developed through scientific research within the INTECRAL project based on research that started during an economically boost phase. It was observed that in these particularly difficult times, positive developments are slowing down, halted, or even reversed. Particularly in critical periods, it is shown how important changes toward more resilient landscapes are when considering the buffer functions of rural areas for livelihoods of directly and indirectly affected people. The measures developed to support the conservation and restoration of landscapes have to be supported, consequently followed, and further improved and adapted.

31.3 Strategies and Tools to Enhance the Resilience of Rural RJ

The management of agricultural land is an important key to halt and reverse land and ecosystem degradation. On the one side, agricultural contribution to the gross domestic product of the heavily urbanized state (97% urban population according to the Demographic Census 2010 – IBGE 2010) was only 0.54% in 2015 (CEPERJ 2017); on the other side, land use management is critical for the highly demanded provision of ecosystem services. Pasture holds the largest share of land cover with almost 50% of the total land cover of RJ (IBGE 2016). An observed trend has been the decrease of cropping area, very seriously affecting the sugarcane sector (Chap. 6).

The high share of family farmers in the agriculture sector (Chap. 3) bears risks and opportunities at the same time: economical pressure and lack of funds for investment make it difficult for farmers to adapt to economic and climatic challenges which resulted in the degradation processes described above.

In the sugarcane sector in the Northern Region of RJ, the reduction of production land has accelerated under recent decline of productivity. This trend is followed by the lack of potential to adopt non-burning harvesting practices and irrigation technologies. The development of the innovative harvesting technology prototype to be adapted to local conditions beyond the end of the INTECRAL project in fall 2017 was seen as a glimmer of hope by farmers and researchers in major sugar mill in Campos de Goytacazes. Future research developed together by researchers and farmers' cooperatives will be essential to conclude and release the technology from the plots to the field.

In contrast to large-scale farmers, the family-managed small holdings demonstrate awareness of the threats of ecological degradation for their own livelihoods. Thus, the response to the adaptation methods offered within INTECRAL and RIO RURAL was cautious but increasingly positive throughout the implementation of pilot installations and pilot technology demonstration in the northwestern region of RJ.

The introduction of native shrubs and trees for restoration (Chap. 13) and a protein bank as a specific productivity-increasing method (Chap. 7) out of a set of options from silvopastoral systems for dairy cattle systems could prove that the introduction of woody crops in pasture management is significantly more drought tolerant than pastures. Within INTECRAL project, both measures were successfully implemented, even though drought periods occurred in 2014–2015 and again in 2016–2017. Especially the protein bank allows reaching and maintaining higher productivity levels than *Brachiaria*-based grassland systems. Where access to water and financial resources are available (as it was the case in the RIO RURAL program), the combination of irrigation with rotational grazing areas also allows to fundamentally increase productivity per hectare as well as per cow. Intensification measures in dairy pasture systems might in the long run lead to the reduction of utilization of low-productivity lands, such as the hilltops and steep slopes, since they cannot supply the nutrient density required for the higher productive cows evolving from this development. Another important component for increasing productivity of dairy system is management practices highlighting animal welfare and health, as shown in Good Dairy Practices and Good Milking Practices Guidelines (Chap. 8). With a simple set of hygienic measures, also the quality of the product will improve, another important factor for increasing the return of farming through the sale of the main product, i.e., milk. The duration of the INTECRAL project was not sufficient to demonstrate productivity increases, but from international experiences, cautious estimations would suggest productivity increases from 10–30% by implementing these measures which are more intensive and also contribute to the conservation of land resources at the same time (Chap. 5).

For farmers as for landowners in general, the National System of Units of Conservation (SNUC; Law 9.985/2000) provides the opportunity to convert private land into a protected area, a so-called Reserva Particular do Patrimônio Natural or RPPN in short (Chap. 11). In the future, a more systematic planning of RPPN establishment could be realized using the spatial data from the CAR initiative. For decision-making at implementation level, it would be important that RPPN datasets and datasets of other conservation units (UCs) as well as datasets of protected areas defined by the Forest Law (RLs – legal forest reserves and APPs – Permanent Protection Areas, Chap. 10) are brought together and made available for corridor design. This development could lead to a state-wide strategy for RPPN establishment that would be the more feasible if one common database existed, irrespective of the type of RPPN, i.e., an RPPN at federal, state, or municipal level. The great number of vestiges found in some forest fragments proved that in the past, different kinds of land use could be replaced by secondary forest after they had been abandoned. All this points to the fact that the enormous area calculated as necessary for restoration measures according to the Forest Law (forest deficit) can be successful in the future. Restoration measures should include the use of native species, to counter biodiversity erosion of Atlantic Forest species (Chap. 13). Even CITES-listed species like the pau brasil, the national tree of Brazil, should be explored for their use in restoration as well as their cultivation potential for economic utilization (Chap. 14). Small islands of native tree species would also enhance connectivity between forest fragments and positively influence the efficiency of the regional protected area network. When low-productivity lands, such as the hilltops and steep slopes, are less utilized for livestock cultivation and this goes hand in hand with a well-developed regional protected area network, a milestone toward a more resilient rural RJ will be reached.

Achieving water security in RJ is a major challenge and prerequisite for sustaining livelihoods, securing human well-being, fostering socioeconomic development, and stabilizing natural ecosystems, as pointed out in Chap. 15 by Ribbe et al. In the past decades, water security in RJ was mainly associated with maintaining adequate water quality and protection from water-related hazards, such as floods, landslides, and waterborne diseases. However, the droughts from 2014 to 2017 showed that water availability is equally important, in particular against the background of the predicted increase of climatic extreme events. The 2014–2017 droughts, however, also revealed obvious deficiencies in water infrastructure and governance, even though the Brazilian water law promotes an integrated approach to water management.

The INTECRAL project emphasized on various aspects of water management and hydrology to improve the current situation. One important part of the project was the ecohydrological modeling in various watersheds. Due to the higher interannual variability induced by climate change, it is important to understand how climate affects the interaction between the timing of extreme rainfall events, hydrological processes, vegetation growth, soil cover, and soil erosion. In this context Santos et al. (Chap. 16) selected the Muriaé River Basin, aiming to develop a consistent procedure for homogenizing climatic data, to analyze the hydro-climatic time series needed for hydrological studies, e.g., trend analysis and ecohydrological modeling, and to contribute to a better understanding of the long-term hydroclimatic patterns in this region. The region experiences water shortages during dry seasons as well as severe floods during rainy seasons. Therefore, it is plagued by high erosion rates on pasturelands. In Chap. 17, Künne et al. explained how ecohydrological modeling can contribute to a better understanding and prediction of ecohydrological processes on a high temporal and spatial scale. The model was applied and validated in three mesoscaled watersheds within the hinterland of RJ. Further, the authors present the evaluation of the impact of future climate development scenarios on hydrology to quantify changing flow conditions under future climate changes within the region. Both chapters provide important ecohydrological information crucial for water resources management in RJ. While Santos et al. showed the importance of hydroclimatic data analysis and how climatic trends are visible in recent data, Künne et al. implemented a method and applied an ecohydrological model. This model was the basis for further evaluation of the impact of downscaled regional climatic data on hydrology. These spatially explicit scenario impact-analyses can form the basis of detailed risk maps and other risk assessment products, hence guiding future disaster risk strategies of RJ.

Complementing water resources management, Cardona et al. (Chap. 18) dedicated their work to sewage treatment, one of the top water pollution drivers in rural RJ and provided an insight into a case study on decentralized sanitation and reuse (DESAR) solutions for small urbanized rural settlements conducted in the Rio Dois Rios basin. The authors argue that DESAR systems contribute to a reduction of costs and improvements of nutrient recovery capabilities of sanitation development in rural communities. They suggest that eco-technologies for wastewater treatment combined with in situ sludge stabilization should be promoted in small urbanized rural settlements in RJ and beyond. Finally, Penedo et al. (Chap. 19) presented a tool to assess land use impacts on surface water quality based on a case study in the Guapi-Macacu River basin. With their tool the authors support practitioners and decision-makers with a sound scientific approach to ensure better water management in Rio Dois Rios watershed which provides drinking water for more than two million people.

Besides compromising the ecosystem services of food and water security, continuous overexploitation of natural resources coupled with climate change resulted in the increase of vulnerability to floods and landslides (Chap. 20). This is particularly the case in urban areas, where 96% of the state's population resides. Several recent disasters, most notably the floods and mudslides mega-disaster that occurred in 2011 in the Serrana Region, forced a change in policy from just confronting disasters toward the development of disaster risk management. Even though the direction preceded by policy and stakeholders is the right one and many advances have been made, there is still a lack of communication between institutions, methodological planning to harmonize different initiatives, and success monitoring (Chap. 21). Additionally, disaster risk reduction (DRR) needs to be addressed by the empowerment of local population and collective decision-making using participatory methods. A case study carried out by Bustillos et al. (Chap. 23) in households of water-related risk areas of Nova Friburgo, São Fidelis, and Rio de Janeiro municipalities showed that the lack of trust on the public institutions and decision-makers is still evident. Hence, participatory stakeholder involvement and communication networks need to be supported to increase trust and to improve transdisciplinary knowledge on strategies for DRR. Promising approaches for DRR are ecosystem-based adaptation (EbA) and ecosystem-based disaster risk reduction (Eco-DRR), aiming at the sustainable management, conservation, and restoration of ecosystems to reduce the impacts of natural hazards. Forest protection, afforestation, and restoration are among the most powerful measures, combining efforts in nature conservation and DRR. In RJ, the restoration of APPs on steep slopes, hilltops, river margins, and headwaters together with RPPNs (Chaps. 11 and 12) can directly contribute to risk reduction and climate change adaptation. However, such synergies with biodiversity conservation require a better coordination between disaster protection and environmental planning authorities. As EbA and Eco-DRR measures are mostly effective at the long run and characterized by a mid- to long-term cost-benefit ratio, there is little acceptance by decision-makers, even though the potential for EbA and Eco-DRR is high in RJ (Chap. 22).

Progressive land degradation, especially caused by inappropriate pasturing at slopes with soils naturally susceptible for erosive forces, such as heavy rainfalls and compaction (Chap. 24), is a widespread phenomenon to be observed almost throughout the entire state. In particular, the highly deforested and hilly northwest Fluminense region moves potentially toward a non-resilient landscape due to its predominance of unsustainable cattle farming and related degradation combined with extended dry seasons and pronounced heavy rainfalls in the wet season, causing both droughts and floods. But there are possibilities to combat pasture degradation with several approaches at varying scales. Sustainable pasture management measures, such as rotational farming and silvopastural systems (Chap. 7), relieve pasturing pressure from degraded slopes and liberate them for recovery and even afforestation without compromising the economic situation of farmers. Furthermore, bioengineered rehabilitation measures, such as hedge terraces, erosion palisades, and geotextiles, were successfully implemented in pilot areas (Chaps. 24 and 25). Such interventions are, of course, cost- and labor-intensive and only successful at a very local scale. Hence, a prioritization of degraded areas taking into account the severity of degradation and effectiveness in terms of conservation and ecosystem services recovery helps in allocating limited financial resources efficiently. For the selection of reforestation sites, Naegeli et al. (Chap. 26) successfully applied a GIS-based tool predicated on ecological principles by integrating data on pasture degradation and landscape fragmentation. From that perspective, degraded pastures are opportunity areas for afforestation increasing forest cover and connectivity. Hence, sustainable pasture management combined with rehabilitation measures and an improved strategy to enhance forest cover and connectivity could support a multitude of ecosystem services, reduce the risk of flood disasters, and maintain rural livelihood.

Yet how can more sustainable agricultural measures, pasture management, water management, and biodiversity conservation be reconciled? May et al. (Chap. 27) highlighted creative approaches that use a mix of policy interventions and provision of funds from several resources that overcome concerns regarding the fiscal feasibility of sustaining payments for environmental services over time. As immediate measures May et al. suggest (1) promoting greater coordination and cooperation between government agencies responsible for different although intimately interrelated public policies in the agricultural, water resource, and conservation themes; (2) establishing priorities of action that go beyond specific government interests to support concerns of the general society; and (3) investing in production of technical and scientific material that documents the experiences carried out as a basis for evaluation and improvement. Chapter 28 by Rodríguez Osuna et al. provided an overview in how far the RIO RURAL program promotes sustainable agriculture and at the same time enhances climate change mitigation and adaptation. As a main outcome, the authors prove that the costs of certifying RIO RURAL's promoted agricultural and forestation practices are higher than the opportunity costs small farmers face in the state. Therefore, Rodríguez Osuna et al. suggest a bundling approach to carbon, where multiple benefits of sustainable agriculture are measured and certified including water, food systems, as well as social and cultural benefits. Moreover, they point out the need for better coordination among environmental protection and agricultural production institutions to achieve these simultaneous goals and suggest the development of metrics to assess environmental services considering the agricultural landscape of RJ aiming at the development/promotion of markets for such services.

Chapter 29 by Torres Zárate and Dornberger focused on alternative income generation in rural areas by making use of innovation resources. The authors explained that most of rural households in their study region either base their income on conventional on-farm activities or on alternative off-farm employment but have no strategies for income diversification. However, a business idea contest that was conducted in the frame of INTECRAL and RIO RURAL showed that innovation resources that are available should be used to create a new rurality perspective including ecotourism and increased regional food quality. Noriega et al. (Chap. 30) took a closer look at regional marketing strategies for sustainable production. The authors came up with the idea of regional branding in combination with PES and the promotion of organic farming. From their perspective, the introduction of a regional sustainability label can be a step toward a more complex certification scheme which is at present not feasible and available for family farmers.

31.4 At What Point Are We Now?

For the next decades, climate models predict an increased occurrence of extreme climatic events to both the wet and the dry side (Chap. 20). Particularly for the northwestern region, the impact of both more dry periods and more heavy precipitation events is particularly drastic: areas whose inhabitants, mostly family farmers, are used to sufficient rainfall with occasional dry phases are now confronted with longer dry spells leading to water scarcity, limited water use, and use conflicts. In valley areas, the dominant rainfed agriculture in rural RJ has potential for adaptation by converting into irrigation agriculture, which would force farmers to invest into equipment and additionally would put pressure on water availability. For livestock farming, additional fodder to overcome extended shortage periods will be needed. This aim can either be achieved through intensive fodder conservation, e.g., silage and hay production or through importing purchased feed from other regions (Chap. 8). Even if this was possible, depending on the number and length of dry periods, the economic success of family farms in rural RJ might be at risk in the long run. Dry periods are also accompanied by an increase of fire events, and burnt hills are occurring more frequently.

As a consequence, with ongoing overexploitation of natural resources, together with increasing extreme climatic events, rural RJ heads for a socio-ecological tipping point: ongoing overexploitation, and thus increasing demand on ecosystem services, triggered by extreme climatic events disrupt ecosystem functioning in such a severe way that the system will change into a different state and strongly affect rural livelihoods. Whether and when this point will be reached depends on the selection of appropriate measures to countersteer this accelerated trend: further reducing overexploitation of natural resources.

31.5 The Way Forward

In order to foster a resilient rural RJ, it is crucial to understand and communicate the interactions between resources use and extreme climatic events. For the assessment of the ecosystem state, including its ecosystem functions and services, the development of standard protocols as well as of metrics and thresholds is needed. In addition to integrating such elements into existing management strategies, there is also a high need for awareness building on environmental issues in the light of increasing extreme climate events. The main difficulty is that the rural as well as the urban populations have little experience regarding the handling of droughts, since this is a comparatively new phenomenon in RJ. Therefore, a more progressive agenda that includes agro-environmental, forest and protected area governance, in which land use policy is more coherently addressed and linked with water resource management, would unlock opportunities to enhance the state's capacity to perform a more integrative management strategy that conciliates environmental and productive processes in the light of increasing extreme climatic events (Chap. 27). One option would be to further improve RIO RURAL's strategy specifically addressing the new challenge of increasing occurrence of extreme climatic events. A resilient rural RJ would be able to provide the ecosystem services increasingly demanded by climate change-stricken degraded landscape. Instead of developing a rural area that is perceived as far away from the attractive urban areas, RJ has the potential to evolve into a rural-urban continuum with lives in well-connected urban, peri-urban, and rural areas.

References

CEPERJ – Fundação Centro Estadual de Estatísticas, Pesquisas e Formação de Servidores Públicos do Rio de Janeiro (2017) Projeto Contas Regionais do Brasil - Produto Interno Bruto – PIB do Estado do Rio de Janeiro – 2015. http://www.ceperj.rj.gov.br/ceep/pib/pib.html. Accessed 12 Dec 2017

George M, McGranahan G (2010) Brazil's early urban transition: What can it teach urbanizing countries? In: Series on Urbanization and Emerging Population, vol 4. International Institute for Environment and Development IIED and Brazilian Association of Population Studies UNFPA, London and New York

Gragnolati M, Jorgensen OH, Rocha R, Fruttero A (2011) Growing old in an older Brazil: implications of population aging on growth, poverty, public finance and service delivery. The World Bank, Washington, DC

IBGE – Instituto Brasileiro de Geografia e Estatística (2010) Sinopse do censo demográphico 2010, Tabela 16: População nos Censos Demográficos, segundo os municípios das capitais – 1872/2010. http://www.censo2010.ibge.gov.br/sinopse/index.php?dados=6&uf=00. Accessed 12 Dec 2017

IBGE – Agência IBGE Noticias (2016) IBGE mapeia a cobertura e o uso da terra no estado do Rio de Janeiro. https://agenciadenoticias.ibge.gov.br/agencia-noticias/2013-agencia-de-noticias/releases/9513-ibge-mapeia-a-cobertura-e-o-uso-da-terra-no-estado-do-rio-de-janeiro.html released 28/07/2016. Accessed 12 Dec 2017

Index

U. Nehren et al. (eds.), *Strategies and Tools for a Sustainable Rural Rio de Janeiro*, Springer Series on Environmental Management,
https://doi.org/10.1007/978-3-319-89644-1

R

The manufacturer's authorised representative in the EU is Springer Nature Customer Service Centre GmbH, Europaplatz 3, 69115 Heidelberg, Germany. If you have any concerns regarding our products, please contact ProductSafety@springernature.com

Printed and bound by CPI Group (UK) Ltd, Croydon, CR0 4YY
30/09/2026
02225658-0014